TECHNOLOGY POLICY AND DEVELOPMENT

International Development Resource Books
Pradip K. Ghosh, editor

Industrialization and Development: A Third World Perspective
Urban Development in the Third World
Technology Policy and Development: A Third World Perspective
Energy Policy and Third World Development
Population, Environment and Resources, and Third World Development
Health, Food, and Nutrition in Third World Development
Economic Policy and Planning in Third World Development
Development Policy and Planning: A Third World Perspective
New International Economic Order: A Third World Perspective
Foreign Aid and Third World Development
Multi-national Corporations and Third World Development
Economic Integration and Third World Development
Third World Development: A Basic Needs Approach
Appropriate Technology in Third World Development
Development Cooperation and Third World Development
International Trade and Third World Development
Disarmament and Development: A Global Perspective
Developing South Asia: A Modernization Perspective
Developing Latin America: A Modernization Perspective
Developing Africa: A Modernization Perspective

TECHNOLOGY POLICY AND DEVELOPMENT

A Third World Perspective

Pradip K. Ghosh, *Editor*

Foreword by Gamani Corea, Secretary-General of UNCTAD

Prepared under the auspices of the Center for International Development, University of Maryland, College Park, and the World Academy of Development and Cooperation, Washington, D.C.

International Development Resource Books, Number 3

Greenwood Press
Westport, Connecticut • London, England

Library of Congress Cataloging in Publication Data

Main entry under title:

Technology policy and development.

(International development resource books, ISSN 0738-1425; no. 3)
Bibliography: p.
Includes index.
1. Technology—Developing countries. I. Ghosh, Pradip K., 1947- . II. Series.
T49.5.T446 1984 338.9'009172'4 83-22771
ISBN 0-313-24139-2 (lib. bdg.)

Library of Congress Catalog Card Number: 83-22771
ISBN: 0-313-24139-2
ISSN: 0738-1425

First published in 1984

Greenwood Press
A division of Congressional Information Service, Inc.
88 Post Road West, Westport, Connecticut 06881

Printed in the United States of America

10 9 8 7 6 5 4 3 2 1

Copyright Acknowledgments

Reprinted with the permission of the Population Council from "The urban prospect: Reexamining the basic assumptions," by Lester R. Brown, *Population and Development Review* 2, no. 2 (June 1976): 267–277.

Reprinted with the permission of the Population Council from "Food for the future: A perspective," by D. Gale Johnson, *Population and Development Review* 2, no. 1 (March 1976): 1–19.

Reprinted with the permission of the Population Council from "Population, development, and planning in Brazil," by Thomas W. Merrick, *Population and Development Review* 2, no. 2 (June 1976): 181–199.

Reprinted with the permission of the Population Council from "Observations on population policy and population program in Bangladesh," by Paul Demeny, *Population and Development Review* 1, no. 2 (December 1975): 307–321.

Reprinted with the permission of the Population Council from "Population policy: The role of national governments," by Paul Demeny, *Population and Development Review* 1, no. 1 (September 1975): 147–161.

Reprinted with the permission of the Population Council from "Asia's cities: Problems and options," by Kingsley Davis, *Population and Development Review* 1, no. 1 (September 1975): 71–86.

Articles from *Impact of Science on Society,* Vol. XXIX, no. 3, © Unesco 1979. Reproduced by permission of Unesco.

Articles from *Impact of Science on Society,* Vol. XXIII, no. 4, © Unesco 1973. Reproduced by permission of Unesco.

Articles from *Impact of Science on Society,* Vol. XXV, no. 1, © Unesco 1975. Reproduced by permission of Unesco.

Articles from *Impact of Science on Society,* Vol. XX, no. 3, © Unesco 1972. Reproduced by permission of Unesco.

Articles from *Impact of Science on Society,* Vol. XXIII, no. 2, © Unesco 1973. Reproduced by permission of Unesco.

Articles from *Impact of Science on Society,* Vol. XXV, no. 3, © Unesco 1975. Reproduced by permission of Unesco.

Articles from *Impact of Science on Society,* Vol. XXIV, no. 2, © Unesco 1974. Reproduced by permission of Unesco.

Articles from *Scientists, the Arms Race and Disarmament,* © Unesco 1982. Reproduced by permission of Unesco.

Articles from *The Use of Socio-Economic Indicators in Development Planning,* © Unesco 1976. Reproduced by permission of Unesco.

Articles from *Methods for Development Planning: Scenarios, Models and Micro-Studies,* © Unesco 1981. Reproduced by permission of Unesco.

Article from *Socio-Economic Indicators for Planning: Methodological Aspects and Selected Examples,* © Unesco 1981. Reproduced by permission of Unesco.

To
Gunnar and Alva Myrdal

"THE ULTIMATE OBJECTIVE OF DEVELOPMENT MUST BE TO BRING ABOUT A SUSTAINED IMPROVEMENT IN THE WELL-BEING OF THE INDIVIDUAL AND BESTOW BENEFITS ON ALL. IF UNDUE PRIVILEGES, EXTREMES OF WEALTH AND SOCIAL INJUSTICES PERSIST, THEN DEVELOPMENT FAILS IN THE ESSENTIAL PURPOSE."

– UNITED NATIONS GENERAL ASSEMBLY RESOLUTION 2626 (XXV), 24 OCTOBER 1970.

Contents

LIST OF TABLES

PART I

TECHNOLOGY GAPS BETWEEN RICH AND POOR COUNTRIES

THE TECHNOLOGICAL GAP BETWEEN DEVELOPING AND DEVELOPED COUNTRIES

TECHNOLOGICAL CHANGES IN DEVELOPING COUNTRIES SINCE THE SECOND WORLD WAR

PART II

STATISTICAL INFORMATION AND SOURCES

LIST OF FIGURES

PART I

TECHNOLOGICAL SELF-RELIANCE OF THE DEVELOPING COUNTRIES: TOWARD OPERATIONAL STRATEGIES

THE SOCIO-ECONOMIC ICEBERG AND THE DESIGN OF POLICIES FOR SCIENTIFIC AND TECHNOLOGICAL DEVELOPMENT

PART II

STATISTICAL INFORMATION AND SOURCES

The Third World

Afghanistan
Republic of Afghanistan
Algeria
Democratic and Popular Republic of Algeria
Angola
People's Republic of Angola
Argentina
Argentine Republic
Bahamas
Commonwealth of the Bahamas
Bahrain
State of Bahrain
Bangladesh
People's Republic of Bangladesh
Barbados
People's Republic of Barbados
Benin
People's Republic of Benin
Bhutan
People's Republic of Bhutan
Bolivia
Republic of Bolivia
Botswana
Republic of Botswana
Brazil
Federative Republic of Brazil
Burma
Socialist Republic of the Union of Burma
Burundi
Republic of Burundi

Cambodia
Democratic Kampuchea
Cameroon
United Republic of Cameroon
Cape Verde
Republic of Cape Verde
Central African Empire
Chad
Republic of Chad
Chile
Republic of Chile
Colombia
Republic of Colombia
Comoro Islands
Republic of the Comoros
Congo
People's Republic of the Congo
Costa Rica
Republic of Costa Rica
Cuba
Republic of Cuba
Dominican Republic
Ecuador
Republic of Ecuador
Egypt
Arab Republic of Egypt
El Salvador
Republic of El Salvador
Equatorial Guinea
Republic of Equatorial Guinea
Ethiopia
Fiji
Dominion of Fiji
Gabon
Gabonese Republic
Gambia
Republic of the Gambia

Ghana
Republic of Ghana
Grenada
State of Grenada
Guatemala
Republic of Guatemala
Guinea
Republic of Guinea
Guinea-Bissau
Republic of Guinea-Bissau
Guyana
Cooperative Republic of Guyana
Haiti
Republic of Haiti
Honduras
Republic of Honduras
India
Republic of India
Indonesia
Republic of Indonesia
Iran
Imperial Government of Iran
Iraq
Republic of Iraq
Ivory Coast
Republic of Ivory Coast
Jamaica
Jordan
Hashemite Kingdom of Jordan
Kenya
Republic of Kenya
Kuwait
State of Kuwait
Laos
Lao People's Democratic Republic
Lebanon
Republic of Lebanon
Lesotho
Kingdom of Lesotho
Liberia
Republic of Liberia
Libya
People's Socialist Libyan Arab Republic
Madagascar
Democratic Republic of Madagascar
Malawi
Republic of Malawi
Malaysia
Maldives
Republic of Maldives
Mali
Republic of Mali
Mauritania
Islamic Republic of Mauritania
Mauritius
Mexico
United Mexican States
Mongolia
Mongolian People's Republic
Morocco
Kingdom of Morocco
Mozambique
People's Republic of Mozambique
Nepal
Kingdom of Nepal
Nicaragua
Republic of Nicaragua
Niger
Republic of Niger
Nigeria
Federal Republic of Nigeria
Oman
Sultanate of Oman
Pakistan
Islamic Republic of Pakistan
Panama
Republic of Panama
Papua New Guinea
Paraguay
Republic of Paraguay
Peru
Republic of Peru
Philippines
Republic of the Philippines
Qatar
State of Qatar

Adopted from THE THIRD WORLD: PREMISES OF U.S. POLICY by W. Scott Thompson, Institute for Contemporary Studies, San Francisco, 1978.

Rhodesia
Ruanda
Republic of Ruanda
Samoa
Sao Tome and Principe
Democratic Republic of Sao Tome and Principe
Saudi Arabia
Kingdom of Saudi Arabia
Senegal
Republic of Senegal
Seychelles
Sierra Leone
Republic of Sierra Leone
Singapore
Republic of Singapore
Somalia
Somali Democratic Republic
Sri Lanka
Republic of Sri Lanka
Sudan
Democratic Republic of the Sudan
Surinam
Swaziland
Kingdom of Swaziland
Syria
Syrian Arab Republic
Tanzania
United Republic of Tanzania
Thailand
Kingdom of Thailand
Togo
Republic of Togo
Trinidad and Tobago
Tunisia
Republic of Tunisia
Uganda
Republic of Uganda
United Arab Emirates
Upper Volta
Republic of Upper Volta
Uruguay
Oriental Republic of Uruguay
Venezuela
Republic of Venezuela
Vietnam
Socialist Republic of Vietnam
Western Sahara
Yemen
People's Democratic Republic of Yemen
Yemen
Yemen Arab Republic
Zaire
Republic of Zaire
Zambia
Republic of Zambia

Countries which have social and economic characteristics in common with the Third World but, because of political affiliations or regimes, are not associated with Third World organizations:

China
People's Republic of China
Cyprus
Republic of Cyprus
Israel
State of Israel
Kazakhstan
Kirghizia
Korea
Democratic People's Republic of Korea
Romania
Socialist Republic of Romania
South Africa
Republic of South Africa
South West Africa
Namibia
Tadzhikistan
Turkmenistan
Uzbekistan
Yugoslavia
Socialist Federal Republic of Yugoslavia

Abbreviations

ADC	Andean Development Corporation
AsDB	Asian Development Bank
ASEAN	Association of South-East Asian Nations
CARIFTA	Caribbean Free Trade Association
DAC	Development Assistance Committee (of OECD)
ECA	Economic Commission for Africa
ECE	Economic Commission for Europe
ECLA	Economic Commission for Latin America
ECOWAS	Economic Commission of West African States
EDF	European Development Fund
EEC	European Economic Community
EFTA	European Free Trade Association
ESCAP	Economic and Social Commission for Asia and the Pacific
FAO	Food and Agriculture Organization of the United Nations
GATT	General Agreement on Tariffs and Trade
GDP	gross domestic product
GNP	gross national product
IBRD	International Bank for Reconstruction and Development (World Bank)
IDA	International Development Association
IDB	Inter-American Development Bank
IFC	International Finance Corporation
IIEP	International Institute for Educational Planning
ILO	International Labour Office
IMF	International Monetary Fund
LAFTA	Latin American Free Trade Association
ODA	official development assistance
OECD	Organisation for Economic Co-operation and Development
OPEC	Organization of Petroleum Exporting Countries
UNDP	United Nations Development Programme
UNEP	United Nations Environment Programme
UNESCO	United Nations Educational, Scientific and Cultural Organization
UNHCR	Office of the United Nations High Commissioner for Refugees
UNITAR	United Nations Institute for Training and Research
UNICEF	United Nations Children's Fund
UNIDO	United Nations Industrial Development Organization
WFP	World Food Programme
WHO	World Health Organization

Foreword

I am pleased to know of the International Development Resources Book project. The 20 resource books which are published under this project, covering the whole spectrum of issues in the fields of development economics and international co-operation for development, and containing not only current reading materials but also up-to-date statistical data and bibliographical notes, will, I am sure, prove to be extremely useful to a wide public.

I would like to commend the author for having undertaken this very ambitious and serious project and, by so doing, rendered a most valuable service. I am confident that it will have a great success.

Gamani Corea

Secretary-General
United Nations Conference on Trade and Development

Preface

Stimulus for the publication of an international resource book series was developed in 1980, while teaching and researching various topics related to third world development. Since that time, I have built up a long list of related resource materials on different subjects, usually considered to be very important for researchers, educators, and public policy decision makers involved with developing country problems. This series of resource books makes an attempt for the first time to give the reader a comprehensive look at the current issues, methods, strategies and policies, statistical information and comprehensive resource bibliographies, and a directory of information sources on the topic.

This topic is very important because within the framework of the current international economic order, developing an effective technology policy is envisaged as a dynamic instrument of growth essential to the rapid economic and social development of the developing countries, in particular of the least developed countries of Asia, Africa and Latin America.

Much of this work was completed during my residency as a visiting scholar in the Center for Advanced Study of International Development at Michigan State University. Suzanne Wilson and Mary Ann Kozak, students at the University, provided assistance with the project. I am thankful to the M.S.U. Sociology department for providing necessary support services and Dr. James T. Sabin, Vice President, editorial of Greenwood Press who encouraged in pursuing the work and finally agreeing to publish in book form.

I would also like to gratefully acknowledge the encouragement given to me by Dr. Denton Morrison to pursue this project and to Dr. Mark Van de Vall who has been an inspiration to me since my graduate school days.

Finally, preparation of this book would not have been completed without the contributions from Paul Streeten, H. J. Rush, Denis Goulet, William L. Eilers, Andre Danzin, Miguel Wionczek, Robin Clarke, Mario Kamenetzky, Charles Weiss, Jr.

and Robert H. Maybury, Constantin V. Vaitsos, Keith Marsden, David A. Phillips, Jorge A. Sabato, Landing Savane, and many experts of the UNCTAD Secretariat and the UNIDO. I am also gratefully indebted to Journal of Economic Literature, U.N. Documents and World Bank Publications for the much needed annotations, and a very special thank to Tom and Jackie Minkel for typing assistance.

For additional resource materials on related topics of international development, please consult the following resource books published by the GREENWOOD PRESS (Westport, Connecticut, London, England).

RESOURCE BOOK NO. 1. INDUSTRIALIZATION AND DEVELOPMENT: A THIRD WORLD PERSPECTIVE
RESOURCE BOOK NO. 2. URBAN DEVELOPMENT IN THE THIRD WORLD
RESOURCE BOOK NO. 3. TECHNOLOGY POLICY AND DEVELOPMENT: A THIRD WORLD PERSPECTIVE
RESOURCE BOOK NO. 4. ENERGY POLICY AND THIRD WORLD DEVELOPMENT
RESOURCE BOOK NO. 5. POPULATION, ENVIRONMENT AND RESOURCES, AND THIRD WORLD DEVELOPMENT
RESOURCE BOOK NO. 6. HEALTH, FOOD AND NUTRITION IN THIRD WORLD DEVELOPMENT
RESOURCE BOOK NO. 7. ECONOMIC POLICY AND PLANNING IN THIRD WORLD DEVELOPMENT
RESOURCE BOOK NO. 8. DEVELOPMENT POLICY AND PLANNING: A THIRD WORLD PERSPECTIVE
RESOURCE BOOK NO. 9. NEW INTERNATIONAL ECONOMIC ORDER: A THIRD WORLD PERSPECTIVE
RESOURCE BOOK NO. 10. FOREIGN AID AND THIRD WORLD DEVELOPMENT
RESOURCE BOOK NO. 11. MULTI-NATIONAL CORPORATIONS AND THIRD WORLD DEVELOPMENT
RESOURCE BOOK NO. 12. ECONOMIC INTEGRATION AND THIRD WORLD DEVELOPMENT
RESOURCE BOOK NO. 13. THIRD WORLD DEVELOPMENT: BASIC NEEDS APPROACH
RESOURCE BOOK NO. 14. APPROPRIATE TECHNOLOGY IN THIRD WORLD DEVELOPMENT
RESOURCE BOOK NO. 15. DEVELOPMENT CO-OPERATION AND THIRD WORLD DEVELOPMENT
RESOURCE BOOK NO. 16. INTERNATIONAL TRADE AND THIRD WORLD DEVELOPMENT
RESOURCE BOOK NO. 17. DISARMAMENT AND DEVELOPMENT: GLOBAL PERSPECTIVE
RESOURCE BOOK NO. 18. DEVELOPING LATIN AMERICA: A MODERNIZATION PERSPECTIVE
RESOURCE BOOK NO. 19. DEVELOPING SOUTH ASIA: A MODERNIZATION PERSPECTIVE
RESOURCE BOOK NO. 20. DEVELOPING AFRICA: A MODERNIZATION PERSPECTIVE

PART I

CURRENT ISSUES, TRENDS, ANALYTICAL METHODS, STRATEGIES AND POLICIES, COUNTRY STUDIES

Introduction

This resource book has two multifaceted purposes. Firstly, to document and analyze the current trends in the development of an effective technology policy of the third world countries--and to evaluate the progress made by them during the past decade in attaining long term objectives of a sustained economic growth and improvement in the quality of living future populations.

We are all very much familiar with the problems of third world countries, usually described by Latin America (excluding Cuba), the whole of Africa, Asia (excluding its socialist countries, Japan and Israel) and Oceania (excluding Australia and New Zealand). They are plagued by poverty, very high rates of population growth, low growth rates of gross domestic product, low rates of industrialization, extremely high dependence on agriculture, rate of unemployment, and uneven income distribution. Although the expression "third world countries" no longer has a clear meaning, a majority of the international development experts would consider the poor developing countries to belong in the third world irrespective of their affiliation as aligned or non-aligned characteristic.[1]

Secondly, major purpose of this volume is to provide the researchers with the much needed knowledge about the different sources of information and available data related to technological change and transformation in the third world countries. Technology policy in the developing countries has raised many complex issues. While these issues are largely dependent on national policies and priorities, their solution is of international concern.

The pace and pattern of technological development have varied widely among the developing countries partly because of the differences in the availability of natural, human and capital resources and in factors such as size and location, and partly because of differences in objectives, strategies and policies of technical change that countries have pursued. The issues affecting strategies and policies differ considerably at the present time from those that were important a decade ago and policy design is thus now more complex and difficult than before.

Technology has become a focus of major interest and a topic of significant debate over the past decade, particularly with regard to those technologies that are relevant to development.

The less developed countries with all their handicaps of a late start have one critical advantage and that is related to the accumulated and growing scientific and technological knowledge. But to use this knowledge creatively for solving their own problems, they need to develop competence in the field of technology choice through an institutionalized technology policy. Sophisticated techniques of planning or project evaluation are not--cannot be--a substitute for such a technology policy; the former in fact have relevance only with such a policy.[2]

This volume examines the experience of a substantial number of the third world countries in implementing development plans during approximately the first half of the 1970's and draws some general conclusions for policy action during the years ahead.

Attention has been focused on some of the major problems faced by hard core developing countries and policy issued posed by those problems. However, the development needs of hard-core developing countries are very large and call for much greater interest and attention from world community than has been the case so far. A systematic attack on the acute problems of countries facing extreme poverty and underdevelopment should therefore now be at the center of the policies designed to usher in a new international economic order.

It is hoped that this resource book will be of use not only to those directly involved in the formulation and implementation of development policies but also will help to acquaint a wide reading audience with the thrusts planned by developing countries for accelerated progress. In addition, the intercountry comparative analysis may be of use to planners and policy makers in developing countries, especially from the viewpoint of harmonizing national plans in order to strengthen economic co-operation with respect to technology policy among interested countries.

The plan of the reading materials in Part I of the book and the selection of the nineteen pieces represents a specific orientation, or bias. They represent current international issues and trends affecting technology policy in third world development, analytical methods, strategies and policies for technology development and selected third world country studies.

Part II includes statistical information and a descriptive bibliography of information sources related to technology and development in the third world countries.

Part III is a select bibliography of books, documents and periodical articles published since 1970, relevant to technology development of the developing countries. Annotations for the different titles have been compiled from The Journal of Economic Literature, International Social Sciences Index, U.N. Documents Index, World Bank Publications, Finance and Development, Book Publisher's promotion brochures and the IMF-IBRD Joint Library Publications.

Part IV consists of a directory of information sources. This section is in four parts; directory of United Nations

Information Sources, listing of bibliographic sources, titles of selected periodicals published around the world and a directory of institutions involved in research relevant to technology policy problems in third world countries.

NOTES

[1]Rodwin Lloyd, "REGIONAL PLANNING PERSPECTIVES IN THIRD WORLD COUNTRIES," in TRAINING FOR REGIONAL DEVELOPMENT PLANNING: PERSPECTIVES FOR THE THIRD DEVELOPMENT DECADE, ed. by Om Prakash Mathur, UNCRD, 1981.
Thompson, W. Scott, THE THIRD WORLD: PREMISES OF U.S. POLICY, Institute for Contemporary Studies, San Francisco, 1978.

[2]V. V. Bhatt, "Financial Institutions and Technology Policy," WORLD DEVELOPMENT, Vol 8, pp. 813-22.

Technology Gaps Between Rich and Poor Countries

PAUL STREETEN*

I. THE CAUSES OF INTERNATIONAL INEQUALITY

The growing inequality in the international distribution of income is of relatively recent origin. It is a phenomenon of less than 200 years. It started towards the end of the eighteenth century when inventors combined engineering skills and entrepreneurial ingenuity, while a social and technical revolution in agriculture released the men and the food for industrialisation. Its twin causes are (a) the appearance of a new condition which permits the income and production of some countries to grow at much faster rates than those of others and (b) the existence of obstacles to the spread of the benefits from the fast-growing to the slow-growing countries. We do not have far to seek for this permissive condition. It is neither the discovery of natural resources nor, as was thought at one time, the accumulation of capital as such. Accumulation of capital, without improvements in knowledge, could not have brought about the substantial increase in income per head that occurred. The ceiling that this condition sets is determined by the stock of scientific knowledge (know-why and know-that) and technology (know-how), by its application to production and organisation in industry, trade and agriculture and by its commercial exploitation.[1] It is the continuing interaction between a succession of scientific, technological and industrial revolutions and the human, economic, political and social conditions that create and use these revolutions, that sets the upper limit to economic growth.[2] According to the assumptions of economic theory, the fruits of economic progress tend to be widely spread and to 'trickle down' through competition, specialisation

From **SCOTTISH JOURNAL OF POLITICAL ECONOMY**, November, 1972, (213-230), reprinted by permission of the publisher.

and public policy, from the rich to the poor. This has been largely true within already advanced countries as they progressed further. Whether competition and specialisation worked for a trickling down or a polarisation of benefits, the policies of the modern state have, on the whole, worked towards reducing domestic inequalities. But it has not been true between different countries nor within many underdeveloped countries. Clearly, no international government and therefore no international redistributive machinery exist. Other reasons why the spread has not occurred internationally will be considered later.

What matters in this context is neither the proportionate difference (of say, income per head or product per worker), nor the differences in magnitudes at the start, but the growth factor.[3] A simple example illustrates the point. Call A_1 and A_2 the initial magnitude of the gross national product per head in two countries.

The initial ratio of these two magnitudes R = ---; the final ratio

(F) after a number of years (t) will depend on the initial ratio (R) and the respective rates of growth (r_1) and (r_2).

F = = R = R.k. where k =

The following table shows k for different annual growth rates over 5, 10 and 20 years.

	t = 5	t = 10	t = 20
r_1 = 2%, r_2 = 1%	1.05	1.10	1.22
r_1 = 20%, r_2 = 10%	1.55	2.39	5.70

It can be seen that, if two countries started with equal income, one growing twice as fast as the other, a difference of 10 per cent. between 20 per cent. and 10 per cent. would make income in the faster growing country 5-70 times that in the slower one after 20 years. If, with the same proportional advantage, the difference were only 1 per cent., i.e. one growing at 2 per cent., the other at 1 per cent., the income ratio of the two countries would hardly have been altered. If the faster country started with 50, the slower with 100, and if the former grew at 20 per cent., while the latter at 10 per cent., the fast would have caught up with the slow within ten years. But if the fast grew at 2 per cent. and the slow at 1 per cent., the latter would still be substantially ahead after twenty years. In fact, over the last 200 years, different countries' growth rates of product per head varied widely, between nearly zero and over 25 per cent. per decade. The post-war period has witnessed an even wider spread of growth rates.

Some growth rates of product per head per decade over long periods for different countries are given in the table below.

II. THE TRANSFER OF KNOWLEDGE

If we are interested in the wide spread of the benefits of science and technology (S & T) and in policies which promote this, we must ask ourselves whether the knowledge is transferable from country to country or whether it is tied to a particular place. Where knowledge is transferable, we must ask how best it can be transferred, and to what extent different methods of transfer are substitutes or complementary. If they are substitutes, the question is whether transfer should be carried out through local subsidiaries of international firms, or through imports or through licenses, joint ventures or collaboration agreements, or through hiring experts, or whether it is not better to rely solely on an indigenous scientific capability, by training students at home or possibly by sending them to be trained abroad, in order to build up indigenous institutions later. If they are complementary, the question is what are the best combinations and what is the appropriate phasing?

In order to answer these questions, we have to construct a typology of S & T. The following questions are relevant:

(i) Is the knowledge physically transferable or is it tied to a particular locality?
(ii) Is it freely available or do patents or other property rights impose a cost on those wishing to acquire it?
(iii) Is the knowledge of the process or product stable or changing?
(iv) Is it separable from other activities of the firm, such as using sources of supply or seeking market outlets or is it inextricably tied up, through feedbacks and 'feed-forwards', with knowledge or information drawn from these other activities? Is it, in other words, an integral part of the whole system or parts of the system of the firm's activities?

The answers to these questions will determine the most effective way of acquiring the knowledge. Thus, if the knowledge is transferable, free and separable, the solution is to look it up; if the property rights are attached to it, it may have to be bought; if it is changing or integrally linked to other activities, direct foreign investment may be invited. The problems arising from this last type are more fully discussed later.

A good deal of S & T is freely available, though even then 'absorptive capacity' is required in the sense that there must be people willing and able to understand and apply the knowledge. Without a receptive indigenous S & T capability and a social structure adapted to receiving its fruits, even freely available

Table I
RATE OF GROWTH PER DECADE OF PRODUCT PER HEAD (PER CENT)

England and Wales—United Kingdom	
1. 1700 to 1780	2·0
2. 1780 to 1781	13·4
3. 1855-59 to 1957-59	14·1
France	
4. 1841-50 to 1960-62	17·9
Germany—West Germany	
5. 1851-55 to 1871-75	9·2
6. 1871-75 to 1960-62	17·9
Netherlands	
7. 1900-04 to 1960-62	13·5
Switzerland	
8. 1890-99 to 1957-59	16·1
Denmark	
9. 1870-74 to 1960-62	19·4
Norway	
10. 1865-74 to 1960-62	19·0
Sweden	
11. 1861-65 to 1960-62	28·3
Italy	
12. 1861-65 to 1898-1902	2·7
13. 1898-1902 to 1960-62	18·7
United States	
14. 1839 to 1960-62	17·2
Canada	
15. 1870-74 to 1960-62	18·1
Australia	
16. 1861-65 to 1959/60-1961/62	8·0
Japan	
17. 1879-81 to 1959-61	26·4
European Russia—U.S.S.R.	
18. 1860 to 1913	14·4
19. 1913 to 1958	27·4
20. 1928 to 1958	43·9

Source: Kuznets, 1967, pp. 64-5.

and communicated knowledge remains unused.

There is, of course, no hard and fast line between knowledge that is freely available and knowledge that is tied to individuals or institutions. Like a recipe, anyone can look it up (if he knows that it is worth looking and where to look), but everyone is not an equally good cook. 'Knowing that' is different from 'knowing how'.

III. OUTPUT GAPS AND TECHNOLOGY GAPS

When we say that there is a technology gap between rich and poor countries we may mean no more than that output per worker is greater in rich countries than it is in poor. Alternatively, we may wish to draw a distinction between an output gap and a technology gap, so that it would be possible, in principle, to have low output per worker and an advanced technology, or high output per worker and a backward technology. The distinction can be illuminated by separating different influences on output per worker.

Output per worker can be analysed as consisting of two components: output per unit of capital, multiplied by capital per worker.[4] Productivity may be high either because workers are backed by a lot of capital or because the productivity of the capital with which they work is high, or both. Knowledge and skills embodied in workers can be regarded either as a form of capital or as a way of raising labour productivity for a given amount of capital. Aggregate differences in productivity per worker may again be due to four factors:

(i) Sectors with different productivities may have a different relative weight in the economy of different countries. Thus richer countries have a larger industrial sector in which productivity is higher than in agriculture in poor countries.

(ii) Within a sector similar enterprises vary substantially with respect to their productivity. The proportion of enterprises following the best practice (with respect to engineering or organization) will vary and will determine the average productivity in the sector and hence in the economy.

(iii) The 'best practices' themselves will vary from country to country and differential standards will mean that some societies show in their 'best practice' firms in similar sectors higher productivity than others.

(iv) Differences in productivity may be due to economies of scale.

This fourfold division refers to output per worker and must again be subdivided in the first three cases into differences of output per unit of capital and those of capital per worker. Either capital or workers may embody technical and organizational knowledge.

Technology gaps may then be of one of two types: they may be due to imperfections in communication and transferring existing technologies (call this the Communications Gap) or they may be due to absence of appropriate technologies (Suitability Gap).[5] A

Communications Gap means that existing knowledge is only partly or imperfectly communicated. It may be responsible for any of the output gaps (i) to (iii). It may lead to a relatively small share of high productivity sectors, or to a low proportion of enterprises practising 'best practice' techniques or to the 'best practice' techniques themselves being of a lower standard.

IV. SUITABILITY GAP

But the Communications Gap cannot explain all the factors responsible for the output gap, because the adoption of 'best practice' techniques developed in industrial countries by less developed countries is likely to involve such a high ratio of capital and sophisticated skills to unskilled labour and such a large scale of production (factor iv) that, even with the best communications, poor countries simply do not have the resources. Compared with the rich countries, poor countries have to provide about three times the number of jobs with about one twentieth of investible resources per worker. These are rough orders of magnitude that show that only about one sixtieth of the investible resources per worker of the rich countries is available for the creation of jobs in poor countries, if only additions to the labour force are to be employed. The ratio is even lower if jobs are to be created for all those already unemployed and underemployed. This alters the scale of the whole problem of job creation and makes historical comparisons with the adaptation of existing technologies by industrializing Japan or Germany quite irrelevant.

This is partly what is meant by saying that rich men's technologies are inappropriate for the poor countries. It is for this reason that we must identify the second type of technology gap, the Suitability Gap. It is a gap in the development of the appropriate technology and of the resources devoted to the discovery of such a technology. The problem here is that techniques with low ratios of physical and human capital to labour but high ratios of output to capital just do not exist (perhaps do not exist any longer), so that the low capital-labour ratios, inevitable for a large proportion of the labour force, lead to extremely low workers' productivity. In many countries many workers have zero capital to work with: they are 'non-employed', rather than 'unemployed'. Resources devoted to closing the Suitability Gap would raise the productivity of labour-intensive techniques and would thereby help to close the output gap.

Inappropriateness is only partly a matter of differential capital availability. In many underdeveloped countries not only labour but also capital is underutilized. Technologies developed in the industrial West are also inappropriate because physical and social conditions differ. Many agricultural and some industrial technologies have been developed for countries in the temperate zone, whereas most underdeveloped countries are in the tropics. Again, technologies depend upon economic, social and political features that may be quite different in the two types of country. Even such a simple implement as a spade is not suitable for bare-foot diggers. Adaptation requires ingenuity and invention. It is not enough to add a simple wooden platform to the top of the spade, for earth would adhere to it. The same is true with much greater force of more complicated production processes which may depend upon large markets, hierarchical management structures, impersonal administration, or labour relations or cultural attitudes which differ in the country to which the technology is transferred.[6]

'Scientific discoveries, and particularly their practical counterparts in inventions and technical improvements, are often the solution to a specific problem in a specific country adapted to the resources it possesses.'[7] To the extent to which this is true, the impression of universal availability of scientific knowledge is an illusion. The agricultural revolution met the needs of agriculture in eighteenth century England with plentiful land and a temperate climate. The improvements of the USA were adapted to a large market of mass consumers and less suitable for smaller countries with more differentiated markets. The factory organization in Western Europe and North America with its system of hiring and firing labour is adapted to a set of human relations quite alien to other cultures.

It is still often said that there is no lack of knowledge as to how to develop, only a lack of will to apply it. This view combines two deep-seated nineteenth century biases. One consists in the abstraction of economic theorizing from scientific innovation and from technical progress. In its nineteenth century guise the assumption is that the 'state of the arts' is constant; in its twentieth century guise that a simple transfer of existing technologies to underdeveloped societies is possible. The other ideological element is to put all the blame on the inability or unwillingness of the people in underdeveloped countries to make use of existing facilities. This comes near to the prejudice about the 'ignorant and idle native'. But to anyone who has worked in this field, it is clear that even the adaptation of existing knowledge to the different physical and social needs and conditions of less developed countries requires creative and imaginative innovation comparable to new discovery, so that there is no hard and fast line between the Communications Gap and the Suitability Gap.

But in many areas there is not even a technology which can be adapted. The technology has to be invented; in some cases, possibly, reinvented, for technical knowledge decumulates as well as accumulates, and things once known and now forgotten may be useful; but invention or discovery there has to be.

The absence of technologies appropriate to the low income countries is the result, partly of the high incomes and the corresponding pattern of demand in advanced countries, and partly of the high ratios of saving and investment per head, to which these high incomes give rise. They are therefore doubly inappropriate: over a wide range of products it is nearly true to say that what the rich consume, the poor do not produce, and what the rich produce, the poor do not (or should not) consume.

The situation may be even worse, because the technologies and products for the rich may compete for scarce skills and capital with the technologies for the poor. Fast private transport, space exploration and catering for sophisticated consumers' choices may prevent or impede the development of technologies for efficient public transport or public health services, for low-cost construction, for mass literacy or for an improved basic diet. In addition, the technologies that meet the most sophisticated demand are also more exciting intellectually, more glamorous professionally, and more closely linked with advanced science. Hampshire (1971) put this point vividly in a review of _Snow's Public Affairs_.

> 'So faith in technology may result in ever better methods in Massachusetts General Hospital and ever greater medical poverty in outlying places, which lose their few remaining doctors to the centres where progress is made: or, more frivolously, the journey from London to New York becomes shorter and shorter, and from Oxford to Cambridge longer and longer, until after a few more years of 'advance', the first will be shorter than the second. The super-highway and aerodrome come near the village just as the train and bus services are phased out. The older technologies had their corresponding social forms; as they are replaced by the new, those who live at the bottom of the scale of opportunity are left out of the social progress...Just as much modern architecture is fun for the architects, but leaves the mere populace depressed; so much very advanced technology is fun for the technologists, and its few beneficiaries, but leaves the world's villagers just where they were before. As a social scientist, Ludd was inclined to oversimplify; but we still do not have an adequate theory to put in the place of his.'

V. RELATION BETWEEN THE TWO GAPS

The common factor in the two technology gaps is that the acquisition of technology is costly though the method of acquisition differs for the two gaps. Imperfect communications and other obstacles to transfer raise the costs of acquisition. But for bridging the Suitability Gap, non-availability rather than high cost is the problem. The implications for analysis and policy of high cost are different from those of non-availability.

The two types of technology gaps, Communications Gap and Suitability Gap, have different causes and call for different measures. Closing one may even open up the other. The Communications Gap can be narrowed by improving communications, compiling catalogues, establishing clearing banks, easing patent laws, facilitating licensing, encouraging training of engineers abroad, changing the balance of bargaining power, etc., as well as by better scientific and technological education and training. But the more and the better advanced technology is communicated, the less likely it is that a new and more appropriate indigenous technology will be developed. Better communication reduces both the opportunity and the will to invent an indigenous technology. closing an economy off against foreign influences and particularly against the multinational enterprise, may be the correct policies if we wish to close the Suitability Gap.[8] Attempts to close the Suitability Gap by devoting resources to the development of an appropriate technology may be frustrated by the increasingly well-communicated foreign technology, producing 'better' goods more 'efficiently'.[9]

At a deeper level, the question arises whether societies developing with different methods and making different products can be properly compared for the purpose of 'gap' calculations. A poor country neither can nor should fully and uniformly adopt the technologies of a rich country, not only because it has different factor endowments, different social and physical (including climatic) conditions and different patterns of demand, but also because it has different objectives. To develop special technologies appropriate to the economic, social and physical conditions of the poorer country, involves accepting permanently some kind of differences between the two societies. Whether these differences are properly understood as 'gaps' or just as varieties is a matter partly of semantics and partly of valuations. The measurement of output gaps is difficult enough when it is remembered that these will depend upon how different components of measured output are weighted and how unmeasured costs and benefits are treated. The difficulties are similar and even greater when we attempt to define and measure technology gaps.

VI. A POSSIBLE OBJECTION

A recent trend in the location of labour-intensive processes by vertically integrated multinational firms in low income countries may appear to contradict what I have been saying so far. Especially in electronics and electrical components, but also in garments, gloves, leather, luggage, baseballs, watches, motor car parts and other consumer goods, and in electrical machinery, machine tools, accounting machines, typewriters, cameras, etc., processes that require much labour and initially limited capital and skills have been located in South Korea, Taiwan, Mexico, Hong Kong, Singapore, the West Indian islands, Mauritius. In one sense, the doctrine of comparative advantage seems to be vindicated, though in a manner quite different from that normally envisaged. It is foreign, not domestic capital and management that are highly mobile and combined with the plentiful immobile domestic labour. Specialization is not by commodities but by factors of production: the poor countries specializing in low-skills and labour, leaving the rewards for capital, management and knowhow to the rich. Cost advantages are not passed on to consumers in lower prices and the profits accrue to the parent company. The continued operation of this type of international specialization depends upon the continuation of substantial wage differentials (hence absence of trade union action to push wages up), continuing access to the markets of the parent companies (hence stronger pressure from importing interests than from domestic producers displaced by the low-cost processes and components including trade unions in rich countries) and continuing permission by host countries to operate with minimum taxes, tariffs and other regulations. The bargaining power of host countries in this situation is likely to be weak and the question is whether such a division of gains between parent and host, between the investment 'package' and labour, remains acceptable. The gains to the host country are confined to the wages of those employed if the alternative is unemployment and the foreign exchange minus the profits remitted and the capital repatriated. There may, in addition, be linkages, but these may be positive or negative. While such a strategy has attractions for some countries faced with labour surpluses and foreign exchange shortages, and poorly endowed with natural resources, the potential gains may not be considered worth the social risks and social costs, including a form of dependence and of dualistic development, that they bring with them.

VII. OBSTACLES TO THE SPREAD OF BENEFITS

Why, then, is there such a wide and widening gap between potential and realizations, between the S & T of the most advanced and of the least advanced countries? Why are there technology gaps? There are (at least) three reasons for the gaps.[10]

First, scientific attitudes and institutions in many less developed countries are weak or absent and there are obstacles to the use of S & T in the social structure of these countries. They often lack a fully rational, experimental and scientific outlook and are not adapted to the introduction or acceptance of systematic change. This explains partly why no use is made of so much S & T that already exists and is available. (This is relevant to the Communication Gap.) The absence of institutions refers not only to institutions directly concerned with S & T, such as research and training institutes. All social institutions may bear on the acceptability of S & T. Thus an antiquated system of land tenure may prevent the use of modern seeds or irrigation to improve crops. A system of sharing earnings with family members may prevent the growth of an innovating entrepreneurial class. Bureaucratic red tape may discourage business men from innovating. A system of social values may attach low status to engineers and entrepreneurs. There is a vicious circle: S & T are necessary for development, but equally, development, with the disruptions that it brings, is necessary for the growth of S & T. Only those who already have an indigenous S & T capability can successfully absorb new S & T. It can be very hard to break out of this mutually reinforcing low-equilibrium trap.

The second reason is the heavy concentration of expenditure on research and development (R & D)[11] in developed countries and the wrong orientation and emphasis of this expenditure there: wrong, that is, from the point of view of solving the problems of the poor countries and contributing to a wider sharing of the benefits of S & T. The orientation of most R & D expenditure is either irrelevant to these problems (like space research) or positively detrimental to their development efforts (like the concentration on improving temperate zone agriculture at the expense of tropical agriculture of the $1,000 million per year research on synthetics which knocks out many exports of the developing countries). (This is relevant to the Suitability Gap.) Of the total expenditure on R & D, which is an important and relatively easily measurable part of S & T, only 1 per cent is specifically directed at the solution of the problems of the poor countries.

The wrong orientation has, in turn, three results. First, it leads to a misorientation of efforts in the less developed countries themselves, because standards and interests are prescribed by what goes on in the most prestigious centres in rich countries. This has been called the internal brain drain: the diversion of talent inside the developing countries to problems irrelevant to their development.

Secondly, it leads to the (external) brain drain, so that the scarce professional manpower, educated and trained in poor countries, contributes to augmenting the S & T of the rich countries. The flow of scientists and engineers from poor to rich countries now exceeds the flow of technical assistance from rich to poor. The inability of the developing countries to find employment for these people is both cause and effect of their loss of professional manpower.

Thirdly, the wrong orientation leads to a composition of the stock of knowledge that may be harmful to the development efforts of the poor, like the concentration on synthetics already cited. Another instance of the wrong direction of existing R & D is the industrial technology, developed in conditions of labour scarcity and capital abundance. There is also the concentration on <u>products</u> that meet the needs of high-income, labour-scarce societies, while gaps remain in the range of products that meet the needs of low-income societies in which labour is abundant.[12] Transfer of inappropriate technologies and products aggravates the large and growing underutilisation of labour from which all underdeveloped countries suffer and it increases inequality in these countries. This is especially relevant to the obstacles to closing the Suitability Gap. As a result, the international technology gaps are matched by domestic technology gaps inside the poor countries. The existence of such gaps within poor countries would not matter if the countries proceeded, like Japan or China in their different ways, 'on two legs', so that modern technology co-existed with traditional and intermediate technology, each supporting the other. But in many countries the modern technology destroys the traditional one and eliminates opportunities and incentives to invent a substitute. International inequalities in job opportunities and income distribution are thus matched by internal inequalities. The modern technology, with its sophisticated products, distributes the lion's share of incomes to profit earners and the small, employed labour aristocracy. These groups in turn provide the market for the high technology products, make the investment profitable and constitute vested interests for its perpetuation. Inequality is both the result and the cause of the misapplied high technology.

In these ways, the small share of 2 per cent of total world R & D that poor countries spend on R & D is (through the internal brain drain) rendered less productive than it otherwise would be and the large share of 98 per cent is either irrelevant or harmful. Of the 98 per cent, 45 per cent is spent on defence and space and another 7 per cent on atomic energy. Even with the most generous allowance for unexpected spin-offs, this a fantastic waste of world resources.

It is important not to make false claims. Statisticians have found no correlation between R & D expenditure and aggregate economic growth. Indeed, there is evidence of a negative correlation between growth and the proportion of GNP spent on R & D. The negative correlation may be due partly to the fact that countries that spend a lot on R & D also spend a high proportion of it on defence, space and allied projects and partly to the fact that much R & D aims at product innovation, not at higher productivity. The links between R & D, S & T and growth are subtle and complicated. In addition to the fact that much innovation is not S & T and that much technology is not the result of R & D, the effective application of innovations usually requires capital and the right type and attitude of people. Large expenditures on R & D can remain unproductive for lack of capital or of engineers and business managers, or absence of a receptive social climate. Another reason for the lack of correlation between the proportion of GNP devoted to R & D and growth is the international spread of the fruits of R & D. Countries with little R & D can acquire the results from those who commit more resources.

On the other hand, at the industry level R & D expenditure appears to yield high rates of return. There can be no doubt that problem-orientated applied research pays off handsomely. Quite small sums devoted to research into the new varieties of rice, wheat and maize showed how the identification of a problem and research directed at its solution can yield high returns. It also, incidentally, illustrates the obstacles put into the path of the spread of progress by the nation state. Only non-governmental institutions like the Ford and Rockefeller Foundations could devote money to a purpose that would benefit others than those in whose country the work was done. The much larger sums of national aid have to be used for purposes that the recipient nation states conceive as in their own interest, which tends to be rather narrowly interpreted.

The third reason for the gap is the prevalence of obstacles to access by less developed countries to whatever relevant and useful S & T exists in the world. Communications in this area are poor, institutions absent or weak and the wheels of the transfer mechanism are badly oiled. This is relevant to the Communications Gap.

VIII. WHY HAS THE MARKET NOT CLOSED THE GAP?

We now turn to policies. The first question that occurs to an economist is to ask why has the market system not provided incentives for the appropriate direction and utilisation of S & T? Though underdeveloped countries are poor, they are potentially large and growing markets. Why have there been so few inventions of low-cost, simple, agricultural or industrial machinery? Why has there not been more progress in low-cost construction or transport? Why do those industrial countries that have a comparative advantage in manufacturing industry protect, often at high cost to themselves, their agriculture, instead of exchanging low-cost machinery and durable consumer goods (say a 10 pound refrigerator) for the agricultural exports of underdeveloped countries? Henry Ford announced in 1909 that his aim was to produce and sell a cheap, reliable model 'for the great multitude' so that every man 'making a good salary' could 'enjoy with his family the blessing of hours of pleasure in God's great open space'. The mass production of the model T Ford ushered in a major industrial and social revolution, the cars of which have, incidentally, destroyed the 'great open space'. Why has no one initiated a corresponding revolution to raise and tap the purchasing power of the world's teeming millions? Insufficient foresight in the face of still small markets (small in terms of purchasing power) and over-estimation of risks or a divergence between private (including political) and social risks may be part of the explanation but it cannot be the whole.

It is easier to see why the market in complex, specialised, often secret or patented, modern technology is different from the market for turnips or even for land. Technical and managerial knowledge and its commercial and industrial application cannot easily be assimilated to the treatment of the conventional factors of production: land, labour, and capital, for at least five reasons.

In the first place, knowledge, although clearly not available in super-abundance, is not scarce in the sense that the more we use of it in one direction the less is left over for use in another, or the more I use it, the less is left for you. The use of knowledge is subject to indivisibilities and average costs diverge widely from marginal costs. The result of this is that it is much cheaper for the multinational firm to use what it already has: the existing but inappropriate technology developed in high-income, labour-scarce countries, than to spend money on developing a new technology, more appropriate for the conditions of the developing countries.

Secondly, there is the well-known difficulty of appropriating the fruits of efforts devoted to increasing knowledge and the need either to treat it as a public good or to erect legal barriers to appropriation by others, in order to create and maintain incentives for research and invention. This leads to the divergence of social from private benefits and costs.

Third, knowledge is, in a sense, substitutable for other

productive factors, so that an improvement in technical knowledge makes it possible to produce the same product with less land, labour or capital, or with more capital but a more than proportionate decrease of labour or land, or a better product with the same amount of other factors. But its costs fall under those of either labour (especially trained employees) or capital (purchase of patents or research laboratories or equipment or intermediate products or other assets embodying the knowledge). As a result, the market for knowledge is normally part of the market for these inputs. If the owners of the inputs that embody knowledge command monopoly power, they can exercise this power over the sale of the knowledge component of the whole package.

Fourth, the accumulation of knowledge is only tenuously related to expenditure on its acquisition. Indeed, useful knowledge can be accumulated without any identifiable allocation of resources for this purpose and, conversely and more obviously, large resources can be devoted to research without any productive results. There is, in the nature of discovery, uncertainty about the outcome of efforts devoted to inventions. This uncertainty cannot be removed by insurance, for insurance would also remove the incentive for research. A common way of reducing it is through diversification of research activities. Only large companies are capable of this. In a private enterprise system the large multinational firm has an enormous advantage in reducing the risks attached to research (Arrow, 1962, pp. 609-26).

A fifth and even more fundamental difference lies in the absence of the justification of the common assumption about the 'informed' buyer. Where technology is bought and sold, as it often is, through the purchase of an asset (or through admitting direct private foreign investment), the underdeveloped recipient country as 'buyer' of the technology is, in the nature of things, very imperfectly informed about many features of the product that it buys. The common assumption about an informed buyer choosing what suits him best is even less justified here than is usual. If the country knew precisely what it was buying, there would be no need--or considerably less need--to buy it. Knowledge about knowledge is often the knowledge itself[13] (Vaitsos, 1970). Part of what it buys is the information on which an informed purchase would be based. As a result, the recipient government will be in a weak position _vis a vis_ the investing firm when it comes to laying down terms and conditions. Excessive 'prices' paid by recipient governments for capital equipment or imported components and technologies inappropriate from the country's point of view, or acceptance of excessively onerous conditions must therefore be the rule rather than the exception in a market where information embodied in equipment is bought by ignorant buyers.

The five features characteristic of the market for technical knowledge--(i) indivisibility, (ii) inappropriability, (iii) embodiment in other factors, (iv) uncertainty and (v) impossibility to know the value until the purchase is made--go some way towards explaining the absence of a free market in which the low-income countries could buy this knowledge.

The situation is quite different from that of an 'equilibrium price' reached in a competitive market. It is more like that of a bilateral monopoly or oligopoly where bargaining theory applies. There is a vast gap between the incremental cost to the owner of the technology of parting with it and the value to the country or firm wishing to acquire it. The cost to the seller is either zero, since the investment has already taken place, or the small amount required to adapt it to the circumstances of the developing country. The value to the buyer is the large amount that he would have to spend to start inventing and developing from scratch and to 'go it alone'. The final figure in the range between these two limits is determined by bargaining strength, which is very unequally distributed.

Here again, international inequality and internal inequality in the poor countries reinforce one another. Unequal income distribution is both effect and cause of inappropriate technologies and products. It is an effect because capital-intensive methods and products raise the share of profits and of rewards for sophisticated skills and reduce that of unskilled labour; and markets for sophisticated, differentiated products require a small elite with high incomes. And it is a cause, because the existence of a market for differentiated luxuries deprives enterprises of an incentive to produce low-cost, most appropriate products for a mass market. Henry Ford had the advantage not only of imagination but also of relatively high real wages.

IX. FOUR POLICIES FOR CLOSING THE GAP[14]

Policies for closing the gap suffer, as we have seen, from the dilemma that measures that narrow the Suitability Gap *may* widen the Communications Gap and *vice versa.* There may, therefore, be alternative routes. Nevertheless, a carefully selected combination of measures for narrowing both gaps is possible. But they have to be seen as a package and have to be adopted together. The adoption of any one of these policies without the others may be worse than not doing anything. The first three types of policy and the need to strengthen bargaining power are relevant to the Suitability Gap; the first also and the fourth policy to the Communications Gap.

First, the less developed countries must build up their indigenous S & T by raising their expenditure on R & D above the present 0-2 per cent of their GNP. It is, of course, true that R & D, even when devoted to the discovery of labour-intensive, low technology processes and products, is itself a skill-intensive high-technology type of activity. And it has been argued that countries in which these resources are scarce should rely on borrowing, buying and adapting the fruits of R & D from abroad. But the ability to borrow, buy and adapt wisely itself depends, as I have argued, on an indigenous scientific capability. R & D abroad is therefore no substitute for R & D at home. While the total percentage is not meaningful, and while opportunities of acquiring and adapting technologies from abroad should not be neglected, the percentage provides a guideline, if combined with the other three policies. At present, not only is the ratio very low, but much of it is spent on research irrelevant to development.

Second, a higher proportion of development aid should be devoted to supporting S & T. Not all of this need or should be intergovernmental aid. As we have argued, non-government aid can have special virtues.

Third, the advanced, industrial countries should devote a larger proportion of their total R & D expenditure (the Pearson Commission recommended 5 per cent), to R & D that is directly relevant to the problems of the developing countries. Priority areas have been identified and include extension of the Green Revolution to millet, sorghum, tuber crops, etc., pest control, desalination, the development of salt-resistant plants, the use of sea water for irrigation, birth control, cyclone warning and weather control, solar power, production of edible proteins, appropriate industrial technology, use of tropical hardwoods for pulp and paper production, use of indigenous building materials, exploitation of the ocean and sea beds, irrigation, new forms of transport (United Nations, 1971).

Here again, precise delineation of what is and what is not beneficial to the developing countries is not possible. Medical research on tropical diseases clearly is directly relevant, desalination is a useful spinoff of research not directly relevant. But by any standards, directly relevant research is now a very small proportion of total R & D expenditure in developed countries.

Fourth, there should be improved access to what is available. An international technology transfer bank has been proposed that would, on the one hand, reduce the risks of those acquiring and wishing to sell S & T, and, on the other hand, reduce the costs for those wishing to buy it. But the basic task goes much deeper than overcoming imperfections of the market: it is to bring about a more equal balance of bargaining power.

Bargaining power is, in turn, partly a function of information and partly of solidarity between countries. Access could be greatly improved by adding to informed knowledge of the buyer, by assisting him in bargaining with the multinational owners of technology, and by joint, rather than competitive action of several buying countries. Potential buyers of knowledge _within_ an advanced industrial country are in a much stronger position than less developed countries, because the government in the former shoulders a large proportion of the costs of acquiring knowledge, through subsidies, tax deductions and the direct supply of information. No similar mechanism of insurance, subsidy and diffusion is at work internationally. The governments of developed countries are also in a stronger position than those of underdeveloped countries in bargaining with the multinational firm, because their officials and businessmen can match its information and skills.

How then can we remove this handicap? Since bilateral assistance in bargaining is bound to be suspect, in view of the presumption that parent countries will support their own parent firms, this is eminently an area for multilateral assistance, at any rate until international political institutions have caught up with the multinational company.

SUMMARY

We have started by asking the question: why are there international income inequalities? We have ascribed them to differences in applied productive knowledge. This raised the question: why can productive knowledge be communicated and diffused within an advanced nation but not between nations or within underdeveloped nations? What are the obstacles to the international diffusion of benefits?

We have found these two areas: obstacles to communication and absence of suitable technologies. The obstacles to communication can again be divided into those due to costs of transfer and those due to intentional restrictions or the exercise of monopoly power. but even perfect communication would not meet the need for quite different technologies from those developed in high-income countries.

Measures that reduce the Communications Gap might make the Suitability Gap wider and vice versa, but a set of integrated actions attacking both gaps has a chance of success. Technical knowledge cannot be marketed like other products or factors because it possesses peculiar features: (i) indivisibility, (ii) inappropriatability, (iii) embodiment in other factors, (iv) uncertainty and (v) impossibility to know its full value until bought. Policies for closing the two gaps are interdependent, so that the pursuit of any one in isolation might make matters worse. What is needed is a set of integrated actions, attacking both the Communications Gap and the Suitability Gap. Transfer must be supplemented by indigenous capability; adaptation by invention and innovation.

NOTES

[1]'Technology' in this lecture is used not in the sense in which the engineer might use it but in the economist's sense: it means the same as the production function which specifies the relation between inputs of all factors of production, including skills and organisation, and outputs. As a result, the discussion in this lecture is unaffected by such common assertions as 'there is no technological gap; there is only a management gap'; or 'the technological gap is entirely a research and development problem'.

[2]Although the bulk of this lecture is devoted to a discussion of the growth of scientific knowledge and its application, attitudes such as those towards entrepreneurs and the cultural, social and political institutions that provide the stimulus for the growth

of knowledge and that makes it possible to absorb it are equally important, though less visible, less well understood and therefore often neglected. Understanding the atom is child's play compared with understanding child's play.

[3]The argument was put forward by Simon Kuznets in a seminar at the Johns Hopkins University in 1956. It can also be found more subtly developed than here, in Kuznets (1954) and (1967).

[4]The relation between income per head and the various factors bearing on the technology gap is this:

$$\frac{\text{Income}}{\text{Population}} = \frac{\text{Output}}{\text{Labour Force}} \cdot \frac{\text{Labour Force}}{\text{Population}}$$

$$= \frac{\text{Output}}{\text{Capital}} \cdot \frac{\text{Capital}}{\text{Labour Force}} \cdot \frac{\text{Labour Force}}{\text{Population}}$$

[5]I am indebted to Mrs. Taya Zinkin for coining this term.

[6]At the same time, there are strong pressures from the underdeveloped countries themselves to seek only the most modern, most sophisticated and often most capital-intensive equipment, even where simpler and lower-cost items are available. Even, or perhaps especially, the smallest countries feel they must have a Boeing 707 if they have an airline or go without it. Many countries prohibit the import of used or obsolete equipment. Sterling International, a Californian pulp, paper and board exporting firm, reported that there had been a change from manufacturing lavatory paper with perforations with a sawtooth edge to perforations in a series of straight lines, like a Morse code. Machinery to make the former cannot be converted to the latter and this made all sawtooth perforating equipment obsolete in North America and available for next to nothing. The firm shipped out three such machines to plants which they had established in Hong Kong, Singapore and Beirut. In each case, marketing pressures in the areas compelled them to re-equip the plants to produce the more elaborate perforation adopted in North America. To have less elaborate perforation than the Americans, even in such private places as lavatories, was regarded as undignified.

[7]Kuznets (1954), p. 244.

[8]Hagen (1962) considers Indonesia, India, China and Japan. The country with the greatest degree of contact with the West is Indonesia, where the Dutch were present for 300 years. Next comes India where the British gradually expanded their footholds; then China, where trade along the coast created enclaves from which trade with the interior was forced on the country; last Japan where the Tokugawa enforced a policy of no contact with the West except through a small Dutch trading group. Yet, Japan started to grow first and made rapid progress; China is well on the way; India comes next and Indonesia last. The order of

economic advance is the reverse of the order of contact (Hagen (1962), pp. 24-5).

[9]In discussing ways of building up local technical capability, Charles Cooper (mimeo) writes: '...governments may simply refuse to allow local companies to sign licence agreements and demand predominantly local development of the technology....Japan, for example, has been able to follow restrictive policies in certain cases where local technical capability is highly developed. It appears that the Indian government has also limited technical agreements from time to time and _de facto_ stimulated some local development in various fields of process technology.' See also (1970).

[10]The following owes much to Singer (1970). pp. 18-41.

[11]R & D, though clearly related to, is not the same as S & T. Many technological innovations are done by small inventors or are the by-product, intended or unintended, of other activities or are imported. Again, innovation overlaps with, but does not coincide with, narrowly interpreted S & T. Some of the most important innovations relate to changes in organisation, marketing, accounting, personnel management, etc.

[12]Furtado (1964) has emphasized the inappropriateness of many _products_ introduced into developing countries, in contrast to the usual emphasis on inappropriate processes. If the costs of thermal and other pollution, and the use of irreplaceable, exhaustible raw materials were to be fully taken into account, the appropriateness of many sophisticated products would become more doubtful even for rich, industrial societies. In any case, appropriate costing would lead to a different international division of labour and location of production.

[13]Arrow (1962) writes: '...there is a fundamental paradox in the determination of demand for information: its value for the purchaser is not known until he has the information, but then he has in effect acquired it without cost.'

[14]These are the policies recommended in the paper by the Sussex Group, except that the argument for the fourth policy is strengthened. The targets of 0-5 per cent of GNP for R & D by less developed countries and of 5 per cent of aid for R & D by developed countries have been adopted by the Pearson Commission and are part of the international strategy of the UN Second Development Decade.

REFERENCES

ARROW, K. J. (1962). Economic Welfare and the Allocation of Resources to Invention. The Rate and Direction of Inventive Activity: Economic and Social Factors. National Economic Research. Princeton University Press.

COOPER, C. (mimeo). The Mechanism for Transfer of Technology from Advanced to Developing Countries.

COOPER, C. (1970). Instrument Industry in India. Bombay.

FURTADO, C. (1964). Development and Underdevelopment. University of California Press, Berkeley.

HAGEN, E. E. (1962). On the Theory of Social Change. Tavistock Publications.

HAMPSHIRE, S. (1971). The Observer Review, 31st October.

KUZNETS, S. (1954). Economic Change. Heinemann.

KUZNETS, S. (1967). Modern Economic Growth, Rate, Structure and Spread. Yale University Press.

SINGER, H. (1970). Science and Technology for Development--Proposals for the Second U.N. Development Decade. United Nations, Sales No. E.70 I.23.

UNITED NATIONS (1971). World Plan of Action for the Application of Science Technology to Development. U. N. POublication, Sales No. #.71 Ii.A.18.

VAITSOS, C. V. (1970). Bargaining and the Distribution of Returns in the Purchase of Technology by Developing Countries. Bulletin of the Institute of Development Studies, Vol. 3 October 1970.

The Technological Gap Between Developing and Developed Countries

UNCTAD SECRETARIAT*

The construction of a viable framework for the technological transformation of the developing countries should begin with a proper understanding of the technological gap among nations. This is a subject on which there is a wide diversity of views. At one extreme it is maintained that the gap is very large and growing. At the other extreme is the view that it is not so large and that the bases are being built up for narrowing it rapidly. Unfortunately, there has been very little systematic analysis on which to draw any firm conclusions.

Of course, each of these extreme views, and any intermediate one, is conditioned by the information and perceptions on which it is based. For example, if the number of patent holdings, or per capita expenditures on R and D or the number of persons engaged therein, or the current technological situation in the most advanced sectors (such as computers and nuclear and space technology), are taken into account, the gap would appear to be very wide, and perhaps even growing. However, if the current levels of productivity in different sectors (a combined result of embodying technology in tools and persons), or the technological inputs actually required for the implementation of the projects and programmes of the national development plans of developing countries, are taken into account, the gap would appear to be not so wide.

Each of these extreme views understandably offers its own policy prescriptions. Those who consider the gap very wide emphasize the excessive degree of dependence and the impossibility of lessening it in any foreseeable future, and hence pay the greatest attention to external factors, considered either as hindering or as stimulating domestic development. The potential for national action is then severely constrained. Those who view

From **PLANNING THE TECHNOLOGICAL TRANSFORMATION OF DEVELOPING COUNTRIES**, (UNCTAD Secretariat), 1981, (19-22), reprinted by permission of the publisher, U.N. Publications, N.Y.

the gap as not so wide generally maintain that the task is to pursue the building up of national capacity so that maximum advantage may be taken of all available technologies, wherever they have been developed, and to embody them in the instruments of production and in the labour force in order to achieve an accelerated rise in productivity. They urge a rapid expansion of output of producer goods, training of technological personnel and promotion of design and engineering capabilities, so as to achieve as quickly as possible the objective of much greater technological and economic independence.

These views on the size of the technological gap are so far apart that there is need for a systematic examination of the bases on which informed judgements could be made. This chapter begins with an indicative measurement of the over-all technological gap. The next section considers the gap in the major economic sectors. The last section sets out some preliminary conclusions, highlighting the dynamics of technological transformation. Because of the weakness of the data used and the preliminary nature of the analysis, the conclusions are to be treated as only a point of departure for more systematic work on the subject in the future.

I. INDICATIVE MEASUREMENT OF THE OVERALL TECHNOLOGICAL GAP

As shown above, the absorption of technology in any country can best be measured by the manner in which man is able to produce more goods and services through the embodiment of technology in the instruments of production and in the skilled personnel using them. Output per capita, or output per head of economically active population (these are functionally related and generally move in parallel), can then be used as a rough and ready measure of the degree of absorption of technology.

As may be seen from table 10, per capita GDP in 1975, converted into dollars at current rates of exchange, was $460 in the developing countries and $5,130 in the developed market-economy countries, an approximate ratio of 1:11. Expressed in terms of GDP per head of economically active population, the comparable figures were 1,230 for the developing countries and $12,290 for the developed countries, a ratio of 1:10, or slightly lower than when total population is considered.[1]

Given a technological gap of this size (about 1:10) in 1975, it is relatively simple to work out the general dynamics of the technological transformation of developing countries required to reach the 1975 technological level of the developed countries. It would take 80 years to reach the 1975 level with an annual growth of per capita GDP of 3 per cent, 60 years with an annual growth of 4 per cent and 50 years with an annual growth of 5 per cent. These relationships would, of course, vary in accordance with the size of the gap (i.e. depending on whether the 1975 level, or any other past level or one assumed for the future, was sought to be attained), and with the assumption concerning the annual rate of and developing countries, 1975 [a]

Sector	GDP by sector: Total			Per head of total population [b]			Per head of economically active population [b]		
	Developed market-economy countries (billions of dollars) (1)	Developing countries [c] (billions of dollars) (2)	Ratio (1)/(2) (3)	Developed market-economy countries (dollars) (4)	Developing countries [c] (dollars) (5)	Ratio (4)/(5) (6)	Developed market-economy countries (dollars) (7)	Developing countries [c] (dollars) (8)	Ratio (7)/(8) (9)
Agriculture	179	185	0.96	230	100	2.3	5 190	430	12.2
Industry	1 844	337	5.5	2 320	180	13.0	14 230	2 370	6.0
of which:									
Mining	57	47	1.2	70	30	2.9	17 810	9 220	1.9
Manufacturing	1 154	168	6.9	1 450	90	16.3	13 640	1 870	7.3
Consumer goods [d]	374	81	4.6	470	40	11.0	10 600	1 290	8.2
Intermediate goods [e]	219	40	5.5	280	20	13.1	18 720	3 250	5.8
Capital goods [f]	561	47	11.9	710	30	28.2	14 960	3 070	4.9
Electricity, gas and water	102	16	6.4	130	10	16.0	36 430	7 270	5.0
Construction	241	51	4.7	300	30	11.2			
Transport and communications	290	54	5.4	370	30	12.6	12 610	2 650	4.8
Services	2 056	340	6.0	2 590	180	14.4			
Total	4 079	862	4.7	5 130	460	11.2	12 290	1 230	10.0

Sources: UNCTAD, *Handbook of International Trade and Development Statistics, Supplement 1977* (United Nations publication, Sales No. E/F.78.II.D.1), and United Nations, *Statistical Yearbook 1977* (United Nations publication, Sales No. E/F.78.XVII.1).

[a] Estimates of GDP by sector were derived by applying to the regional GDP figures (available in the UNCTAD source) the estimated percentage shares of sectors in each region. The latter were estimated on the basis of the sectoral weights used for the construction of GDP and world industrial production indices (1970 = 100), brought up to 1975 using the sectoral and subsectoral indices; the relevant figures are found in the United Nations source mentioned above. Total economically active population and the proportion in agriculture were taken from FAO, *Production Yearbook*, 1970 and 1977, and the proportion in industry was estimated on the basis of sectoral weights shown in industrial employment index tables in the United Nations *Statistical Yearbook* and the estimates made by ILO on manufacturing employment for certain years, in *Labour Force Estimates and Projections, 1950-2000*, vol. V, *World Summary* (Geneva, 1977). Economically active population in other sectors was derived as a residual. Because of these crude methods of estimating, and also because the country groupings vary slightly for the different series, the table provides only rough orders of magnitude.

[b] Rounded to the nearest $10.

[c] Not including China and other socialist countries of Asia.

[d] ISIC divisions 31, 32, 33, 34 and 39.

[e] ISIC divisions 35 and 36.

growth or the period of time within which it was sought to accomplish the transformation.

Indeed, the dimensions of time and pace discussed above would seem to overstate rather than understate the real position. They are based on the use of per capita income comparisons at current exchange rates--a practice that was very widespread in the 1950s and 1960s. It is now generally recognized that such comparisons can give a highly misleading, in fact a very distorted, view of reality.

A more systematic analysis of differences between countries in the more recent period has contributed to an improved understanding of the limitations of such comparisons of per capita output. To begin with, the size of the labour force, the number of working hours and levels of unemployment or underemployment vary from country to country. Moreover, the varying extent to which output is actually marketed, the fact that certain activities are not covered in the measurement of GDP (particularly the major contribution of the unpaid labour of women in the home, which is at best marginally reflected in present methods of income estimation), differences in relative prices, and the limitations of exchange rates as a tool for measuring such differences among countries, all combine to overstate the differences as reflected in superficial comparisons of per capita income or output among countries. Any effort at an adequate understanding of the technological gap must therefore keep these limitations in mind and devise means of taking them into account.

The influence of one important limitation--the use of current exchange rates--may be examined here in some detail. Recent work on comparisons of gross product in terms of purchasing power parity shows the major distortion that results when comparisons are based on current exchange rates alone. It has been estimated, for instance, that the deviation from real purchasing power resulting from the conversion of the value of national output in 1970 at current exchange rates was as high as 3.49 for India, 2.32 for Colombia and 1.91 for Kenya. In other words, the ratio of the purchasing power of the GDP of India to that of the United States of America in 1970 works out at only 1:14, whereas when comparisons are made on the basis of current exchange rates the ratio is as high as 1:50.[2]

If allowance is made for the distortions resulting from the use of current exchange rates, it would seem that the real technological gap would be narrower than the ratio of 1:10 mentioned above. The per capita availability of products is not, of course, a complete indicator of the standard of living of a society, or of its well-being in the widest sense. The latter depends not only on the size of national income and how it is distributed but also on the stock of accumulated assets, such as durable consumer goods, housing, social infrastructure and technology.

II. TECHNOLOGICAL GAP BY INDIVIDUAL SECTORS

The general ideas on the technological gap described above should be regarded only as a first step towards an informed understanding. Moreover, they tell little about the situation in different sectors of the economy, which may vary widely. An attempt has therefore been made to present in table 10 information on the sectoral composition of GDP per head of total population and per head of economically active population in different sectors. These are rather crude global estimates, based on simplified methods, and represent no more than broad orders of magnitude. None the less, several observations that have a certain policy significance can legitimately be made.

In the first place, a comparison of total GDP, broken down by major sectors of economic activity, shows that the differences between the developing countries and the developed market-economy countries are relatively small with respect to agricultural and mineral output. Thus, in 1975 the developing countries had an agricultural output somewhat above and a mineral output somewhat below that of the developed market-economy countries. The differences were much larger for the other sectors. The developed market-economy countries produced 6.9 times as much in the manufacturing sector as did developing countries, the differences in other sectors being a little smaller.

Secondly, since the total population of the developing countries is higher than that of the developed market-economy countries, it is more meaningful to compare per capita availabilities by sector. Agricultural output per head of population in the developed market-economy countries was only 2.3 times higher than in developing countries, a ratio for manufacturing (over 16) and capital goods (28) (see column 6 of table 10). The comparison for the service sector is less meaningful since the very large differences in the current levels of wages and salaries between the two groups of countries do not necessarily reflect differences in productivity.

The third observation is perhaps more pertinent to the measurement of the technological, or productivity, gap. Strikingly, the technological gap, measured in productivity per economically active person, is the smallest in mining and the largest in agriculture, the ratios being 1.9 and 12.2 respectively (column 9 of table 1). The small difference in mining is accounted for by its being mostly an export-oriented activity, often externally owned and controlled. The very large difference in agriculture is clearly a reflection of how this vast sector, where most of the labour, particularly the female labour, in the developing countries use mainly their muscle power, aided with little else but neolithic tools, has remained virtually untouched by the impact of technological advance. In comparison, the ratio for manufacturing was 7.3, a little higher (8.2) in the consumer goods industries, and lower (about 5.0) in the capital goods industry. These productivity differences are obviously a combined reflection of the technological intensity (inputs of more productive tools and more skilled labour force and management) and of the spread of technology among the sectors.

These rough indicators give not only a measure of the technological gap but also an indication of the main lines along which the technological transformation of developing countries could proceed. They are not blueprints but can serve as a broad frame of reference.

III. TECHNOLOGICAL DYNAMICS OF TRANSFORMATION

The indicators of time and pace required to reach levels as high as those prevalent in the developed market-economy countries in 1975 are summarized in table 2. Crude as they are, they give an insight into the extent of the rise in output that would be called for if the populations of the developing countries were to be supplied with a basket of goods and services adequate not only for meeting their basic needs but also for furnishing them with the amenities associated with the process of development. Much more systematic work is needed at the national level, with an analysis by sector, before valid indicators can be constructed to outline the dynamics of technological transformation in the developing countries.

The indicative estimates show that if the 1975 level of per capita availability of goods and services in the developed market-economy countries were to be set as an objective for the developing countries, its attainment within 50 years would require an annual per capita growth of output in the agricultural, mining, manufacturing and service sectors of 1.4 per cent, 2.2 per cent, 5.7 per cent and 5.4 per cent respectively. The time span would obviously be shorter if these estimates were to be adjusted for the factors that tend to overstate the present gap. Moreover, it would be shorter still if higher growth rates were achieved through rapid technological development.

This approach is intended to serve only as a frame of reference, and it shows that the task is not a hopeless one or one that would take centuries to accomplish. It is a manageable task, which could be accomplished within 25 or 50 years, with annual rates of growth of per capita GDP of 10 per cent and 5 per cent respectively, which are not impossible of achievement.

It should be emphasized that there is nothing sacred in the productivity level reached in 1975 in the developed market-economy countries. Real per capita productivity in these countries had in fact more than doubled since 1950. Depending upon cultural patterns, social requirements, resource endowments, and sectoral employment and productivity levels, any of the per capita GDP or productivity levels in the developed market-economy countries in the period 1950-1975 could be an objective for the technological transformation of a developing country.

It might be added that the developed countries themselves are now in search of an appropriate development path and may therefore change their objectives of economic and technological advance in any of the sectors for which data have been given here. Indeed, the need for fresh thinking on patterns of development is as urgent for these countries as it is for the developing countries. Each developing as well as each developed country will have to set its own targets for the future, taking fully into account its specific endowments and requirements.

The task of attaining such a transformation will be particularly formidable for the 39 countries and territories that constitute the hard core of the development problem and account for about three fifths of the population of all developing countries. Their average per capita GDP was below $200 in 1975, and they included some of the rather large developing countries and the least developed among the developing countries (see table 2). Their difficulties of transformation have been aggravated by the economic recession of recent years. The growth of their export earnings, from both primary products and manufactures, has declined at the same time as their import bill for capital goods and intermediate products, including energy, has risen very sharply. The foreign exchange requirement of these countries would require an even more careful planning hitherto, with increasing emphasis on reducing the costs of imported technologies and beginning to substitute domestically developed technologies.

An entirely new and unprecedented pattern of national action, founded on the rapid strengthening of the domestic technological capacity of developing countries, will be called for to accelerate their technological transformation. It is urgently necessary to supplement such action by removing international constraints, affording these countries the widest and freest possible access to the existing world storehouse of technologies, and facilitating the rapid diffusion, adaptation and assimilation of those technologies in production structures. This requires a bold programme of action aimed at accelerating the technological transformation of the developing countries and complemented by new forms of international and interregional co-operation.

NOTES

[1]The socialist countries of Eastern Europe and of Asia are excluded from consideration in this chapter because of the gaps in data and conceptual differences in measurement.

[2]See I.B. Kravis _et al., A System of International Comparisons of Gross Product and Purchasing Power_ (Baltimore, Md., The Johns Hopkins University Press, 1975). The authors conclude that the size of the exchange rate deviation tends to decline with a rise in the per capita level of GDP, expressed in dollars at current exchange parities (p. 5). Thus, current exchange rates tend to

exaggerate the differences at the extremes of per capita income levels. For a discussion of the same problem, see S. Kuznets, Modern Economic Growth . . ., op. cit., p. 385, where it is estimated that the ratio between Asia and the United States of America would be 31:1 at current exchange rates, but only 10.5:1 when United States price relatives are used.

Sources of Technology in Development

H. J. RUSH*

The accumulation of literature concerning technological change attests to the growing recognition of this factor as an integral part of contemporary economic and social activity.[2] Reliance on technology to perform a wide range of functions has come to affect virtually every sphere of human activity (Dickson, 1974), for a large and increasing proportion of the world's population. As the dependence on technology becomes institutionalized by the choices made by influential forces within society concerning the direction of technical change, certain possibilities are opened up while constraints are simultaneously placed on others (Singer, 1977; Dickson, 1974). Realization of this has focused considerable interest on the role which the choice and source of technology plays in understanding the cause and the resolution of the difficulties (i.e. unemployment, environmental degradation, resource supply, inequality of income, etc.) which planners are concerned with.

The importance of technology, ascribed to in many discussions of development, is that it can contribute to economic growth, through the incremental and radical innovation of skills, machines and organizations. Innovation,[3] as Jones (1971) points out, can allow for a more efficient use of resources; greater productivity can be achieved with existing resources and resources that were previously unobtainable or considered of low value can be made available. The direction of this growth is determined by the 'rigidities' which exist in the available technology and what David (1975) has called the 'myopic' decisions whose objectives solely consist of the minimization of current and private costs of production. One result of the diffusion of technology should be improvements in sectoral industrial efficiency (Jones, 1971). But determining whether technology is used efficiently, and what

---

From **METHODS OF DEVELOPMENT PLANNING**, (161-173), reprinted by permission of the publisher, UNESCO Press, 1981, Paris.

criteria are used to judge efficiency, are the development and application of various forms of knowledge including managerial skills and technical know-how. Whereas some people see the contemporary forms of such skills as the only rational means of realizing the potential of the technology (what we might call the deterministic model), others would argue that these forms of knowledge themselves reflect and reproduce a wider set of social choice (what we might call the social-construct model).

It is often pointed out in the economic literature on innovation that the process of learning, in the form not only of structured training but also the more informal learning-by-doing, makes an important contribution to increasing this efficiency and therefore to sustaining growth. In learning-by-doing, as much as in formal training, what is learnt is to some extent predetermined. Clearly what is learnt depends upon what is done, and what is done -- the operations of production -- reflects earlier decisions about both what products are to be produced and which techniques and organizational arrangements are to be used to produce them (David, 1975). The course and location of learning (among persons), the direction of future technical knowledge, is influenced in part by present technical choices, the division of labour in the economy and workplace, and the social and economic context in which these exist.

Technical choices can therefore be instrumental in determining the nature of economic growth and thus in achieving various development goals. The development strategy chosen reflects the answers to such questions as: 'Who has the knowledge necessary for sustained growth?' 'Within which sectors of the economy is this knowledge concentrated?' 'Who controls investment and resources?' And 'What criteria of development are used?' Implicit in these questions is the need to develop an analysis of the political economy of technical choice.

Analysis of the political economy of technical choice is defined as the study of distribution of control over resources, the use of this control and where it comes from. The actual choice of technology is made by various actors; or decision-makers, often with different and divergent objectives influencing their decisions.[4] Singer (1977) lists among these decision-makers: the transnational enterprises, which strive to maximize world-wide profits; the large national firms, whose aim is to maximize profits at the local level; national governments, whose goals vary but may include modernization, redistribution of wealth, or increased employment; and the family firms and small firms, whose principal concern is to maximize family consumption (Singer, 1977). Other actors might include political parties, foreign States, military forces and international organizations, local communities, workers and unions. As these actors are all operating within a context constituted by class relationships one could extend the goals of, for instance, national governments as the preservation of the power of the dominant groups. The distribution of influence which each of these actors has over the available resources, and the coalitions which form between different actors, will be a significant factor in the innovation and diffusion of techniques and the direction in which development takes place.

Stewart (1977, p. 110-11) points out that there is a characteristic distribution of benefits associated with each technique in the current world system. On a general level, advanced industrial technologies typically concentrate these benefits in the monopoly enterprises of the advanced countries themselves, through the ownership of the equity and the receiving of profits through the sale of techniques. Benefits are also acquired within the Third World/peripheral countries for the minority of the population who work in or own part of the 'modern' sector and consume its products.

The use of advanced industrial techniques in the Third World often requires the adoption of methods of management and organization employed in the so-called 'advanced' countries (Stewart, 1977). This may be a deliberate importation of work relations or simply the conscious or unconscious acceptance of these relations because of a lack of opportunity to develop alternative social organizations or production around the imported technology. As mentioned above, this is alternatively seen as providing increased efficiency, or as implanting the social values associated with 'modern' technologies. Both of these views perceive a widening gap between people in the 'modern' and 'traditional' sectors of many Third World nations; whether or not there will be a 'diffusion' of entrepreneurial skills and patterns of work organization between these sectors or whether one actually depends on holding down the other, is a matter of dispute. As such a diffusion has largely failed to take place -- indeed gaps between and within countries and sectors have grown, in terms of productivity and income differentials -- a re-evaluation of wisdom concerning both the choice of technique and the source of technology has taken place.

This paper will concentrate on the question of the source of technology and the institutional process within, which Singer (1977) calls the 'supply determinants' of a country's technological mix. These have been classified as the acquisition of existing ('off the shelf,') technology, the adaptation of existing technology (widening the shelf), and the generation of new indigenous technology (building new shelves). The ability of planners to meet future criteria for appropriate choice of technique will be influenced by the characteristics of these three sources of technology.

I. EXISTING TECHNOLOGY

The majority of innovation activity incorporated in modern manufacturing and agricultural technique has been developed in the industrially advanced nations. Some view the existence of 'off the shelf' technology as a distinct advantage to less-developed countries -- it is seen as a means of avoiding the considerable expense of the structures necessary for creating such technology (cf. Singer, 1977). The examples of France, Germany, Russia and the United States in importing the Industrial Revolution from Great Britain are held up as illustrating the benefits of this historical 'late-comers' advantage.

There are, however, considerable disadvantages associated with this strategy, including the enormous financial expense, the unsuitable nature of the technology compared with factor endowments, the disruptive impact on indigenous scientific capability, and the loss of control over decisions internal to development (Singer, 1977; Stewart, 1977). Furthermore, the imperfections in today's technology markets, the different requirements for absorbing the technology, the inability of agriculture to provide the required surplus, the lack of large external or internal markets, and the different development objectives necessitated for late-comers make comparisons with experience of previous industrialization of little direct relevance in many Third World countries.[5]

Quite apart from being in an advantageous position, owing to the vast amount (and rapid turnover) of 'on the shelf' technology, the substantial information requirements necessary in making an appropriate choice place Third World Countries in a difficult position. In such conditions of uncertainty and because of dependent links already established with technology suppliers, many Third World firms have, at the present time, little actual choice, or incentive, to look further than the 'off the shelf' technologies which are suggested to them (Singer, 1977).

The main concern in the literature has been with the role, often negative, of transnational enterprises in the 'transfer' or 'commercialization' of technology. Sufficient evidence exists, however, which indicates that the restrictions and control placed by transnationals on the technology they 'provide' are basically similar to those restrictions placed by most suppliers of technology, regardless of size and global interests. Neither level of analysis is intrinsically better than the other; however, as the literature concerning the type of enterprise which exercises control is generally well known, we focus here on the nature of the control itself.

Cooper and Hoffman (1978) outline three forms of control which are employed in technology transactions. The first and most straightforward is the 'direct' sale of technology. Here firms (or States) purchase techniques and technology directly from capital-goods producers and technical consultants (Cooper and Hoffman, 1978):

> Firms in developing countries buy in this form, either because it will give them a competitive advantage in their product markets through a degree of monopolization or in order to survive against competitors who have obtained the technology before them.

A second form of control referred to is 'process packaged' sale of technology. Included in this package is the 'innovated component' for which the seller has monopoly control, along with additional non-monopolized components designed to complete a system (Cooper and Hoffman, 1978):

> Suppliers try to maintain the system by engineering the non-innovative components to match up precisely with the innovated one. This differentiation of non-monopolized components makes it difficult for a buyer to make price comparisons for any component in the line.

The third category of technology control is 'project packaging'. Here the seller, often a transnational firm, is engaged in both 'production of goods as well as the production of the technology'. The seller, therefore, has a greater interest in the way this technology will be employed and will attempt direct control of its uses.

The commercial acquisition of technology does not come cheaply to the recipient firms or countries. UNCTAD has estimated that by 1980 as much as one-third of world export receipts will be spent on capital repayment for foreign technology (Stewart, 1977). Some of the more identifiable means of extracting payments are listed by Huepe (1977). These include profits on equity invested, royalty payments, technical licensing fees, interest payments, import payments for capital goods, and transfer pricing. Indirect costs to the recipient nations occur in the way the aforementioned are employed -- what might be termed 'creative accountancy', from the perspective of the transnational enterprise. These payments have very different private versus social costs. All have an effect on the international distribution of income; however, this is rarely of direct concern to the recipient firm. An affiliate of a transnational, for example, may not be concerned with the maximizing of profits locally; where as a national firm would rarely enter into an agreement if profits were not to be expected.

As Cooper and Hoffman (1978) point out, 'fairness in international distribution of gains is a matter in which governments' perception of economic advantage may take a different view from private firms'.' A cause for concern resulting from the import of existing technology, which illustrates these differences in perception, is the potential displacement of labour in some industries as small-scale producers are unable to compete with newer technologies appropriated by larger firms. There may thus be a concentration of economic power, in addition to a possible negative effect on the recipient country's GNP.

As important as the financial cost to Third World nations are the 'foregone benefits', relating to restrictive clauses often attached to the transfer of technology. Such practices include: (a) export restrictive clauses limiting sales to a country or region, with volume restrictions aimed at maintaining a high price; (b) restrictions on channels of distribution; (c) tie-in clauses which limit the sources of raw materials, intermediate goods and spare parts; (d) research restrictions which place limits on extending or adapting the technology ;(e) training and employment restrictions against local personnel; and (f) package licensing preventing the purchase of technology from alternative sources (cf. Huepe, 1977).

The business-practice clauses associated with importing existing technology are likely seriously to restrict the choices open to firms or States, regardless of the criteria selected or the motivations of those making the choice. In addition, the likelihood of Third World nations developing their own national capacity, in which more appropriate technology might be developed, can be severely reduced. Local entrepreneurs, when faced with a choice between local and foreign technology, will rarely select the former. The view of foreign technology as 'modern' and 'efficient', the ability at times to have an agreement for monopoly use, and the speed at which it is delivered, the availability of information and promotional material, as well as bribes, all influence their choice. With the use of local capability seen as increasing the risk to the individual private firm, the choice of foreign technology is often viewed as rational. The missed opportunities for local suppliers to improve their capabilities, through learning-by-doing, and to reduce unit costs through production is seen as a heavy 'public' social cost to pay. In the long run it may also prove to have 'private' economic costs as well (cf. Cooper and Hoffman, 1978; Stewart, 1977).

Although constraints are thus placed on the Third World's technological development through the import of existing technology, this does not mean that more appropriate choices cannot be made from 'off the shelf'. Nor is it to suggest that individual local firms have not improved their ability to bargain and receive better conditions. However, the choice of technologies which are appropriate, as Westphal (1974) emphasizes,

> can only be made by individual producers (i.e. entrepreneurs, managers, production and design engineers, technicians, marketing specialists and others involved in implementing the choices made) familiar with the specific technological alternatives and the characteristics (market demand, supply conditions, etc.) of the industry in which they operate.

This requires increased quantity and improved quality of information about existing technology, including information about its sources and the environment (economic and physical) in which it will be used. Enabling local actors to obtain such information may well require government intervention. But this alone may still not be enough for development goals to be pursued. These can necessitate the expansion of the 'shelf' to include more appropriate techniques developed by a science and technology which reflects and is sensitive to local conditions and social objectives.

II. ADAPTATION OF TECHNOLOGY

The introduction of pre-existing technology into Third World nations has often been seen as being unsuitable. A number of reasons are cited, including technical ones and problems of consistency between foreign technology and development objectives. Part of the solution to this problem has been the adaptation (incremental changes in existing technology) of technology to local conditions in order to make it more 'appropriate'. The development of the skills necessary for the adaptation of technology, skills which may be different from those required in making appropriate choices, can enable both the appropriate choice of technique through recognition of what is adaptable from 'shelf' technology, and the development of a national capability for creating appropriate technology.

Although in a highly protected oligopolistic market there may be little incentive to adapt technology (Huepe, 1977), and restrictive-business-practice clauses may at times prevent such adaptation, in many instances adaptation does take place (Katz, 1976). Incremental changes are made necessary by the various differences between advanced industrialized nations and the Third World, such as environmental conditions, the availability of certain skills and management and work-force attitudes, composition of local raw materials, infrastructural facilities, the size of the market, and preferences in product characteristics (Giral and Morgan, 1974, in Huepe, 1977). These varying conditions influence the form of adaptation required, which can be categorized as being to the product or to the process.

Adaptation in either of these categories requires a certain amount of local capability. Entrepreneurial and engineering abilities are essential skills in identifying the potential for adaptation (Morowetz, 1974). However, these skills are in themselves insufficient. A further requirement in implementing the changes is the existence of a capital-goods production capacity. Without a capital-goods sector Third World countries have to import not only their technical machinery but also the 'nature and direction' of technical progress (Stewart, 1977). This is the case because decisions taken in the capital-goods sector will affect not only the capital-goods sector but all industries to which it supplies machines (Stewart, 1977, p. 152).

> The capital goods sector does more than simply enable ideas to be realised -- it also is a major initiator of change. The major source of market expansion for machinery producers lies in the replacement of existing machines and there is considerable technological feedback with developments at one stage stimulating and often requiring developments elsewhere.

The existence of a capital-goods sector, while required in developing a flexible engineering capacity for adaptation and development (Cooper, 1976), will not necessarily lead to 'appropriate' technical progress. As an overwhelming proportion of existing technology has been designed to meet the needs of a minority of the population, adaptation may only result in meeting those same needs more efficiently. Certain types of capital-goods capability are also more likely to be successful than others. Huepe (1977) notes, for example, that batch production may be more easily adapted than continuous process and with less expense.

Concentration on capital goods required by small-scale producers, particularly mechanical and simple electrical engineering, may well prove more advantageous for Third World countries than large-scale industries which require more costly research and development investments (Cooper, 1976) and possibly less flexible engineering skills.

The associations historically identified between the capital-goods sector and heavy industries, with long gestation periods, do not necessarily have to pertain (Cooper, 1975). Apart from the possibilities in the small-scale industries (or perhaps in co-operation with them) the capital-goods sector can play an important role in the adaptation of second-hand techniques, where available, and the 'upgrading' of 'traditional' techniques.

As Cooper and Bell (1975) point out, techniques can become second-hand for various reasons. These include the decline of an industry, replacement of worn equipment, and obsolescence through new technical developments. Techniques discarded for either of the first two reasons are unlikely to be of value to Third World countries, although some industries in decline in western countries (e.g. railways, construction) may have plenty of life left in Third World countries. But old techniques replaced by newer ones in advanced industrial nations are not necessarily inefficient in all circumstances (Cooper and Bell, 1975). Generalizations on the use of second-hand techniques are unreliable and difficult to make. Specific country-by-country case studies are necessary in order to clarify which techniques are available and at what price, in what situations they might be useful, what risks are involved and what further infrastructure would be required (Singer, 1977). It does, however, deserve further study as a form of adaptation which can contribute to the creation of a national technical capacity.

A final form of adaptation to be mentioned in this paper, which also deserves further investigation, is 'upgrading', an alternative form of adaptation based on improving techniques developed in the 'traditional' or 'informal' sectors (Singer, 1977). It has been suggested that techniques in these sectors are 'appropriate' techniques because they both reflect the factor endowments of Third World nations and have been designed to solve the problems that the majority of people in these sectors face in meeting their basic needs (Singer, 1977). If the required links were made, for example between capital-goods producers and innovative co-operative groups, upgrading could provide what Jequier (1977) has termed 'first generation' forms of appropriate technology with a dynamic element that they have often lacked.

III. GENERATION OF NEW INDIGENOUS TECHNOLOGY

As is illustrated by any examination of the capital-goods sector, it is at times difficult to draw boundaries between the adaptation of existing technology and the generation of new indigenous technology. Because innovative activity is found in most stages of the production process the skills required in both will overlap to a certain extent. There does exist, however, considerably greater difficulty in the development of new generating capability, owing to political, organizational and financial barriers. In addition, while there is a growing realization that such a capability is essential in determining the nature and direction of technical progress in Third World nations, there is little guarantee that this capability will lead to the use of more appropriate technology than have adaptation or 'off the shelf' purchases.

Cooper and Hoffman (1978) point out that the use of locally generated capability requires complex intervention into local industry, as was the case in the development of industrial nations. Even where government awareness does exist, it is at times seen as a more efficient use of scarce resources to attempt to control the cost of foreign technology than it would be to develop the links between local technical capacity and production. There is also likely to be strong resistance to such attempts at intervention by firms which see little private gain in taking what they see as high risks or by foreign firms which often have much to lose in the way of control over production and in turn over their profit remissions. In addition the 'expert syndrome' in the international community, and the orientation of the local scientific elite towards methodology geared to the solving of international rather than local problems, are rarely likely to offer any satisfactory alternative advice other than that found in existing technology.

As foreign technology is likely to remain an important source of technical progress in Third World countries for some time, there is a strong need for the import and use of this technology to be more discriminating. Unless this is achieved, it is unlikely that local technical capacity will be able to provide a relevant input to development. Furthermore, as Singer (1977, p. 17) notes, the longer the development of a national capacity is postponed 'the more resources required for selection and adaptation will increase and compete with resources needed for creating new technology'.

Included in the cost of reliance on existing technology over a national capacity is the substantial foreign exchange required -- often a scarce resource in Third World nations. In addition to foreign-exchange costs, researchers have recently begun to examine the costs in terms of health, safety and environmental degradation; not to mention cultural imperialism, reinforcement of colonial/neo-colonial consciousness and importation of western work relationships. Although the cost of local technology is often higher than that of imports there do exist important economic and social gains associated with developing a technology-generating capacity. These gains, which are difficult to quantify or have long-term impacts only, are not always -- or indeed, rarely -- included in the costing analysis. These 'externalities' include the acquisition of skills through learning-by-doing (Cooper and Hoffman, 1978), a variable which we have already described as an essential component in determining the direction of technical progress.

As a result of the costs of importing technology, and the

reality that 'appropriate' technology is either not always available 'on the shelf' (because it does not exist) or cannot be supplied to the particular market which requires it, the components of the innovative and production process have begun to be examined in detail for ways in which a national capability can be applied. By classifying the 'elements of technology', through the type of skills required in each stage of the production process, Cooper (1976) develops a useful framework. These elements, not all of which are applied in every industrial project, include (Cooper, 1976, p. 21):

> Feasibility and pre-project analytic skills; skills required in preliminary choice of production methods and in initial design; detailed engineering skills; skills involved in choice, acquisition and construction of process and product technology (including machine construction); engineering construction skills; commissioning and initial operation skills; operational skills, including maintenance and repair.

Many of these skills are employed in the innovation process and some of the skills already exist in Third World nations. This investment in selected 'elements' can make an important contribution to the development of a national technical capacity which is linked to production in a meaningful way. Cooper notes that the difficulty lies with determining in which categories the social return to development of local capabilities will be the highest. He defines the factors which this will depend upon as (Cooper, 1976, p. 23):

> (a) Differences between the various 'elements of technology' in terms of complexity, degree of specialization and sophistication of the skills required to provide them; (b) differences between industries in terms of the complexity and scale of technology; (c) differences arising from different methods of transferring technology internationally; (d) problems arising from the small scale of national markets for particular types of technical skills; (e) problems arising from inappropriateness of some types of skills to economic and social objectives in less developed countries -- especially employment and income distribution objectives;[6] (f) differences in initial endowment of technical skill, between less developed countries.

A thorough analysis of each of these factors must be conducted on a case-by-case basis and requires information which is not always available. Such an approach must also be grounded in a serious analysis of the social forces operating within society. Without such a combination of the empirical with the theoretical, no basis for developing general criteria can be proposed. In addition to the factors mentioned above, and the development of a national technical capability (to some extent this does already exist) which can create 'appropriate' technology, attention to these factors must be sensitive to the problems of the majority of the population as well as being linked to the production system.

In order to achieve this, the process of innovation and production must include the perceptions of the problems and the solutions by those whose lives they have an impact on. Over and above people's perceptions, an understanding of the exploitation components of the social structure that we want to replace is necessary. This must be accomplished on a basis which regards the indigenous population as more than merely passive, although astute, observers of their environment.

As Howes (1978) emphasizes, 'the depth and breath of local knowledge warrants some more fundamental role in the development process'. He points out that the 'selectivity' of the scientists and the technology resulting from their perceptions present a potential danger because of the rigidity of their training and experience. There is considerable evidence which indicates that the 'dynamic processes of experimentation', which are required for innovation, 'do operate in supposedly "traditional" settings'. This requires that science be linked with indigenous expertise or, if necessary, that new sciences be developed with different methods from those historically developed in the North. The value in such developments would be that technology in the Third World would then become an indigenous cultural phenomenon (Herrera, 1973). This in turn would permit development along paths other than those modelled directly on the North, enabling the expression of different and diverse cultures and perhaps without the social costs of Northern, capitalist industrialization.

IV. CONCLUSION

The choice and source of technology has been viewed here as instrumental in determining the nature of economic growth, the direction of which is conditioned by the distribution of resources between those actors who are responsible for making the choice. In any concrete situation there exists a distribution of benefits associated with each technique; historically for the most part this has accrued to the industrially advanced countries and to a small minority in Third World countries. And because of this distribution, often evident in the restrictive business-practice clauses associated with the transfer of technology, attempts at an international level to moderate and influence transfer of technology or to improve domestic innovative capability have been relatively ineffective. While technology transferred from central to peripheral countries may be considered by some as 'appropriate' in terms of maximization of overall economic growth it has achieved little in the way of redistribution. In recognition of this ineffectiveness technological self-reliance has become the stated aim of many Third World nations.

The creation of technology through the development of national capabilities can be a time-and-resource-expensive undertaking. For this reason, as Hoffman and Unger (1978) point out, it is neither feasible nor advisable that this objective should imply autarky.

The problem as stated by these authors is one of 'achieving an appropriate balance between foreign and local supply of the required technological inputs'. This balance will change over time as local capabilities are developed and replace imported capabilities.

Bearing this in mind it is necessary, in scenarios of self-reliance, to examine, among other factors, the time-scale on which this would occur, the processes of decoupling sectors from the world economy and the reaction by other countries following different directions of development. As might be expected, planning activities of this kind run up against a number of conceptual and measurement difficulties. In order to incorporate these technological issues it is necessary to go beyond a simple North-South view of international relations. In the South, for example, some discrimination is required between at least three different categories of countries: large, newly industrialized countries (e.g. Argentina, Brazil, India); medium-sized economies (e.g. Egypt, Kenya, Nigeria, Pakistan); and small economies which could be further differentiated into those relying on primary products (such as Guatemala) and those relying on 'runaway' industries or subcontraction (e.g. Singapore, Taiwan).

While adequate work on these issues is relatively recent, a considerable amount of information on capability development already exists, both at the individual-firm level and in country case studies which can be employed in developing and implementing a strategy of self-reliance. The flexibility of technological policy is considerable, and the differing requirements and resource availability of each country will ensure that no two countries will follow exactly the same path. Of great importance to all those concerned with creating a future society is that the goals of development are clearly stated, debated and their implication understood. This in itself, however, is not sufficient. There must also be an understanding of existing social relations and the role technology could play in reproducing or transforming these relations, if goals such as self-reliance are to be achieved.

NOTES

[1]The main arguments summarized in this review are discussed in depth in Stewart, 1977; Singer, 1977; Huepe, 1977; and Cooper and Hoffman, 1978.

[2]Among academics it has been the economists, perhaps those closest to the policy-makers, who have produced the lion's share of the publications concerning technology. Other social sciences, particularly sociology and social psychology, might be expected to have distinctive and important contributions yet to make.

[3]We adopt Freeman's (1975) definition of innovation being any changes in technology introduced for the first time.

[4]These are not necessarily heads of firms. The decisions made at that level have either been narrowed down or made, often by engineers, at lower levels.

[5]The very use of the term 'late-comers' may also be inappropriate as it implies blame on those arriving 'late' rather than the transport system that has structurally disadvantaged them.

[6]In addition to the employment and income distribution objectives, the primarily descriptive definition of working skills must be extended to include 'control' skills with the element of control by the work force becoming a central objective. See Acero below (pages 189-200) for a fuller explanation of 'control' skills.

References

ACERO, L. 1978. Working Skills as a Critical Issue for Self-Reliance. Proceedings of the Joint Workshop on Analytic Techniques for Long-term Development Issues, Institute of Development Studies, University of Sussex, 20 November to 2 December 1978.

COOPER, C. 1975. Draft Proposal for a SPRU/IDS Research Programme on Innovation and Alternative Technology in Less Developed Countries. Science Policy research Unit (SPRU), October. (Mimeo.)

_____, 1976. Policy Interventions for Technological Innovation in Less Developed Economies. Washington, World Bank, January, (Mimeo.)

COOPER, C.; BELL, M. 1975. Industrial Technology and Employment Opportunity: A study of Technical Alternatives for Can Manufacture in Developing Countries. IDS/SPRU, University of Sussex, September. (Mimeo).

COOPER, C.; HOFFMAN, H. K. 1978. Developing Countries and International Markets in Industrial Technology. SPRU/IDS, University of Sussex, July. (Mimeo).

DAVID, P. 1975. Technical Choice, Innovation and Economic Growth. London, Cambridge University Press.

DICKSON, D. 1974. Technology and the Contruction of Social Reality. Radical Science Journal, January.

FREEMAN, C. 1974. The Economics of Industrial Innovation.

Harmondsworth, Penguin Books.

GIRAL, J.; MORGAN, R. 1974. Appropriate Technology for Chemical Industries in Developing Economies: Report of a Summer Research Training Project in Mexico. Techos, Vol. 3, No. 2, April -June.

HERRERA, A. 1973. La Creacion de Technologia como Expresion Cultural. In: K. Heinz-Stanzick and P. Schenkel (eds.), Ildis, Quito.

HOFFMAN, H. K.; UNGER, K. 1978. (Paper.) International Peace Research Association Seminar, Sweden, July 1978. (Mimeo.)

HOWLES, M. 1978. The Uses of Indigenous Technical Knowledge in Development. IDS, University of Sussex, April. (Mimeo.)

HUEPE, C. 1977. Technological Options for Developing Countries. Proceedings of the Conference on Technical Co-operation and Development, CEFA Bologna, Italy, November 1977. (Mimeo.)

JEQUIER, N. 1977. Appropriate Technology for Basic Needs: The Criteria of Appropriateness. OECD Development Centre on International Action for Appropriate Technology, Geneva, 5-9 December 1977.

KATZ, J. 1976. Creacion de Technologia en el Sector Manufacturero Argentino. Programme BID/CEPAL sobre Investigacion en Temas de Ciencias y Technologia. Buenos Aires, Naciones Unidas, Comision Economica para America Latina.

LALL, S. 1973. Transfer-Pricing by Multinational Manufacturing Firms. Oxford Bulletin of Economics and Statistics. Vol. 35, No. 3, August.

MOROWETZ, D. 1974. Employment Implications of Industrialisation in Developing Countries. The Economic Journal, Vol. 84, No. 335, September.

SINGER, H. 1977. Technologies for Basic Needs. Geneva, ILO.

STEWART, F. 1977. Technology and Underdevelopment. London, Macmillan.

VAITSOS, C. 1974. Intercountry Income Distribution and Transnational Enterprises. Oxfrod, Clarendon Press.

VAITSOS, C., et. al. 1976. Technology Policy and Economic Development. Ottawa, International Development Research Centre. A summary report on studies undertaken by the Board of the Cartagena Agreement for the Andean Pact Integration Process. (IDRC-061e).

WESTPHAL, L. 1974. Research on 'Appropriate' Technology. In: L. J. White (ed.), Technology, Employment and Development. Selected

<u>Papers Presented at Two Conferences Sponsored by the Council for Asian Manpower Studies.</u>

Can Values Shape Third World Technology Policy?

DENIS GOULET*

I. INTRODUCTION

Can values shape technology policy in developing countries, or are values themselves shaped by the policy? This is the central question posed in these pages. One must also ask: whose values and which category of values are decisive for policy? The arena of technology policy under consideration likewise needs to be specified--policy for a nation, for some region or locality, or for a particular sector of economic activity? It should not be imagined that policy decisions are preceded by value-free "scientific" prescriptions. On the contrary, technology is itself a highly conflictual arena in which competing visions of what is good or socially desirable clash. A few prior definitions are in order, however, if we are to understand the issues under dispute.

(a) Value

In everyday usage the term "value" is applied to any preference, norm, belief, or meaning which exercises some claim upon our conscience, our esthetic sense, or our respect for truth. To get beyond this broad notion, however, and to craft a precise definition of value is a task which has long stymied philosophers, theologians, social scientists, historians, poets, and legal scholars. For present purposes I must nonetheless submit a working

From **JOURNAL OF INTERNATIONAL AFFAIRS**, Vol 33, Spring 1979, (89-109), reprinted by permission of the publisher.

definition: a value is "any object or representation which can be perceived by a human subject as habitually worthy of desire." Each element in this definition is essential. Value resides either in an existing reality ("object") or in some image of a reality which may or may not exist ("representation"). And unless a value is amenable to being grasped by a perceiving subject ("can be perceived"), no intelligibility will attach to it. Most importantly, values are not mere preferences: they contain a quality of "oughtness" ("worthy of desire") which is habitual, not ephemeral. Hence projected images of a better life and a preferred future are values no less than are moral obligations to respect the rights of one's fellow humans. It would be futile to engage here in fine debates over criteria for classifying values, or to decide whether hierarchy or some alternative pattern of relative weighting must always exist among values. It suffices to note that the values discussed in this article are those which are shared by human communities--a local population, any identifiable cultural unit, a national society, or the entire human race.

An "identifiable cultural unit" is any group of persons who claim a common identity, who share meanings they deem important, and who agree on the traits which distinguish members from outsiders. In the past, members of a common culture[1] have usually agreed on standards for deciding what was important or not to the survival of the identity they derived from their culture. This traditional agreement is now threatened, however, by encroaching modern ideologies and technologies.

(b) Technology Policy

The main object of inquiry here is national policy within developing countries; hence technology choices within developed countries or in international arenas are treated only incidentally. Even in countries where technology policy has not been explicitly formulated by planners or politicians, a de facto policy can nonetheless be inferred from the patterns of technology choice, application, and utilization observed therein.

National technology policy for development is necessarily broad in scope;[2] so are the instruments which can be deployed to implement policy.[3] Policy choices range from the educational strategy for training the requisite personnel, to incentive systems (comprising a mix of material and moral rewards for socially beneficial efforts), vehicles for disseminating technological attitudes throughout the working population, legislation governing the conditions under which proprietary technology is to be imported and utilized, ways of promoting new circuits of technological demand and supply, aids given to indigenous research activities, modes and scales of technology "appropriate" for diverse purposes,[4] and criteria for technological collaboration to be pursued with various actors in international arenas. Third World policy makers in growing numbers now seek a pattern of selective linkage and de-linkage with rich country institutions deemed most beneficial to them. Obviously TNCs (transnational corporations), development banks, and other powerful institutions operating from the "metropoles" of the world emit both positive and negative impulses to "peripheral" weaker countries. Consequently, when they negotiate with suppliers of technology, many developing countries look for ways of "maximizing" the beneficial while "minimizing" the harmful effects of "transfers."[5]

Not surprisingly, only a few developing country governments have framed comprehensive technology policies. Fewer still successfully harmonize technology policy with broader development strategies. Nonetheless, the implicit criteria operative in their technology choices reveal value priorities adopted by their societies. In this general sense, therefore, an inquiry into the degree of consonance between the central values of a society and its "technology policy" can be fruitfully conducted.

A third term still remains to be clarified--"development."

(c) Development

On-going development debates presently emphasize such themes as meeting basic human needs[6] as a first priority, promoting national and local self-reliance in the transfer of resources, experimenting with planning from the bottom up,[7] and redistributing the benefits of economic growth.[8] But no one knows how to assign relative importance to development's economic, cultural, and political dimensions.[9] Nevertheless, it is obvious that economic development, institutional modernization, and planning efficiency are, ultimately, means to a larger end. That larger end is the actualization of all human potentialities, personal and communal. In 1945 Lord Keynes referred to economists as "the trustees not of civilization but of the possibility of civilization."[10] On such a normative vision is founded the holistic view of development I have proposed in an earlier essay and which I here endorse anew:

> What is conventionally termed development--dynamic economic performance, modern institutions, the availability of abundant goods and services--is simply one possibility, among many, of development in a broader, more critical sense. Authentic development aims at the full realization of human capabilities: men and women become makers of their own histories, personal and societal. They free themselves from every servitude imposed by nature or by oppressive systems, they achieve wisdom in their mastery over nature and over their own wants, they create new webs of solidarity based not on domination but on reciprocity among themselves, they achieve a rich symbiosis between contemplation and transforming action, between efficiency and free expression. This total concept of

> development can perhaps best be expressed as the 'human ascent'--the ascent of all men in their integral humanity, including the economic, biological, psychological, social, cultural, ideological, spiritual, mystical, and transcendental dimensions.[11]

All who ignore such a broad notion of development risk falling into a "reductionist" stance which rules out sound policy. Reductionism is the posture which "reduces" (practically if not theoretically) a larger reality to a small facet of its totality. Development planners and agents are often prone to be reductionists--to treat development *as if* it were nothing other than growth in output, the "rationalization" of public institutions, or the efficient allocation of scarce resources. Ultimately, however, development poses to all societies the need to choose which images of the good life and which principles for promoting justice they will favor. This is the reason why the relationship between human values and technology policy needs to be elucidated.

(d) A Vital Distinction: Instrumental and Non-Instrumental Handling of Values.

Politicians and planning experts readily profess their regard for the values of the population affected by their policies. Too often, however, they treat these values in purely "instrumental" fashion, manipulating them as mere aids or obstacles to reaching objectives which they have already defined (and judged beneficial!) outside the framework of those values themselves.

One example of the instrumental use of religious values in fostering development is seen in the policy of Prime Minister Mohammad Daoud Khan of Afghanistan during the decade 1953-63.[12] Daoud, aware of the powerful hold Islam had on his country's largely illiterate population, shrewdly harnessed religious symbols to promote a secular pattern of development. He employed liberal young Koranic lawyers to undermine the traditional authority of older religious leaders and to win support for abolishing the wearing of the veil by women (burqa). Anthropologist Louis Dupree judges that "[P]robably the most important even in 20th-century Afghanistan was the promulgation of a new Constitution, one which used Islam as a positive, dynamic weapon in helping create a nation-state...the government continues its policy of using certain important religious places to further the cause of nationalism and the creation of a nation-state."[13]

A similar "instrumental" use of values is made by many advocates of popular participation in development planning.[14] Many change agents, notwithstanding their professed respect for the wishes of their intended beneficiaries, nonetheless seek above all to rally peasants to support a decision already adopted by planners or technicians. Peasants are consulted only after the basic contours of a problem have been diagnosed, *after* an array of possible options has been laid out, and after some alleged cost-benefit analysis has isolated a preferred solution. Yet the key to genuine participation is to engage people at an early moment in the overall decisional sequence. To act otherwise is to handle the value of participation "instrumentally." In contrast, non-instrumental change catalysts first enter inside the value framework of the affected people on its own terms; only afterwards do they specify the development goals to be pursued. The crucial point lies in recognizing that a community's identity, cultural integrity, and meaning system are themselves the matrix out of which emerge the goals of authentic development for that community. Consequently, values are not treated as a means to be mobilized in pursuit of development ends which are themselves legitimized outside the boundaries of those values.

Ideologically, historically, and culturally diverse communities are now searching for alternative designs of development. They have come to regard economic growth, institutional modernization, and productive efficiency as instruments of human development to be critically and selectively utilized, not as final goals dictated by experts. A brief illustration of critical selectivity is offered by Daniel Lerner, author of the land-mark study The Passing of Traditional Society (1964). Lerner has returned to his earlier research sites and now writes that:

> The most important change that I want to make is not to call the whole process 'modernization' any more but rather 'change'...I would think of the factors not as indicators of modernity but as 'propensity to change' or readiness to try new things...here is another index that I have been working on and I have called it 'ambivalence.' This concept especially refers to those who are conflicted, who have some feeling for the worldly set of goals, but who are also sensitive to traditional values.[15]

To speak in more concrete terms, and in another geographical setting, dwellers in the fishing village of Kuala Juru in northwestern Malaysia have decided not to accept the offer of their state government to move inland to a new industrial site where employment and social amenities would be guaranteed.[16] Earlier industrialization programs sponsored by the national government had created massive pollution which destroyed their livelihood as traditional fishermen along the coast. Nonetheless, the villagers decided that: "We are people of the seas and will stay on in the village. The skill of fishing is something precious to us, handed down from our forefathers. From the very day we were born, the sea and the water have always been in our blood. The sea is our life-line, our ricebowl, and our link with nature." The values cherished by these villagers led them to repudiate conventional "modern" patterns of development and create one in harmony with their own identity and desired relations with the natural environment. But their decision required arduous changes in their traditional way of living. They had to give up individual fishing and adopt cooperative conch-farming; furthermore, cash incomes had to be distributed according to new socially recognized priorities. Kuala Juru's example is already spreading: six other villages in adjacent areas have sought advice from the pioneer village to learn how to make the transition from traditional fishing to communal conch-farming. Like Kuala Juru, these villages have also been deprived of their ancient form of livelihood by industrial pollution and the introduction of modernized fishing vessels. They too are searching for new forms of livelihood more congenial to their values than that offered by salaried work in factories.

II. THE "VITAL NEXUS": WHEN IS IT POSSIBLE?

National technology policy is best formulated and implemented when planners create a "vital nexus" which links, in triangular fashion, their nations' priority social values to its development strategy, and to practical criteria for making technology choices.[17] The contention that the vital nexus is crucial rests on several assumptions drawn from the concrete workings of technology circulation systems. The first assumption states that modern technology powerfully reinforces specific conceptions of rationality, efficiency, and stances toward the forces of nature. This is to say that technology is not value-neutral; on the contrary, it simultaneously creates and destroys values. The same is true of development processes themselves--they are not value-free but give birth both to beneficial and to destructive effects. Consequently, it is vitally important for any society to identify the priority values (or "images" of good life and the just society) it wishes to foster in its pursuit of a certain style of development as well as in its utilization of technologies. Moreover, the mechanisms through which technology is channelled across national boundaries reinforce or weaken certain social interests. And why is this so? Because technology is simultaneously a resource (a means for creating new wealth), an instrument of social control (a tool used by dominant social groups to serve their interests), and a shaper of decision making patterns exhibiting varying degrees of elitism or democratic participation. Therefore, unless priority social values are clearly stipulated and normative directions in one's development strategy specified, many destructive values will impinge upon weak and vulnerable societies. This damaging effect is traceable both to the very nature of modern technology and to the social interests served by prevailing mechanisms of technology circulation.[18] Countries which lack a technology policy, or which adopt one not consistent with their central value preferences and general development objectives, condemn themselves to having a bad policy.

Yet even a cursory observation reveals that in most developing countries the "vital nexus" simply does not exist. One is tempted to prescribe that the nexus simply be created. But such advice is useless; in many countries it is extremely difficult, if not impossible, to create the linkage. The crucial question then becomes: upon what foundations can a sound technology policy be formulated? No reply is possible unless we first answer the larger question: under what conditions is a "vital nexus" possible?

Leaders Not Rulers[19]

Many leaders in developing countries embody the values of Western industrial nations more faithfully than those of their own people. Numerous reasons converge to explain this phenomenon: ideological mimetism, various demonstration effects, the education history of the first generation of leaders, the colonial heritage, the lack of alternative "modern" role models, and a sense of embarrassment related to status over purely indigenous values. Yet is is impossible for national policy to reflect authentic values if political, intellectual, and technical leaders do not stay in close touch with their populace. Accordingly, when speaking of the "vital nexus" one is not referring to modern values which Westernized leaders may try to impose upon their people or induce them to accept. Nevertheless, it is apparent that even leaders who are sensitive to the aspirations of their own people nourish certain ambitions for their societies. And at least some of these ambitious cannot be satisfied without adopting certain modern technologies. For this reason revolutionary leaders usually insist that the "liberation" of the oppressed calls for new styles of learning, of problem solving, and of social organization, all of which help communities of struggle go beyond their indigenous traditions.

Leaders, in short, need to listen to their masses in formulating desired social values. Proximity to, and consultation with, the people clearly presupposes a certain leadership ethic and patterns of reciprocal dialogue if it is to succeed.[20] To state the matter differently, a "vital nexus" cannot thrive if too great a chasm separates the values held by leadership groups from those cherished by the population at large. The dramatic speed with which the power base of the Shah of Iran was eroded testifies to the dangers inherent in promoting values which the common people repudiate or suspect

What To Do When National Values Conflict

The values which national societies seek to promote cannot easily be identified. Especially in large heterogeneous countries like India or Nigeria, no single constellation of values commands the loyalties of all citizens. At times, the only values which can rally all segments of society are those which lie on the outer edges of the ethnic or religious values from which different population groups derive their particular identity and cultural integrity. The fragile basis for a national consensus then becomes precisely those values essential for constructing a pluralistic nation around common humanistic ideals like social justice, the abolition of misery, and the forging of a modern nation on the twin pillars of common geography and history. To illustrate, most Malays are Muslims whose basic values flow from Koranic interpretations of life, death, and destiny. Chinese and Indian Malaysians, in contrast, draw on different philosophical sources for their central values. Accordingly, a Malaysian constellation of values serviceable for national policy purposes would, of necessity, focus on the humanistic themes common to all ethical heritages.[21] Nevertheless these common values need to be perceived as relevant to modern challenges.

To summarize, the "vital nexus" is possible in two circumstances:

-where a high degree of homogeneity among specific values exists in the society at large; or

-where, lacking homogeneity, a relative consensus can nonetheless be built around common humanistic social values needed to build up the nation or pursue sound development.

What is a "Viable" Development Strategy?

The "vital nexus" approach to technology policy presupposes the adoption of a "viable" development strategy. The nexus is precluded if no coherent development strategy exists, or if strategy amounts to little more than a collection of rhetorical flourishes rather than effective norms for action in such diverse arenas as investment, fiscal incentives, guidelines for exports and imports, and employment. Trouble also arises when a coherent development policy does exist, but when it assigns greater importance to economic growth than to equitable distribution of growth's benefits or to open access to the assets upon which growth depends. A given development strategy may also perpetuate dependency on outside economic actors, a condition viewed by some governments as quite acceptable, by others as intolerable. The advocacy of the vital nexus obviously presupposes a prior qualitative judgment about the development strategy whence a potential technology policy would derive its directions.

Professor A. K. N. Reddy of India's Institute of Science does not hesitate to make such a judgement. As the principal author of a recent report on "Methodology for Selection of Environmentally Sound and Appropriate Technologies,"[22] Reddy declares that the only valid kind of development must take as its objectives:

-the satisfaction of basic human needs (material and non-material), starting with those of the neediest, in order to achieve a reduction of inequalities between and within countries;

-endogenous self-reliance through social participation and control; and

-harmony with the environment.

He further stipulates that "the advancement of differing development objectives of industrialized and developing countries requires the establishment of a New International Economic Order, for it is only such an order which can make these differing objectives compatible with each other" (Report cites, Paragraph 1.12). Most importantly, Reddy postulates an intimate correlation between technology and the value-inspired choice of a specific conception of development. He concludes that "it must not be assumed that all available technologies (however modern they may be) and all future technologies (likely to emerge in the guise of 'technological progress') are necessarily consistent with development objectives." (Paragraph 1.13).

From the unavoidably normative character of development choices, it follows that different qualities of "vital nexus" can be discerned in the technology policy of different nations. If in one case the dominant objectives are rapid industrialization, the enrichment of middle classes, the creation of large technical infrastructures, and the acquisition of advanced military capabilities, the explicit and coherent "vital nexus" will be at the service of spurious development. The nexus leads to sound technology policy only when the social values and the development strategies whence that policy is derived are themselves just and sound. Obviously, leaders committed to an alienating conception of development cannot be induced to adopt a technology policy which will produce effects other than those implied in their initial options. Thus the former Brazilian president Emilio Garrastazu-Medici, could acknowledge that "Brazil's economic miracle was good for the economy, but bad for the people." Two questions remain fundamental at all times: what kind of development is being sought? and, at what social cost?[23] It needs to be remembered, however, that certain preferred technologies reinforce the dominance of certain social values; and conversely, certain preferred social values affect the criteria adopted for making technology choices.

Experimenting with Different Technologies

The vital nexus is not a static construct which can determine, once and for all, the criteria for appropriate technology choices. On the contrary, the nexus flourishes best in societies willing to experiment with different scales and modes of technology. Although some technologies favor given social constellations of values, other reinforce quite different values. Massive reliance on automobiles instead of on bicycles for urban transportation creates quite a different kind of metropolis. One has only to evoke images of Mexico City and Peking to see the difference. Similarly, the small cottage industries favored by Gandhi in rural India imply a different pattern of working relationships, of mutual help, and of trading complementarities than those associated with gigantic factory complexes. The lesson is clear: a society's readiness to experiment with different mixes of technological inputs is a favorable factor which fosters a range of social values which are not highly congenial to modern scales of technology. Several writers[24] have recently called our attention to "diseconomies of scale" attaching to conventional Western technologies--ecological destruction, wastefulness of scarce resources, and a high level of political dependency. It is equally essential to weigh the cultural and psychological diseconomies attendant upon dehumanizing labor imposed upon large numbers of workers.[25] High productivity is only a relative good: it needs to be balanced against other goods such as a creatively satisfied work force, equitable access to resources and to jobs, and the subordination of competitive instincts to collaborative needs.[26]

Although consumer patterns may display great uniformity, or identical industrial plants fill every empty landscape, different societies remain possible because the meaning and relative importance attached to these outward realities can vary. Upon returning from a trip to the Soviet Union several years ago, the Italian novelist Alberto Moravia meditated for the first time on the intimate link tying a society's spiritual values to the material objects with which it surrounds itself. Moravia concluded that it is catastrophic for any system of production to eliminate variety in consumer goods. More significantly, he discovered that variety can symbolize, in addition to wastefulness, prodigality, or even the liberation of the person, the affirmation by a people of its creative spirit.

> The borderline separating the work of art from a product of handicraft, or from the product of light industry cannot be traced with certitude. One can even affirm in this regard that the identical creative spirit lies at the origins of a monument, a novel, a rug or a crystal vase. On the other hand, we can locate the exact dividing line between these objects and a tractor, a truck or any other product of heavy industry. In the former category what is expressed with more or less talent is taste, artistic sense, and imagination; whereas in the latter what is revealed is rational utility. The former objects manifest the profound diversities in traditions, national genius, and particular characters. The second category, on the other hand, is based on the precepts of universal necessity.[27]

This text suggests the great cultural importance to be placed on a society's choices of the consumer goods it produces, whether by personal artistry, handicraft, light or heavy industry. Any society's choice of goods will affect its selection of technology. Because different work styles call for different kinds of technology, if a nation desires to maintain its cultural distinctiveness, it must also select technologies consonant with its day-to-day values. Indeed it is in the arena of daily problem solving that new cultures are forged and old ones displaced.

A nation unwilling to experiment with varied technological "optimum" scales and modes will be exposed, in full vulnerability, to the powerful conditioning effects channelled by mainstream modern technologies. In thus exposing itself, it may well disqualify itself from establishing the kind of vital nexus which promotes sound social values, viable development, and good technology policy.

Freedom from Outside Influences

Outside domination effects are, in some circumstances, quite overwhelming. This appears to be the case in Paraguay, a country whose technological decisions have been shaped largely in response to Brazilian pressures favoring the massive generation of hydro-electricity.[28]

Nations which are too dependent on outside forces cannot create a healthy vital nexus because they lack minimum zones of freedom not only in technology policy, but also in development strategy and, ultimately, in setting the profile of their social values. Canadian economist Kari Levitt laments that American business influence on Canada's national life has led to a "silent surrender."[29] Even within national boundaries, specific regions may be hindered from experiencing autonomous decision making, in technology no less than in spheres of economics and politics, by the dominance of outsiders. This happens almost everywhere in the Third World: cities dictate the patterns of development allocations, usually at the expense of rural areas.[30]

The choice of technologies suited for integral development often presupposes a revolution in development priorities[31] and development "styles."[32] The underlying issue remains, of course, how successful poor nations can be in defining their own development needs and in choosing their own social institutions to meet them. The answer states, in part, that they can never achieve development autonomy if they submit to dominant patterns of technological circulation, which reinforce the use of technology as a commercial weapon in a competitive arena. Important "humane" values have deliberately been kept external to commercial calculations of "optimality" in technology. As a result, many technologies are not ecologically sound or socially equitable because they were never designed so as to "internalize" these values. The "basket of consumer goods" which mainstream Western technologies were meant to produce postulates high purchasing power in the hands of large numbers of people. When these technologies are introduced into social settings where the masses have very low buying power, the goods produced are destined disproportionately to the satisfaction of the wants of middle and upper classes, not to the priority needs of poorer classes. There is, manifestly, nothing inherent in technology *per se* producing such an effect. Distortions arise because specific technologies originate in a concrete social matrix and reflect the value preferences and interests of those groups in that social matrix who have created or exploited the technology in question. To state the matter in naked terms, modern technology was never invented to achieve development, but rather to confer advantages to certain categories of producers or warriors over others. If technology is to become an instrument of human development, it must be re-invented under the aegis of values opposed to those which presided at its first incarnation.

From this outline of five prerequisites for a "vital nexus," the key to good technology policy, one conclusion emerges sharply: very few countries possess all these pre-conditions. Can other countries, nonetheless, articulate a salutary technology policy for themselves?

III. THEN THE "VITAL NEXUS" IS MISSING

What of developing countries which lack the requisite conditions for setting up a coherent "vital nexus:" are they condemned to having no guidelines for formulating their technology policy? Not necessarily: for even when full success is unattainable, the vital nexus as an image of the desirable can help planners achieve relative approximations of the ideal. Several measures can be taken to improve technology policy.

Applying the "Sabato Triangle" in Key Sectors

Sound diagnostic tools make policy formulation easier by facilitating the analysis of structural failings in the present system. Such tools point to cures for failure's causes rather to mere cosmetic tinkering with symptoms. Many developing countries are powerless to exert control over their technology imports because their technological demand and supply circuits are not properly integrated to their internal priority needs. In such settings, one useful principle for screening and evaluating technological options is the "Sabato Triangle," a construct which draws its name from Jorge Sabato, an Argentine physicist, metallurgist, and policy scholar. The triangle provides a simple model to guide technology policy.[33] Sabato urges creating practical linkages among technology researchers, economic producers, and development policy makers. His underlying metaphor is the triangle, a familiar geometric figure composed of three inter-connected vertices. The top vertex represents governmental decision makers, whereas the second and third vertices stand for producers and technological research workers, respectively. Sabato argues that many valuable technological outputs in developing countries are wasted because they are not converted into the concrete inputs needed by producers. His solution is to have each category of actors in the technological arena, particularly, the policy makers, devise incentive systems which activate reciprocal flows of information with each of the other two. Incentives are likewise needed to reward new initiatives in demanding or supplying technology. Speaking concretely, factories and mines should have access to, and influence on, laboratory and university researchers. Government planners, in turn, should be able to influence the technologies which manufacturers will employ and which researchers will devise. Unless circulatory flows connect all three elements of the triangle, Sabato warns, potentially available technology will not be incorporated to national development efforts.

During public lectures delivered in Argentina in 1973, Sabato offered two historical examples of the workings of the triangle. His first illustration recalls the invention of the stirrup in the early Middle Ages. The stirrup, it will be recalled, was a technological breakthrough which transformed horses, as mounts for cavalrymen, into devastating weapons of war. Rulers immediately started placing increased demands for more horses and for more land on which to raise horses. Under prevailing land tenure systems, however, these new demands could only be met by expropriating more land from churches and feudal lords. Thus, what at first seemed to be a mere technological invention turned out to be a potent agent of change in social structure. Pre-World War II Germany provides Sabato with his second example. Because Germany was rich in zinc but poor in copper, Hitler ordered his industrial researchers to find ways of making automobile carburetors out of zinc instead of copper. Thus goaded, Germany's research and productive actors in the "triangle" proceeded to invent and produce zinc carburetors on a mass scale.

As these cases show, if the Sabato triangle functions vigorously technology can contribute directly to development. On the other hand, if the triangle is absent or weak--that is, if the vertices are not joined or if infrastructures are deficient--many technologies, whether locally produced or imported will make but slight contributions to development. The triangle is a useful conceptual tool not only for joining locally generated technology to local production, but also for making a more rational selection of foreign technologies and for obtaining a more effective incorporation of these within existing local productive structures.

Sabato urges governments to set up triangular flows--of information and incentives--in each sector of activity essential to their national economies. Such action usually requires new government initiatives for endowing research institutions with specific resources or integrating government agencies to new information networks. It can also mean redirecting existing research capacities in order to meet production needs. New circuits in information flow, and new rules for bargaining with suppliers of technology, must also be instituted.

The worth of the Sabato model lies in its great simplicity, practical applicability, and flexibility. Even poor countries unable to erect triangles in all sectors of economic life can at least start doing so in one or two vital sectors or product lines:

Bolivia in tin, for example, or Sri Lanka in tea. Within a few years national technology capabilities can be strengthened and new reciprocities between demanders and suppliers created.

The Sabato triangle can also help Third World nations whose firms import most of their technology from transnational corporations. By using the triangle, government policy makers can better identify and manage the conflicts of interest which pit suppliers against purchasers of technology. More importantly, it suggests to planners the measures needed to launch new circuits of technology demand and supply. These measures range from technology registry to ceilings on servicing payments, to restrictions on unused patents and new rules for creating indigenous R & D laboratories.

Keep Mistakes Small and Reversible

Adepts of the "small is beautiful" philosophy plead the merits of "appropriate" technologies suited to poor local conditions. Their critics retort that "small is not necessarily powerful," a crucial point inasmuch as one major objective sought by Third World governments aspiring to modern technology is, precisely, power. Obviously there is no single scale, mode, or level of technology which is appropriate to all development purposes. Even very poor countries may need a whole array of differently scaled and designed technologies so as to meet different goals. A Haitian central banking official once explained to this author that it was precisely because Haiti constituted such a fruitful field of application for small and simple technologies (like family solar cookers and building materials made from waste materials) that certain complex co-ordinating activities--in this case, rapid transfers of credit monies from Port-Au-Prince to small farmers in isolated mountain villages--required "high" technological instruments such as computers. The obvious point is that a national technology policy dictates "appropriate" mixes of technologies located at various points on the twin spectrums of size and complexity.

A strong case can nonetheless be made for sinning on the side of "small and reversible" mistakes in many Third World sites. Most modern technologies are so costly to install and maintain, so complex to repair, and so inflexible in their requirement of standardized intermediate inputs that their adoption by poor societies risks being very wasteful. In technology policy as in development planning itself, it is the better part of wisdom to seek solutions to problems in experimental ways which respond flexibly to the lessons learned from on-going experience. For this if for no other reason, technology policy ought to be biased in favor of multiple small innovations over a few larger ones. Lessons gleaned from the massive introduction of "miracle strains" of wheat in several Asian countries suggest that ecological vulnerability to new blights was needlessly courted--all because a single technological improvement was applied indiscriminately on a vast scale.

Developers who seek to launch technological practices in ways which keep cultural and social damage to a minimum should experiment with a variety of alternative technical solutions. Otherwise the value sacrifices exacted of societies may be seen as excessive only after irreversible commitments (in the form of capital, creation of a support structure, and training of personnel) have been made to a given solution. In several Third World countries experiments are now under way to determine what are the least expensive and the most equitable ways of harnessing technology to the needs of poor villagers. Consultative processes between specialists and the general population are essential to adjudicate competing claims. In Indian villages where animal dung is undergoing bio-gasification treatment so as to produce electricity, for instance, the key equity question is how to bring electricity to families too poor to own cows and, therefore, unable to supply dung. Where electricity has multiple uses--household lighting and cooking, the pumping of water in village wells, and the running of machines in small workshops--how is the distribution of technology's benefits to be made "appropriate" to the full range of villagers' needs? Continued experimentation is necessary if we are to gain sound answers. Flexible experimentation is rendered impossible, however, by initial commitments of large-scale capital, design, and training to single levels of technology.

Collective Self-Reliance

The recent United Nations Conference on Technical Co-operation among Developing Countries (TCDC) signals the coming of age of the concept of technical collective self-reliance among poorer countries. In the words of one observer,

> the Buenos Aires Conference sanctioned the integration of a new South-South dimension into the international development system, thus creating the potential for expanding and transforming the existing system of technical co-operation into a truly international co-operative of development skills, experience and knowledge for the benefit of all mankind.[34]

Developing countries wishing to dilute the strength of their vertical links to rich country suppliers of technology can now explore horizontal avenues for collaborating with poor countries like themselves. Collective self-reliance is not a new principle; the experience gained by Andean Pact members in practicing collective action is clearly useful to other nations in the Third World. That experience reinforces the belief that horizontal technical cooperation is one important measure which LDCs can adopt in order to progress toward a more mature and comprehensive technology policy. Concrete examples of what can be accomplished with TCDC have already appeared.[35] A brief mention of two cases can help illustrate the principle at issue.

(a) First Example: Rural Development Technology: "Las Gaviotas."[36]

The "llanos" (plains) region covers 300,000 square miles in eastern Columbia. At present this area is sparsely populated (containing some 100,000 inhabitants), a condition traceable to the low fertility of land and its resultant low agricultural productivity. Efforts were launched nine years ago by a Colombian community organizer to study the special needs of these tropical plains and to settle them more densely. These efforts, later supported by the national government, aim at finding ways of generating employment without disturbing the delicate ecological balance of the region. This path to development, in turn, is seen as a way of creating alternatives to rampant urban migration coming from poor countrysides.

A full service center called "Las Gaviotas"[37] is now in operation and takes care of the educational, medical, transportation, and supply needs of local farmers. Las Gaviotas has developed and field-tested simple machines designed to help isolated agricultural workers labor more efficiently: a water dam, a solar heater, a bicycle-pedalled yucca grinder, a hydraulic turbine, a windmill, and a manual pump. The Center's founder, Paolo Lugari Castrillon, explains that "We are trying to create communities of people who 'think tropically,' who know how to live in these tropical plains without destroying them."

Las Gaviotas has begun to share the fruits of its experience with an analogous organization in Ecuador, and with technology centers in India and Papua New Guinea.

(b) Second Example: Education Technologies for Teacher Training in Arab Countries.[38]

Most Arab countries suffer from a critical shortage of qualified teachers at a time when their student populations are growing rapidly. Classes must often be taught by instructors with sub-par qualifications or with heavy teaching loads which undermine their effectiveness. The problem is especially acute in schools for Palestinian refugees located in Jordan, Lebanon and Syria. A survey conducted in 1964 revealed that 90 percent of the 4,768 teachers working in UN programs in these three countries were professionally or academically unqualified.

Faced with the problem, several UN organizations launched new programs to improve the quality of teaching. Thanks to a regional cooperative program based on experience gained in Palestinian schools, the techniques they have selected are now spreading to nine Arab countries.

Early seminal work was performed by the Institute of Education, founded in Beirut in 1964. Valuable lessons were learned from experiments with various multi-media techniques ranging from programmed instruction to special radio and closed-circuit TV programs. These techniques are now integrated into comprehensive approaches which also include tutorial guidance, demonstration teaching, debate panels, role-playing and simulation. The educational technologies tested in these experimental years are now being adopted by Bahrain, Iraq, Jordan, Lebanon, the People's Democratic Republic of Yemen, Sudan, the Sultanate of Oman, Syria, and the Yemen Arab Republic.

Attempts are made to adjust the approach to the specific needs of each country. Thus in Jordan where government policy calls for universal education through the ninth grade, teachers of grades one to nine are all enrolled in the same course. And in Sudan, whose national policy assigns a vital role to community education, teacher-training programs strongly emphasize community development. Throughout the program, training is provided in such ways that teachers are not withdrawn from their normal teaching duties.

As these examples suggest, the creative adaptation of technologies originally tested in developing countries can help planners in other developing countries to harmonize technical solutions with their own local values. TCDC also encourages critical reflection on development priorities in varied settings. At least on these two counts, collective self-reliance is a useful instrument in making progress toward the eventual creation of a vital nexus.

New Incentive Systems

Managers of public and private firms in developing countries usually have strong reasons for buying technology from international suppliers. They seek the punctual delivery of reliable technological goods. And because technology is a powerful weapon in their marketing strategy, they are reluctant to use alternative local technology supplies, if only because foreign technology commands greater prestige in the market-place. Firm managers likewise place great importance on favorable financing terms when they purchase technology. For all these reasons, government policy cannot easily wean them away from their usual sources of technology supply. For many Third World firm officials, the single most compelling pressure upon them for buying local technology is the shortage of foreign currency tied to the inability of securing easy import credits. Accordingly, Third World governments wishing to reorient technology demand need to provide concrete incentives for users of technology if these are to seek other sources of supply.

Government policies often rely on negative incentives to achieve their goal of greater "indigenization" of technology use. Negative incentives take such forms as restrictive legislation on technological imports, compulsory registry of licensing contracts, ceilings on technology payments to outsiders, and the requirement that local consumers of technology prove that indigenous technology is unavailable to them for a particular task. In the absence of positive incentives, however, purely restrictive measures cannot succeed. Unless local suppliers of technology can satisfy the aspirations of firm managers on the demand side--their need for reliable and timely delivery, easy marketability, and credit facilities--restrictive legislation usually leads to circumvention, whether by black-market operations or by the disguising of *de facto* foreign purchases. Wherever vulnerable Third World economies are powerfully linked to markets responsive to the control of outside forces, it will prove very difficult *in the short term* to reverse established technology supply and demand circuits. But even in adverse circumstances like these, governments can provide supportive incentive systems which at least facilitate the creation of the vital nexus at the micro level.

To illustrate, agriculture ministries can give credits to small farming units, individuals and cooperatives, to develop new technologies using inexpensive materials available locally. Ministries can at least confer legitimacy on the goal of maximum technical self-reliance at local levels by favoring small-scale production units aimed at the satisfaction of priority food needs. They can likewise provide technical advice to researchers and problem solvers trying to create new sources of technology supply.

The rationale for this sort of government policy is twofold. First, the policy goals of fostering more indigenous technologies in harmony with societal values and with the larger goals of development strategy need to be pursued wherever leverage exists to do so, especially when large numbers of people are directly affected by choices made. New circulatory flows governing indigenous supply and demand of technology can also be set up in industrial and service sectors, but greater freedom of action and a higher probability of initial success are found in the rural sector. A second reason for pursuing leverage points at the micro level is the vital importance of building up a critical mass of relatively self-reliant producer units at the grass roots. Small innovations can be multiplied and the on-going correction of errors can be realized more cheaply in local projects, especially in rural settings. More importantly, the dialogue between technical problem solvers and the populace which is essential to a true vital nexus can best be achieved in small-scale operations.

In short, countries lacking the knowledge, the trained people, the supportive structures, and the political strength to achieve a fully coherent *national* technology policy can nonetheless make gradual progress toward such a policy by applying the vital nexus at more modest levels, a locality or selected sectors of activity.

Choosing Among Conflicting Goals

The Third World's technological aspirations center around four goals:

(a) gaining access to the full range of actual and potential technologies;

(b) enjoying a fair price structure in international technology markets;

(c) purchasing technology in ways which favor the optimum use of local resources (materials, human skills, and capital) instead of buying "packages" imposed upon buyers on "all or nothing" terms by outside technology suppliers; and

(d) lessening their dependence on outside technology innovators, who enjoy a near-monopoly on R & D activities.

No poor country can make technology choices in a political vacuum: existing linkages with specific markets often limit their range of freedom. Thus Senegal cannot abruptly shift its sale of ground nuts to other than its traditional French and European markets. Parallel constraints weigh upon Egypt for cotton, on Brazil for coffee, and (to a lesser extent) on Hong Kong for manufactured items. Moreover, the four goals just listed are not readily compatible among themselves. The quest for a high degree of technical autonomy may dictate sacrifices in access to the full gamut of existing merchandise. Likewise, a policy of making optimum use of local materials, skills, and finance capital--with the correlative decision to limit technological imports to what Andean Pact specialists call "modular" or "core" technology--can weaken the negotiating pressure poor countries can apply to the world pricing system in force. The reason is that many transnational suppliers will simply refuse to sell to purchasers who "break up the package."

The four general goals just listed are attainable only at a price. That price may well be the temporary acceptance of scant progress on one front while efforts are concentrated on achieving greater gains in other arenas. Nevertheless, the careful trading-off of gains among several arenas is a useful tactic which can make it easier to establish the vital nexus. Successful trade-offs are impossible, however, unless a prior screening has been made of a society's value priorities and development strategies. If, for example, the creation of new jobs is a high priority in development strategy; if, moreover, most jobs are to go to the poorest among the population, then the decision to maximize the use of local materials, of untrained non-specialists, and of local currency in technological activities already constitutes the third element in the vital nexus. Merely to deliberate on the weighting to be attached to each of the four general goals of itself carries policy makers several steps along the road to formulating the vital nexus.

In this paper several measures to be taken by Third World countries unable to make a complete or consistent application of the vital nexus have been outlined. Any measure which induces policy makers, in consultation with the affected populace, to assess the relative importance of clustered social values (especially when this gauging is wrought in the light of technological possibilities), enhances the chances of formulating a sound technological policy. No doubt a wide panoply of good policy instruments is also needed: updated information systems, infrastructure for suitable training, fiscal incentives rightly applied. But the very process of debating value preferences and social costs in technical arenas heightens the ability of policy makers to discern the technological scales and modalities most consonant with their society's deepest needs.

IV. SOME LESSONS ABOUT VALUES

Many analysts doubt that a respectful stance toward local values can lead to good technology policy in developing countries. They argue that these values impede "progress," reinforce passivity, and legitimize resistance to necessary innovations. This view is self-defeating, however, because the development process aims at helping people grow in maturity, in the ability to solve their problems, and in gaining greater control over their physical, social, and psychological environments. Therefore, it is impossible to achieve true progress or sound development if the central values cherished by people are disdained. It is from these values that human communities derive their identity, their cultural integrity, and their sense of meaning. Any form of technological "progress" which runs roughshod over these will betray its own promise; indeed sensitive regard for a society's key values is essential to the formulation of sound technology policy. Unless policy makers respect values ON THE TERMS OF THE COMMUNITIES HOLDING THEM, they risk promoting technologies which will destroy cultural life and shatter the fabric of social conviviality. More than any other instrumentality of modern life, technology ought to become what Illich calls a "tool of conviviality,"[39] an aid to living well together in society.

The assumption that indigenous traditional values reinforce inertia in non-industrial societies sins gravely against the facts. As many scholars[40] note, traditions need to display great vitality in the face of new challenges if they are to survive. Consequently, the proper strategy for proposing (and not imposing) developmental change is one which builds on the latent dynamism found in each society's "existence rationality." Until proponents of modern technology overcome their bias in favor of a single mode of rationality--namely, their own--they cannot fruitfully dialogue, even about necessary technological solutions, with people who adhere to other modes of rationality.[41] To the charge that this stance is "inefficient," it must be replied, along with Brazilian educator Paulo Freire,[42] that time spent by technical problem solvers in discussing the mutual impingement of technique and the central values of the affected populace is timed gained "in the process of human development."

Value systems labelled "traditional" vary widely one from another: Aboriginal populations in Australia are not like "Montagnards" of Vietnam's central highlands and Andean peasants do not resemble African tillers of the land. These rich variations mean that cultural communities possess different coefficients of adaptation to the challenges posed to their value systems by modern technology. Elsewhere I have written that "technology is perhaps the most vital arena where cultures and sub-cultures will either survive or be crushed. Their absorptive capacity will be tested in this arena."[43] There is no doubt that the advent of modern technology is an irreversible historical phenomenon, but the "technological imperative" needs to be countered by an explicit "policy imperative." One of the main tasks of policy is precisely this: to subordinate technology to essential human values. The late Max Millikan, a noted econometrician and development planner, wrote in 1962 that planning ought to be "the presentation of certain key alternatives to the community in ways which will help shape the evolution of the community's value system.[44] It is not only the community's value system which must evolve, however: the very manner in which problem solvers frame policy alternatives must also change.

This essay argues that values ought to shape technology policy, not the inverse. The vital nexus briefly described here is simply one way to assure the primacy of values. This nexus is as crucial for good social policy in the United States and other "developed" countries[45] as it is in the Third World. Ultimately, what is needed is a wisdom to match our sciences. Paradoxically, many nations still labelled "less developed" may be closer to gaining that wisdom than those currently enjoying hegemony in domains of technology. Not only must individual technologies be assessed[46] for their impact on human values, but the entire technological universe itself. For without such an assessment, no valid technology policy is possible.

NOTES

[1]As used here the term "common culture" includes sub-cultures, that is, cultural units existing within the confines of a larger unit possessing a broader basis for identity, whether on grounds of nationhood, language, religion, kin, geography, etc.

[2]On arenas of technology policy see Denis Goulet, The Uncertain Promise: Value Conflicts in Technology Transfer (New York: IDOC/North America, 1977), pp. 167-170.

[3]On instruments see Francisco Sagasti, Science and Technology Policy Implementation in Less-Developed Countries (Ottawa: International Development Research Centre, 1976), 2 volumes.

[4]For a discussion of policy issues related to "appropriate" technologies see P. M. Henry, A. Reddy, and F. Stewart, A New International Mechanism for Appropriate Technology (The Hague, Netherlands: August 1979).

[5]Denis Goulet, "The Suppliers and Purchasers of Technology: A Conflict of Interests," INTERNATIONAL DEVELOPMENT REVIEW, Vol. XVII, No. 3 (1976/3), pp. 14-20. Cf. Goulet, "The High Price of Technology Transfers," Interciencia, Vol. 2, No. 2 (March/April 1977), pp. 81-86.

[6]Denis Goulet, Strategies for Meeting Human Needs," NEW CATHOLIC WORLD, September/October 1978, pp. 196-202.

[7]Albert Waterston. "A Viable Model for Rural Development," FINANCE AND DEVELOPMENT, Vol. 11, No. 4 (December 1974), pp. 22-25. Cf. ibid., "Regional Planning: From Its Past to Its Future," AFRICA; INTERNATIONAL PERSPECTIVE, No. 1 (December 1975-January 1976), pp. 17-21.

[8]Hollis Chenery, et al., Redistribution With Growth (London: Oxford University Press, 1974). Cf. Irma Adelman, "Redistribution Before Growth: A Strategy for Developing Countries," University of Maryland, Department of Economics and Bureau of Business and Economic Research, Working Paper 74-14, 1978.

[9]On this see Goulet, "The Challenge of Development Economics" COMMUNICATIONS AND DEVELOPMENT REVIEW, Vol. 2, No. 1 (Spring 1978), pp. 18-23.

[10]Cited in Benjamin Higgins, Economic Development: Problems, Principles, and Policies, revised ed. (New York: W. W. Norton, 1968), p. 3.

[11]Denis Goulet, "An Ethical Model for the Study of Values," Harvard Educational Review, Vol. 41, No. 2 (May 1971), pp. 206-207.

[12]See Louis Dupree, "The Political Uses of Religion: Afghanistan," in Kalman H. Silvert, ed., Churches and States: The Religious Institution and Modernization (New York: American Universities

Field Staff, Inc., 1967), pp. 195-212.

[13]Ibid., pp. 207, 209.

[14]John M. Cohen and Normal T. Uphoff, "Rural Development Participation," in ALS International Agriculture, Cornell University, Vol. 5, No. 1, January 1978. This is a newsletter of the New York State College of Agriculture and Life Sciences.

[15]"Modernization Revisited, An Interview with Daniel Lerner," COMMUNICATIONS AND DEVELOPMENT REVIEW, Vol. 1, No. 2 & 3 (Summer-Autumn 1977), p.5.

[16]Details of this case are reported in several unpublished papers authored by Lim Teck Ghee of the Centre for Policy Research, Universiti Sains Malaysia in Penang, Malaysia. For other examples see Denis Goulet, "Development as Liberation: Policy Lessons from Case Studies," International Foundation for Development Alternatives, Nyon, Switzerland, IFDA DOSSIER, No. 3 (January 1979), pp. 1-17.

[17]Goulet, The Uncertain Promise: Value Conflicts in Technology Transfer (New York: IDOC/North America, 1977), pp. 42-50.

[18]The more fundamental question--Why are developing societies so vulnerable to value damage arising from the adoption of certain "modern" modes of problem solving?--is analyzed in Goulet, "The Cruel Choice: A New Concept in the Theory of Development (New York: Athenum, 1971), Chapter Nine and Appendix No. 3.

[19]On the difference between the qualities found in "leaders" as opposed to those of "rulers" see Paul T. K. Lin, "Development Guided by Values, Comments on China's Road and Its Implications," in Saul H. Mendlovitz, On the Creation of a Just World Order (New York: The Free Press, 1975), pp. 259-296.

[20]For the efforts of leaders in one new nation see Goulet, Looking at Guinea-Bissau: A New Nation's Development Strategy (Washington, D. C.: Overseas Development Council, 1977).

[21]See Chandra Muzaffar, Aliran: Basic Beliefs (Penang, Malaysia: Aliran Movement, 2nd edition, 1978).

[22]Unpublished report entitled "Methodology for Selection of Environmentally Sound and Appropriate Technologies," United Nations Environment Programme, n.d., 54 pages.

[23]On social costs see Goulet, "The High Price of Social Change," CHRISTIANITY AND CRISIS, Vol. 35, No. 16 (October 13, 1975), pp. 231-37.

[24]Ezra J. Mishan, The Costs of Economic Growth (New York: Praeger, 1967), ibidem, The Economic Growth Debate: An Assessment (London: George Allen & Unwin, Ltd., 1977); E. F. Schumacher,

Small is Beautiful: Economics As If People Mattered (New York: Harper Torchbooks, 1973); Dennis Clark Pirages, ed., The Sustainable Society (New York: Praeger, 1977); and Hazel Henderson, Creating Alternative Futures: The End of Economics (New York: G.P. Putnam's Sons, 1978), Cf. Fred Hirsch, Social Limits to Growth (Cambridge, Mass.: Harvard University Press, 1976); also Harlan Cleveland and Thomas W. Wilson, Jr., Humangrowth: An Essay on Growth, Values and the Quality of Life (Princeton, N.J.: Aspen Institute for Humanistic Studies, 1978).

[25]See Charles M. Savage, Work and Meaning: A Phenomenological Inquiry (Springfield, Va.: National Technological Information Service, U.S. Department of Commerce, 1973), Document No. PB-224-333.

[26]Arthur M. Okun, Equality and Efficiency: The Big Trade-Off Washington, D.C.: Brookings Instituion, 1975).

[27]Alberto Moravia, Un Mois en U.R.S.S. (Paris: Flammarion, 1958), p. 165. Cited in Barnard Cazes, "Galbraith ou du Bon Usage des Richesses," in Cahiers de I'Institut de Science Economique Appliquee, No. 111 (March 1961, Series M), p. 47.

[28]Jose Bernabe, "Exportando Energia," AMERICAS (Vol. 25, No. 5), May 1973, pp. 25-27.

[29]Kari Levitt, Silent Surrender (New York: St. Martin's Press, 1970).

[30]On this see Michael Lipton, Why Poor People Stay Poor: Urban Bias in World Development (Cambridge, Mass.: Harvard University Press, 1977).

[31]On this see What Now: Another Development (Uppsala, Sweden: The Dag Hammarskjold Foundation, 1975).

[32]On development styles see Marshall Wolfe, "Development: Images, Conceptions, Criteria, Agents, Choices," in ECONOMIC BULLETIN FOR LATIN AMERICA, XVIII, Nos. 1 & 2, 1973. Cf. Wolfe, "Development Under Question: The Feasibility of National Choice Between Alternative Styles," in CEPAL REVIEW, First Semester 1976, United Nations Document No. 76.11.6.2.

[33]Jorge Sabato and Natalio Botana, LaCiencia y la Technologia en el Desarrollo de la Sociedad, edited by Amilcar O. Herrera, Santiago, Chile: Editorial Universitaria, 1970, pp. 59-76. Cf. Jorge Sabato, "Quantity Versus Quality in Scientific Research: (1) The Special Case of Developing Countries," IMPACT, Vol. 20, 1970, pp. 183ff.

[34]Eduardo Albertal, "The TCDC Conference: Hard Work and Good Timing," UNDP NEWS, November-December 1978, p. 2.

[35]See, for example, the packet of documents prepared by the United

Nations Conference on Technical Co-operation Among Developing Countries (30 August to 12 September 1978, Buenos Aires, Argentina) and entitled Case Studies on TCDC.

[36]Case Study No. 1 in Packet mentioned in previous note.

[37]From the Spanish name for "seagulls," after the fresh-water seagulls, well-liked by the people, which nest in the many webs of creeks and small rivers that cover the "llanos."

[38]Case Study No. 4 in Packet, op. cit.

[39]Ivan D. Illich, Tools for Convivality (New York: Harper & Row, 1973).

[40]Lloyd and Suzanne Rudolph, The Modernity of Tradition (Chicago: The University of Chicago Press, 1967; Mirrit Boutros Ghali, Tradition for the Future (Oxford: the Alden Press, 1972); Robert E. Gamer, The Developing Nations: A Comparative Perspective (Boston and Bacon, Inc., 1976).

[41]The epistemological issues underlying this statement would themselves require lengthy treatment. For a useful recent discussion of the range of programs posed by the issue in domains of research, see Simposio Mundial de Cartegena, Critica y Politica en Ciencias Sociales, el debate Teoria y Practica (Bogota: Punta de Lanza, 1977), 2 volumes.

[42]Source: Paulo Freire, Education for Critical Consciousness (New York: The Seabury Press, 1973), pp. 91-164, section entitled "Extension of Communication."

[43]Goulet,The Paradox of Technology Transfer," THE BULLETIN OF THE ATOMIC SCIENTISTS, Vol. XXXI, No. 6 (June 1975), p. 46.

[44]Max F. Millikan, "The Planning Process and Planning Objectives in Developing Countries," in ORGANIZATION, PLANNING AND PROGRAMMING FOR ECONOMIC DEVELOPMENT, US Papers Prepared for the UN Conference on the Application of Science and Technology for the Benefit of the Less Developed Areas Vol XIII (Washington D.C.: US Government Printing Office, 1962), p. 33.

[45]For a discussion of the myths hidden in the term "developed countries," see Goulet, "The United States: A Case of Anti-Development," MOTIVE, January 1970, pp. 6-13. Cf. Mal-Developpement Suisse-Monde (Geneve: Centre Europe-Tiers Monde, 1975). Commission des Organisations Suisses de Co-operation au Development. Cf. Amilcar O. Herrera and others, Catastrophe or New Society? A Latin American World Model (Ottawa: International Development Research Centre, 1976).

[46]See Denis Goulet, "Capabilities for Technology Assessment: A Concept Paper," THE INDIAN & EASTERN ENGINEER, 1979.

Technological Changes in Developing Countries Since the Second World War

UNCTAD SECRETARIAT

The period 1950-1975 was a golden age of growth for the international community. Both developed and developing countries attained economic growth rates--or technological absorption--far in excess of those achieved in any other comparable period since the Industrial Revolution. The trend was even more pronounced for the developing countries, most of which emerged from colonial dependence during this period. For the first time, and in a vastly altered world, they began to take steps to achieve their economic and social transition to modernization as rapidly as possible.

This chapter reviews broadly the nature and the degree of the change attained in the developing countries since 1950. The data base for giving a concise picture for the period as a whole and for individual developing countries, covering all aspects of their development, is inadequate. The review therefore concentrates on pinpointing the broad sweep of change, often using only illustrative data. It is hoped that this preliminary presentation will help create a conceptual framework within which the problems of the technological transformation of the developing countries in the period ahead can be properly assessed. Only then can a beginning be made in assessing the real causes of the change and the nature of future constraints.

The chapter begins with a summary presentation of the scale, spread and structure of economic change in these countries; in a limited sense, the presentation parallels the main thrust of the first chapter. It is followed by a somewhat closer look at the main elements in building up their technological capacity, including reference, by way of illustration, to a few technological indicators. But before going further the essential features of the technological dependence of developing countries should be recalled.[2] Despite the accelerating pace of scientific advance,

From **PLANNING THE TECHNOLOGICAL TRANSFORMATION OF DEVELOPING COUNTRIES**, (UNCTAD Secretariat), 1981, (7-18), reprinted by permission of the publisher, U.N. Publications, N.Y.

particularly in the twentieth century, all developing countries, though in varying degrees, are in a state of technological dependence. This is accounted for by a series of asymmetries: (i) in commodity patterns--the types of consumer goods produced in poor countries are strongly influenced by those consumed in the rich; (ii) in the means of production--the ability of poor countries to produce capital goods is very limited; (iii) in technical knowledge--there are gaps concerning both communications and suitability: the third world does not know what is widely available elsewhere, and there is little appreciation of the kind of expertise required; (iv) in skills--the fundamental difference is that there are very many more opportunities in advanced countries of learning by doing, which is just as important as it was at the beginning of the Industrial Revolution; (v) in trade--there are often restrictions on exports from developing countries that use an imported technology in their manufacture, and imports are frequently tied to the country of origin of the technology; (vi) in finance--most developing countries have limited and uncertain access to capital; (vii) in control--the main decisions are generally taken in a foreign country; (viii) in initiative--stemming from the inability to make technological decisions.

I. MAIN ELEMENTS OF TECHNOLOGICAL CHANGE IN DEVELOPING COUNTRIES, 1950–1975

1. The Pace of Technological Transition

Technological transition is characterized by the diffusion of knowledge of the productive process, resulting in increased output of goods for final consumption. A more precise indicator would be the change in output per head of population. Between 1950 and 1975, the total real output of developing countries tripled or quadrupled, rising at a rate of 5 per cent per year.[3] This growth rate was nearly five times as high as that for the first half of the twentieth century and higher than even the historical growth rate in the developed countries (see paras. 8 and 9 above).

At the same time, there was a marked success in reducing death rates, particularly through control of infant mortality. Population growth therefore accelerated--to a rate of about 2.5 per cent per year--with the result that per capita output grew slowly, by 2.5 per cent per year. In sharp contrast, this per capita growth rate could not have been more than 0.5 per cent in the 25 years preceding 1950.[4]

Obviously, in a group comprising some 103 countries varying widely in size, level of development and output, natural resource endowment, year of obtaining independence and type of economic, social and political policies pursued, there are bound to be

TABLE 1

Selected indicators of change in developing countries, 1960-1975

	Number of countries	*Population 1975* Number (millions)	Share of total (percentage)	*Gross domestic product 1975* Total (billions of dollars)	Per capita (dollars)
Developing countries with per capita income in 1975:					
1. Over $800	26	324	17	474	1 465
2. Between $300 and $800	38	416	22	210	506
Subtotal (1 + 2)	*64*	*740*	*39*	*684*	*923*
3. Below $300	*39*	*1 161*	*61*	*186*	*160*
of which:					
India, Indonesia, Pakistan, Bangladesh	4	881	76	140	158
Others	35	280	24	46	166
Total developing countries	103	1 901	100	870	458
Total developed market-economy countries	28	795	100	4 079	5 140

Source: UNCTAD, *Handbook of International Trade and Development Statistics, Supplement 1977* E/F.78.II.D.1); *Handbook of International Trade and Development Statistics, 1979* (United Nations publication

TABLE 2

Changes in the shares of major sectors in output in developing countries, 1950-1975 [a]

(*Percentage*)

	Agriculture [b]	Industry [c]	Services [d]
Developing countries			
1950	36.8	26.4	36.8
1955	34.2	28.8	37.0
1960	33.0	31.9	35.1
1965	27.6	34.5	37.9
1970	25.3	37.0	37.7
1975	22.1	38.7	39.2
Developed market-economy countries			
1950	7.3	42.8	49.9
1955	6.7	44.5	48.8
1960	6.2	44.1	49.7
1965	5.2	45.9	48.9
1970	4.6	46.7	48.7
1975	4.5	45.1	50.4

Sources: Statistical Yearbook 1968 (United Nations publication, Sales No. E/F.69.XVII.1) and *Statistical Yearbook 1976* (United Nations publication, Sales No. E/F.77.XVII.1).

TABLE 3

Structure of industrial output in developing countries, developed market-economy countries and socialist countries of Eastern Europe, 1960 and 1975
(Percentage)

Branch of activity	*ISIC code*	*Developing countries*		*Developed market-economy countries*	
		1960	*1975*	*1960*	*1975*
All industry	2-4	*100*	*100*	*100*	*100*
of which:					
Mining and quarrying	2	25	19	7	4
Electricity, gas and water	4	4	7	5	8
Manufacturing	3	71	74	88	88
of which (as a percentage of total manufacturing output):					
(*a*) *Consumer goods*	31-34, 39	*57*	*46*	*39*	*32*
Food, beverages and tobacco	31	23	8	12	11
Textiles, wearing apparel, leather	32	18	17	9	8
Wood products and furniture	33	6	3	4	4
Paper, printing and publishing	34	2	5	4	7
Other	39	8	13	10	2
(*b*) *Intermediate goods*	35-36	*19*	*24*	*15*	*19*
Chemicals, petrochemical and coal products, rubber products	35	14	19	11	15
Non-metallic mineral products	36	5	5	4	4
(*c*) *Capital goods*	37-38	*25*	*29*	*45*	*48*
Basic metals	37	8	6	8	7
Metal products	38	17	23	37	41

Sources: Statistical Yearbook 1966 (United Nations publication, Sales No. E/F.67.XVII.1); *Statistical Yearbook 1977* Sales No. E/F.78.XVII.1).

TABLE 4

Gross domestic capital formation, gross domestic savings and external resources in developing countries, 1970-1975
(Percentage of GNP) [a]

	Gross domestic capital formation	*Gross domestic savings*	*External resources* [b]
Developing countries with a per capita GNP in 1975:			
Over $800	22.8	24.4	−1.6
$300-$800	21.0	17.5	3.5
Under $300	16.7	13.7	3.0

Source: "Domestic savings in developing countries: note by the UNCTAD secretariat" (TD/B/C.3/153), October 1978, table 1 (UNCTAD secretariat estimates, based on United Nations and other international sources).

differences in the ability to incorporate technological knowledge in the productive system. Some of the relevant data are shown in table 1.

As may be seen from that table, in some 64 countries, with over $300 per capita output in 1975 and accounting for 39 per cent of the total population of developing countries, the growth rate of GDP averaged 6.2 per cent per year between 1960 and 1975. Their per capita output rose by 3.4 per cent annually. On the other hand, in the other 39 countries, with just over three fifths of the total population of developing countries, expansion was much smaller, their total real output increasing by 3.6 per cent per year, and per capita output by only 1.3 per cent. This group includes most of those that may be considered as confronted with the hard core of the development problem. Among them are some of the largest countries in Asia and most of the least developed among the developing countries.

2. Changes in Origin of Output by Major Sectors

The structure of output changed only marginally in the developing countries in the 100 years or so preceding 1950. Since then, the changes have been rather marked for the group as a whole. As shown in table 3, the most important structural change took place in the relative shares of agriculture and industry and related activities. The share of agriculture in GDP fell from 37 per cent in 1950 to 22 per cent in 1975, and that of industry (including mining, manufacturing, electricity and gas, construction, and transport and communications)[5] rose from 26 per cent to 39 per cent--a near reversal of relative positions. In comparison, the share of the service sector remained relatively constant, at 37-39 per cent. Changes in the sectoral composition of employment were in the same direction, though less pronounced.[6]

The reversal of the shares of agriculture and industry followed from differences in their growth rates during the period 1960-1975: industrial output expanded by 7 per cent per year, with a rise in employment of 4.4 per cent and an implicit rise in productivity of 2.6 per cent. The corresponding growth rates for agriculture were lower: 2.7 per cent, 1.3 per cent and 1.4 per cent respectively.[7]

As in the case of overall output growth, there were significant variations among countries in the extent of the shift from agriculture to industry. In developing countries with per capita incomes over $800, the shift was much greater than in those with lower incomes. For instance, the share of agriculture in the first group had fallen by 1975 to 10-15 per cent, whereas in the other two income groups, with per capita incomes below $800, it remained as high as 25-40 per cent. The important point to note, however, is that in 50 developing countries, the share of agriculture had fallen to just over 10 per cent, or the level prevailing in most developed market-economy countries (other than

TABLE 5.

Indicators of capital goods production in selected countries, 1974

	Value added in capital goods production[a]		
	Share in total manufacturing (percentage)	*Ratio to consumer goods production*[b]	*Annual growth rate (in real terms)*[c] *1965-1974 (percentage)*
Developed countries			
Czechoslovakia	50.4	1.92	7.6
Germany, Federal Republic of	47.0	2.58	3.8
Japan	45.5	1.65	14.0
Sweden	42.3	1.40	5.3
Italy	40.9	1.54	6.0
United States of America	39.2	1.16	3.2
United Kingdom	37.2	1.10	1.0
Developing countries			
Chile	45.2	1.24	0.1
Mexico	33.9	0.99	10.2
India	32.7	0.87	2.7
Brazil	31.9	0.84	15.5
Iran	27.8	0.61	30.0
Turkey	24.6	0.65	7.0[d]
Venezuela	23.9	0.57	7.5
Peru	22.6	0.43	15.5
Republic of Korea	21.5	0.42	32.0
Egypt	20.2	0.33	6.2
Colombia	17.1	0.34	7.0[d]
Malaysia (peninsular)[e]	16.9	0.39	23.9
Nigeria	15.7	0.26	30.9[d]
Zambia[e]	15.0	0.28	14.0
Indonesia[e]	14.6	0.20	43.8
Ghana[f]	14.3	0.30	1.7
Philippines	13.7	0.25	1.8
Tunisia[e]	12.8	0.25	13.5
Mozambique[e]	12.3	0.20	4.6
Bangladesh	11.1	0.15	22.0[d]
Ecuador	10.8	0.16	24.0
Jamaica	10.7	0.18	—
United Republic of Tanzania[e]	10.4	0.17	21.5
Zaire	10.2	0.16	—
Malawi[e]	6.6	0.11	20.3
Libyan Arab Jamahiriya	4.4	0.07	30.1[d]
Dominican Republic	3.2	0.04	10.6[d]

TABLE 6. **Selected aspects of educational development in developing countries**[a]

	Literacy rate[b] (percentage)	Percentage of adult population that entered[c]			Number of students enrolled in universities and other institutions of higher education (thousands)
		Second level	Post-secondary		
A. *Developed countries*[d]					
1960	—	..	..		9 544
1974	95[b]	..	..		24 831
B. *Developing countries*[e]					
1960	—	..	..		2 112
1974	38[b]	..	..		8 453
Kenya					
1962	20	2.0	0.3	1965	2
1969	24[f]	2.8	0.8	1974	11
Iraq					
1957	14	0.9	0.7	1965	28
1965	24	1.7	0.9	1974	79
Algeria					
1954	19	1.2	0.4	1965	8
1971	26	2.2	0.3	1975	42
India					
1951	20	1.5[g]	0.5[g]	1965	1 054
1971	33	3.9	1.1	1974	2 230
Syrian Arab Republic					
1960	30	1.8	0.5	1965	33
1970	40	4.4	1.1	1974	64
Zambia					
1963	29	1.3	—	1965	—
1969	47	2.0	0.6	1974	5
Indonesia					
1961	39	1.4	0.1	1965	140
1971	57	5.1	0.5	1975	278
Brazil					
1950	49	1.0	0.9	1965	156
1970	66	5.9	2.0	1974	955
Mexico					
1950	57[h]	2.9	1.1	1965	133
1970	74	4.1	2.6	1974	453
Thailand					
1947	52	3.0	0.2	1965	36
1970	79	4.4	1.1	1975	78
Colombia					
1951	62	5.7	0.9	1965	44
1973	81	18.4	3.3	1974	149
Republic of Korea					
1955	77	5.3	1.5	1965	142
1970	88	21.8	5.6	1975	297
Argentina					
1947	86	3.7	1.2	1965	247
1970	93	7.8	4.0	1975	597

Sources: UNCTAD, *Handbook of International Trade and Development Statistics, Supplement 1977* (United Nations publication, Sales No. E/F.78.II.D.1); and UNESCO, *Statistical Yearbook 1976* (Paris, 1977).

[a] Countries arranged in ascending order of literacy rate in the later year shown.

[b] Unless otherwise stated, the rate refers to population over 15 years of age covered by census or survey. The rates shown for developed and for developing countries as a whole are very rough estimates.

[c] Adult population in most cases refers to those over 25 years of age.

[d] Developed market-economy countries (including Israel) and socialist countries of Eastern Europe (including Romania).

[e] Excluding China and other socialist countries of Asia.

[f] Of persons over 25 years of age.

[g] Of all age groups.

[h] Of persons over 6 years of age.

the United States of America and the United Kingdom), in the mid-1950s and in all the socialist countries of Eastern Europe in the mid 1960s.

The differences in degree of structural change in the two broad groups of countries--with per capita incomes of over $800 and under $800 respectively derived primarily from the varying rates of growth in industry. The rate of agricultural growth in both groups was much the same, ranging from 2 per cent to 3 per cent for the period 1960-1975. Much greater differences are to be noted in the expansion of industrial output. To illustrate, in the absence of comprehensive data, the former group included a number of countries with an annual industrial growth of over 10 per cent--notably, the Republic of Korea (22.5 per cent), Iran (15.7 per cent), Singapore (15.3 per cent) and Brazil (11.0 per cent) whereas industrial growth in most of the countries in the latter group was less than 5 per cent per annum.

II. STRUCTURAL TRANSFORMATION OF INDUSTRY

The rapid growth of industrial output in the developing countries was accompanied by a drastic change in the structure of their industry, as shown in table 3. The most striking feature of the industrial structure of the developing countries in 1975 was the large share of mining in total industrial output--nearly one fifth. Slightly less than one third of the world's mineral output is produced in the developing countries, which export most of it, often in wholly unprocessed or semi-processed form, to the developed countries. This is obviously an area where a big gap remains to be filled in beginning the processing of mineral output and moving on to the manufacture of mineral-based products.

An examination of the structure of industrial output of the developed market-economy countries and the socialist countries of Eastern Europe shows that a broad measure of similarity exists between them. Capital goods account for roughly one half of the total, their share being slightly higher in the socialist countries. Intermediate goods production accounts for slightly less than 20 per cent in both. Broad similarities are visible even when individual subsectors are compared.

The structure in the developing countries is sharply different. Consumer goods industries still continue to account for nearly one half on industrial output, though this proportion declined considerably during the 15 years from 1960 to 1975. The greatest weakness in their industrial structure remains in the capital goods sector, or the materials in which advanced technology is embodied and which form the basis for raising the productivity of the work force. It should be noted, however, that a significant change occurred in the relative share of metal products from 17 per cent in 1960 to 23 per cent in 1975.

III. BUILDING UP THE TECHNOLOGICAL CAPACITY OF DEVELOPING COUNTRIES

The central economic difference that distinguishes pre-Stone Age man from any other in succeeding ages lies in the capacity to produce goods and services. This capacity is embodied in the tools and instruments and acquired skills with which he participates in the production process. While the ancient history of Africa, Asia and Latin America abounds with examples of highly developed civilizations, the vast changes, partly described in chapter I, which revolutionized the process of production in the developed countries simply by-passed the third world. Except for a few countries, and even then in only isolated segments and on a small scale, both these embodiments of technology--skills and tools--were largely absent in the third world. Whatever little embodied technology was used in developing countries was imported from the metropolitan countries. Productivity per person increased only marginally, if at all, up to the mid-1950s.

In the period 1950-1975, a beginning was made in filling this gap. Detailed information on this subject is scanty, but an attempt is made here to give an impression of the changes that have taken place in the ratio and volume of capital formation, the import and domestic manufacture of capital goods, skill formation and expenditure and manpower devoted to research and development (R and D).

1. Changes in Capital Formation and Domestic Savings

In the early 1950s, gross domestic capital formation as a proportion of GDP in developing countries was only 10-12 per cent. As these countries achieved independence and adopted planning as an instrument of policy, there was a considerable increase in this share. Thus in 1960 there were only 10 developing countries in which it was above 20 per cent, whereas by 1975 there were 46; in the developing countries as a whole it was about 20 per cent in the period 1970-1975 (table 4), or nearly the same as in the developed market-economy countries. Gross domestic capital formation, at over $170 billion in 1975, was some six times higher in real terms than 25 years previously.

The share of gross capital formation varies, of course, from country to country. As shown in table 5, it was much higher (21-23 per cent) in developing countries with a per capita GNP of more than $300 in 1975 than in those below that level, where it was about 17 per cent. It is striking indeed that most of the capital formation was financed from gross domestic savings, with external resources contributing less than one fifth of the total even for those with a per capita GNP of less than $300.

TABLE 7. **Scientists and engineers engaged in R and D, and expenditures on R and D** [a]

	Scientists and engineers engaged in R and D			*Expenditure on R and D*				
	Period	*Total (thousands of FTE)* [b]	*Per 10,000 of total population*	*Fiscal year beginning*	*Total (millions of dollars)*	*Per capita (dollars)*	*Percentage of GNP*	*Annual percentage increase (at current prices)*
A. *Rough totals* [c]								
Developed market-economy countries	[c] 1973-75	1 390	43 [d]	[c] 1973-75	80 900	111	2.0	..
Socialist countries of Eastern Europe	[c] 1973-75	1 440	82 [d]	[c] 1973-75	39 400	114	4.0	..
Developing countries [e]	[c] 1973-75	210	1 [d]	[c] 1973-75	1 900	1	0.3	..
B. *Selected developing countries*								
India	1973	97.0 [f]	—	1972	256	0.5	0.4	18.1 (1969-1972)
Indonesia	1975	12.2	0.9	1975	47	0.3	0.2	80.5 (1972-1975)
Argentina	1974	8.1	3.2	1974	184	7.4	0.5	26.9 (1968-1974)
Brazil	1974	7.7	0.8	1974	346	3.4	0.4	10.6 (1973-1974)
Egypt	1973	6.9	1.9	1976	85	2.2	0.7	4.1 (1973-1976)
Republic of Korea	1974	6.3	1.9	1976	128	3.6	0.7	20.7 (1969-1976)
Thailand	1974/75	6.1	1.5	1968	14	0.4	0.2	..
Mexico	1974	5.9	1.0	1973	102	1.8	0.2	18.8 (1970-1973)
Philippines	1965	5.6	1.8	1973	32	0.8	0.3	15.1 (1965-1973)
Iran	1972	4.9	1.6	1972	47	1.5	0.2	..
Pakistan	1973/74	4.2	0.6	1973	15	0.2	0.2	..
Ghana	1975	3.9	3.9	1971	21	2.4	0.7	..
Venezuela	1973	2.7	2.6	1973	67	6.0	0.3	43.6 (1970-1973)
Sudan	1974	2.7	1.6	1973	9	0.5	0.3	..
Nigeria	1970/71	2.1	0.4	1970	33	0.6	0.3	11.5 (1966-1970)
Peru	1975	2.1	1.3	1970	25	1.9	0.4	..
Cuba	1969	1.9	2.2	1969	92	11.0	2.2 [g]	..
Bangladesh	1973/74	1.6	0.2	1974	24	0.3	0.3	..
Iraq	1974	1.5	1.4	1974	25	2.3	0.2	67.0 (1971-1974)
Colombia	1971	1.1	0.5	1971	10	0.5	0.1	..

2. Imports and Domestic Manufacture of Capital Goods

The data on gross capital formation give only a rough impression of the changes that have taken place, since they include construction (involving inputs of intermediate products and labour), inputs of intermediate products, capital goods (means of production) and skills. If the technological capacity of a country is to be assessed, a much greater role has to be assigned to the means of production needed for translating the general concept of technology as "capability" into actual "capacity" to produce goods. It is really the balance between import dependence for capital goods, and domestic capability to produce them, that can serve as the critical indicator for this purpose.

The value of imports of machinery and transport equipment by developing countries (SITC sect. 7) increased by 13 per cent per year between 1955 and 1975. Deflated by the index of the unit value of exports of manufactured goods from developed market-economy countries, this implies a sixfold increase in volume, or an annual rate of 9.4 per cent.

The rapid increase in imports of capital goods was accompanied by an equal, if not more rapid, increase in the domestic manufacture of these goods. If value added is taken as an indicator of the latter activities, domestic manufacture of capital goods--basic metals (ISIC 37) and metal products (ISIC 38)--grew at an annual rate (in real terms) of slightly more than 10 per cent during the period 1960-1975. Changes in the ratio of imports to domestic manufacture of capital goods (SITC 7), calculated on the basis of current prices, may be taken as a rough indicator of changes in the degree of dependence of developing countries on imports for the supply of capital goods. This ratio decreased slightly, from 1.4 to 1.3, between 1960 and 1970, and then increased to 1.5 in 1975.[9] If the value added in metal products manufacture (ISIC 38) alone is taken as an index of capital goods production, the ratio declines from 2.0 in 1960 to 1.7 in 1970, followed by an increase to 1.9 in 1975. The picture given by these ratios is not very conclusive. It is probable that the increase in the ratios between 1970 and 1975 resulted on the one hand from the rise in prices of capital goods imported by the developing countries due to inflation and, on the other hand, from reduced domestic manufacture in the developing countries, reflecting the global economic difficulties of the 1970s. However, detailed study would be required to determine the impact of the global economic situation in the 1970s on the development of capital goods industries in the developing countries. The ratios none the less serve to show that for the period 1960-1975 as a whole the domestic manufacture of capital goods was increasing, or at least keeping pace with imports of these goods.

The type of capital goods imported by a particular developing country depended to a large extent on its capacity to produce them domestically. In the relatively more industrialized developing countries, where such capacity had been built up, imports of capital goods consisted largely of those embodying sophisticated

TABLE 8

Selected indicators of technological diffusion, 1960-1976

	Fertilizer consumption (kg/ha) [a]	*Per capita energy consumption (kg of coal equivalent)* [b]	*Per capita electricity consumption (kWh)*	*Number of commercial vehicles in use per 10,000 of population*	*Railway freight traffic (net ton-kilometres per capita)* [c]
Developing countries					
1960	6	211	97	26	124
1976	23	426	305	46	209
(1976 as percentage of 1960)	(383)	(202)	(314)	(177)	(169)
Socialist countries of Asia					
1960	13	552 [d]	102	—	—
1976	51	719	189	—	—
(1976 as percentage of 1960)	(392)	(130)	(185)	(—)	(—)
Developed market-economy countries					
1960	67	3 995	2 596	333	1 916
1976	114	6 388	6 077	1 189	2 270
(1976 as percentage of 1960)	(170)	(160)	(234)	(357)	(118)
Socialist countries of Eastern Europe					
1960	26	2 893	1 305	41 [e]	5 428
1976	98	5 251	4 032	140 [e]	10 034
(1976 as percentage of 1960)	(377)	(182)	(309)	(341)	(185)

Sources: For fertilizer consumption: FAO, *Annual Fertilizer Review, 1977* (Rome, 1978); for energy consumption and electricity consumption: *World Energy Supplies 1950-1974* (United Nations publication, Sales No. E.76.XVII.5) and *World Energy Supplies 1972-1976* (United Nations publication, Sales No. E.78.XVII.7); for commercial vehicles in use and net ton-kilometres of railway freight traffic: United Nations, *Statistical Yearbook 1966* (United Nations publication, Sales No. E/F.67.XVII.1) and *Statistical Yearbook 1977* (United Nations publication, Sales No. E/F.78.XVII.1).

[a] Fertilizer consumption is expressed in terms of effective nutrient content of nitrogen, phosphate and potash per hectare of arable land and land under permanent crops. Data refer to the fertilizer year and those for 1960 to the average for 1961-1965.

[b] Consumption of commercial energy.

[c] Freight in net ton-kilometres on railway lines (except railways entirely within an urban unit), including fast and ordinary goods services but excluding service traffic, mail, baggage and non-revenue governmental stores.

[d] Average for 1959-1961.

[e] Excluding USSR, for which data are not available.

technology. In the less industrialized countries where there was some, but less, capacity for capital goods manufacture, attempts were made to undertake an increasing proportion of assembly work on standardized capital goods on the basis of imported components, while at the same time relying on imports of a wide variety of fully assembled capital goods with a large technological content. In the least industrialized countries, imports in fully assembled form were practically the only means of introducing capital goods into domestic production activities--and this, quite often, through reliance on the inflow of external capital.

Table 5 provides an overview of capital goods manufacture in selected developing countries and a comparison with the situation in some developed countries. It is evident from the table that, first, capital goods manufacture already accounts for an important share of manufacturing output in a number of developing countries. The share in terms of value added in 1974 was already above 30 per cent in Chile, Mexico, India and Brazil, and in six other countries (Iran, Turkey, Venezuela, Peru, Republic of Korea and Egypt) it was above 20 per cent. The ratio of value added in capital goods production to that in consumer goods production was closely correlated to the share of capital goods production in total value added in manufacturing. For Chile, Mexico, India and Brazil, this ratio reached the level of that of some of the developed countries.

Secondly, the growth of capital goods production in the developing countries was considerable. As may be seen from table 6, the annual growth rate of value added in capital goods production (in real terms) for all but a few developing countries was either equal to or more rapid than in the developed countries, except Japan. In a number of developing countries with a relatively high share of capital goods production in manufacturing, such a Brazil, Iran, Peru and the Republic of Korea, the growth rate for the 1965-1974 period was even more rapid than in Japan. It is also worthy of note that a number of developing countries with a relatively small share of capital goods production, such as Tunisia, Bangladesh, Ecuador, the United Republic of Tanzania, Malawi and the Libyan Arab Jamahiriya, recorded a relatively high growth rate, though needless to say that was partly a reflection of a late start, from a low base.

The expansion of domestic capacity to manufacture capital goods in developing countries has far-reaching implications. First, it strengthens the scope not only for "reproducing" domestically the technology that is imported from abroad in embodied form but also, and perhaps more important, for producing the technology that is most suited to the specific needs of these countries. Secondly, it broadens the options open to planners and policy-makers in these countries in planning and managing investment activities, allowing the possibility for foreign exchange savings on a given investment project and thereby reducing the impact of the foreign exchange constraint on the overall investment programme. Thirdly, the development of the capital goods production capacity in developing countries has in fact increased the scope for trade in these goods among the developing countries themselves.[10]

Two important qualifications have to be made, however,

concerning this rough picture of capital goods manufacture in developing countries. The first is that in some of the relatively more advanced developing countries the development of the capital goods sector relied quite heavily on the financial, technological and managerial input of transnational corporations through direct foreign investment.[11] The extent to which the participation of these corporations contributed to the strengthening of the national capacity varied from country to country. The second qualification relates to the impact of the extensive and wide-ranging manufacture of capital goods on a small scale characterized by simple repair work on imported machinery, which is not above analysis was based.[12] An in-depth analysis of these questions is needed to provide a more comprehensive and more accurate picture of capital goods manufacture in the developing countries and its contribution to the technological development of those countries.

3. Skill Formation and R and D Resources

The building up of the technological capacity of any country starts with education. Developing countries have made significant progress in this area in the 25 years under review. Total school enrolment in these countries grew at an annual rate of 5.6 per cent during the period 1960-1974. Particularly significant was the increase in enrolment in universities and institutions of higher education, reaching an annual rate of more than 10 per cent over the same period. As shown in table 7, the size of such enrolment in 1974, involving nearly 8.5 million students, came close to the level of that in the developed countries in the late 1950s. One out of every four students enrolled at the university level in the world is now from a developing country (excluding China and other socialist countries of Asia, for which data are not available).

One direct result of the general spread of education has been the improvement in the literacy rate. Although the rate varies widely among different developing countries--for many countries in Africa it is less than 30 per cent, while for Latin American countries it is usually more than 70 per cent--there has nevertheless been a universal improvement, as shown in table 7. For the developing countries as a whole, it is estimated to be currently slightly less than 40 per cent, in contrast to about 95 per cent in the developed countries.

Improvement in the literacy rate implies the strengthening of the skill base in the broadest sense. A much more direct impact of education on skill formation in the developing countries occurs through improvement in the average level of education attained by the labour force. Table 6 also contains such data for selected developing countries expressed as the proportion of adult population entering schools at the secondary and post-secondary level. For the former, the proportion ranged from about 2 per cent to about 20 per cent, and for the latter from about 0.5 per cent to about 6 per cent, and in the more recent period. However, there again, the dominant trend over time has been an improvement, often a doubling or even tripling of the proportion over a decade or so. The picture can also be completed by information on the occupational composition of the labour force. In selected developing countries for which data were available, the proportion of professional, technical and related workers in the total economically active population has tended to rise over time.[13]

Skill formation is also closely related to capacity to undertake R and D activities, that is, capacity to adapt and generate technology. Table 7 gives the latest available information on the number of scientists and engineers engaged in R and D in selected developing countries and establishes a comparison with developed market-economy countries and the socialist countries of Eastern Europe; however, per 10,000 of economically active population, the gap widened to more than one fortieth and one eightieth respectively. R and D manpower (full-time equivalent) per 10,000 of (total) population in individual developing countries, which ranged from 0.2 to 3.9, was much smaller than the corresponding figure in most developed countries, where the ratio was more than 10. However, in a number of developing countries the absolute size of such manpower had reached a significant level in terms of manning requirements for plant, particularly in large countries such as India, Indonesia, Argentina, Brazil, Egypt and the Republic of Korea.

Table 7 also contains some data on expenditure on R and D, which may be regarded as a general indicator of the level of activity in that area.[14] A study of the data would suggest that the gap in R and D activity between developed and developing countries was much wider than suggested by the data on scientists and engineers engaged in such activity. R and D expenditure per capita in developing countries was well under 1 per cent of that in developed countries. Expressed in terms of GNP, total expenditure on R and D was estimated to be about 0.3 per cent for developing countries compared with 2 per cent for the developed market-economy countries and 4 per cent for the socialist countries of Eastern Europe.

Among developing countries, much of the R and D activity took place in a relatively small number of countries; six countries--India, Argentina, Brazil, Egypt, Republic of Korea and Mexico--accounted for roughly 60 per cent of the total expenditures on R and D of developing countries, and spent a generally higher proportion of GNP on R and D than other developing countries. Limited time-series data on R and D expenditure for these countries also seem to suggest a relatively high rate of increase in their expenditures on these activities. In most of the other developing countries, R and D activities were only beginning to take shape, so that the relevant figures recorded were still low or not yet available.

As shown above, most of the indicators of skill formation and R and D resources demonstrate improvements and positive gains achieved by developing countries, though to varying degrees. It should be noted, however, that these countries also face the problem of putting to productive use the skills and other human capacities that are created. The record of developing countries in this respect has been far from satisfactory, as unemployment and underutilization of skills created are prevalent in many of them. Moreover, even when the skills are utilized, they are not necessarily geared to increasing output. While there are many reasons for this situation, it should be noted that the quantitative expansion of skilled personnel has not been integrally linked to the development of opportunities for using them adequately. Thus, one of the main tasks involved in altering the technological profile of developing countries is to tackle the problems of both skill formation and improved utilization within the framework of national development.

IV. DIFFUSION OF IMPROVED TECHNOLOGY: EXAMPLES

Another aspect of the technological profile of the developing countries relates to the diffusion of improved technological practices. While there are countless instances of such diffusion, they do not lend themselves to any systematic accounting, partly because of lack of data and partly because of the absence of appropriate criteria for judging the rate and magnitude of diffusion. There are, however, some universally accepted technological practices whose diffusion can be "compared", not only among countries or groups of countries, but also over time. Table 8 shows five such indicators of technology diffusion: fertilizer consumption per hectare, per capita energy consumption, per capita electricity consumption, number of commercial vehicles in use per 10,000 of population, and net ton-kilometres of railway freight traffic per capita.

The table is self-explanatory, and only some general remarks may be called for. Comparing the developing countries with the developed market-economy countries, there remains a sizeable gap in the level of these indicators of technological diffusion. However, with respect to all of them, except the number of commercial vehicles in use per 10,000 population, the rate of change for the developing countries over the period covered was much greater than the corresponding rate for the developed market-economy countries. Notwithstanding the varying rates of diffusion among the countries, they appear to indicate that improved technological practices were being spread in the various economic and social sectors of the developing countries to narrow the gap between them and the developed countries.

V. EVOLUTION OF TECHNOLOGY POLICY AND PLANNING

Technology planning should perhaps be regarded as the final stage in a process that starts with national action to control imports of technology, leads on to the formulation and adoption of a coherent and integrated technology policy, with a corresponding network of co-ordinated institutions, and culminates in technology planning in the framework of development planning.

1. National Action to Control Technology Imports

The first attempts by developing countries to control technology imports go back little more than a decade. In India, in 1969, the guidelines on policies and procedures for foreign collaboration agreements provided for a selective approach to technology imports, with limits on the equity capital that could be held by foreigners (exceptions being made when the technology sought could be obtained in no other way), limits on royalties, and prohibition of restrictive clauses. In Brazil, imports of foreign technology embodied in machinery are controlled by the Department of External Commerce (CACEX) of the Bank of Brazil. Other transfers of foreign technology are controlled by the National Institute of Industrial Property (INPI), which was established in 1971 under the Ministry of Industry and Commerce. INPI is concerned only with contractual agreements, i.e. payments and conditions. It is not concerned with evaluation in terms of employment, foreign exchange or suitability of the technology; its small staff consists of lawyers and economists. It has no link with the organs responsible for controlling foreign investment or with those engaged in developing domestic technology.

In accordance with Decision No. 24, adopted on 31 December 1970 by the Commission of the Cartagena Agreement, the five member countries of the Andean Group (Bolivia, Chile, Colombia, Ecuador and Peru), later joined by Venezuela, established a common regime for foreign capital, trade marks, patents, licensing agreements and royalties. Decision No. 24 established norms to be followed by each country within the basic objectives of the Andean Pact: harmonization of economic development, and in particular of industrial development, and promotion of trade among members. It was left to each country to determine its own institutional framework, and the Decision has undoubtedly had some impact in this respect. But the main emphasis seems to have been on reducing excessive charges for imported technology; Colombia's success in reducing payments made via transfer pricing in the pharmaceutical sector has frequently been cited. Changes in basic economic policy in some countries, which have weakened the Andean Pact, have also had their effect in limiting progress towards the setting up of comprehensive technology policies and institutions. Chile has withdrawn from the Pact and Peru has moved towards a more liberal regime with respect to foreign investment.

In Argentina, legislation was enacted in 1972, inspired to some extent by Decision No. 24 of the Andean Group; this led to the registration of technology transfer contracts, but without evaluation. The following year foreign investment laws were introduced, but there was no provision for linking control of foreign investment with that of imported technology. By this time responsibility for registering technology contracts resided in the National Institute of Industrial Technology, a step in the direction of relating imports of technology to technologies produced domestically. Prior to 1973, the role of the registry had been advisory, but a law passed in that year, which became operative the following year, reinforced its powers by providing for royalty ceilings and restrictions on foreign trade marks. A change of government led to a much more liberal approach to foreign investment and technology. The registry was transferred to the Ministry of Industry and has become less active; there have also been further changes in legislation, so that control of imported technology has become more formal in character.

In Mexico, the Registry began operations in 1973, and was to become a model for the approach to control of technology imports in developing countries. The focus is on registration of all contracts, elimination as far as possible of restrictive practices, limitation on payments for technology, evaluation of technology and establishment of administrative guidelines to ensure that imported technology makes a positive contribution to national development. But there has been insufficient linkage and interaction among the different Mexican institutions concerned with technology transfers, in particular the Registry for the Transfer of Technology, the Registry for Foreign investment and the Industrial Property Office. New arrangements have recently been announced in this respect, but the context of a more open approach to foreign investment.

Screening of imported technology has gradually evolved in the Philippines since the early 1970s, but the system is still somewhat ill-defined. The Central Bank plays a certain part through a ceiling on royalties, a maximum rate in trade mark agreements, prohibition of export restrictions and a fixed duration for technology transfer agreements. The Board of Investments has worked out guidelines based on three major considerations: essential need for the technology by the industry concerned and the national economy; cost of the technology; and restrictive practices. A technology transfer and development unit in the Department of Industries undertakes technical, economic and legal evaluation of agreements with a view to the unpackaging of technology and the inclusion, where possible, of local technology. The Republic of Korea has set up a technology transfer centre within the Korean Institute of Science and Technology; the main emphasis at present is on controlling the high rate of growth of payments for foreign technology and diversification of sources of such technology, obtained at present mainly from one country.

By the middle of the 1970s several of the more advanced developing countries, i.e. those referred to above, could be said to have significantly improved their bargaining position as technology importers. Most of these countries--and some others--had by then a variety of research and development institutions. Yet the linkage between imports of technology and promotion of domestic capability was--and remains--non-existent or ineffective. Even with regard to the acquisition of foreign technology there is little linkage among the various institutions directly or indirectly involved in the drawing up and implementation of technology policy, i.e. those concerned with development planning and policy, with finance, with foreign investment, with the preparation and appraisal of projects in different economic sectors (especially in industry), on with education and training.[15]

2. Towards Technology Policies

By its resolution 87 (IV), on strengthening the technological capacity of developing countries, the United Nations Conference on Trade and Development recommended that developing countries should establish national centres for the transfer and development of technology and decided that an advisory service on transfer of technology should be established within UNCTAD. A new impetus was thus given to the evolution within developing countries of more integrated and effective technology policies and institutions, and a beginning was made in providing them with assistance to that end.

UNCTAD preparatory missions, some of which have already led to follow-up action, have been to several countries: Afghanistan, Burundi, Egypt, Ethiopia, Iraq, Somalia, Sri Lanka, Thailand and Venezuela.[16] Further missions are being considered by the governments concerned, which include those of Angola, China, the Democratic People's Republic of Korea, Ghana, Guyana, Ivory Coast, Malaysia, Nicaragua, Nigeria, Nepal, Sierra Leone, Sudan, the United Republic of Cameroon and the United Republic of Tanzania. If the general themes arising in connection with the acquisition, adaptation and domestic generation of technology by developing countries are necessarily the same, the specific policies and, still more, the institutions established differ according to national circumstances. Thus Ethiopia and Iraq, which are building socialist economies, are setting up national centres with considerable executive powers. Other countries envisage centres having more of an advisory and co-ordinating role, as is the case with Sri Lanka and Venezuela.[17]

3. From Technology Policy to Technology Planning

In the mid 1970s five developing countries--Brazil, India, Mexico, Pakistan and Venezuela--drew up and published science and technology plans.[18] As will be seen from their titles, each plan is concerned with both science and technology. However, the definitions used, even of the terms "science" and "technology", are not very explicit.

The plans indicate the method used in drawing them up, in terms of the country's development plan or strategy, but the relationship between the technology plan and the development plan or strategy is tenuous. This is hardly surprising in view of the fact that four of the five countries have not proceeded very far with development planning as such. It follows that there are only limited arrangements for the implementation of the technology plan: in other words, only a few countries have even begun to work out comprehensive and integrated policies for the transfer and development of technology and to establish and link together the corresponding institutions.

NOTES

[1]Owing to limitations of data, the discussion in this chapter generally leaves out the socialist developing countries of Asia (China, Democratic Kampuchea, Democratic People's Republic of Korea, Mongolia and Viet Nam).

[2]See the report by the UNCTAD secretariat, "Technological dependence: its nature, consequences, and policy implications" (TD/190), chap. I, reproduced in Proceedings of the United Nations Conference on Trade and Development, Fourth Session, vol. III, Basic Documentation (United Nations publication, Sales No. E.76.II.D.12).

[3]The rate sustained for these 25 years was indeed higher than most economists and planners considered possible in the early 1950s. Arthur Lewis, for example, after noting that some development plans set targets of economic growth at over 4 per cent per year, stated: "but 4 per cent is so difficult to attain that it is really quite an ambitious target". See W.A. Lewis, "Some reflections on economic development", Economic Digest (Karachi), vol. 3, No. 4, winter 1960, p. 3.

[4]The rate may have been even lower; see the study by the UNCTAD secretariat, Energy Supplies for Developing Countries: Issues in Transfer and Development of Technology (United Nations publication, Sales No. e>80.II.D.3), annex A, table A-2.

[5]Output in the transport and communication sector tends to rise in line with industrial output and has therefore been included in industrial output in table 3 for analytical convenience.

[6]Employment in agriculture fell from 71 per cent in 1960 to 63 per cent in 1970 and rose in industry from 9 per cent in 1950 to 14 percent in 1970; (See ILO, Labour Force Estimates an Projections, 1950-2000, vol. V, World Summary) (Geneva, 1977).

[7]These growth rates for total output in industry and agriculture (though not for productivity) in developing countries were greater than those for the developed market-conomy countries.

[8]According to the Statistical Yearbook 1977 (United Nations publication, Sales No. E/F.78.XVII.I), these countries were: Africa: Algeria, Botswana, Egypt, Gabon, Guinea, Ivory Coast, Kenya Liberai, Libyan Arba Jamahiriya, Mauritania, Mauritius, Morocco, Nigeria, Tunisia, United Republic of Tanzania, Upper Volta, Zaire, Zambia; Asia Iran, Iraq, Jordan, Lebanon, Oman, Philippines, Republic of Korea, Saudi Arabia, Singapore, Syrian Arab Republic, Thailand; Latin America: Argentina, Barbados, Brazil, Costa Rica, Ecuador, El Salvador, Guatemala, Guyana, Honduras, Mexico, Panama, Paraguay, Venezuela; other: Malta, Tonga, Hong Kong and Martinique are also included in these figures. To this list should be added, by virtue of their membership of the Group of 77, Romania and Yugoslavia.

[9]The ratios were calculated on the basis of unpublished data of the Statistical Office of the United Nations on the value added in developing countries.

[10]See Trade in Manufactures of Developing Countries and Territories: 1977 Review (United Nations publication, Sales No. E.80.II.D.2), part one, "Recent trends and developments in trade in manufactures and semi-manufactures of developing countries and territories". chap. II, sect. 2.

[11]Transnational Corporations in World Development: A Re-examination (United Nations publication, Sales No. E.78.II.A.5), annex III, tables III-59 to III-63.

[12]See, for example: J.T. Thoburn, "Exports and the Malaysian engineering industry: a case study of backward linkage", Oxford Bulletin of Economics and Statistics, vol. 35,No. 2, May 1973; C. Child and H. Kaneda, "Links to the Green Revolution: a study of small-scale, agriculturally related industry inthe Pakistani Punjab". Economic Development and Cultural Change (Chicago, Ill.), vol. 23, No. 2, January 1975; ILO, "Technology diffusion from the formal to the informal sector: the case of the auto-repair industry in Ghana" by A. N. Hakam, working paper WEP 2-22/WP 35, July 1978.

[13]According to data compiled by the UNCTAD secretariat on 26 developing countries on the basis of information contained in ILO, Year Book of Labour Statistics, and United Nations, Demographic Yearbook (various issues).

[14]It should be remembered that R and D expenditure includes in most cases both capital and current expenditures. The level of R and D activity is in general better reflected in capital expenditures, for obvious reasons. Current expenditures in developed countries tend to push up the total level because of the relatively higher wages and costs of services.

[15]See the report by the UNCTAD secretariat, "Action to strengthen the technological capacity of developing countries: policies and institutions" (TD/190/Supp.1), para. 26, reproduced in Proceedings of the United Nations Conference on Trade and Development, Fourth Session, vol. III, Basic Documenation (United Nations publication, Sales No. E.76.II.D.12).

[16]Reports on these missions, available or forthcoming, are: "Transfer and development of technology in Afghanistan" (UNCTAD/TT/AS/3); "Transfer and development of technology in Burundi" (UNCTAD/TT/38); "Transfer and development of technology in Egypt" (UNCTAD/TT/AS/7); "Transfer and development of technology in Ethiopia" (UNCTAT/TT/AS/4); "Transfer and development of technology in Iraq" (UNCTAD/TT/AS/2); "Transfer and development of technology in Somalia" (UNCTAD/TT/39); "Transfer and development of technology in Sri Lanka" (UNCTAD/TT/5).

[17]For a description of different approaches to the problem of establishing integrated policies for the transfer and development of technology, and a summary of the recommendations of the UNCTAD missions referred to, see UNCTAD, Handbook on the Acquisition of Technology by Developing Countries (United Nations publication, Sales No. #.78.II.D.15 and Corrigendum), chap. IX.

[18]Brazil, Presidency of the Republic, Plano Basico de Desenvolvimento Cientifico e Tecnologico, 1973/1974 (Brasilia, 1973); idem, Basic Plan for Scientific and Technological Development, 1976-1979 (Brasilia, 1976); India, National Committee for Science and Technology, Science and Technology Plan, 1974-1979, 2 vols. (New Delhi, 1974); Mexico, Consejo Nacional de Ciencia y Tecnologia, Plan nacional indicativo de ciencia y tecnologia, 1976-1982 (Mexico City, 1976); Government of Pakistan, Scientific and Technological Reserach Division, Prospects for National Science and Technology Policies, 1976-1981 (Karachi, 1976); Venezuela, Consejo Technology Policies, 1976-1981 (Karachi, 1976); Venezuela, Consejo Nacional de Investigaciones Technologicas, Primer Plan Nacional de Ciencia y Tecnologia, perfodo 1976-1980 (Caracas, 1976). (These plans are summarized and analysed in their economic and historical context in document TD/B/C.6/29 and Corr. 1, chap. 111.)

Technological Self-Reliance of the Developing Countries: Toward Operational Strategies

UNITED NATIONS INDUSTRIAL DEVELOPMENT ORGANIZATION

I. TECHNOLOGICAL DEPENDENCE OF THE THIRD WORLD AND LIMITATIONS OF THE PRESENT TECHNOLOGY SYSTEM

A. Technology and development

Technology is one of the prime motive forces of development. Whether the need is more food, better education, improved health care, increased industrial output or more efficient transportation and communications, technology plays a decisive role. It consists of a system of knowledge, skills, experience and organization that is required to produce, utilize and control goods and services. Technology is critical to development because it is a resource and the creator of new resources, is a powerful instrument of social control and affects decision-making to achieve social change.[1]

Technology is not neutral; it incorporates, reflects and perpetuates value systems and its transfer thus implies the transfer of structure.[2] Technology is both an agent of change and a destroyer of values. It can promote equality of income and opportunity or systematically deny it. It follows that technology not only influences society but also that society imposes limits on the choice and development of technology.

Because the technologies adopted by the developing countries not only shape national development options but also affect, directly and indirectly, the economic structure of the industrialized countries, it is not surprising that technology is of concern to both rich and poor countries. Inevitably, it has

From **TECHNOLOGICAL SELF RELIANCE OF THE DEVELOPING COUNTRIES TOWARDS OPERATIONAL STRATEGIES**, No. 15, 1981, (3-46), reprinted by permission of the publisher, U.N. Publications, N.Y.

become one of the major areas of negotiation for the establishment of a new international economic order.

B. Technological dependence of the third world

Although aware of the great importance of technology for their development, the developing countries are unable to exercise real choice in designing effective strategies for their technological transformation. The growth of the international economic system has resulted in a profusion of institutions and mechanisms that maintain developing countries in conditions of dependence and that lead to ever-widening disparities between the richest and poorest nations.

Industrial production in the industrialized world has been accompanied by a process through which the sources of new technology have been concentrated and a handful of enterprises and government agencies dominate and control most of it. For example, in the United States of America the top 50 corporations and the government research agencies in the fields of defence, energy, space and health accounted for more than three quarters of the $38 billion spent on research and development in 1976. A few hundred people in the highly industrialized nations are able to make decisions on who is going to get which part of these new technologies at the world level and under what conditions.[3]

Nowhere are the disparities between the industrialized countries and the third world more marked than in the crucial field of technological development; the dependence is almost total (table 1). Developing countries possess only 12.6 per cent of global stocks of scientists and engineers engaged in research and development (R and D), of which 9.4 per cent are concentrated in a few countries of Asia. Developing countries account for only 2.9 per cent of global expenditures on R and D and 3.3 per cent of global exports of machinery and transport equipment. There are no readily available data for services but there is little reason to suppose the picture would be much different. Approximately 95 per cent of developing countries' imports of machinery and transport equipment come from developed countries.[4]

In general, technological dependence arises when most of a country's technology comes from abroad and the greater the reliance on foreign technology and the more concentrated the source, the greater the dependence. For the developing countries, the major source is a small number of industrialized countries. A country dependent upon a single source for all of its foreign technology can thus be considered more dependent than one that obtains its technology from a range of countries. For some technologies, sources may be widely spread; for others, for specific industries, they may be highly concentrated. At present, the United States is the world's main supplier of technology, being responsible for 55-60 per cent of the world's technology flow.[5]

Third world countries cannot offset the direct costs of

technology imports with the proceeds of technology and exports of manufactures.

Many developing countries suffer "double dependence" in that they need not only to acquire the elements of technical knowledge but also to import the capacity to use this knowledge in investment and production.

Technological dependence seriously undermines the attempts that might be made by a developing country to strengthen its own capacity for scientific research and technological development. It does so in two ways: it inhibits processes of "learning-by-doing", essential for the development of scientific capacities; and it tends to devalue the activities of local scientific and technology institutions, making them either irrelevant or poor copies of those in the industrialized countries.

C. Limitations of the present technology system

The international system with its built-in automatic mechanisms that maintain dependencies imposes severe constraints on the exercise of the technological options open to the developing countries. Some of the most severe of these constraints are the costs of technology transfer; the role of transnational corporations in the transfer process; the relevance of the technology transferred; and the restrictions imposed by the international industrial property system.

Costs of technology

During the last two decades there has been a rapid growth of technology exchange between enterprises in different countries and an emergence of industrial technology as a highly marketable commodity. Trade in technology rose from approximately $2.7 billion in 1965 to over $11 billion by 1975, mainly in the form of lump-sum payments, royalties and fees. Most technology trade has taken place between enterprises in the industrialized countries, with the highest outflow coming from the United States, followed by Switzerland, the United Kingdom of Great Britain and Northern Ireland, the Federal Republic of Germany, the Netherlands, France, Belgium, Italy and Japan. The payments made by the developing countries for transactions in technology are estimated at $1 billion in 1975 which is less than 10 per cent of the total value of such transactions. Of this amount, about 50 per cent was paid by Latin American countries, particularly Brazil and Mexico, and about 35 per cent by Asian countries. Payments to United States enterprises from developing countries accounted for $316 million in 1965, increasing to $845 million in 1975.

In a perfect market, competition would reduce to marginal the cost of acquiring technology; the market in technology, however, like so many others of importance to the developing countries, is imperfect, with great monopoly advantages for the seller because of secrecy and the protection of patents and trade marks. The technology (whether in the form of pure knowledge or embodied in foreign investment or machinery) is transferred under terms that are the outcome of negotiations between buyers and sellers in situations frequently approximating monopoly or oligopoly. The final returns and their distribution largely depend upon the relative power of the bargainers and an unfavourable outcome is probable for the dependent countries.

UNIDO has estimated that the trade in technology by developing countries, in terms of fees, royalties and other payments for technical know-how and specialized services, could increase from approximately $1 billion in 1975 to over $6 billion in 1985.[6] This is approximately 15 per cent of the total trade in technology which, if the growth in the period 1965-1975 is maintained in the period 1975-1985, is likely to be approximately $40 billion by the mid 1980s. Most of the payments made by developing countries would be for technology and know-how imported from the industrialized countries and would represent payment outflows by the third world as a whole. The figure can, however, be considered an under-estimate since it takes no account of underpayments through the manipulation of transfer prices or the cost of technology transferred implicitly via sales of product and payments for foreign personnel.

The indirect costs of technology acquisition, which take the form of restrictions on sources of input and access to market outlets, are held to be many times higher than direct costs. Rough estimates indicate that indirect and hidden costs could be from $6 billion to $12 billion a year, equivalent to 2-4 per cent of the national income of the developing countries.

Although the total cost of the technological dependence of the third world cannot be accurately stated, if allowance is made for the transfer of inappropriate technologies and the long-term influence of technologies that undermine the development of endogenous capabilities, it could well be as high as $30 billion to $50 billion a year.

Role of transnational corporations

Transnational corporations have been responsible for approximately 80-90 per cent of the technology transferred to the developing countries and much of the third world has been dependent upon transnationals for acquiring and expanding their technological development capability.[7] This has mainly involved "contractual transfers", the main way of acquiring the technologies required for such science-intensive industrial sectors as chemicals, pharmaceuticals and electronic components.

TABLE 1. SELECTED INDICATORS OF TECHNOLOGICAL CAPACITIES

Indicator	*Developed market economy countries*	*Eastern Europe (including USSR)*	*Developing countries*		
			Africa	*Asia*	*Latin America*
R and D scientists and engineers, 1973 (percentage of world total)	55.4	32.0	1.2	9.4	2.0
R and D expenditures, 1973 (percentage of world total)	66.5	30.6	0.31	1.63	0.94
Share of exports of machinery and transport equipment, 1976 (percentage of world total)	86.9	9.5	0.04	2.6	0.68
Developing country imports of machinery and transport equipment, 1971 (percentage of total)	90.3	4.2	5.1		

Source: UNIDO, *Industry 2000, New Perspectives* (ID/237), tables 7 (1) to 7 (4).

TABLE 2. ESTIMATED FOREIGN-CONTROLLED SHARES OF THE PHARMACEUTICAL INDUSTRY–SELECTED COUNTRIES, 1975

Country and groups of countries	*Share of sales (percentage)*
Saudi Arabia	100
Nigeria	97
Belgium	90
Colombia	90
Venezuela	88
Brazil	85
Canada	85
Australia	85
Indonesia	85
Mexico	82
Central American Common Market (1970)	80
India	75
Iran	75
Argentina	70
United Kingdom	60
Italy	60
South Africa	60
Finland (1971)	50
Sweden	50
France	45
Portugal (1970)	44
Turkey (1974)	40
Norway (1971)	36
Germany, Federal Republic of	35
Switzerland (1971)	34
Greece	28
Egypt (1971)	19
United States	15
Japan	13

Source: For details of sources see "Transnational corporations and technological development" (ID/WG.301/12), p. 35.

The use of transnationals as a major source of technology has given rise to many problems. Confusion over basic values and social priorities has often led to the uncritical purchase of technologies and techniques and this has in many cases proved detrimental to genuine development. The indiscriminate extension of the technologies and the productive systems employed by transnationals into the third world has resulted in much destruction of traditional technologies as well as social problems. It has been clear that the development fostered by transnationals, especially those involved in consumer goods industries, is not always responsive to social needs, particularly those of the poor. Because of their necessity to continually expand and grow, transnationals must have an increasing number of responsive buyers. Since their capacity to sell largely determines their profits, they must inevitably produce for those who can afford rather than those in need. Thus, they have become linked to and dependent upon the affluent sectors of poor societies because these are the principal consumers. When allowed to operate on the principle of artificially stimulated demand, and thus waste, it is inevitable that they tend to repeat the patterns of Western market economy societies through a type of technology very often inappropriate to the needs of third world countries.

The market power of transnational corporations largely determines the availability and pattern of technology transfer in advanced science-based sectors where technology ownership is mainly concentrated in a few large enterprises. In these sectors, owing to increasing R and D costs, the economies of scale involved in technological innovation and commercialization, and high costs of market failure, large companies have become the major source of technological development and, consequently, the owners of improved and new technology. Similarly, in sectors where fast technological change reduces the product life cycle, for example, drugs, scientific instruments and electronics, the importance of technological advantage makes control over the technology within the corporate system the major motivating factor in its commercialization. In these sectors, the transfer of technological know-how is confined largely to wholly or majority-owned subsidiaries. Where foreign minority ownership is unavoidable, effective control over technology use is sought through management or service contracts. The diffusion of technology in these advanced sectors and the participation by competitive firms is frequently limited, on the one hand, by a large degree of cross-licensing, patent pooling and other forms of technology sharing arrangements between the leading transnationals and, on the other, by intra-firm technology flow. Such barriers have been particularly prevalent, for example, in the chemical industry and in the manufacture of heavy electrical and telecommunications equipment.

Despite the regulatory measures instituted by several developing countries and the increased availability of technological alternatives in certain sectors, the role of transnational subsidiaries and affiliates in most developing countries is important because of their dominant position in several sectors, for example, the pharmaceutical industry (table

2). In most countries of Africa and Latin America, mineral industries remain largely under foreign ownership or control even though domestic participation, often through state enterprises, has been increasing in recent years. State participations and control have been most marked in the petroleum industry, but in both petroleum and other resource-based industries, transnational corporations have continued to exercise significant control through the supply of technology and services. In several developing countries, even relatively low-technology consumer goods production has remained under the control of foreign subsidiaries. In the case of middle-technology and high-technology industrial sectors, transnational subsidiaries and affiliates exert dominant influence even in such countries as Brazil, India and the Republic of Korea where significant domestic entrepreneurial capability exists. In several service sectors, including merchandising, transnational subsidiaries and affiliates continue to play a decisive role in many developing countries.

Transnational corporations have generally contributed little to the development of technological infrastructure in the developing countries. Rather, they have sought to minimize the value-added of their production in a developing country. This procedure has frequently been aggravated by the excessively high prices at which some technological know-how is supplied by so-called "tie-in" clauses. Since proprietary and non-proprietary knowledge is transferred by a transnational corporation partly in embodied form or as know-how from the parent company, there is little interest or initiative for R and D activities by subsidiaries and affiliates. Since affiliates obtain only those elements that have already been commercialized in the home market, the R and D function has been completed for the specific technology at the parent company. New technologies, including improvements, are developed in the parent company, which is close to the home market and has an advanced scientific and technological infrastructure. Centralization of technology generation at the parent company also helps to ensure control over proprietary technology. Local R and D activity in developing countries is often confined to the adaptation and local testing of products that are not available in the industrialized countries, such as certain drugs, or that are produced locally, such as tea processing. The absence of R and D in the host country renders the affiliate dependent on the parent company for the flow of technological improvements. This inevitably becomes a major element of control. Technology ownership can be similarly the controlling element in the case of joint ventures, particularly when technology transfer includes patented know-how brands or trademark names.

The lack of local R and D activity and the resultant low demand for scientific and research personnel hinders the development of indigenous engineering and design capabilities necessary for the creation of technology and the effective adaptation and absorption of foreign technology. Moreover, in the absence of local R and D the affiliate has little technological linkage with local scientific and research institutions, which would promote technological research capability and diffusion.

Given the above, it is inevitable that transnational

corporations have provided but little employment in the developing countries. Estimates vary from 1.6 million to 2.5 million for 1967 for all industrial sectors combined, approximately 0.3 per cent of the third world's total active population. Even when generous allowances are made for the indirect employment created, this figure is insignificant.

The transfer of technology from the parent company to foreign affiliates, because it takes place as a purely internal process, provides opportunities to manipulate the prices of goods and services supplied by one part of the enterprise to another. Internal transfers have been used, for example, to shift profits and to control "free" cash. The extent to which transnationals manipulate transfer prices appears to depend upon the gains vis-a-vis the costs, in terms of the effort and risk involved. Manipulated prices are most likely where large corporations trade in large quantities in products encountering little or no competition. The potential for such manipulation appears likely to increase as a result of the continued concentration of economic power in the hands of transnationals and the increasing importance of intra-corporate transactions in their total trade, and in particular of the continued diversification of their activities on a horizontal, vertical and conglomerate basis.

The problem of transfer pricing is one that confronts both industrialized and developing countries. However, inducements to manipulate prices may be greater in the countries of the third world. This is owing to, among other things, import controls, limits on dividend remittances and royalty payments, and the desire by transnationals to achieve, for a variety of reasons, a higher return on investments. Intra-firm trade probably exceeds 50 per cent of the international trade transactions of the developing counties, and case studies show that the extent and range of intra-firm imports of developing countries are large, especially in such industrial sectors as chemicals, pharmaceuticals, electrical machinery and rubber.[8]

Transfer price manipulations can seriously prejudice a developing country's possibilities for economic development: they can, for example, have adverse effects on competition and the balance of payments, domestic capital formation, tax revenues of individual developing countries, and local industrial structure.

Transnationals are expressions of a system whose values and orientations have given them their characteristics and stimulated their unprecedented expansion.[9] Transfer pricing is a function of the corporate system and it may prove exceedingly difficult to change transfer arrangements without first changing corporate structures. Transfer pricing may well be an area in which the power of transnationals is greater than that of governments to control it, although government supervision, especially of the industrialized countries, has so far been minimal.

Relevance of the technology transferred

Much of the technology developed in the industrialized countries has little direct relevance to the problems confronting many developing countries because it is not geared to the satisfaction of basic human needs; more than 50 per cent of world investment in science and technology is directed towards the production of evermore sophisticated weapons and armaments and about two thirds of the remainder towards marginally increased consumption of non-essential goods.[10] Research on problems directly relevant to the third world probably accounts for little more than 1 per cent of the total research expenditures of the industrialized countries.

Most of the world's ready-made technologies are optimally suited to the industrialized countries. The introduction of inappropriate Western technologies has had a wide range of consequences for the developing countries, particularly those noted below.

Although some Western technology has undoubtedly contributed to economic and social progress, in many developing countries it has been instrumental in increasing the gap not only between the rich and poor, but also, in earnings and social status, between men and women. This is especially so with some of the "modern" technologies introduced by transnationals. Although these enterprises can be powerful engines of growth, their activities are not inherently geared to the goals of development and, in the absence of proper government policies and, in some cases, social reform, tend to increase rather than reduce inequalities in poor societies.

Technology is not a neutral factor in social and economic development; various types of technologies can be used to promote different types of development and to reinforce patterns of privilege and power. Western technologies have been used by developing country elites to strengthen their power position at a time when self-reliant development calls for increased participation and the decentralization of decision-making functions.

One of the reasons why Western enterprises have sought to establish affiliates in the developing countries is to escape the increasingly strict pollution control legislation in their home countries. Some developing countries have implicitly acquiesced to the use of their "open spaces" as pollution havens although it is increasingly being realized that development that is in harmony with the environment can contribute, especially at the local level, to the satisfaction of basic needs and the promotion of self-reliance.

TABLE 3. SHARE OF PATENTS REGISTERED BY NON-RESIDENTS IN SELECTED DEVELOPING COUNTRIES

Country	*1965*	*1970*	*1975/76*
Argentina	...	77.7	69.2
Bolivia	...	89.5	86.7
Chile	91.5	93.8	89.5
Colombia	93.4	80.8	78.6
Ecuador	...	96.3	86.7
Ghana	100.0	100.0	100.0
Hong Kong	...	98.6	98.8
India	90.2	83.1	82.6
Iran	93.1	92.7	96.8
Kenya	100.0	100.0	100.0
Korea, Republic of	38.7	25.1	32.5
Morocco	93.5	94.5	93.5
Philippines	96.0	96.5	87.6
Tanzania, United Republic of	100.0	100.0	100.0
Tunisia	95.6	99.3	91.6
Venezuela	94.5	92.0	84.5
Zaire	100.0	100.0	92.5
Zambia	...	99.4	98.0

Source: Based on *Industrial Property* (Geneva, WIPO, 1977).

TABLE 4. SHARE OF PATENTS REGISTERED BY FOREIGNERS IN CHILE

Year	*Percentage*
1937	65.5
1947	80.0
1958	89.0
1967	94.5
1976	90.0

Sources: C. V. Vaitsos, "Patents revisited: their function in developing countries", *Journal of Development Studies,* vol. 9, No. 1 (October 1972); *Industrial Property* (Geneva, WIPO, 1977).

International industrial property system

International patenting does not itself generate technological dependence. It is, however, a means of regulating the application of technological knowledge in different countries and by different types of enterprises. The international patent system thus exerts considerable influence on who gets industrialized, and the method, conditions and cost.

Patents confer on the owner a monopoly of production and distribution of products in a specified territory for a given period of time. Of the current 3.5 million patents, only approximately 200,000 (6 per cent) have been granted by developing countries. Of these, moreover, five out of six are held by foreigners (table 3) and only one sixth (1 per cent of the world total) are held by nationals of the developing countries. Most of the developing country patents held by foreigners are held by large corporations headquartered in five developed market economy countries: the Federal Republic of Germany, France, Switzerland, the United Kingdom, and the United States. Approximately 90 to 95 per cent of the patents granted by developing countries to foreigners are not actually used in production processes in those countries; instead, the overwhelming majority are used to secure import monopolies. In some cases, utilization rates fall below 1 per cent of the registered patents. In Peru in 1975, the patent utilization rate was below 0.5 per cent.[11] It is thus evident that international patent practices have come to represent a reverse system of preferences granted to foreign patent holders in the markets of developing countries.

Even when foreign-held patents are actually used in production processes, the agreements entered into by developing countries concerning patent use through foreign investment or licensing arrangements frequently contain high royalty payments and charges for the technology, restrictive practices and in some instances abuses of patent monopolies, either explicitly embodied in the contractual agreements or implicitly followed by subsidiaries and affiliates of transnational corporations, which impose heavy indirect or "hidden" costs by overcharging for imported inputs. The foreign exchange burden of these costs, which are much larger than direct costs, applies to all developing countries regardless of whether they have national patent laws.[12]

The extensive patenting activity of Western transnational corporations has undoubtedly suppressed the development of local innovative activity and has contributed to a continuous decline in the share of locally owned patents. This trend is demonstrated in the example of Chile (table 4), which is characteristic for most developing countries for the past decades. The insignificantly low share and economic importance of locally held patents tend to demonstrate that the present industrial property system has hampered the development of indigenous technological capabilities.

Summary

The suppliers and purchasers of technology have different motivations and hence obey different criteria.

The suppliers seek lucrative, free and diverse markets in which they can generate revenues on their investments in R and D. In carrying out their activities they exploit international financing contracts, and become involved in measures against competitors and in countering national pressures over ecological and working conditions. The purchaser of technology, on the other hand, wants to master the imported technology, exploit it to develop his country and remain competitive domestically and internationally.

In this conflict of interests, the weaker partner, which is often the developing country, is destined to lose. The technology supplier is able to take advantage of a wide range of weaknesses of the recipient country, including a lack of capital and of appropriate skills and information.

The transfer of technology from the industrialized countries has enabled some developing countries, particularly the more privileged groups within them, to benefit from some of the advances made in the field of science and technology in the past two centuries. The transfer has allowed these countries to use this technology without themselves having to go through the difficult and costly process of developing it. Technology transfer has introduced high-productivity techniques, and also, in many cases, inspired the desire for technical change. Whereas there are some benefits from the present system of transfer, there are none from the dependence that the process of technology transfer, development and concentration has created.

Technological dependence has many dimensions; at its simplest, it results from the fact that a handful of rich countries are the source of almost all the industrial technologies that are currently being applied, and that the virtually sole supplier of technologies has been transnational enterprises, which are motivated by business (profit-maximization) rather than development (social welfare) considerations.

At another level, technological dependence is one aspect of the general pattern of dependence resulting from the operation of the international economic system and the institutions and mechanisms that govern the relations between rich and poor countries. The international economic system is a stratified system of power relations. Because it has a structure that helps to determine who decides and who controls, it is a system of domination.

Rather than reducing the technological dependence of the developing countries, the international system actively and persistently reinforces it. The transfer of technology may facilitate the expansion of industrial output in the developing countries; it does not, however, necessarily further the ability of these countries to produce that output, or, more precisely, does not give them the capacity to adapt and modify existing technology or to evolve new technologies.

Technological dependence can be seen as both cause and effect of general dependency relationships.[13] It leads to foreign investment, loss of control and the introduction of alien patterns of consumption and production. This creates an enclave economy, dependent on the advanced countries for inputs, markets, management, finance and technology. This creates, in its turn, a society in the image of the advanced countries, requiring further imports of technology to satisfy new demands and to enable the industries to survive and expand. There is a vicious circle in which a weak technology system reinforces dependence, and dependence perpetuates weakness. Some of the elements of this vicious circle are shown schematically in figure I.

The international system and the channels through which technology is transferred to the developing countries thus contain many elements that are incompatible with the attainment of many of the objectives of the New International Economic Order, including the industrialization target contained in the Declaration and Plan of Action of the Second General Conference of UNIDO held in Lima in 1975. The use of existing channels for the attainment of this target--an increase in the third world share of industrial output from its present level of under 10 per cent to 25 per cent by the year 2000--could, according to some estimates, increase by fivefold to eightfold the costs of technological dependence. As noted earlier, these approximate costs may already be from $30 billion to $50 billion a year.

The present technology system thus maintains the developing countries in a situation of dependence and frustrates efforts that might be made to develop an indigenous technological

II. TOWARDS TECHNOLOGICAL SELF-RELIANCE: ISSUES AND IMPLICATIONS

A. Towards technological self-reliance

The new approach to the technological transformation of the third world must aim at reducing the technological dependence of the developing countries by strengthening their autonomous capacity for technological change and innovation. This approach, a movement away from a "flow" concept to a "stock" concept, must necessarily be supported by determined efforts to restructure the prevailing legal and juridical environment with the aim of developing new sets of internationally agreed norms for the benefits that arise from the international transfer of technology. Such restructuring should focus on the formulation of appropriate codes that can be used to control transfers and the activities of transnational corporations and on the reform of the industrial property system in order to create the necessary conditions for strengthening the endogenous technological capabilities of the developing countries and for reducing their dependence.

Technological autonomy, although a particularly important component of self-reliant development, cannot, under the present global circumstances, mean technological independence. Just as national self-reliance may require selective participation in the international system, enhanced technological autonomy may necessitate selective technological delinking from the world market, however difficult this might prove. Some scientists in the third world argue that the poor countries should cut themselves off from Western science and technology, that traditional cultures "must be protected from the onslaught of Western patterns of consumption and those consumer goods that represent the omnipresence of technologies". By the same token, the developing countries "should reject all Western offers of technological assistance".[15]

Whereas structural disengagement could in some instances have an unexpectedly stimulating effect on the development of local technological capabilities, it appears to be an option open to only large developing countries. Even for these, however, it will be difficult to ignore the fact that the international economic system is the dominant system that governs the behaviour of subsystems. The Union of Soviet Socialist Republics has been seeking ways in which it can strengthen its links with the market economies, and China is also carefully exploring such possibilities. Clearly then, disengagement is bound to be difficult for most developing countries. Besides, autonomous capacity for technological development does not mean that a country must reinvent the wheel. Rather, it means that it should have the capacity to do so if it had to, possibly in circumstances beyond its control, and have the capability to improve upon wheels invested elsewhere.

Technological self-reliance is defined as the autonomous capacity to make and to implement decisions and thus to exercise choice and control over areas of partial technological dependence or over a nation's relations with other nations. It follows that technological self-reliance can be effectively pursued only when a nation understands the nature and extent of its technological dependence and possesses the will and self-confidence to seek to overcome it and to maintain its cultural identity. Technological self-reliance must thus be conceived in terms of the capacity to identify national technological needs and to select and apply both foreign and domestic technology under conditions that enhance the growth of national technological capability. Enhanced technological capacity appears to be an essential precondition for developing countries to deal with their economic and social problems.

B. Dimensions of technological self-reliance

There is an enormous variety of technological situations in

the third world and generalizations concerning strategies may have little relevance. The decisions that must be taken by developing countries will be determined by such considerations as factor endowments, cultural patterns, national aspirations, present levels of development and industrialization (sectors, products, processes, functions performed in productive operations), geographic location, size of markets, and so on.

In seeking to exercise their limited choices, developing countries will be constantly confronted with complex problems that defy quick solutions. A major problem will almost inevitably concern the science and technology system. The behaviour of this system is conditioned by the larger social system of which it forms part. A science and technology system has various components such as institutions (scientific and technological) and production facilities, physical facilities and human skills all of which are embedded in, and which affect, a pattern of values. The technology system performs several functions starting with a specification of its "outputs" (products and services) and proceeding to choice of technologies in the usual sequence of prefeasibility, feasibility studies, engineering design, implementation, management, marketing and R and D. As the components in the technology systems of some nations are more developed than those in others, there are differences in the ability to perform various functions.

There appears to be fundamental differences in the science and technology systems in industrialized and developing countries. In the industrialized world--whether as a result of an internal cumulative process, as in the case of Western Europe, or of a transplant that grew its own roots, as in the case of United States and Japan--the evolution of scientific activity has led directly to, or is clearly linked with, advances in production techniques. In the developing countries, knowledge generating activity is often, for various reasons, not related in any significant way to productive activities. Industrialized countries might thus be described as possessing an endogenous scientific and technological base, and developing countries as having an exogenous scientific and technological base.[16] The process of interaction between science and production is complex and took place among considerable social upheavals, and concurrently with the emergence of capitalism as the dominant mode of production.[17]

The technology systems in most developing countries are characterized by "dualism": the existence of a modern, urban enclave linked to the international market place, usually producing for and adapting to the needs of the industrialized countries, within a traditional, rural setting that contains know-how accumulated over centuries. The modern sector frequently operates independently of the traditional sector. Moreover, the modern sector has traditionally been associated with technological progress while the traditional sector has been undervalued and underdeveloped.

The modern sector generally employs imported technologies that bring with them skill requirements, use of materials, organizational styles, and technical traditions that are alien to the local environment and the traditional sector. Furthermore, the technological capabilities associated with modern production are

Figure I. The "vicious circle" of technological dependence

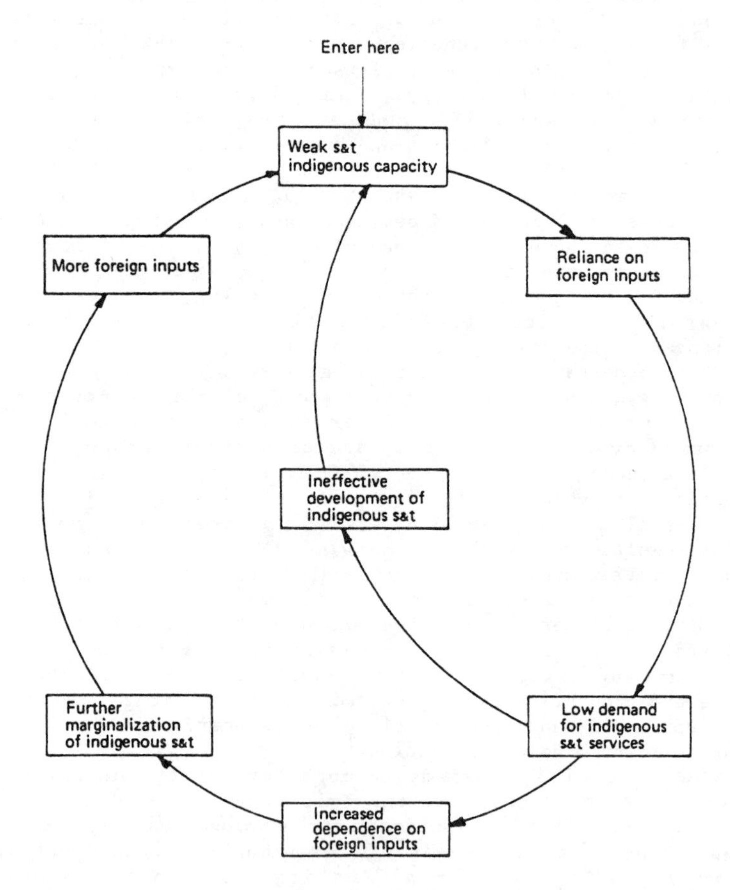

Source: "The structure and functioning of technology systems in developing countries" (ID/WG.301/2), p. 43.

Note: s & t = science and technology.

expanded primarily through new technology imports, which means that the technological traditions, developed slowly and cumulatively, become increasingly neglected and even eliminated. This has inevitably led to a reduction in the variety of indigenous technological responses. The situation described is represented graphically in figure II. Science and technology systems in developing countries are thus frequently underdeveloped. The existence of individual components, sometimes, like science and technology institutions, artificially created, does not constitute a system. There can be a viable system only when the components are linked through feedback effects that form closed loops for the effective exchange of experience gained, the transmission of new demands for better performance and innovation, as well as the provision of better facilities for meeting these new demands. The effective functioning of feedback loops and linkage arrangements implies the existence of a decision-making capability that can mobilize the system and harness it for the purposes of national development. This, in turn, requires the existence of political leadership convinced of the importance of the technology system. In some developing countries all the components of the technology system do not yet exist, in others the linkages are weak, ineffective and, sometimes, non-existent. Furthermore, in some countries, decision-making capabilities need to be strengthened and political leadership convinced of the role that the science and technology system can play in the attainment of development goals.

The success achieved by some developing countries in the development of their technological capacities is difficult to generalize. It is interesting to compare, for example, the "models" of, say, India and the Republic of Korea, two countries that have developed an indigenous base. The rationale of the Korean model is export orientation, foreign investment, foreign technology and foreign management know-how transfer in the first phase, leading to what might be called outward-oriented dependence. Foreign investment and control together with selective delinking take place. In the second phase, on the basis of the expertise gained during the first phase, the model becomes more international and delinking takes place. The rationale of the Indian model is to block rather than to pormote the foreign ownership of productive activities and foreign control and domination of the economy. The emphasis has been on the internalization of skills and institutional structures and the acquisition of the self-confidence required to meet the nation's needs. The development of such self-confidence was a basis for the subsequent entry into the world market through nationally owned companies.[18]

There is no simple method for defining the best course of action for the development of indigenous technological capabilities. Too much emphasis, for example, on meeting the basic needs of the poor masses through small-scale village technologies may result in a nation being permanently relegated to second-class social, economic and technological status. On the other hand, over-emphasis on so-called industrializing industries combined with aggressive acquisition of modern "high" technology may result in greater dependence at a qualitatively higher level. Clearly, there are no magic formulae or quick solutions. Rather, there is a need

Figure 2. **The relations between science, technology and production in industrialized and third world countries**

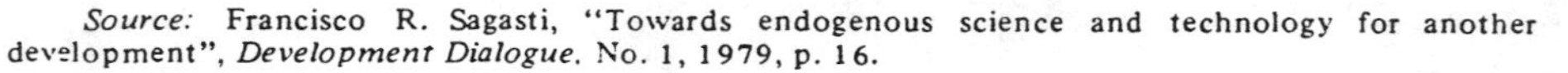

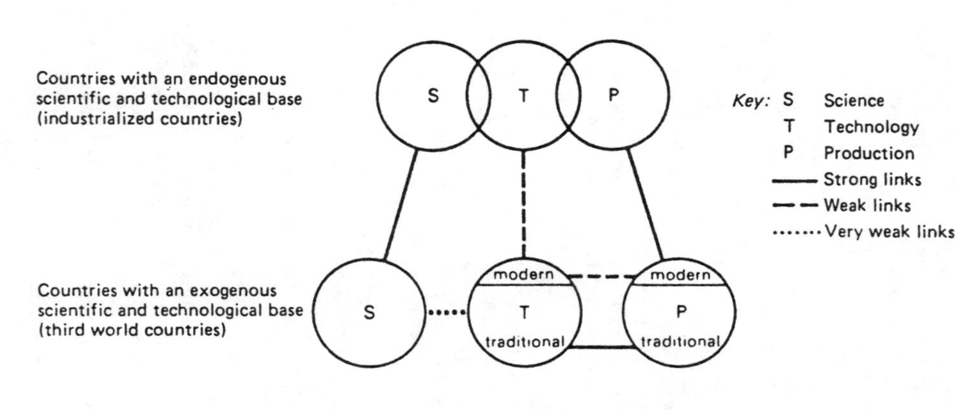

Source: Francisco R. Sagasti, "Towards endogenous science and technology for another development", *Development Dialogue*, No. 1, 1979, p. 16.

for a planned series of trade-offs through the introduction of a technological component in the national development strategy, and thus, for the gradual building-up of the institutions that make this approach feasible. This implies the inculcation of what might be termed a technological culture.

There will also be many problems of a more specific nature. Some of the situations that typically confront a developing country as it embarks upon the process of strengthening its technological autonomy are listed below:

(a) Many production decisions incorporate "wrong" types of products, for example, through the imitation of foreign consumption patterns;

(b) Much technology development has been left to the personal inclinations of researchers, developed during training courses frequently reflecting Western-oriented curricula and programmes;

(c) Much imported technology is not understood and there is little tradition of taking imported technology apart as a first step in its adaptation and absorption, and its subsequent substitution by local technology;

(d) There are contractual obstacles to the development of such understanding, for example, prohibitions on further use and patent systems that obstruct local adaptation;

(e) There are contractual limitations on the expansion of the use of acquired technologies, for example, export prohibition clauses;

(f) There are cases of unnecessary purchases of goods or technical processes tied to the necessary acquisition of knowledge and know-how;

(g) There is a lack of criteria for the effective selection of technology and confusion concerning the goals to be employed in evaluation such as efficiency, global output, development of initial skills, employment generation, income redistribution;

(h) The development of national technological institutions is unrelated to the development of productive units and processes;

(i) Purchasing policies in the public and private sectors are either ill-defined or non-existent;

(j) There is a lack of consultancy, extension, information and other link-up facilities.

The types of problems will obviously differ from country to country, which emphasizes the need for specificity in approaches aimed at strengthening national technological capacities.

C. Elements of technological self-reliance

The core problem of a strategy aimed at promoting technological self-reliance is twofold. It involves, on the one hand, the selection and management of foreign inputs, and on the other, the stimulation of indigenous supplies of technology. The first task requires the existence of a well-developed capacity to select and acquire technology from a variety of sources, and,

Figure 3. The process of technology selection and adaptation

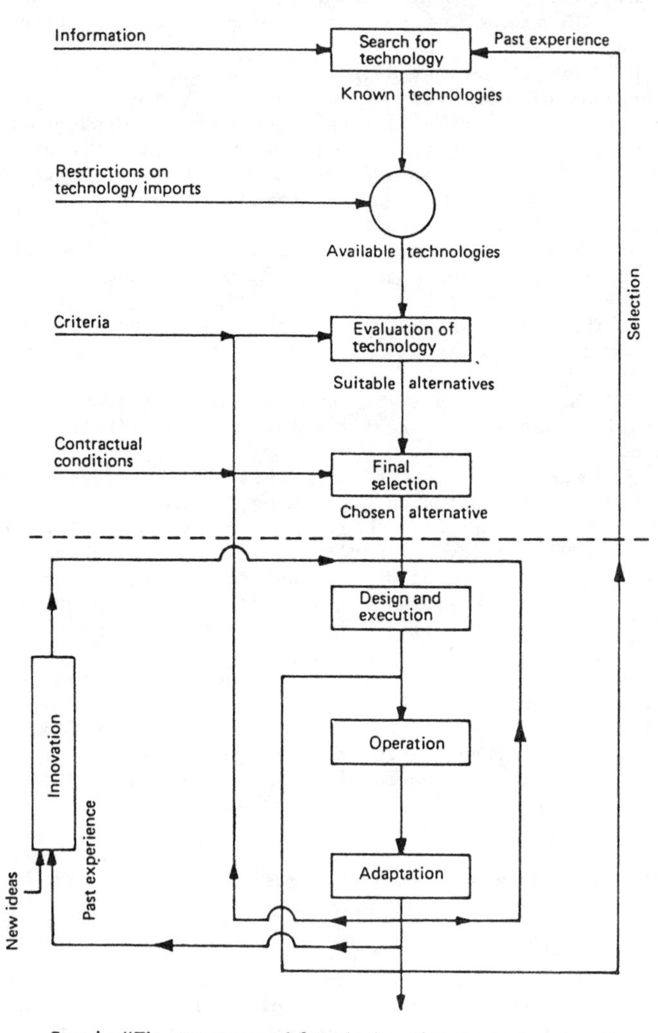

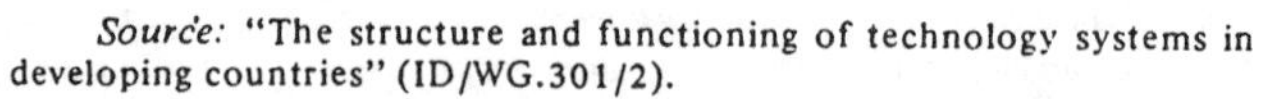
Source: "The structure and functioning of technology systems in developing countries" (ID/WG.301/2).

since none will usually have been tailored to local needs and conditions, to adapt the imported technology and its products to ensure that they can be absorbed by and can operate effectively in their new environment. The second task is to initiate an autonomous process of technological innovation and development, which requires the mobilization of the technology system.

Selection and acquisition

The technologies chosen by developing countries should obviously be appropriate, that is, they should contribute most to the economic and social objectives of development. In general, three sets of factors should be considered in determining whether a technology is appropriate, namely, development goals, resource endowments and conditions of application. Development goals can include growth of employment and output through more effective use of local resources; formation of skills; reduction in inequalities in income distribution; meeting the basic needs of the poor; improvement of the quality of life in general; and promotion of self-reliance. Resource endowments can include the availability and costs of local manpower; the level of skills and local management capabilities; availability and costs of water and energy; and natural resources. Some of these are more or less fixed while others can be influenced in the short- or long-term. The conditions of application include a number of economic and non-economic factors, such as the level of infrastructure, climate, natural environment, the social structure of the population, traditions, cultural and educational background as well as the location of industry, the size and demand of the foreign and domestic markets and the foreign exchange situation.[19] It follows that appropriate technology is not synonymous with traditional village technologies, with labour-intensive and small-scale production. Depending upon the circumstances, the most appropriate technology could be capital-intensive involving large-scale production. The appropriateness of a technology can be defined only in the economic, social and ecological context. It is meaningless to try to devise a set of appropriate technologies as such, although certain basic generalizations concerning appropriateness of sets of resource endowments and of conditions of application appear possible.

The scope for choosing a technology that is appropriate in the sense described above varies according to sectors. Agriculture, construction, and the service industries are often considered to offer more opportunities for exercising choice than some manufacturing sectors. Within the industrial sector, alternative technologies in some branches might be abundant; in other branches, especially those with sophisticated, modern technologies, they may be very limited and available only in the form of a "package". The choice of sectors depends upon a number of factors that are not influenced by the choice of technology, such as the natural resources available and the size and growth of markets, and in many cases can only be made if the sectors are described by the technology used. Once the sectors have been selected, the choice of technology can only be operationalized at the product or process level, and sometimes even at a more disaggregated level. If a product is specified in great technical detail, there may be little scope for the application of alternative technology or inputs. The choice of technology is thus not a simple choice of a capital-labour ratio or between labour- and capital-intensive production, but much more complex.

Policies aimed at promoting self-reliance in manufacturing industries should thus be based upon a systematic review of sectors and branches and an identification of product specific patterns of dependence. The forms and growth patterns, quantitative and qualitative, of technological dependence should be differentiated accordingly. Only through the application of branch and product specific criteria will a developing country be able to identify "frontier technologies" and those areas in which selective technological delinking appears possible.

Whatever choices developing countries make, they cannot afford to isolate themselves from fiercely competitive international markets nor can they cut themselves off from the mainstream of industrial and technological innovation and development. Many of the "industrializing" industries lend themselves to large-scale production and the manufacture of many products often needs to be undertaken through large-scale units based on the most modern technology. Developing countries will require a mix of technologies with different degrees of sophistication and a modern sector to increase their productivity, enhance their competitive position in international markets, and serve as an important source of technological innovation. As noted above, the problem will be of expanding the modern sector without increasing technological dependence.

While the technology introduced into a developing country must recognize the prevailing factor proportions, it is not necessary that it slavishly correspond to the resources available. If it were to, the mix of factors would be frozen and the deficiencies reproduced indefinitely. for instance, in China the guiding principle has been the introduction of the pioneer technology that entails the highest organic composition in as many units as possible, regardless of the fact that the shortage of capital prevents its immediate spread over the rest of the branch.[20]

Development requires both industry and agriculture. Industry normally grows faster than agriculture and the development of the agricultural sector requires an increasing number of industrial inputs. Development also requires the production of a range of consumer goods to meet the basic needs of the people and a range of capital goods, without which an economy cannot expand. There are various reasons why a country should seek to produce capital goods at an early stage in the development process. Some capital goods, for example, are required to produce industrial and agricultural consumer goods. The capacity to import, however much it can be increased, will frequently be limited in the face of increasing needs; and the developing countries cannot always import the kind of capital goods they most need--and where they can, they frequently must do so on onerous terms. The production of capital goods also promotes learning-by-doing and it is usually in the capital goods sector that innovation and technological development most rapidly gather force.

The effective selection of technology requires both information and evaluation.[21] Enterprises in developing countries, with the exception of a very few large enterprises, do not generally possess technological information, and more important, they often do not know where it can be obtained. As a result, industrial and technological decisions are taken on the basis of inadequate information. Where information does become available, the ability to evaluate such information for purposes of decision-making is often lacking.

Information centres have been established in many developing countries, in several cases with the assistance of UNIDO or UNESCO.[22] These information centres are either independent institutions or part of research institutes or other institutions. They are sometimes part of sectoral centres. The extent to which information available to all such institutions is fed into the decision-making processes varies significantly from country to country. The institutions vary in their organization and structure from a library or a mere collection of books to extension and consultancy facilities. Technological information is often a relatively undeveloped component of their activities. Processed technological information of practical value in decision-making requires trained personnel with access to information from all over the world. They will often need to be not only information specialists, but also persons with techno-economic backgrounds.

In developing countries, the evaluation of a project from an economic and technological point of view suffers not only from want of information but also from lack of capabilities and adoption of the relevant criteria.[23] Entrepreneurs make private cost-benefit analyses of their own. The banks and financial institutions also make such analyses of a project's economic viability. Several developing countries have attempted to upgrade their capabilities through institutions where evaluation is a major function. In some countries, UNIDO has assisted in establishing industrial studies, development centres or investment production centres that facilitate the building-up of evaluation capabilities. Such evaluation, however, does not always deal directly with the choice between alternative technologies. Technology is often viewed as a constant and not as a variable. Many developing countries do not appear to have made a systematic examination of the implications of the choice of technology and the criteria to be applied in such a choice.

For the acquisition of technology, the capabilities required are to specify the technological services required and to negotiate the terms and conditions. Entrepreneurs have not, with notable exceptions in the more advanced of the developing countries, significantly built up such capabilities. This compounds their weak bargaining position. Guidelines for negotiation, model contracts, and investment promotion institutions serve to assist the entrepreneurs in this regard.[24] In addition, government regulations on the import of technology help not only governmental authorities but also enterprises.

Such regulations exist, however, in only about 20 developing countries.[25] The reasons for this are many. Some countries have not been aware of the value of regulations and others have perhaps made a conscious decision not to have such regulations at present. Among the latter category are countries that regard themselves as not having reached a stage of development at which such regulations are necessary. Yet others are faced with financial and managerial resource constraints, and they believe the climate for foreign investment would be upset by the regulations on technology imports.

Even where the capability for acquisition is built up through regulatory institutions, the direction of such regulations varies considerably. With few exceptions, government regulations were introduced only in the 1970s. Such regulations have, however, generally been concerned with limiting the size of payments and avoiding restrictive clauses. They have also helped to build up indigenous technological capabilities by not allowing restrictive clauses that might have an adverse effect on such capabilities, and, more importantly, by not allowing the import of technology when technology is indigenously available. However, monitoring and follow-up of imported technology have not yet become the strong points of the regulatory agencies. Nor do they appear to have contributed significantly to disaggregating the technology packages offered for import or developing sectorwise technology policies based on an assessment of the state of the art in the respective industrial sectors.

Adaptation

All countries import technology. Most import more than they export. An industrialized country is generally able to import the technology it requires from another industrialized country and, because it has its own technological infrastructure, to adapt it to its own requirements. As domestic infrastructure is often weak and sometimes totally lacking in developing countries, they are generally less able to adapt technology imports to their own needs.

The technologies traditionally imported by developing countries are optimally suited to the factor endowments of the rich exporting countries. Adaptation is the process of matching alien technologies to local factor endowments, social customs and values and national development objectives. It may necessitate, for example, the scaling down of the technology to the size of the local market or matching it to the local skills available which, in some cases, may require increasing the unskilled labour force. Adaptation is also the means of linking imported technology with domestic R and D.

Adaptation is therefore--in that it is consonant with the strengthening of the capacity for effective acquisition and mastery of foreign technology as well as the building-up of an effective research and development system--an essential element of attempts to foster technological self-reliance.

Effective adaptation requires skilled manpower, which has had at least several years' experience in related production. This condition is not generally met in all but the more advanced developing countries. The adaptation of technology can, however, be undertaken by engineers and technicians within an enterprise, by industrial research institutes, and by consulting engineers.

Developing countries generally do not appear to have initiated incentives designed to promote adaptation. However, the environment of a protected market does not encourage the process of adaptation.

In only a few developing countries, such as Brazil, India, the Republic of Korea and Yugoslavia, have consultancy engineering capabilities been created in any significant measure. In several other countries consultancy engineering firms have come into being but their experience and versatility are limited, often confined to local consultants of equipment suppliers or other consultants from abroad.

In general, most developing countries are found to lack technological service capabilities. Such services range from macro-level industrial planning to micro-level project identification, feasibility studies, plant specifications, detailed engineering designs, civil construction and machinery installation, and the commissioning, start-up and operation of plants. The most significant gap, even in fairly industrialized developing countries, is in detailed engineering and design and in sectoral consultancy services through nationally owned units. This makes the disaggregation of imported technology packages extremely difficult and creates a lack of infrastructure, resulting in an undue dependence on foreign design and engineering services. This, in turn, has a negative impact on the pattern of investment for particular projects, on the requirements for capital goods and equipment, and on subsequent plant operations and management. In other developing countries, the gaps in consultancy services are even more marked and extend to almost the entire range of services indicated above.[26]

Absorption

The process of selecting and adapting technologies requires careful thought as to their subsequent absorption and diffusion, which in turn requires consideration of who is to use the technology and of the constraints on its application.

As noted in chapter one, technologies incorporate and reflect value systems and embody social and cognitive structures. They contain intrinsic characteristics that cannot be altered by narrowly defined processes of adaptation. Some of these characteristics (such as the degree of complexity of the technology, its scale, spatial extension, energy, material requirements, transformation, and skill, manpower and knowledge content) have the greatest possible bearing on the possibilities for diffusing and absorbing the technology.

The concept of social carriers of technology, developed by Edquist and Edqvist, is useful for identifying some of the problems associated with technology absorption.[27] They suggest that effective absorption and diffusion is dependent upon the existence of a social entity or category, a "social carrier", which has an interest in applying that technology. A social carrier could be an individual farmer who changes his pattern of production as a result of the introduction of an improved plough. The Indian "mistri" is another example of a social carrier who plays an important role in the adaptation as well as absorption and diffusion of technologies. The carrier can also be an institution. It might be, for example, an enterprise or agricultural co-operative that can develop and promote the use of new machinery and agro-technologies.

A developing country might choose to initiate a domestic network of air services using modern jet aircraft. These will need to be imported and there is nothing that can be done in the way of adaptation. The use of the aircraft is dependent upon the existence of airports and a complex system of air traffic control. A nation cannot have aircraft without a wide range of supporting and often very expensive infrastructure and services. Similarly, if a new agricultural technology is to be introduced, there must be peasants or peasant organizations that can acquire the inputs needed (seeds, implements, fertilizers, pesticides etc.), organize the labour (own or hired) and distribute the products. If these requirements cannot be met, there will be little point in attempting to introduce the technology.

A social carrier must have an objective interest in choosing and applying a specific technology. This objective interest must coincide with a subjective interest, i.e. the objective interest must be consciously felt or perceived as an adequate goal by the carrier. In order to function as a carrier, the social entity must further have some degree of social, economic or political power to be able to materialize its objective and subjective interests.

Every technology must thus have a social carrier in order to be absorbed and diffused. For a large-scale industrial technology in a developing country, the social carrier might be the government and its planning authorities, an international organization and a transnational corporation, alone or in different combinations. An agricultural technology may be "carried" by individual farmers or by the leadership of agricultural co-operatives and associations.

UNIDO has suggested that in seeking to strengthen endogenous capacities for technology adaptation and absorption, developing countries should pay particular attention to:

Industrial sectors and manufacturing processes

The assimilation of design know-how and related R and D efforts

The further development of technology and its incorporation into the production process

The development of special skills[28]

Human resource development is a particularly important aspect of technology absorption. The base for adaptation and absorption, as indeed of technological development, is provided by qualified engineers and scientists, middle-level technicians and skilled labour. As regards engineers and scientists, the situation in most developing countries is marked by both paucity in numbers and less than full utilization of their capabilities. Educational facilities for this purpose are generally lacking and the university traditions and curricula are not such as to promote their capabilities in, or association with, applied research and production activities. Several relatively small developing countries also lack the necessary scale of requirements to have fully fledged technical institutions of various types. Another phenomenon in certain developing countries is that of "brain drain" involving the export of much needed technological manpower. Generally speaking, requisite educational policies and manpower plans for projected requirements are still at an initial stage in developing countries.

The process of both adaptation and absorption would be greatly facilitated should technology contracts specify in detail the number of persons to be trained and the nature of training to be provided by the transferer of technology. The number of persons trained as part of technology contracts varies not only with the nature of technology and type of contract (including whether foreign investment is associated or not), but also with the countries of origin of the technology suppliers.

Development

The process of technological innovation is not well understood. It involves very much more than the creation of national technology centres, improved access to foreign patents and know-how or even the availability of capital to exploit them. The appropriation of knowledge appears to be the mainspring of innovation and this requires scarce resources of skill that have a high opportunity cost.

A capacity to innovate is the result of complex relationships between available capital, skills, information, communication and scientific infrastructure. It is not an autonomous process but rather a consequence of patterns of social, economic and cultural interactions. It requires, for example, high levels of cooperation between government and industry; that capacity for science and technology be linked with the productive and educational systems; integration of fiscal and trade policies; and mobilization of the creative energies and problem-solving capacities of a nation's population. It also requires the existence of social carriers that have an interest in, and possibilities for, introducing and spreading technologies thereby engendering innovation; and conscious policies designed to trace the unexploited knowledge and technologies in local communities, for example, of small farmers and women. All this implies national self-confidence and the exercise of political will. Against this background, it may prove impossible for some developing countries to embark upon their technological transformation without a corresponding and parallel social and political transformation.

One of the keys to innovation and to the mobilization of the national technology system is linkage, i.e., the creation of institutional devices that facilitate continuous intercommunication and mutual assistance between representatives of the educational system, enterprises, employers' associations, organized labour, and development agencies. The intention to communicate and coooperate must be inculcated at national, regional and local levels and permeate everyday thinking. Attitudes and motivation throughout society will determine the climate for innovation.

It is doubtful whether all countries in the third world can bring about and maintain the process of technological innovation. More than 50 developing countries, many island states among them, have very small populations. These countries lack technicians and experience. Because their markets are small, and because it is difficult to create necessary industrial complexes, they lack the capacity to produce industrial goods. The solution to such a problem is generally assumed to lie in specialization through international trade and the promotion of regional coooperation to enhance collective self-reliance. However, free trade does not always work to the advantage of small developing countries. Moreover, it implies a commitment to levels of cooperation never achieved by the industrialized countries.

Technology development calls for well-directed programmes designed to promote research activity. At present, in only a few developing countries do industrial establishments have research and development units of their own, and even these have a limited record and with very little horizontal transfer. In general, whatever research takes place in developing countries is by and large government funded through industrial research institutes or universities. This expenditure, which does not exceed 0.4 per cent of the gross national product (GNP) of the developing countries, is often spent on basic rather than applied research, involving programmes not necessarily drawn up as a result of clearly defined industry-related priorities. In some developing countries, voluntary agencies and institutions are attempting to promote appropriate technology in one or more specific sectors. But they tend to be small, lacking in government support and isolated from the mainstream of industrial activities. They have thus not generally been able to make any significant impact on the technological development of the countries concerned. The number of developing countries that have activities for invention promotion or patent registration is also small, thus hardly providing any impetus to the innovative capabilities of the local population. The approach to industrial research is itself generally more Western-oriented than inward looking, contributing very little to the technological advancement of locally used technologies and to the solution of the problems of the rural areas.

Manpower and financial constraints affect the process of technological innovation and development in a large number of other, more detailed, ways. The commercialization of research findings, for example, is dependent upon the existence of such services as product and process development, pilot plant, plant design and installation, process adjustment, advice on manufacturing operations, quality control, product and process improvement. With a few exceptions, such skills and services are lacking in developing countries. The number of processes commercialized by industrial research institutes in developing countries is not significant. Except for the least developed countries, most developing countries have one, and often more, research institutes. Some countries even appear to have too many research institutes to function in an effective and co-ordinated manner. The research institutes established have been of various types but, by and large, they belong to the categories of government controlled, autonomous, state-aided or quasi-government institutes. There are both single purpose and multi-purpose institutes and single sector and multi-sector ones. At one end of the spectrum are institutes providing quality control and testing services in a single sector of industry and, at the other end, are multi-sector institutes with services extending to applied research, pilot plants and extension and consultancy.

The limitations of such institutes in developing countries have been well documented.[29] Principal among the causes of their ineffectiveness appear to be:

(a) A structure that is too ambitious, and executive officers with inadequate training and experience;

(b) Failure to assess the applied research and development needs of the nation and industry prior to formulating programmes, building infrastructure and equipping laboratories;

(c) Operational shortcomings, including inept management, wrong type of staff, poor staff remuneration, and lack of business orientation, staff mobility, priority-based research, commercialization efforts, package of services to industry, guarantees for technology development, and motivation to undertake contract research;

(d) Weak contacts and co-ordination with industry and government, lack of adequate funding and the indiscriminate importation of technology.

Even if these constraints were removed, some of the more general problems involved in encouraging the process of technological innovation and technology development, those rooted in social and economic structures, might still remain. Improvements in organizational design, for example, provide no guarantees that technological development will be transmitted to rural areas, to the vast majority of the population or lead to an improvement in the general level of technological awareness and capability of the population (as distinct from a number of scientists and engineers). These and similar problems have yet to be adequately considered by either the developing countries or international bodies.

D. Technology policy and technology planning

The above considerations indicate that while developing countries are becoming increasingly aware of the challenge of developing technological capabilities, their responses have varied. The elements of such capabilities and the factors that influence them are so numerous and varied that policies and actions have generally been compartmentalized and uncoordinated. Adequate methodologies for the formulation of technology policies and plans have yet to emerge.

Technology policies and plans, however, are a matter of high priority. It will no doubt prove impossible to promote technological self-reliance without recourse to planning and the preparation of policies that are linked to strategies of national development. Indeed, given the pervasive influence of technology and its motive force, planning in the area of technology may, in many cases, prove more important than planning for investment.

Technology policy is not synonymous with technology planning but is a basic function of government aimed at creating a framework in which decisions concerning technological choice can be made and implemented. Technology planning implies the existence of a formally constituted and internally consistent set of goals, objectives and instruments. Whereas all developing countries should seek to formulate technology policy within which basic choices can be made, the preparation of comprehensive technology plans may be both beyond the scope of, and unnecessary for, countries with limited regulatory and supervisory capabilities and where institutional continuity is a problem. Past experience with both technology policy formulation and technology planning show that the state of the art is at a rather elementary stage.

For the majority of developing countries, the need to develop a technology planning capability will no doubt become increasingly urgent but experience has so far been disappointing. It was only in the early 1970s that such countries as Argentina, Brazil, India, Mexico, the Philippines, the Republic of Korea and those of the Andean Group set out to control technology imports.

In the mid 1970s, the first technology plans, prepared by Brazil, India, Mexico, Pakistan and Venezuela, appeared. The importance afforded technology by the developing countries is evidence by the fact that by 1977 the number of countries exercising governmental control of technology imports had, according to UNIDO estimates, increased to approximately 30.

As suggested earlier, the experience gained so far indicates that while regulations and programmes have helped to build up technology institutions and to strengthen the bargaining position of the developing countries as technology importers, they have gone little further than the review and approval of technology supply arrangements at the enterprise level, and that the linkages between technology imports and the promotion of national capabilities are generally ineffective. Problems associated with technology absorption and adaptation have so far generally received little attention. Even where technology plans have been prepared, the relationship between these plans and national development strategies is weak.[30]

In discussing technology planning, it must be acknowledged that there has been, in many respects, a growing disenchantment with the idea of comprehensive planning as advocated and described in traditional textbooks. Indeed, the truth is that few people today have the same sort of blind faith in planning that was prevalent at the end of the 1950s and the beginning of the 1960s. Even in centrally planned economies, such as China and the Union of Soviet Socialist Republics, attempts are being made to correct apparent rigidities in planning and to increasingly liberalize the operation of the economy. In non-centrally planned economies there are only a handful of countries that have medium-term plans that play a role in the process of allocation of resources. The trend of de-emphasizing the importance of comprehensive plans has continued because of the many difficulties encountered, not so much in the formulation phase of such plans, but rather in their implementation. Discrepancies between planned and actual figures are all too common. Planning is difficult because, in spite of the calls for increased self-reliance, the economies of most countries have become more instead of less open to the world economy. Also, there have been considerable increases in the flow of financial resources, as shown by the growth of the external debt of developing countries, and payments for the transfer of technology account for an ever-expanding portion of trade in services. Further, sudden fluctuations in the prices of basic inputs and commodities have meant the transfer of inflationary pressures from one country to another.

In setting out to plan its technological future, a nation is seeking to control and manage something that is pervasive and that does not recognize sectoral distinctions and ministerial responsibilities. Of all the things that man might set out to plan, technology is undoubtedly one of the most elusive and difficult. Nowhere has a nation yet demonstrated a real capacity to control its technological future. Even in the motherland of planning, the Union of Soviet Socialist Republics, the Director of the Institute of Economics of the Academy of Science has been quoted as saying that "the planning of scientific and technical progress . . . is the weakest link in the whole complex of economic planning and in the whole system of national incentives for production".[31]

Yet without technology planning a country will find it difficult to decide whether the technological inputs into national development efforts ought to be imported or obtained from domestic sources. Also, it will not be possible to ensure that the technological inputs are appropriate from the viewpoints of resource use, employment creation, income redistribution, need satisfaction and environmental effects. In general, systematic progress towards the strengthening of endogenous capabilities and the substitution of appropriate domestic technologies for imported ones will be impossible without the existence of a broadly planned framework over a long period within which individual development projects can be fitted.

In formulating a technology plan, developing countries should seek to create a framework for effective interaction between government, private enterprise and institutions for science and technology. They will need to give careful consideration to such matters as the needs, resources and socio-economic objectives of the country; the promotion of a social climate that encourages the application of technology in different sectors and at different levels; the formulation of measures designed to stimulate local technological capabilities; the setting-up of machinery for the selection and assessment of technologies and techniques; the selective import of know-how and its adaptation to local requirements; the development of technology packages involving frontier technologies and sets of technologies; and the development of manpower for the management of technology. Above all, the environment created should, at one level, inspire the confidence of industry and research, of engineers, technologists and scientists and, at another, seek to mobilize the creative problem-solving capacities of ordinary people at the local level.

The effective exercise of a technology function and of a technology planning capability implies the existence of scientific and technological intelligence, or the capacity to appropriate and utilize knowledge. Technological intelligence is an essential component of an anticipatory intelligence, or the capacity of a nation to identify its relevant strengths and weaknesses, to understand and analyse threats and opportunities of different kinds and to translate the resulting knowledge into policy and action. It is doubtful whether any of the world's nations, developed or developing, have yet developed a real social intelligence, although several countries, notably Japan, have demonstrated a technology intelligence capability.

E. Constraints on technological self-reliance at the intermediate level

It has been stressed that the technological dependence of the third world is but one aspect, though a crucial one, of the general pattern of dependence into which the third world is locked. It is the institutions and mechanisms that underlie the functioning of the international economic system that generate dependence. Many of the system's mechanisms are not the result of conscious design. They work automatically, but once in operation they persistently aggravate the fundamental inequalities between rich and poor nations.

The international economic system is a complex mixture of dynamic forces, of active and potential conflict. It is characterized by unequal specialization and exchange reflected in an inequitable international division of labour. The system, with its tendencies towards the internationalization of capital and the transnationalization of production, has inherent forces that tend towards the marginalization and fragmentation of the developing countries. Within this system, modern science and technology are becoming ever more hierarchical, centralized and specialization oriented. Scientific innovation and technology development are dominated by transnational structures, military industrial complexes, a near global network of agro-business, and a network of universities and research institutions, all of which are highly interpenetrated and mutually reinforcing.

Against this background, strategies designed to enhance national self-reliance are imperative. It must be questioned, however, whether they are really feasible for all but a handful of developing countries. Such strategies inevitably impinge on the profits and perceived interests of the rich nations and are thus unlikely to receive their support. Radical circles have warned that self-reliance will only have real meaning for the developing countries when they have freed themselves from the system that maintains their underdevelopment.[32]

Even if possible for some developing countries, technological self-reliance may prove beyond the reach of many small and economically and politically vulnerable developing countries. This questions the validity of theory, especially classical economic theory: whether, for example, the technological transformation of the third world based upon strategies of national and collective self-reliance is wholly comparable with the attainment of an international division of labour based upon industrial and agricultural comparative advantage.

F. Constraints on technological self-reliance at the national level

The concept of technological self-reliance is, like other concepts before it, in danger of becoming co-opted by vested interests within the existing international order. Some of the arguments underlying self-reliance tend to become distorted and used to reinforce the power of entrenched interests in the developing countries. Some elites in developing countries have a tendency to use such arguments to increase their independence from rich country interests without demonstrating a readiness to share any of the advantages that might result from increased self-reliance.

So far only a handful of countries have been able to incorporate in a meaningful way the concept of self-reliance in their national development strategies. Not many countries have found it easy, or an absolute necessity, to disentangle themselves from the complex webs of commercial, financial and technological relations that link them to the outside world in a sort of "external-reliance" and that, in many cases, maintain and feed their dependence.

Movements to organize labour, to mobilize peasants and to create conditions at the local level conducive to increased self-reliance are sometimes systematically suppressed in developing countries. In such situations it is difficult to see how a redistribution of Western science and technology and the strengthening of domestic technological capabilities would serve to improve the conditions of the poor and under-privileged masses. A central question, already raised in this paper, is whether it is possible for all developing countries to embark upon their technological transformation without a corresponding and parallel social and political transformation. This raises questions concerning the conditions required for, and the nature of the social transformation most conducive to, self-reliant development and selective technological delinking.

Even where favourable conditions exist, it is questionable whether something as pervasive as technology can be planned and whether nations can choose their technological future. Planning requires consensus on the goals and objectives of development, a commodity frequently in very short supply. And without clarity on the nature of the internal development to be pursued, it will be difficult to answer questions concerning technology development and innovation.

When the conditions are right, a good deal may be achieved in technology planning in a comparatively short period. Many developing countries are already involved in the process of strengthening their science and technology institutions and there is evidence that many developing countries may be able to increase their self-reliance in a wide range of consumer and capital goods industries in the near future.

This will be an important start. However, self-reliance, if it is to have real meaning, must be defined to include more than the production of goods and services and more than the building-up of science and technology institutions. It must ultimately be seen as a strategy that builds development around individuals and groups through the mobilization and deployment of local resources, material and non-material, and indigenous effort. In this sense, self-reliance transcends the application of techniques. Rather, it contributes directly to the formation of new value systems and to a direct attack on poverty, alienation and frustration as well as to the more creative utilization of productive factors. Self-reliant development, with its emphasis on local rather than imported institutions and technologies, is thus a means whereby a nation can reduce its vulnerability to events and decisions that fall outside its control.

All developing nations should be able to strengthen their technological capabilities, especially their capacities to control the inflow of foreign technology. Not all developing countries may be able to do so, however, within the framework of meaningful strategies of national self-reliance, with their emphasis on the mobilization of indigenous resources and knowledge.

It follows that a diversity of starting points necessarily implies a diversity of responses. In fashioning their strategies for national self-reliance, developing countries will no doubt experience a greater need for the systematic exchange of relevant information and experience than for a generalized, universally valid approach.

III. TOWARDS OPERATIONAL STRATEGIES

A. Objectives of technology policy

Technology policy can only be formulated on the basis of clearly defined development goals and objectives and in terms of decisions concerning the type and volume of goods and services that need to be produced and the resources to be mobilized and deployed. In this context, the production of the "right" goods with the "wrong" technology could in some respects be considered preferable to the production of the "wrong" goods with the "right" technology.

The technology policies of the developing countries are likely to be guided by a common goal, namely, the desire to exercise greater control over their social, economic and industrial development through the promotion of technological self-reliance, a precondition for meeting the basic material needs of their poor and under-privileged masses. Policies should address the problem of controlling and managing foreign technology inputs on the one hand and of stimulating the development of indigenous supplies of technology on the other. This implies the effective integration of two main streams: the "flow stream", with its emphasis on the selection and acquisition of foreign technology and its subsequent adaptation, absorption and diffusion; and the "stock stream", with its emphasis on the development of endogenous technological strengths and the promotion of the capacity to innovate.

The emphasis in the past has been firmly on questions relating to the transfer or flow of technology, the development of stocks having received only scant attention. It will be the task of technology policy to harmonize flows and stocks. Attempts at harmonization will need to recognize, however, that the two streams are not independent or mutually exclusive, but rather interactive at different levels. It may also be necessary to tackle the problems associated with each stream within different timeframes. The development of the capacity to control foreign technology inflow might be afforded short-term importance. Without such a capacity, policies aimed at fostering endogenous technology development and the capacity to innovate are likely to be continuously undermined.

The exercise of a national technology function obviously requires that the national science and technology system must function properly. However, these systems are, for a variety of reasons, frequently underdeveloped in developing countries. Typically, technological capacities are not strongly linked to industrial production, and the modern sector, usually export-oriented, frequently operates independently of the traditional sector. It will be one of the key tasks of policy to come to terms with these problems: to link the conduct of technological activities and the development of technologies with the growth of production; and to systematically and selectively recover the traditional technological base, weaving modern methods into the traditional tapestry of a developing society. If this is achieved, the technology system will be better able to react to stimulation and a revision of inputs within realistic periods.

Experience gained in developing countries indicates that these and similar problems can best be tackled when science and technology policies are formulated and implemented separately. These cannot be categorically differentiated with any clarity since they overlap to a great extent,[33] yet there is a difference in emphasis, which is of great importance to developing societies. Science is essentially attitudinal and science policy has the objective of encouraging the acquisition of scientific and technological understanding that may, or may not, be of use in the development of knowledge directly applicable to the pursuit of economic and social goals. The objective of technology policy, on the other hand, is to stimulate the generation of the scientific and technological knowledge to be applied in the solution of well-defined problems in certain areas of production and in social welfare. Although both science and technology policies are concerned with the generation of scientific and technological knowledge, there is a basic difference in that with technology policy the knowledge concerned is organized, promoted, financed etc. by policy-making institutions with the explicit purpose of using it to serve specific social and economic needs. In other words, technology policy is defined by objectives external to the scientific world as such. Technology policy is oriented towards the finding of acceptable solutions within a given social context and timeframe; since its objectives are essentially production and social welfare and it is not developed in the abstract, it is subject to decisions of a scope much wider than merely solving technical problems.

Moreover, as is well known, scientific knowledge usually flows freely without significant constraints whereas technological know-how is a commodity that is traded on the world market and is vigorously protected.

Separate, but interlinked, policies for science and technology should make it possible to deal more effectively with technology problems and problems with the development of indigenous technological capabilities.

B. A framework for national action

A framework for national action in technology consists of four interrelated steps:

"(a) A broad consensus on the desired mix of appropriate technology and the pattern of national technological capabilities;

"(b) An assessment of the present status of technological capabilities and identification of gaps and shortcomings;

"(c) Strategy formulation in terms of policies, programmes and institutions, together with the financial and manpower resources needed for its implementation;

"(d) A re-assessment of the coherence of ends and means as well as arrangements for co-ordination and monitoring."[34]

The purpose of the framework given below is not to present a step-by-step approach to the formulation of policy but to list what might be termed indicative issues. Its purpose is to foster the awareness that technology is a resource and that there is a continuous need for clarity in the relationship between ends and means in technology policy.

The framework is based upon the three essential pillars of policies, programmes and institutions. Policies by themselves can only act like levers or valves that can be used to channel or to cut off the flow of national resources or energies. The specific orientation of resources and energies is conditioned by programmes of action. Institutions are the instruments that formulate and implement policies and programmes. Excessive reliance on any one of these three pillars at the expense of the other two should be avoided.

Technology mix

The first step toward an effective technology policy requires a broad agreement on the mix of appropriate technology and then on the pattern of national technological capabilities. Though in a general sense technological capabilities will be required whatever the technology mix, clarity is essential for the generation of particular types of capabilities. The latter will in turn be derived from national development objectives. If the benefits of technology are to be spread throughout the population, then its application and the capabilities required should cover a very wide field of national activity. Hence, it can be said for all developing countries that the basic common skills should be generated abundantly and existing technological skills should be upgraded rather than uprooted. Subject to this, the technology mix, and therefore the desired pattern of technological capabilities, may vary for each country. In a labour surplus economy the emphasis may be on labour-intensive industries while in developing countries with a shortage of manpower, labour-saving technologies and skills to operate sophisticated machines may requires special attention. In the case of export-led growth, the technological capabilities of the export industry sector should receive priority. Wherever possible, the desired levels of particular technological skills should be quantified. Broad norms should be adopted bearing in mind that technological skills should be created as an infrastructure ahead of demand rather than as a response to demands as they merge at a particular time.

The selection of the most appropriate technological mix requires the identification of technological needs at both the macro level, in terms of sectoral priorities and the technological inputs for each priority and critical manufacturing sector, and the micro level of individual industrial enterprises. At the macro level, sectoral priorities can normally be identified through national plans and growth strategies. At the technological level, such priorities have to be broken down in terms of requirements of process or production know-how, the supply of technical inputs, provision for technological services, specialized manpower training for management and plant operations and the like. These, in turn, are determinants of, and closely dependent on, the choice of technology from among the various alternatives that may be available. At the micro level, principal technological needs include improvement of productivity, quality control and institutional technical support to industry, including information linkages, which have to be tackled on a national or even regional basis but which relate primarily to the working of individual enterprises.

Sector technological demand should also be identified at the regional level in the case of developing countries. Several regions, particularly in Latin American and in parts of Africa, lend themselves to an effective regional approach to several priority industrial sectors such as fertilizers, petrochemicals and capital goods production. Such identification could be a prerequisite for strengthening the bargaining position of regional industrial units in respect of technology acquisition and the development of regional technological capability.

Assessment of the present situation

An assessment of the present status of technological capabilities and of the effectiveness of national technology systems, aimed at identifying gaps, limitations and deficiencies, has not yet been carried out by many developing countries. It is, however, a prerequisite for the proper formulation of a strategy.

Reviews of existing situations are notoriously static undertakings. It is essential that an assessment of technological capabilities take place in a dynamic and development oriented framework, being cognizant of global and regional technological trends and developments on the one hand and national development aims and ambitions on the other.

An assessment of technological capabilities may include the items listed below.

Technological manpower

The strength of the existing technical and scientific manpower should be evaluated quantitatively and qualitatively, as should likely developments in patterns of deployment and utilization. The extent of brain drain, if any, may need to be assessed. The evaluation of manpower resources should be undertaken keeping reallocation possibilities in mind, since additions to manpower may require from three to five years, short of reversing the brain drain or of bringing in expatriate manpower. The categories of manpower to be assessed include scientists, science graduates, research and development personnel, teachers and engineers (civil, mechanical, electrical, chemical, metallurgical, electronic etc.) who are engaged in production, teaching, consultancy, design and other occupations; middle-level technicians of various types; trained artisans; traditional artisans etc.

Indigenous technologies

Many developing countries have yet to obtain a clear picture of the traditional technologies available to them. Such technologies, developed over centuries and representing accumulated experience, are likely to be appropriate to local conditions and particularly relevant to the problems of rural areas and to the development in these areas of such activities as agro-processing and building materials and construction. The inventory and evaluation of indigenous technologies should take place with a view to identifying the possibilities for their systematic upgrading and improvement through the application of modern science and technology. Research and development institutes in developing countries have an important role to play in the assessment of indigenous technologies.

Sectoral developments

As assessment of the status of technological advance and manpower in specific sectors will need to be made. The sectors should include not only individual industrial sectors, but also technological service capability areas such as consultancy, design and construction. High priority industrial sectors are likely to include food processing and engineering industries as well as the industrializing industries, which allow for the optimal utilization of local natural resources and for the longer-term accumulation of technological capabilities. The assessment of sectoral developments should cover not only large-scale industrial units and technologies, but also small-scale and traditional technologies.

Impact of policy

The effective exercise of a technology function requires a careful assessment of the scope for implementing policy and for government intervention and regulation in the technology market. In making such an assessment, it must be recognized that there is a range of contextual considerations involving social, political and economic structures that constrain policy formulation and implementation and that policies can have an indirect as well as a direct effect on the development of technological capabilities. The technology system operates within the frame of an intellectual climate, a system of values, attitudes and modes of behaviours as well as of current legislation. The direct impact of this on strategies, policies and plans and on the definition of the composition of social demand may be obvious, although difficult to generalize. Less obvious is the indirect impact on the components of the science and technology system of policies governing such areas as taxation laws, import controls, customs duties, the influx of foreign capital and labour. All these will have a profound effect on the operation of the technology system and together constitute what might be termed an implicit science and technology policy.[35] In many areas, implicit technological policies are able to run directly against the explicit technological policies contained in science and technology plans. It is this contradiction that frequently lies behind failures in policy implementation.

Internal diffusion of technology

The state of diffusion of technology within the country and the existence of conditions to promote such diffusion should be assessed. Internal mobility of technical personnel promotes transfer and diffusion and enables the training and transfer of skills to a much larger number of persons than would otherwise be possible. The economic relationships between the urban and rural areas have to be examined to see how their strengthening could contribute to the growth of technological skills in the rural areas. The facilities and instruments available for the promotion of innovation should also be examined.

Technological institutions

An assessment of the capacities of existing institutional infrastructure is essential. This should identify the function performed by institutions, the means at their disposal and their potential for change and development. Technological institutions cannot be construed in the narrow sense of industrial research organizations and the like. The assessment should also cover such institutions as information centres, project formulation and evaluation centres, investment promotion agencies, investment boards, technology regulating agencies, productivity councils, design institutions, consultancy and other technological service agencies, extension centres for small industries, institutions for technological education and research institutes. In other words, the review should include promotional, regulatory and service institutions since their activities will involve implicit policy and impinge in a variety of ways on the process of technological development. In this sense, it may be more appropriate to think in terms of functions and services to be performed rather than in terms of institutions *per se*, since, ultimately, it is there that the major interest lies. This approach requires the specification of such functions and services and its correlation with the potential offered by available institutions.

In assessing existing institutional capabilities, it is essential to go beyond "numbers" (of technical personnel, expenditure incurred and so on) to a qualitative evaluation of the output of the institutions. The possibilities of strengthening the institutions, extending the scope of their activities to include more functions and services, avoiding duplication in their work and ensuring co-ordination should be identified. The place of the respective institutions in the government hierarchy, their involvement in decision-making for industrial and technological development, and the contacts they have with industry and the public are critical factors in assessing their effectiveness. With respect to research institutes, their role in essential technological functions such as extension, pilot plant and commercialization of technologies should also be assessed.

Summary

The above assessment should provide the following: (a) sufficient information and insights to understand ongoing processes at different levels and to identify future possibilities; (b) an understanding of the scope for technology policy and the possibilities for government intervention and regulation in the development of technological capabilities; (c) the possibility of identifying sector and branch specific patterns of dependence, sector and branch priorities and important intersectoral relationships with significant linkages and backward and forward multiplier effects; (d) an understanding of available and needed institutional infrastructure and manpower requirements; (e) an extensive basis for identifying priorities in a range of interrelated areas and of evaluating the advantages and disadvantages associated with technology alternatives at different levels; and (f) the linking of technology policy with national economic, social and industrial development objectives.

Policies and policy instruments

The practical formulation of the strategy in terms of policies, programmes and institutions will vary from country to country in accordance with conditions, requirements and priorities. While specific actions are suggested illustratively in what follows, the emphasis is on providing a framework for action.

Developing countries are able to apply a large number of policy instruments in seeking to attain their technological objectives and to achieve the technology mix deemed most desirable. The effective application of such instruments, however, will require the identification of the structural forces and deficiencies that are likely to invalidate their utilization. One of the themes of this report is that contextual factors may be equally or even more important than individual policy instruments in determining the success of technology policy-making.

Policy instruments can take various forms and be of the explicit or implicit type. They include national laws and regulations for licensing of production capacity of industrial enterprises (as in India) or the defining of new and necessary industries (as in Mexico), controls over majority foreign equity holdings, employment for expatriates, controls over imports, incentives for exports and import substitution, regulatory control over foreign technology, regulations for use of domestic consultancy agencies, and technical services, various forms of financial assistance and incentives for small-scale and rural industries and the like. In most developing countries, several fiscal and regulatory instruments are utilized in combination with one another. A number of governmental and semi-governmental agencies are consequently involved in dealing with one or another policy instrument. One of the criticisms often levelled is the multiplicity of governmental regulations and agencies with which domestic industry has to deal. While adequate co-ordination is undoubtedly necessary and bureaucratic delays need to be minimized, the complex and manifold issues of industrial and technological growth in most developing countries necessitate that governmental agencies play a critical and determinant role in several policy areas. The nature and extent of such a role obviously depends on the circumstances and objectives of each developing country but the nature and magnitude of the problems are such that the free play of market forces may only accentuate existing gaps and problem areas.

As noted above, policies and instruments relating directly to technology have to be viewed within the framework of overall economic and industrial policies. By and large, however, such policies and mechanisms need to be defined in respect of (a) the role of private foreign investment, both existing and new; (b) fields in which foreign technology is considered particularly necessary, including measures designed to ensure adequate flows, such as tax benefits; (c) production and service sectors in which foreign technology should not be encouraged, including technical and management services, merchandising, and internal sales and sectors where domestic capability is either adequate or should be developed; (d) the establishment and development of a regulatory mechanism to regulate such inflow in accordance with prescribed and well-defined guidelines; (e) incentives and measures to encourage domestic technological growth, including tax rebates for R and D expenditure, limited duration of foreign technology agreements etc.; (f) incentives and measures to promote domestic technological services, particularly consultancy and engineering services, including tax relief and regulatory action such as insistence on local consultancy agencies being appointed as prime consultants in selected fields; and (g) financial assistance and support to domestic technology agencies. Such a list of policy measures and instruments relating directly to technology can only be illustrative and not exhaustive and most be formulated in the context of each country or region.

General policy guidelines

In every developing country, technology policy will need to take into consideration selective action. As noted above, the definition of the technology mix that is socially optimized requires the systematic identification of sector- and product-specific alternatives and the careful analysis of the various constraints associated with each of the options. Despite the enormous differences among developing countries, five general guidelines appear to possess a particular relevance in the identification of the most appropriate technology mix:

Effective control of key sectors. Without this there will be little progress in the direction of autonomous decision-making and little influence over the process of accumulation let alone development. Such control is a precondition for the establishment of dynamic inter-industry linkages. It involves control of the market, of essential inputs, of forward and backward linkages as well as of research and development of technologies. The control of key sectors may call for policies of selective nationalization. Such policies should recognize, however, that ownership should not be confused with control and that it is control that counts.

Converging needs with effective demand. In many developing countries, the gap between the needs of society, or more specifically, the needs of the underprivileged majority, and effective demand, i.e., the demand that can enter monetary exchange relations, is dramatically increasing. Decreasing fulfillment of basic needs and overconsumption in some urban growth poles are the familiar symptoms of this trend. A conscious policy to reconcile needs with effective demand thus becomes of utmost importance. This implies three interrelated priority activities: the identification of social needs; the definition of criteria for the adjustment of effective demand to social needs (such as maximizing the basic needs satisfaction of the poor, the productive integration of the labour force, the use of local natural resources and the use of local scientific, technological capabilities and traditional skills); and restructuring the supply side and solving the problem of the choice of product.

Support for agriculture. Especially important is the promotion of self-sufficiency in basic foodstuffs. Support for agriculture, which would help guarantee self-sufficiency in food, is one of the main priorities for development strategies and especially for industrialization strategies. This applies to sectors producing agricultural inputs (implements, fertilizers, pesticides, irrigation equipment etc.), to sectors serving transport and distribution requirements, and to those processing agricultural goods. Possibilities for the application of science, and technology to increase agricultural productivity, improve post-harvest technology, and introduce innovations into plantation industries, fisheries and forestry are considerable.

Social optimization of using and processing resources, including energy resources. Some developing countries still have to develop the preconditions for effective control over the natural resources located within their frontiers, i.e. national capacities to detect, exploit and process such resources. Thus, the utmost importance should be given to activities in this field, which should include a systematic search for areas in which co-operation between developing countries appears feasible. Availability of natural and energy resources should have a determining effect on the contents of industrialization strategy as regards choice of sectors, choice of process and techniques.

The identification and strengthening of industrializing industries. Priority should be given to the identification and promotion of the so-called industrializing industries, i.e. industries that allow for the optimal use of local natural resources, guarantee the fulfillment of basic needs and allow for the long-term optimization of accumulation and scientific-technological capacities. Such a strategy includes, among other things, the development of the engineering machine tool industry, the production of textile and agricultural machinery, and a reorientation of basic industries, processing locally available resources that aim to increase the share of down-stream activities and to foster the integration of the country's industrial and agricultural production. This strategy should include attempts to strengthen local engineering capacities, especially with regard to pre-investment studies, chemical engineering and equipment design and attempts to control technological building blocks and technology life-cycles.

The development of industrializing industries should be related to the growth and development of physical infrastructure, another prerequisite for the process of industrial development. The planning and provision of such physical infrastructure as electric power, transport and communication systems, including railways, roads and shipping, should ensure that such facilities would effectively meet the projected needs of at least those industries that form the spearhead of industrialization efforts.

Acceptance of the general guidelines outlined above may well call for the transformation of the productive system. The transformation suggested would involve the reorientation of production away from mimetic patterns of consumption that favour a diversity of goods for higher-income groups, to a productive structure based on the satisfaction of basic needs and with greater emphasis on collective rather than individual consumption. This revised pattern could substantially reduce the need for imported technologies and lead to an increased demand for local scientific and technological activities. The selection of the most appropriate policy instruments for promoting the development of indigenous technological capabilities will need to take account of the above guidelines.

Levels

Technology policy should address problems and outline options at different levels. National strategies for technological development should be based upon the recognition that the international technology situation and the international division of labour are not static but dynamic. National strategies should thus reflect an appreciation of global and regional trends and developments, a consideration that will become increasingly important as efforts in the direction of collective self-reliance and Technical Co-operation among Developing Countries (TCDC) and Economic Co-operation among Developing Countries (ECDC) are intensified.

As noted above, an essential ingredient of technology policy is decisions concerning sector and branch specific product and process technologies. Such decisions can be articulated only at the enterprise level. The enterprise level is thus of critical importance. Technology choices at this level, however, cannot be left to the discretion of individual entrepreneurs and to market mechanisms. The interest of the nation will not necessarily be compatible with that of an individual or of groups of entrepreneurs. Individual enterprises may well be motivated by profit rather than social welfare considerations. Profit maximization may well encourage them to import foreign technologies under conditions that perpetuate national technological dependence. One of the essential functions of technology policy is thus to guide the actions of entrepreneurs in socially desirable directions. In most cases this will necessitate a system of incentives as well as of regulation and control.

Policies for selected areas

In chapter II, technological self-reliance was defined in terms of the capacity to select, acquire, adapt and absorb foreign technology inputs (regulating the flow stream) and of developing an indigenous base and the capacity to innovate (the development of stocks). We now consider some of the policy options under each of these main headings.

Selection and acquisition of technologies

As regards the technology mix, developing countries appear to have a special need for technologies that meet the following criteria:

"(a) High employment potential, including indirect employment through backward linkages with national suppliers and forward linkages with national processors, distributors and users;

"(b) High productivity per unit of capital and other scarce resources;

"(c) Higher labour productivity in the context of increased employment, that is, the maximization of the productivity of labour in the economy as a whole;

"(d) The utilization of domestic materials, especially of raw materials previously considered of little value;

"(e) A scale of production that is suitable for the local markets to be served (unless exports are involved), with special consideration being given to small, fragmented markets in rural areas;

"(f) Low running costs and cheap and easy maintenance;

"(g) Maximum opportunity for the development, as well as use, of national skills and national management experience;

"(h) Dynamic opportunities for the further improvement of technologies and feedback effect on the national capacity to develop new technologies."[36]

Instruments that can be employed by developing countries to promote the selection of appropriate technology could include, for example:

"(a) Differential direct and indirect taxation (e.g. the exemption or lower taxation for products/enterprises in the small-scale sector or utilizing newly developed or indigenous technologies);

"(b) Differential financial and credit policies (e.g. lower rates of interest and liberal credit for products/enterprises in the small-scale sector or utilizing newly developed or indigenous technologies);

"(c) Industrial policies concerning size of units of criteria for expansion (e.g. certain products could be reserved for manufacture in the small-scale sector; policies discouraging more assembly industries based on imported components);

"(d) Trade policies on import of capital goods or raw materials (e.g. import control; not permitting import of equipment of too large a capacity; phased programmes for the reduction of import content of raw materials and components);

"(e) Policies on foreign investment and import of technology (e.g. discouraging turnkey contracts; not allowing foreign investment or import of technology in specified areas; associating local consultants or research and development institutions in selection)."[37]

Policies aimed at regulating the acquisition of foreign technology should not only cover technology per se, but also equipment (which embodies technology) and foreign investment which is a vehicle of technology and invariably predetermines it). A mechanism for screening technology contracts will be necessary. Such screening could ensure that the technological services required are clearly specified; technology packages are "unpackaged" wherever possible to admit contributions from indigenous technological capabilities; adequate provision should be made for the training of local technicians; there should be no unwarranted restrictions on the further dissemination of the technologies and the technological capabilities involved. Although each developing country may have its own approach towards the extent of production or regulation of foreign technology, the establishment of a screening mechanism will enable the continuous and systematic monitoring of foreign technology inflows, which does not exist in many developing countries at present.

The above suggests that methodologies will need to be developed to evaluate alternative technologies in terms of overall costs and benefits. This could necessitate the defining of numerical values for critical parameters such as labour costs and shadow wage rates, foreign exchange costs and shadow prices and thereafter the application of a discounted cash-flow approach. While the information network should provide the basic information regarding alternative production techniques, the evaluation of alternatives would need to be done by developing country enterprises and the national agency with responsibilities for reviewing arrangements.

Adaptation and absorption of technologies

Policies of technology adaptation and absorption should focus on the process of ridding imported technologies of their rich country "ethnocentricity" and of stamping them with the societal imprint of the importing country. No less important will be the process of upgrading local technologies to improve their productivity.

The adaptation of imported technology may necessitate, for example, the scaling down of the technology to the size of the local market, a process that has already been satisfactorily demonstrated in several fields, including bricks and cement, paper, textiles, packaging, sugar and a wide variety of agricultural equipment. Adaptation will also necessitate the matching of the technology to available local skills which, in some cases, may require maximizing its labour intensity and capital savings.

Since technology adaptation is the means of linking imported technology to national R and D, policies designed to enhance capacities for adaptation and absorption will need to give due consideration to the building-up or enhancement of national R and D capabilities. Technology policy will be required to forge closer links between R and D institutions and industries.

Adaptation to the satisfaction of a technical authority could be imposed as a condition in contracts for the acquisition of foreign technology. The costs of adaptation could receive preferential treatment in taxation. Adaptation to local raw materials and components could be secured through a phased programme of reduction of imported materials and components.

The sequence of main activities involved in the processes of technology selection and adaptation is outlined in figure III. The illustration suggests that there are three decisive inputs into these processes: information, the criteria for selection, and legal and contractual practices, each of which has been discussed above.

Absorption of technology in a narrow sense could be facilitated by policies which insist that foreign technology/investment inflows be accompanied by adequate training of local personnel both in terms of the number of persons trained and the extent of their training. The passing of a National Apprenticeship Act whereby each industrial unit is required to take a certain number of apprentices for training would also enlarge the pool of trained personnel. Free horizontal occupational mobility has also to be ensured, although there are no known direct policy instruments for this purpose. However, general policies that do not unduly restrict the setting-up of new units in the same industry may help incidentally. Policies for attracting the country's technical personnel resident abroad either for permanent settlement or for short-term guidance should also be formulated and implemented, as is already being attempted by a few developing countries.

Long-term policies for the absorption of technology should concentrate on human resource development. Policies that promote a greater involvement of scientists and technicians in the development problems of the country will be needed, including, where necessary, the restructuring of their salaries and responsibilities. This calls for serious reappraisals of educational policies, particularly in respect of the following:

(a) Introduction of a vocational content in the school educational curricula and making such courses available to as large a number of students as possible;

(b) Reorienting the technical courses at the university level so that the awareness of the students of the technological problems in the country, particularly with reference to rural areas, is enhanced;

(c) Educational curricula should include association with

Figure 4. The process of innovation

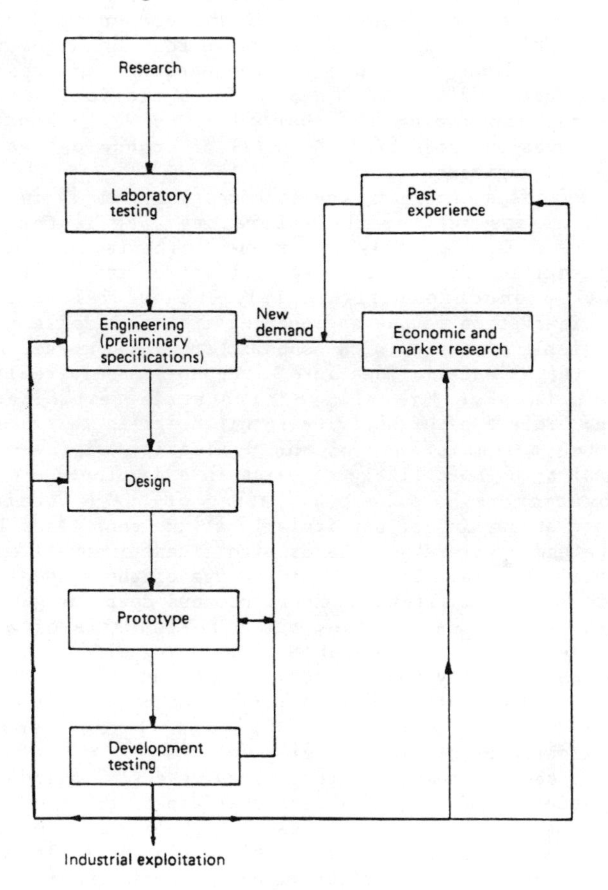

Source: "The structure and functioning of technology systems in developing countries" (ID/WG.301/2).

industry and practical training.

Development of technologies

The development of the capacity to innovate requires much more than the building-up of R and D institutions. In countries where development has been decentralized and community development programmes initiated, experience has shown that local governments, local organizations, agricultural co-operatives and the like, as well as motivated individuals, can be technological innovators. Technological innovation is a bottom-up as well as a top-down process: innovation comes from the users of technology as well as scientists and engineers.

In many cases, however, the technology system is incapable of bridging the gaps between the laboratory, the factory and the market place. This deficiency is due to the lack of integration between scientific and technological activities and the process of industrial production (figure IV). The crucial stage in the process of innovation is the engineering treatment of a new idea. Past experience, coupled with economic study and market research, transforms the scientific idea into a techno-economic reality that can be put through the mill of industrial exploitation and production. This type of activity requires skills that are neither those of the scientist nor of the production engineer, the two professional types most likely to exist in a developing country.

Innovation is not the prerogative of the scientist. The practitioner at any level, particularly at the shop-floor level, as well as the end user are sources of significant innovative ideas of considerable potential. The great advantage of these ideas is that they often reflect first-hand experience and deep insight into the actual needs of the user. They are often capable of producing working models; but considerable engineering effort is needed to transform the basically sound concepts into an economic reality. It is another task of policy to promote the application of such first-hand experience and to facilitate the process of the commercialization of new technologies.

R and D can be promoted through levying a tax on industry and utilizing the proceeds for promotional expenditure. Tax rebates could be allowed on the R and D expenditures of enterprises to encourage them to set up such facilities. In India, it is part of the condition of approval for the import of technologies that the importing organization should set up R and D facilities within the period of the contract so that the need for continuing the import beyond that period is obviated.

To preserve traditional technologies and capabilities, protection could be provided by way of the reservation of lines of manufacture, policies of government purchases etc. The adoption of technologies developed locally (e.g. by research institutes or industrial enterprises) could be encouraged by tax or interest concessions or by liberal conditions of industrial approvals.

For widespread dissemination of technology and for encouraging innovative capabilities, the promotion of self-employment and techno-entrepreneurs should be encouraged as a matter of policy. Concessional financial assistance through financial institutions are important in this respect. Policies of worker participation in production and technology decisions are of help. Patent laws and financial encouragement for innovations and their application are necessary. Special incentive schemes aimed at universities and academic institutions designed to promote innovative activities may also need to be devised.

In the section on technology programmes, a number of special instruments for promoting technology adaptation, absorption and development will be discussed.

Policies concerning transnational corporations

A substantial increase in the flow of technology to developing countries must take place if an adequate pace of industrial growth is to be achieved. Since, in a large number of manufacturing and service sectors, transnational corporations retain oligopolistic control over technology, a considerable proportion of technology acquisition may need to take place through their operation. Technology plans and policies should thus channel the operation of transnationals according to national objectives and priorities.

Policies aimed at regulating the activities of transnational corporations should recognize the conflict between the profit-maximization objective of transnationals on the one hand and the development of national scientific and technological capacities on the other. These conflicting interests can only be harmonized and the negative impact for developing countries reduced by the introduction of a regulatory and monitoring system. Elements of this control function should focus on the extent of the local integration of the foreign subsidiary, including the utilization of technologies appropriate to the country's needs and conditions, the extent to which it is involved in the building-up of indigenous capacities. The exercise of the control function should be guided by the need to secure decision-making autonomy in the host country.

Once technological needs have been defined and the most appropriate technology mix identified, the specific role and the possible pattern of corporate relationships with transnational corporations in various sectors of the economy can be established. In certain branches, particularly high-technology industries, it may be necessary to utilize transnationals both as sources of investment and as suppliers of proprietary technology. In sectors where the domestic industry has the necessary entrepreneurial capability and technological base, technological needs may be served by licensing and other contractual arrangements without foreign capital participation. In certain fields, in order to utilize and enhance domestic innovative capability, it may not be desirable to encourage foreign technology flows, for instance, in sectors where appropriate domestic technology is available or where foreign technology has been adequately absorbed by domestic industrial enterprises.

The technological requirements of linkage industries constitutes an important element of negotiations with transnationals. In the case of mineral industries, for example, technology for down-stream processing stages is an important aspect to consider and the interests of both the host country and the enterprise should be harmonized. Similarly, the extent and nature of domestic integration and the increase in value-added over a specific period should be established in the course of negotiations. The development of domestic marketing and managerial expertise, as well as operational skills, should also be identified as an important responsibility of transnationals in various sectors.

An important aspect of negotiations with transnational corporations is the disaggregation of the technology package. Transnationals tend to aggregate the investment function with the various technology elements, including project engineering, production technology, management and marketing. From the viewpoint of the host developing country, it is important that the package should be unbundled and evaluated in terms of its various elements. Of even greater importance is the possibility of the participation of domestic industry in the supply of inputs and project engineering services. Even if the cost of domestic goods and services tends to be above world market prices in the earlier stages of industrialization, this may be justified in the long-term interests of the development of domestic capabilities. The extent of unpackaging may, however, be limited in certain sectors where transnationals can ensure that the technology is used only by a subsidiary or affiliate under its control or sold only as a complete system and not as separate components. Similarly, where foreign engineering contractors with the skills to combine various inputs are themselves dependent on the technology supplier, the incentive to unpackage may be weak or lacking. In such cases, a great deal may depend on the technical and managerial expertise and contracting skills available in the host country. Some countries have, accordingly, placed great emphasis on the development of domestic capabilities in consultancy services.

Efforts in the direction of unpackaging should obviously aim at maximizing the use of local inputs, especially technological services. Policy guidelines can be prescribed concerning restrictions in the use of foreign personnel, training programmes for domestic personnel at various levels, and enterprise-level R and D. Import restrictions and controls can significantly affect greater technology flow for linkage industries and adaptive use of local materials and parts. Export incentives and insistence on export commitments by the subsidiaries of transnationals can, on the other hand, improve the balance of payments performance of transnationals and achieve better quality production.

It is important that the impact of operations of the subsidiaries and affiliates of transnational corporations on domestic technological development be monitored continuously.

The review process should monitor the path of technological development, the R and D undertaken by the foreign affiliate, and the adaptations performed to suit local conditions and requirements. This review should cover existing subsidiaries and affiliates and also new enterprises in which transnationals are involved.

Special attention may also need to be given to the high costs resulting from the extensive use of foreign brand names and trademarks by transnational corporations. Measures that can be used in this respect include the compulsory use of domestic brand names which, after a certain period, obviates the need for foreign brand names. The diffusion of foreign technology can be facilitated by restrictions on the duration of licensing agreements (usually from 5 to 10 years). The shortening of the period of patent validity below the norms of the international patent system can also be introduced, as has been done by such countries as Brazil and Mexico, and the possibilities for introducing patents in vital sectors can be severely restricted.

Technology programmes

Technology policy will have to be translated into programmes and, eventually, subprogrammes, projects and specific activities.

Development of the engineering and machine tool industry

One of the most important of all industrializing industries is the engineering and machine tool industry. It is the basis for much industrialization, and experience in developing countries has shown that a broad-based industrial structure cannot be sustained without the existence of a growth-oriented engineering sector. The engineering industry is traditionally an important source for the growth and development of technical manpower and a focus for the process of technological innovation; it is thus advisable for all developing countries to assign high priority to its development, and especially to the production of machine tools.

The development of the engineering sector may call for the setting-up of facilities for the production of ferrous and non-ferrous castings, forgings, machine tool and machine shop equipment, fabrication (including weldments and stampings), rolling, bending and pressing facilities, heat treatment and plating and steel rolling mills.

Raw material supplies will be of decisive importance, especially steels, castings and forgings. With respect to steel, construction steel (mild steel), alloy steel and sheet steel are the raw materials essential for engineering products. Whether a developing country should develop its own iron and steel industry depends upon a complex of factors, not the least important of which is the availability of necessary mineral resources. Developing countries that have no supplies of iron ore or coal, do not have abundant power and have not reached a high level of industrial development should import whatever steels are required to develop their engineering industry.

The availability of ferrous and non-ferrous castings depends upon the existence of foundries and forge shops, whose development should thus, where necessary, be afforded high priority. Since casts and forged components are to be made specifically to drawings, they can be more advantageously produced in the country itself.

The decision to develop a national machine tool capability should not be made dependent upon the size of the market. Virtually every artifact is made on machines that are themselves made on machine tools. Even in the least developed countries a machine tool industry can and should be developed. It might, for example, be organized as a cottage industry and involve the production of essential spare parts.

Small and medium-sized enterprises

Special programmes may be required to promote the technological development of small and medium-sized enterprises. An environment that encourages small firm initiative is likely to be more competitive and able to promote an active search for more appropriate technologies. A small firm is usually less inclined towards vertical integration so that it is more likely to rely on small, relatively labour-intensive local producers and suppliers than is a large enterprise. Small-scale industries also have a critical role to play in integrating the agricultural and industrial sectors, a key aspect of development policy.

In some developing countries, transnational corporations receive more privileged treatment than local small and medium-sized enterprises that typically receive little support in dealing with the problems they face. They generally lack, for example, the necessary resources to maintain specialized personnel for technological management and do not even have enough technicians to adequately maintain and supervise ongoing production processes.

The effectiveness of small and medium-sized enterprises could be improved through support programmes involving R and D institutions, industrial extension services and technological service organizations. Governments might seek to develop entrepreneurial skills in small and medium-sized enterprises through programmes aimed at reducing the risks incurred by groups of entrepreneurs in the development of their technological capacities.

Development of a technological service capability

Inadequate technological service capability is a major constraint in most developing countries. Such services range from macro-level project identification, feasibility studies, plant specifications, detailed engineering designs, civil constructions and machinery installation, and plant commissioning, start-up and operations. While the extent of the gap varies from country to country, the most significant gap, even in fairly industrialized developing countries, is in respect to detailed engineering and designing and sectoral consultancy services through nationally owned units. This makes disaggregation of foreign technology packages extremely difficult and also creates a critical gap in infrastructure, which result in undue dependence on foreign design and engineering services which affects the pattern of investment for particular projects, the requirements of capital goods and equipment, and subsequent plant operations and management. In the less developing economies, the gaps in consultancy services are even more marked and extend to almost the entire range of service activities indicated above. The identification of gaps in service capability has to be done on a country-wide basis and for critical and priority sectors in each economy. An appropriate policy package also needs to be prescribed and the extent to which preferential treatment is necessary for national or regional consultancy services, including engineering and designing capability, needs to be identified regarding the use of such domestic capability in a progressive manner at successive stages of industrial growth. It may also be necessary to provide technical and financial support to national consultancy firms undertaking detailed engineering and other technological services, particularly in priority production sectors. Fairly effective steps have been taken by certain developing countries, notably India, in this field and similar action can be emulated in other developing countries, with appropriate adjustments to suit national or regional considerations.

Technological services include the promotion of standardization, quality control, common testing facilities, productivity, metrology and other such general service functions. There are a number of institutions in developing countries in several of these fields. Such institutional activities are usually supported by governments or are financed through universities or research organizations. In many countries, standardization and quality control have made effective progress and have constituted an essential feature of export promotion of non-traditional products. Productivity organizations have also proved useful in identifying specific production problems at the micro-level in several industries though, by and large, there has been limited communication and linkage with production sectors and enterprises.

Industrial extension services

The processes of technology adaptation, absorption and development would no doubt be facilitated by the creation of industrial extension services. Such services, which would parallel those applied in agriculture, could serve to accelerate the growth of manufacturing industry, especially in small and medium-sized enterprises, and, in time, provide an important input into the strengthening of national, research and development activities.

Industrial extension services could be used:

"(a) To identify and resolve, to the extent possible, problems faced in manufacturing. It may be necessary, however, to refer the more complex problems to R and D institutions for advice or resolution;

"(b) To identify new areas for the adaptation and development of appropriate technologies. Such areas might include leather, processed food, metallurgy, forest products and building materials. The work would be undertaken either in the extension centres themselves, or in indigenous R and D institutions, according to needs and resources;

"(c) To familiarize industries within the country with development and improvements in related techniques;

"(d) To train local professionals;

"(e) To provide essential support for future expansion into R and D institutions and assist in the growth of other institutions."[38]

Information networks

Special programmes in the field of information may need to be initiated. There will be a need for an adequate information network that can ensure a flow of fairly detailed data and material on production and technical requirements projected both for the economy as a whole and at the micro-level with specific growth projections and technical requirements of significant production sectors and enterprises. Once the nature and magnitude of sectoral growth projections and technological requirements are defined, the information system should also be able to provide possible technological sources, both indigenous and external and for specific projects and enterprises. At the micro- or enterprise-level, the information mechanism should provide for detailed data flow regarding existing industry in terms of (a) production capacity in various or selected sectors, production techniques employed, utilization of capacity and technological problems encountered; and (b) the nature of the expansion proposed, with its technological implications. The flow of information should also cover the need for new enterprises that may have to be set up to cover critical production gaps in various sectors.

Technical education and training programmes

Technological self-reliance is linked to education and training, which is linked to the process of confidence building. The promotion of technological self-reliance should thus be seen as a learning process in which the emphasis is on thinking in terms of independence and the need for autonomous decision-making.

Education is a basic element of the scientific and technological infrastructure of the country. Because of the complexity and overriding importance of the educational system, however, the design of this system should be left to institutions that deal specifically with it. Education should fulfill the dual function of instilling appropriate values and attitudes and upgrading and developing necessary skills. Technology and education plans should thus be closely interlinked. Training at the production level--that is, in its actual operation in the factory and the agricultural sector that directly influences the adaptation, absorption and diffusion of technology--falls within the framework of technology policy.

The strengthening of technological capacity requires:

(a) Making science education more responsive to the needs of the country and using science and technology effectively in the achievement of national goals;

(b) Stimulating the choice of scientific and technology education and careers to increase the number of scientists, engineers and technicians;

(c) Reinforcing the social status and prestige of technical and technological work;

(d) Emphasizing and stimulating imaginative inquiry and independent learning.

Well-defined training programmes for the development of special skills in specific industrial operations and technological services will be necessary. Short-term training programmes may include, for example, processing technological information; training of managers, entrepreneurs and government officials in the evaluation, negotiation and acquisition of technology; training of research and development personnel in the management of research and development, evaluation of research and development projects, commercialization, extension work, liaison with industry and other related matters. Training programmes and sensitization courses may be necessary for policy-makers in project and technology evaluation and the implications of choice of technology. Special courses may be necessary for technical personnel in such aspects as design, production engineering and productivity. In-plant training programmes for engineers and skilled workers will be essential. While some of the programmes referred to could be organized within the country itself with, where necessary, the help of outside experts, there are others where training may need to take place in an industrialized country or be organized within the framework of TCDC efforts.

Special programmes might also be launched to deal with the problems of brain drain. The aim of such programmes should be to enable experts, technologists and managerial/supervisory personnel to return home, even if for only a limited period, so that their knowledge and expertise can be adequately utilized.

The need for action programmes

The development of national technological capabilities requires concerted action in a wide range of interrelated fields. However, it will not generally be possible for developing countries to do everything at once, even if this were considered desirable. There is thus, as noted earlier, an overriding need for selective action in areas that will lead to an immediate and demonstrable improvement in technological capacities.

One area in which such an action programme could yield substantial results is in bringing technology and production into a relationship of co-operative association and mutual reinforcement after, in many developing countries, decades or even centuries of separation. The main elements of such an action programme could include:

(a) The selection of a small number of sectors or areas of production in which there is considerable scope for the introduction of frontier technologies that could be used to spearhead the process of industrial development. These sectors are generally in the industrializing industries. In other sectors, technology would still play an important role as one of the major factors of growth and production but these would not be afforded such strategic importance. In a third group of sectors or areas of production, there would be no determined efforts to stimulate technology development much beyond current and available levels;

(b) In the high priority sectors and production areas, special efforts should be made to create "integrated technology development, upgrading and application systems", both at the level of the enterprise and of broad policy;

(c) For this purpose, expenditure on technology selection, acquisition, adaptation, absorption, development and application in the selected sectors may need to be expanded to perhaps 10 times the current average level of expenditure for the rest of the economy. National policies and other public and private institutional programmes and instruments should be developed and applied to ensure the desired results;

(d) The supporting services, skills, legislation and regulations required should be gradually expanded to serve as an indigenous basis for promoting the development of other sectors, thus ensuring that there is a general advance not only in technology, but also of social and economic development.

In this manner, real progress could be made in a number of high priority sectors that would facilitate the process of development-centred industrialization. A stronger technological basis could be established and closely integrated with production, management and investment in the country concerned. The process would be inevitably long-term and it could take five years or more before it started to bear real fruit; it is a dynamic process and this dynamism could serve to give real expression to the concept of technological self-reliance. Although the technology used may be imported or traditional, or both, it would always be the result of decisions taken by qualified technologists and technicians who would seek to choose the optimum technology and technology mix within a dynamic and development oriented framework.

The action programme outlined could also utilize the idea of industrial development complexes that could serve as a focus for the implementation of mutually reinforcing sets of proposals. Such complexes might aim at bringing together a group of related enterprises and technologies to form a co-ordinated vertical system with an enhanced capacity for innovation and technology development.[39]

Institutions

Policies and programmes are formulated and implemented by institutions. Their value resides in the fact that they provide a measure of continuity and experience and in due course become the repositories of technological capability. Although they have a decisive role to play in the promotion of technological self-reliance, institutions can only be as good as the policies and programmes that they implement and as effective as the means that they have at their disposal.

Unfortunately, a good deal of the literature on technology policy has given the impression that the implementation of policy calls for the creation of new institutions, or even of a single omnipresent institution in which the "technology function" is centralized. Yet there can be no watertight boundaries for defining all the elements of technology policy and, in its implementation, there is a wide range of "implicit" instruments that, although not usually applied by technology institutions, are able to influence technological development in crucial ways.

Conceivably, if a nation were to set out to build an administration from scratch, all the areas affecting technology development could be separated and grouped together for institutional purposes. But no country does start from scratch and most already have a range of institutions that deal, in various ways, with aspects of technological policy. Developing countries typically possess ministries of science and technology, technology transfer centres, sectoral industry development centres, research institutions of various types, information centres, technology regulation agencies etc., each performing one or more technological functions. The difficulty of attempting to disentangle technology as a separate policy area is not surprising given that the area encompassed by technology is virtually co-extensive with that of economic progress, since technology in part determines productivity and productivity influences incomes. Because one of the principal aims of all policies implemented by all government institutions is the fostering of economic progress, few governments think it productive to separate such policies and to consolidate them institutionally. The same applies to technology policy.

Clearly, then, the process of developing technological capabilities is far too complex to become the exclusive domain of a single institution, and the development of appropriate institutional infrastructure lies in the strengthening of existing institutions rather than in the creation of new institutions. There may of course be both the scope and need for new institutional initiatives but, as a general rule, the creation of new institutions only appears justified where there is a role for them that is demonstrably different from the functions of existing institutions.

A balanced approach to institution building will have to start from the functions, capabilities and services required and a review of how these can be most effectively made available or linked to entrepreneurs on the one hand and government officials and policy-makers on the other. Apart from institutions for technical education and training, at least three basic types of institutional functions may be required. One type relates to technology policy formulation and monitoring at the macro-level and technology screening and evaluation at the micro-level. These functions will generally have to be fulfilled by a government department or agency, suitably located in the governmental hierarchy so that it is able to influence decision-making. Another type of institutional function relates to technological information, evaluation and consultancy assistance to entrepreneurs. This may have to be performed by an agency, governmental or quasi-governmental, which maintains effective relations with government, financial institutions and industry. The third type of function relates to technology development, adaptation and commercialization with facilities for consultancy and extension work. Such a function needs to be exercised by research institutions or technology development centres, which may be single or multi-sectoral, depending on requirements.

Several of these functions will need to be injected into ostensibly non-technological institutions, such as ministries of industry, planning, trade and finance, and financial and banking institutions. Sectoral industry centres should function as technology adaptation and development centres. Technological diffusion, particularly in the rural areas, may require institutional innovations such as those adopted in India, including the creation of small industry centres, district industry centres and polytechnology clinics (i.e. extension and consultancy outposts of research institutes).

Some of the more important steps that a developing country may need to take in the development of its technological institutions are summarized as follows:

(a) Examining whether adequate institutional arrangements exist for the exercise of technological functions;

(b) Strengthening existing institutions or networks of institutions, to ensure that they are able to exercise such functions;

(c) Creating new institutions where existing institutions are unable to effectively exercise their required functions;

(d) Ensuring adequate linkages and co-ordination between the institutions, government and industry;

(e) Providing the institutions with adequate manpower, material and financial resources;

(f) Developing institutions for technical education and manpower training;

(g) Reorienting the programmes of the institutions towards problems of national development, especially those associated with strategies of (technological) self-reliance and aimed at meeting the basic needs of the poor and underprivileged masses;

(h) Instilling an awareness of possible technological impacts into the operations of relevant non-technological institutions;

(i) Creating organic linkages between technology institutions and decision-making for social and economic development;

(j) Providing appropriate encouragement to voluntary agencies and universities so that they can become catalytic agents in the promotion of technological self-reliance.

There is likely to be considerable scope in most developing countries for experimentation in the field of institution building. In countries with a weak technological base, it may be necessary to create national technology centres vested with a broad range of responsibilities. Because of the enormous variety of technological situations in the third world, no single blueprint for such a centre can be hypothesized. Its core functions might, however, be to:

"(a) Assist, within the framework of national, social, economic and political constraints, in the identification of technological needs for a variety of economic activities;

"(b) Assist in the acquisition and analysis of information required on alternative sources of technology from all available sources, domestic and foreign, and its delivery to users;

"(c) Assist in the evaluation and selection of technologies appropriate for the different jobs to be done, with the emphasis on decision-making;

"(d) Assist in the unpacking of imported technology, including assessment of suitability, the direct and indirect costs and the conditions attached;

"(e) Assist in the negotiation of the best possible terms and conditions for the technology to be imported, including arrangements for registration, evaluation and approval of agreements for its transfer;

"(f) Promote and assist absorption and adaptation of foreign technology and generation of indigenous technology, linked specifically to design/engineering, research and development;

"(g) Promote the diffusion among users of technology already assimilated, whether indigenous or foreign;

"(h) Co-ordinate policies in general and evaluate their internal consistency in relation to the transfer and development of technology."[40]

In view of the observations made above, it would be appropriate to determine whether such functions could best be met by a network of existing government agencies and private institutions implementing an interrelated set of policies and programmes at different levels (figure V).

Some developing countries might also find it useful to establish a national centre for development alternatives for the purpose of conducting multi-disciplinary and multi-institutional studies into alternative development systems that are mancentred, need-based, endogenous and self-reliant and into the possibilities for change in and development of the national science and technology system.[41]

Monitoring of technological strategy

Developing countries should continually review and evaluate the process of technological change and the efforts made to strengthen technological capabilities. This evaluation should seek to clarify the relationships between ends and means in technological policy and to assess the effectiveness of the programmes and projects initiated. The evaluation of technology policy and of efforts in the direction of technology planning should be made in the light of national development aims and aspirations and of economic, social and industrial development objectives. If the development of technological capabilities and the promotion of technological self-reliance are to be treated as more than mechanical exercises in manpower projections, then it is essential that the monitoring and evaluation system developed should be a derivative of development strategy and focus on problems related to human resource development and the mobilization of national problem-solving capacities.

Especially important is the monitoring of foreign technology flows and their impact on domestic technological progress in specific sectors as well as on changing technological needs. The absorption and diffusion of foreign technology following its adaptation is also an area that should be subject to continual review.

The institutional arrangement for monitoring and evaluation will vary from country to country. It is essential, however, that the institution or group of institutions vested with monitoring responsibilities should be in the mainstream of technology policy formulation and implementation. It should have sufficient authority to speak instead of being merely represented at, for example, interdepartmental consultations on technology policy. The monitoring authority should have its own budget and be in a position to allocate funds to different bodies for the purposes of policy evaluation and review. It should make and publish periodic reviews of progress in the direction of strengthened technological capabilities and be able to relate such reviews to progress in the direction of development goals. It is essential that the monitoring and review body, irrespective of its exact composition or structure, does not become another department lost in a bureaucracy or in routine and administrative tasks.

IV. ROLE OF INTERNATIONAL TECHNOLOGICAL CO-OPERATION

A. Issue of international technological co-operation

The countries of the third world will be unable to strengthen their technological capacities unless they themselves become the active agents of their own transformation through their own efforts and the application of their own resources and knowledge. The pursuit of increased self-reliance, however, even when aimed at more selective participation in the international economic system, in no way excludes technical co-operation with other nations, both developing and developed.

Since the technology needs and experience of many developing countries bear close affinity and follow similar patterns, co-operation between developing countries will be invaluable in the process of collectively strengthening technological capabilities. Equally, the industrialized countries will remain the predominant suppliers of much modern technology. Co-operation with the governments of and enterprises in these countries will be required to ensure that the transfers that take place contribute to, rather than undermine, national development efforts. In both these areas--technical co-operation between developing countries and between developing and industrialized countries--there is considerable scope for new approaches and fresh initiatives.

B. Co-operation between developing countries

A great deal of attention has been focused in recent years on the potential offered by ECDC and TCDC. It is recognized that they constitute an important framework for the forging of bilateral links of co-operation at the subregional, regional and interregional levels. The Buenos Aires Plan of Action lists the objectives of TCDC as follows:

"(a) To foster the self-reliance of developing countries through the enhancement of their creative capacity to find solutions to their development problems in keeping with their own aspirations, values and special needs;

"(b) To promote and strengthen collective self-reliance among developing countries and through exchanges of experiences, the pooling, sharing and utilization of their technical resources, and the development of their complementary capacities;

"(c) To strengthen the capacity of developing countries to

identify and analyse together the main issues of their development and to formulate the requisite strategies in the conduct of their international economic relations, through pooling of knowledge available in those countries through joint studies by their existing institutions, with a view to establishing the new international economic order;

"(d) To increase the quantum and enhance the quality of international co-operation as well as to improve the effectiveness of the resources devoted to over-all technical co-operation through the pooling of capacities;

"(e) To strengthen existing technological capacities in the developing countries, including the traditional sector, to improve the effectiveness with which such capacities are used and to create new capacities and capabilities and in this context to promote the transfer of technology and skills appropriate to their resource endowments and the development potential of the developing countries so as to strengthen their individual and collective self-reliance;

"(f) To increase and improve communications among developing countries, leading to a greater awareness of common problems and wider access to available knowledge and experience as well as the creation of new knowledge in tackling problems of development;

"(g) To improve the capacity of developing countries for the absorption and adaptation of technology and skill to meet their specific developmental needs;

"(h) To recognize and respond to the problems and requirements of the least developed, land-locked, island developing and most seriously affected countries;

"(i) To enable development countries to attain a greater degree of participation in international economic activities and to expand international co-operation."[42]

TCDC should thus be viewed as a multi-dimensional process that should prove of decisive importance in enabling the third world to free itself from some of the worst forms of technological dependence and domination.

The need for intensified TCDC arises not only from the recognition that the developing countries cannot rely solely on the goodwill and participation of enterprises from the industrialized countries but also from the awareness that the technological needs and experience of the developing countries have, by and large, common elements. The technological capabilities of several developing countries have achieved a level where their know-how, expertise and services, supported by supplies of machinery and equipment, can be effectively transferred to other developing countries at both the government-to-government and firm-to-firm level. The sectors in which such capabilities have been developed extend over the production of a wide range of consumer durables, intermediate products, light and medium engineering goods, and machinery and equipment. In all these areas, developing countries are effectively competing in international markets. Technological service capability, for example in consultancy and engineering services, has also grown considerably in many of these countries and could be extended to other developing countries. Most of the process and production know-how in these countries have been acquired through foreign affiliates, joint ventures and former licensing arrangements, though a large number of locally developed processes and techniques have also been developed. Both these categories of technical know-how can be effectively and appropriately transmitted between developing countries.

Most prospective developing country licensees continue to seek assistance form Western transnational corporations, even for relatively unsophisticated production processes for which there is a fairly wide range of technological choice available in other developing countries. This is partly due to a lack of knowledge of the availability of appropriate technology and technological expertise and know-how in other developing countries and partly to a continuing preference for more sophisticated production techniques used in highly industrialized countries. Closer contacts and greater sharing of knowledge and experience between developing countries would improve this situation.

An important prerequisite to greater technology flow between developing countries is the compilation of a new set of guidelines and principles in respect to technology transfer arrangements between enterprises. Licensor enterprises from developing countries should not impose unduly restrictive contractual conditions on licensees in other developing countries as was often done by technology licensors from industrialized countries. On all critical negotiable issues, such as the extent of foreign holding, duration of agreement, technology remuneration, technical service support and other contractual conditions, new standards and principles should be set and agreed upon, based on a maximum degree of co-operative partnership. A set of model guidelines should be prepared and developing countries should ensure that the guidelines are applied by licensor-licensee enterprises. With the greater degree of control exercised in most developing countries over the production sector, it should be both feasible and practicable that such guidelines, agreed upon at the intergovernment level, should be universally applied in technology and investment cum technology transactions between developing country enterprises.

Other areas in which there is considerable scope for technological co-operation between developing countries include: (a) exchange of information and experience regarding technology licensing and contracts; (b) collective adoption of guidelines governing foreign technology inflow and regulation; (c) joint action in the selection of appropriate know-how in certain sectors; (d) collective bargaining for the licensing of a particular technology for similar projects in more than one developing country; (e) development of joint R and D facilities for selected production sectors; and (f) joint manpower training programmes in selected branches.

An exchange of information and experience on the terms and operation of technology contracts would greatly strengthen the bargaining power of the developing countries because of the greater knowledge and information they would have at their disposal and the extension of the area of technological choice.

UNIDO has suggested that the information that could be shared to great advantage between developing countries could be broadly categorized as follows:

"(a) Available alternative sources of technology;

"(b) Terms and conditions of acquisition of specific technologies;

"(c) Terms and conditions of supplies of raw materials and intermediates;

"(d) Sectoral trends in terms of applicable royalty rates, technological developments etc.;

"(e) Corporate ownership and structures of various suppliers of technology etc.;

"(f) The availability of skilled manpower and expertise in various countries."[43]

Of particular importance is specific information on prices of know-how, engineering, technical services etc.; applicable royalty rates; methods of calculation of running and fixed payments; prices and terms of delivery of raw materials, components, and intermediate products; scope of sales and manufacturing rights; limitations of volume of production/sales; duration of agreements; and parties to agreements.

In respect of policy guidelines governing foreign technology contracts, a similar pattern and approach has already been adopted in most developing countries where such contracts require to be reviewed by regulatory institutions. As pointed out earlier, however, model guidelines should be prepared for other developing countries where such institutional arrangements still need to be established. These could be considered and adopted in the light of each country's specific situation and objectives.

An important field of technological co-operation relates to the joint acquisition of technology and know-how for use in several countries through a process of collective bargaining. Though seemingly difficult, this has considerable possibilities, both for technology suppliers and for recipients and licensees. There is considerable commonality in industrial programming in countries in comparable stages of development and projects in the same field could be undertaken in more than one country about the same time. Such projects could include large-scale industries, such as steel, petroleum, fertilizers and chemicals, and machine building; medium-size plants for textiles, sugar, cement and agro-industries; and small-scale units covering a wide range of intermediate and consumer products. In a number of these cases, the acquisition of foreign know-how on a collective basis for more than one project could be considered. This would enable more detailed evaluation and consideration of technological alternatives and would reduce technology costs in addition to securing better contractual terms. Such an approach towards collective bargaining would have particular significance in contiguous countries, as in the case of the Andean Group or regional country-groups in Africa and Asia. It would also have relevance for countries at a similar stage of industrial growth, such as Brazil, India and Mexico. Significant collective action has not so far been initiated in the acquisition of technology primarily because this issue has been viewed in national terms and left to the initiative of individual enterprises. With the increasing realization of the interrelationships in technological growth, a joint or collective approach to technology acquisition appears particularly important.

The institutional arrangements for the joint acquisition of technology should also be considered. They can take either the form of joint negotiations by a group of developing countries for identified sectors in which the country-groups are interested or the establishment of an international mechanism through which technology can be acquired and transferred to projects in more than one country. The former approach necessitates close collaboration and co-ordination between country-groups and the identification of common technological needs in specific industrial sectors, after which a joint body can be constituted for evaluating, negotiating and acquiring selected technology in the fields identified. The second alternative requires the creation of an appropriate international mechanism through which such joint technology transactions can be channelled. UNIDO could function as such a mechanism.

TCDC should be intensified in respect to consultancy and engineering services and the development of manpower skills, including managerial expertise. Hitherto, linkages in these fields have been established primarily at the enterprise level between licensees and foreign parent organizations and technology licensors from industrialized nations, though some joint training programmes have been undertaken in some developing countries. There is considerable scope for the setting-up of joint consultancy and engineering services, either on a regional basis or between country-groups at a similar stage of industrial growth. The first step in this direction is the greater use of consultancy and engineering services available in certain developing countries by other developing countries, followed by the creation of appropriate national consultancy services in each country or in regional groups.

Co-operation in research and development programmes is another promising field. Such co-operation would help to ensure that R and D is better geared to meeting the developmental needs of the developing countries. Experience of industrial R and D in the institutions set up in developing countries has, at best, been fairly mixed, which emphasizes the need for the systematic sharing of experience and the implementation of joint research activities. Electronics, drugs and pharmaceuticals and non-conventional sources of energy are high priority research areas. They could be followed by joint R and D programmes in agro-industries, leather, chemicals, engineering products and several other sectors of interest to a number of developing countries. It is essential, however, that R and D programmes are directly related to the needs of the production sector and, though the results of industrial research can only be assessed over a relatively long period of, say, from three to five years, such an assessment has to be made in terms of changing costs and benefits. Cost-benefit analyses in terms of utilization of research results by industry are all the more necessary and significant for joint programmes.

While the part played in TCDC by institutions and enterprises is important, governments also have a crucial role to play in defining the nature and extent of TCDC programmes and in monitoring and evaluating their effectiveness. It will be necessary, therefore, for developing countries to arrive at intergovernmental framework agreements that specify the nature, extent and modalities of TCDC and provide a framework within which to conclude bilateral and multilateral agreements in different fields and at different levels.

The Round Table Ministerial Meeting on Industrial and Technical Co-operation among Developing Countries, organized by UNIDO and held at New Delhi in January 1977, detailed specific areas of technological co-operation as follows:

"(a) Co-operation in the field of industrial technology with a view to improving the identification and use of technologies already available in the developing countries, including technical know-how and skills, machinery and equipment, design, consulting and construction capabilities;

"(b) Collaboration in respect of the proposal for a technology bank, which would also include consideration of joint purchase of technology and examination of contracts and agreements already concluded, to provide guidance to others so as to avoid the mistakes and problems relating to the experience of particular technologies in any of these countries;

"(c) Promotion of collective action for negotiating and bargaining for more equitable economic relationships and for acquisition of technology;

"(d) Development of concrete programmes for using engineering and consultancy capabilities available in the developing countries;

"(e) Co-ordination of industrial training programmes to augment the skills considered basic to industrial development programmes;

"(f) Co-operation in the establishment and strengthening of national and regional institutions concerned with industrial and technological development;

"(g) Co-operation in applied research and development in specific sectors, drawing upon machinery and capabilities already available in the developing countries and concentrating specifically on engineering industries, electronics, fertilizers and agro-chemicals, pharmaceuticals, chemical industries and energy."[44]

C. Co-operation between industrialized and developing countries

Bilateral and multilateral technical co-operation between industrialized and developing countries has increased considerably in the past decade or so. Evidence indicates, however, that past patterns of co-operation have not only strongly reflected but also tended to reproduce the disparities in the technological capabilities of the rich and poor countries. Certain types of well-meaning co-operation have tended to be wasteful in their use of resources, ineffective in terms of their contribution and, when they have involved in the transfer of "ready-made" solutions and the conscious and deliberate transmission of value systems, consumption patterns and ways of thinking in the Western industrialized countries, may have contributed to the problems of developing countries.

The efforts of the third world to strengthen its technological autonomy should be underpinned by well-conceived international, multilateral and bilateral programmes of technical co-operation. The reports of many of the conferences held in recent years by United Nations bodies reflect this belief and refer to the importance of science and technology in the development process and to the valuable role that can be played by programmes of technical co-operation.

A study made of the results of such conferences points to areas in which the industrialized and the developing countries have reached a large measure of agreement.[45] These covered balancing of the "flow" and "stock" streams of science and technology, emphasis on "appropriate" technologies, stopping the "brain drain", participation of the developed countries in solving problems of concern to the developing countries, more technical co-operation between the developing countries and the satisfaction of basic human needs.

The consensus implied is stated in general terms and has yet to be put to the test; also the listing takes a predominantly instrumental view of science, technology and co-operation. However, there is evidence of a foundation upon which to construct new approaches to scientific and technical co-operation.

It is not the validity of technological co-operation that should be questioned, but rather the relevance of past approaches. There appears to be scope for new initiatives which, stripped of ethnocentric prejudices, preoccupations and predilections and free of parochial interests, seek to support the third world in its struggle to free itself from some of the worst forms of technological dependence. If mobilized in the right ways, the technological expertise accumulated in the industrialized countries could stimulate a conscious process of technological change in the developing countries.

For some time to come, enterprises and institutions in the industrialized nations will continue to function as the major source of much industrial technology. With the exception of the centrally planned economies, ownership and knowledge relating to industrial technology will be concentrated in individual enterprises in the developed countries and technology flow will continue to take place through various mechanisms, ranging from the supply of capital goods and licensing arrangements to joint ventures and foreign affiliates with varying degrees of foreign ownership. In almost all cases, technology transfer takes place through contractual arrangements between enterprises in these countries and in developing countries and it is the nature and content of the contract that should take the legitimate doubts and aspirations of the developing countries into full account. Given the screening arrangements already operating in many developing countries and those likely to be established, it would be impracticable to expect that unreasonable restrictive provisions should continue to form part and parcel of such contracts, particularly provisions that may not even be legal in the country of the licensor. Representative bodies of technology suppliers and licensors should prescribe and adopt guidelines in technology supply and contracting that are consistent with the requirements of developing countries and licensees from these countries. It is only then that the present mood of confrontation will be satisfactorily resolved and a more appropriate climate created for investment cum technological collaboration at the enterprise level.

There is also a need for a greater flow of technology from a larger number of industrial enterprises in the industrialized countries. The majority of technology transactions between industrialized and developing countries have, in the past, involved Western transnational corporations and large industrial establishments. As noted in chapter I, however, transnationals have not always proved satisfactory agents of transfer. Because they are concerned with world-wide operations they tend to standardize their procedures and processes. As a result, they often become too dominating and inflexible for the development process, especially in small developing countries. There appears to be considerable scope for a range of public-supported initiatives aimed at involving small and medium-sized enterprises in the process of international technological co-operation. Small industries may well prove a more appropriate instrument for the transfer of technology. The smaller scale of operations lends itself to an efficient use of labour-intensive techniques. Capital-labour ratios also tend to be lower in small industries, which suggests that a given amount of investment would generate greater employment if it were allocated to small instead of large industries. Thus, the experience of small and medium-sized enterprises in the industrialized countries is likely to be more relevant to the factor endowments (relative abundance of labour and scarcity of physical capital and skills) of the developing economies.[46]

Small and medium-sized enterprises have not yet been adequately involved in formal transfer programmes because they lack the capacity to enter into negotiations and co-operative arrangements. This suggests a potential area for new initiatives: the governments of the industrialized countries might consider such measures as the coverage of investment risks, the promotion of contacts through the dissemination of information, and financial support to bring interested parties together. Small and medium-sized enterprises may well have a special role to play in the process of technology adaptation, conceivably within the framework of basic needs strategies.[47]

Research co-operation is typically an area in which several ideas, old and new, still have to be fully tested. Special attention could be given to the promotion of twinning arrangements that lead to direct channels of communication between research institutes in industrialized and developing countries and that facilitate the transfer of know-how and provide a framework for various types of training arrangements, longer-term collaboration and more effective involvement of universities in the industrialized countries in technological transformation activities. Efforts might also be made to launch co-operative research programmes aimed at developing new technologies from which both industrialized and developing countries could benefit, such as small-scale energy modules based on "soft" sources, small-scale electrification systems, low energy housing and transport technologies, and rediffusion systems. Similarly, the governments of the industrialized countries, which are generally able to exert considerable direct and indirect influence over national R and D budgets, could seek to promote research into technologies of special interest to the third world. Whatever the institutional framework selected, however, the resulting programmes should involve the active participation of research workers and institutions from the developing countries so that they can benefit from the experience of working on the solution to some of their own problems.

The governments of the industrialized countries might establish programmes aimed at subsidizing the sale of technological knowledge and product and process know-how.[48] Governments are not generally in a position to force their industries to part with their technological property which, in some cases, may have involved exceedingly high R and D costs and which determines their competitive position in both domestic and international markets. Governments of the industrialized countries could possibly seek to overcome some of the constraints involved by agreeing on preferential terms for the sale of various categories of patents and know-how to third world countries, the difference between such terms and those between the industrialized countries being borne by the governments. In this way, the competitiveness of the firms involved would not be disturbed on world markets.

Another approach would be to establish in the donor countries policies that would enable them, as part of their aid policy, to subsidize the sale of technological property to the third world on an ad hoc basis. Such policies could focus on the technologies that are controlled by Western transnationals and that would contribute to strengthening the technological research and development capabilities of the recipient countries. Arrangements might also be made for the joint purchase of technology use rights for several firms in a number of developing countries. This could be organized within the framework of support for TCDC.

Technology is embodied in people. Industrialized countries could consider ways in which their specialists could more effectively contribute to the process of strengthening the technological capacities of the developing countries. In this respect, industrialized countries could establish programmes whereby voluntary technical advisers could assist developing countries in such areas as negotiations with Western transnational enterprises, the creation of industrial extension services, and the setting-up or strengthening of local design and engineering consulting services. Assistance could be organized on a government-to-government or firm-to-firm basis.

The exercise of this responsibility may call for the introduction of such measures as codes of conduct, incentives and sanctions. Support for the efforts being made by developing countries to develop their technological capabilities must also be reflected in support for measures designed to protect infant science and technology in the third world at national and regional levels, for automatic financing mechanisms aimed at promoting human resource development and technological innovation in the third world, and for intergovernmental programmes, especially those launched under the auspices of the United Nations, that are aimed at strengthening the technological capacities of the developing countries.

D. Role of UNIDO

In assisting developing countries in their quest for technological self-reliance, UNIDO could continue:

(a) To assist developing countries in a practical manner in developing and implementing a framework for national action towards self-reliance;

(b) To generate an extensive movement for creation of awareness, sensitization and mobilization of interest and effort;

(c) To develop human resources, thereby strengthening technological capabilities in the widest sense;

(d) To develop technology, both processes and equipment, suitable for use by developing countries;

(e) To promote technological co-operation among developing countries;

(f) To develop and promote the concept of technological self-reliance towards operational strategies.

A large number of developing countries have not yet taken effective and systematic measures for technological development. The urgent need at the national level is to create a framework for national action for technological development in the place of ad hoc and unco-ordinated efforts. Such a framework has been developed by UNIDO whose technical assistance and technological advisory services will be available to developing countries in elaborating that framework and in implementing several programmes arising therefrom.

The creation of awareness, sensitization of issues and mobilization of interest and efforts is of primary importance. The third meeting of the Consultative Group on Appropriate Industrial Technology felt that UNIDO could perform a major service to developing countries by generating a sustained process in this regard.[49] The International Forum on Appropriate Industrial Technology, including the preparations leading to it, was a major sensitizing factor at the international level. It resulted on the one hand in the evolution of a conceptual and policy framework for appropriate industrial technology, endorsed by the Ministerial Level Meeting; and on the other in a detailed expert examination of technology options and issues in a dozen industrial sectors. The conceptual, analytical and empirical basis for a major effort in appropriate industrial technology will have to be further built upon so as to create a movement for technological development between as large a number of developing countries as possible.

The process of sensitization should include the presentation of technological options available to developing countries in specific industrial sectors. The work of the UNIDO Industrial and Technological Information Bank (INTIB) will need to be enlarged and supported by adequate resources. Special attention should be paid to the collection and dissemination of information on technology available from developing countries and also to the possibilities of co-operation in small and medium-sized industry between developed and developing countries as well as between developing countries themselves. Since INTIB by its very nature has to draw upon a large number of sources of information and serve users in all developing countries, its work presents great potentialities for developing international co-operation and could well form the nucleus of a technology bank with wider objectives and functions.

A concept of human resource development that would act as a framework for strengthening the technological capabilities of developing countries should be elaborated through specific programmes of action. Within such a framework the training activities of UNIDO in the industrial and technological fields will have to be intensified. Training programmes for upgrading specialized skills should cater to a variety of personnel such as skilled workers, industrial engineers, production engineers and managers. The skills to be upgraded include those pertaining to the production process and also to other technological functions including information, technology acquisition, technology planning and policy co-ordination, innovation, R and D, and all aspects of policy-making in relation to technology, in other words, the whole spectrum of the process of development and transfer of technology. Special attention should be paid to the promotion of engineering and consultancy services, technological delivery systems, industrial extension services etc.

Efforts to develop new technologies should concentrate on certain areas that will make the maximum contribution to the achievement of technological self-reliance in critical sectors. For certain developing countries this will mean the upgrading of technologies in the processing of materials for export and in adding further value to them. A measure of contribution to industrial output could be obtained from the rural areas if technologies suitable to them are identified or developed and applied. This will also have the effect of industrial dispersal and the expansion of the market. The development of technologies should be based on the application of the results of modern science and technology. For this purpose technological advances should be continually monitored. Attention should also be paid to the identification and development of alternative technologies in the field of energy, in view of the close interrelationship between energy and industrialization and the energy constraints that several developing countries may face.

The role that UNIDO could play in the promotion of technological co-operation among developing countries can best be summarized by the recommendations of the Round Table Ministerial Meeting of Industrial and Technical Co-operation among Developing Countries, referred to earlier.

"1. An information system should be set up that would concentrate on the kinds of information that could broaden the possibilities of co-operation among the developing countries, i.e.

information on : (a) the availability of appropriate technologies; (b) the terms of license or collaboration agreements concluded by developing countries; and (c) the availability of skilled manpower and expertise in various countries.

"2. In consultation with Governments, UNIDO should explore the possibility of expanding and strengthening already establishing R and D institutions in developing countries to make them 'centres of excellence' in specific technical fields and prepare a detailed study of the subject.

"3. UNIDO should review possible constraints, both internal and external, that may affect the setting up of joint industrial projects and market-sharing arrangements.

"4. UNIDO should initiate studies to identify ways of co-operation in the following sectors of industry:

Chemicals
Engineering
Electronics
Energy
Fertilizers and agro-chemicals
Pharmaceuticals

"5. Under the auspices of UNIDO, concrete programmes through which the relatively more developed of the developing countries could assist the least developed countries should be formulated and implemented.

"6. UNIDO should convene round-table ministerial meetings periodically, to be held in developing countries in different regions in co-operation with the host country.

"7. UNIDO should outline projects of co-operation and submit them for consideration to the developing countries. A committee of experts should work out guidelines for collective action" (UNIDO/IOD.133).

The Declaration of the Istanbul Round-Table Ministerial Meeting on the Promotion of Industrial Co-operation among Developing Countries, held in October 1979, reiterated those tasks of UNIDO, including examination by UNIDO of the possibilities of developing INTIB as the nucleus of a "technology bank" (ID/WG.308/4, p. 21 (q)).

Finally, the technological self-reliance of developing countries can be achieved rapidly only through the promotion of major concepts on which action could be based. The first major concept in this connection is that of technological self-reliance itself. What has been presented in this report is a preliminary view of this concept that will have to be complemented by assessments, at the field level, of the actual experience of developing countries and by promotional measures for the adoption of operational strategies by individual developing countries. UNIDO will endeavour to carry forward this concept to the point of integrated and sustained action at the country level.

NOTES

[1]See Denis Goulet, The Uncertain Promise: Value Conflicts in Technology Transfer (New York, IDOC/North America, 1977), pp. 7-12.

[2]See Johan Galtunk, "Development, environment and technology: towards a technology for self-reliance" (TD/B/C.6/23), June 1978.

[3]Francisco R. Sagisti, "Knowledge is power", Maxingira, No. 2, 1979, p. 28.

[4]See Industry 2000, New Perspectives (ID/237), pp. 180-182.

[5]See "Technological cooperation between developing countries including exchange of information and experiences in technology and know-how arrangements" (ID/WG.272/1), p. 3.

[6]See "Towards a strategy of industrial growth and appropriate technology" (ID/WG.264/1), p. 4.

[7]See "technological cooperation between developing countries . . .", p. 3.

[8]See UNCTAD, Intra-firm Transactions and the Impact of Development, UNCTAD Seminar Programme, Report Series No. 2 (May 1978).

[9]The Impact of Multinational Corporations on Development and International Relations (United Nations publication, Sales No. E.74.II.X.S.), p. 162.

[10]See Sagisti, loc. cit., p. 28.

[11]M. A. Zevallos y Muniz, Analisis Estadistico de las Patentes en el Peru (Lima, Consejo Nacional de Investigacion, 1976).

[12]"The role of the patent system in the transfer of technolgy" (TD/B/AC.11/19/Rev.1), p. 64.

[13]Frances Stewart, Technology and Underdevelopment (London, Macmillan, 1977), p. 138.

[14]See Surendra J. Patel, "Plugging into the system", Development Forum, October 1978.

[15]"Separate development for science", Nature, vol. 277, May 1978.

[16]See Francisco Sagisti, "Towards endogenous science and technology for another development", Development Dialogue, No. 1, 1979, pp. 15-17.

[17]For a detailed discussion of the processes described, see Francisco Sagisti, Technology Planning and Self-Reliant Development: A Latin American View (New York, CBS International, 1979), chap. 10.

[18]See Ashok Parthasarathi, "India's efforts to build an autonomous capacity in science and technology for development", Development Dialogue, No. 1, 1979, pp. 58-59.

[19]See "Draft Report", Second Consultative Group on Appropriate Industrial Technology, Vienna, 26-30 June 1978 (ID/WG.279/12), p. 5 See also "Report of the Ministerial Level Meeting", International Forum on Appropriate Industrial Technology (ID/WG.282/123) and Conceptual and Policy Framework for Appropriate Industrial Technology, Monograph on Appropriate Industrial Technology No. 1 (ID/232/1).

[20]See A. Emmanuel, "The multinational corporations and inequality of development", International Social Science Journal, vol. 28, No. 4 (1976), pp. 754-772.

[21]UNIDO has initiated several efforts designed to overcome some of the existing gaps in the area of information. Such initiatives include the Industrial and Techological Information Bank (INTIB) and various publications in the "Development and Transfer of Technology" series.

[22]For example, UNIDO has established over 40 such centres in developing countries.

[23]See Guidelines for Project Evaluation (United Nations publication, Sales No. E.72.II.B.) and Guide to Practical Project Appraisal (United Nations publication, Sales No. E.78.II.B.3.).

[24]See UNIDO, Guidelines for Evaluation of Transfer of Technology Agreements, Development and Transfer of Technology Series No. 12 (ID/233).

[25]The countries include the Andean Group countries, Argentina, India, Malaysia, Mexico, the Philippines, Portugal and the Republic of Korea. The nature the extent of regulations vary from country to country.

[26]"The role and functions of technology regulatory agencies in technological development" (ID/WG.272/7), p. 11.

[27]Charles Edquist and Olle Edqvist, Social Carriers of Science and Technology for Development, Discussion Paper 123 (Lund University, Research Policy Program, Sweden, October 1978).

[28]See "Survey on the impact of foreign technology in selected countries and priority sectors" (ID/WG.275/4/Rev.1), May 1978.

[29]See, for example, ESCAP, Guidelines for Development of Industria-Technology in Asia and the Pacific (Bangkok, 1976), chap. IV and V. See also "Joint UNDP/UNIDO evaluation of industrial research and service institutes; Addendum 1" (ID/B/C.3/86/Add.1).

[30]See, for example, "Technology planning in developing countries"

(TD/238/Supp.1), May 1979.

[31]B. Williams, Technology Investment and Growth (London, Chapman and Hall, 1967), p. 149.

[32]See, for example, Samir Amin, "Self-reliance and the new international order", Monthly Review, July-August 1977; and Harry Magdoff, "The limits of international reform", Monthly Review, May 1978.

[33]See Junta de Acuerdo de Cartagena, Technology Policy and Economi-Development (Ottawa, IDRC, 1975), pp. 7-8.

[34]See, in this connection, "Strengthening the technological capabilities of developing countries: a framework for national action" (A/CONF.81/BP/UNIDO), pp. 19 and 20.

[35]The International Development Research Centre's project on science and technology policy instruments provides ample and interesting examples of "implicit" science and technology policies from several countries in Latin America, the Middle East, southern Europe and Asia. See Francisco Sagisti, Science and Technology for Development: Main Comparative Report of the Science and Technology Policy Instruments Projects (Ottawa, IDRC, 1979).

[36]See Hans Singer, Technologies for Basic Needs, (Geneva, ILO, 1977), p. 32.

[37]See report of the Second Consultative Group on Appropriate Industrial Technology and also reports of the International Forum on Appropriate Industrial Technology.

[38]See Cooperation for Accelerating Industrialization: Final Report by a Commonwealth Team of Industrial Specialists (London, 1978), pp. 30-31.

[39]The development value of such complexes is discussed in "The effectiveness of industrial estates in developing countries" (UNIDO/ICIS.32), May 1977.

[40]See Handbook on the Acquisition of Technology by Developing Countries (United Nations publication, Sales No. E.78.II.D.15), p. 41.

[41]See "Science and technology for development--indigenous competence building" (ID/WG.301/3), June 1979, pp. 12-17.

[42]Report of the United Nations Conference on Technical Co-operation among Developing Countries (United Nations publication, Sales No. 78.II.A.11), pp. 5 and 6.

[43]See "Technical cooperation between developing countries . . .

[44] See "Industrial and technical cooperation among developing countries (UNIDO/IOD.133), October 1977.

[45] J. M. Logsdon and Mary M. Allen, Science and Technology in the United Nations Conferences: A Report for the U.S. Office of Science and Technology (Washington, D.C., George Washington University, Graduate Programme in Science, Technology and Public Policy, January 1978).

[46] See A. S. Bhalla, "Small industry, technology transfer and labour absorption", Transfer of Technology for Small Industries (Paris, OECD, 1974), pp. 107-120.

[47] See Antony J. Doman, The Like-Minded Countries and the Industrial and Technological Transformation of the Third World (Rotterdam, Foundation Reshaping the International Order (RIO), 1979), pp. 76-78.

[48] See Jan Tinbergen, co-ordinator, Reshaping the International Order: A Report to the Club of Rome (New York, Dutton, 1976), chap. 14 and annex 6.

[49] "Draft report", Third Consultative Group on Appropriate Industrial Technology (ID/WG.309/6), September 1979.

Doing More With Less: The Future of Technology in the Third World

WILLIAM L. EILERS*

A new profession of futurists has emerged in our society. The more successful are invariably optimists, who proclaim that we live today in the best of all possible worlds. The pessimists fear that this is true.

The tumultuous events of late -- Iran, then Pakistan and Libya, plus the famines in Cambodia and Somalia -- have many of us asking just where we went wrong and what we can do about our position in the world.

What there is to say about the grave situation in developing countries is not new, but it bears repeating. A quarter of the world's population -- a billion people -- exist in abject poverty. This poverty is so desperate that most of us have difficulty comprehending it. In South Asia alone, the numbers who live in utter deprivation are twice the U.S. population. Children suffer the most. Half of all deaths in these countries are children under 5, with one out of four never reaching that age. More than half don't attend school.

The world population growth in the next 20 years alone will equal the growth in population from the birth of Christ to 1950. Ninety per cent of this growth will occur in the developing countries, adding more than 500 million to their labor force by the turn of the century. The burgeoning population of Third World cities will suffer even greater increases, fueling unemployment, crime, and environmental stress.

Compounding these problems are the immediate crises of increasing energy costs, rising external debt and the threat of industrial recession in the industrial countries. The slowdown we are seeing in our economy, with big steel and the car plants shutting down, means that exports from developing countries are going to contract sharply and will further restrict their ability

From **AGENDA**, 3, June 1980, (24-28), reprinted by permission of the publisher.

to import goods and repay foreign lands. The 1980s will see a disintegrating international economic order; economic growth is going to be extraordinarily difficult to achieve.

In its formative years, the United States had a lot going for it -- abundant natural resources, navigable rivers, access to two oceans, an energetic population, and a workforce with skills and a work ethic. Many developing nations have few resources. Fifteen of the 30 least developed are totally landlocked. Most lack trained people of any kind. Most are in the tropics, where they are plagued with diseases such as river blindness and snail fever, a recent upsurge of malaria, spreading deserts, tsetse flies, locust plagues, monsoons, earthquakes.

The specter of starvation already confronts Cambodia. One-third the population of East Timor is malnourished and has reached the point of being unable to recover. The world's largest refugee population, in Somalia, faces the spread of disease and starvation. India has suffered major crop failures this year. Zambia's disastrous corn crop may be only another harbinger of more serious and widespread famine we may see in the 80s.

Significant international steps are being taken to cope with global problems. Yet, despite these initiatives and vast increases in bilateral assistance in the food field, there are a lot more hungry people today than five years ago and the growth of the world food deficit has accelerated.

Margaret Mead remarked that the United States alone has the power to destroy the world, but not the power to save it. All the donor nations together, including the Communist bloc, cannot save the world without the full cooperation of the developing countries. When we are dealing with issues as diverse as the global refugee problem, exploration of the deep seabed, control of offshore coastal resources, control of nuclear proliferation, protection of the earth's atmosphere and ending international terrorism, we need the cooperation of all developing countries, most of whom share our desire for just solutions of these problems.

Traditionally, we have justified development assistance on humanitarian grounds or have argued that economic aid is vital to national security. But now there is a growing perception that the developing countries have become essential to the health and growth of the American economy. Our economic well-being depends to an ever-increasing extent upon our ties with the Third World, links that are growing faster than with any other group of nations.

Export promotion has become far more significant -- a major policy instrument to promote growth and strengthen the faltering dollar. We export more to the developing countries than to Western Europe and Japan combined. About 35% of our exports go to the developing states -- a third of our wheat, cotton and rice is sold to these countries.

These less developed nations are also the source of valuable imports. Last year the U.S. bought nearly $75 billion dollars, or 42%, of our total imports from them. We rely heavily on them for many commodities critical to industry -- 30% of all raw materials are derived from this source. By the year 2000 only 20% of our primary raw materials will come from domestic sources. Right now we depend on them almost exclusively for tropical products --

coffee, cocoa, nuts and spices, bananas and fibers.

It is estimated that almost half of the unexplored global deposits of gas and oil lie in non-OPEC developing countries. They offer potentially diversified sources of supply for all oil-importing countries.

Last year U.S. firms invested over $40 billion in developing countries, a quarter of our total foreign direct investment. And the rate of return on investment in these countries last year was 66% greater than the rate for investments in developed countries. They proved highly profitable, and last year generated 35% of total income from U.S. foreign investments. And during a time when we are facing massive unemployment from the shutdown of major industries and their suppliers, we should keep in mind that the jobs of 1.2 million American workers depend on exports to developing countries. If these exports continue to grow, developing nations can import more and more U.S. goods and services.

If the argument that it is in our national interest to help the Third World is persuasive, then a question arises: In what ways can this country share its scientific knowledge and technological know-how with these countries to mutual advantage? It is well known that one of the most strident and pressing demands from the developing countries is that the United States share its technology with them, help them to build their own capacity to adapt foreign technology to our needs, and generate their own appropriate technologies. But does modern technology possess the real potential they credit it with?

In any event it is inevitable that, skeptics notwithstanding, new technology will pervade our lives in the 80's. The greater potential of the new technology lies in relatively new fields where large-scale applications have yet to be made. The emphasis will be on doing more with less in a climate of dwindling resources. We all hope that the auto makers are right when they predict that minicomputers in cars may reduce gasoline consumption by at least 15%.

A recent Newsweek article indicated there will be few opportunities for quantum leaps in older industries such as steel or textiles, but for the most part new technologies will fall into five rapidly emerging fields -- microprocessers, fiber optics, superconductivity, space technology and recombinant DNA. As far as developing countries are concerned, a number of more promising fields of technology will come into their own during the next 10 years.

Let's look at microprocessors. Within five years we'll be able to put 256,000 bits of information on a single silicon chip at no increase in price, whereas just 10 years ago only 16 bits could be fitted onto one chip. By 1990 portable microcomputers will be almost as commonplace as handheld calculators are now. This means that many of the functions now performed by human beings in the manufacturing and services sector of industrial economies can be done faster and cheaper by microelectronic devices or by robots controlled by these solid state instruments. This means speed, cost reduction, flexibility, reliability, and capacity for self-diagnosis and correction. One forecast is that while the

French banking industry will expand measurably during the 1980s, employment will decline 30%. Since the occupations involving information now institute the largest category in the work force of most industrial countries, even moderate increases could have a devastating impact on employment. Think of the millions of lower level white collar workers in banks, small businesses and other enterprises in developing countries who will be displaced by the small computer. If you introduce microprocessors in these countries, there is bound to be a severe impact on employment.

The Science Policy Research Unit at the University of Sussex in England believes that job displacement will draw far ahead of job-generating investment in the 1980s. In Third World countries already confronted with high unemployment -- in Iran it is running 25% -- there are likely to be large-scale social crises when they are forced to deal with a rapidly growing work force, widespread poverty and a badly skewed income distribution.

Recombinant DNA holds great promise. We now produce insulin in this country with a new DNA process and hope we can adapt a similar process to produce interferon, a tremendously expensive and little studied substance that could offer a new approach to fighting cancer. Scientists are growing a special type of bacteria that has an enormous appetite to gobble up massive oil spills from supertankers. In Saudi Arabia Chinese scientists are working to manufacture an economical single cell protein that can be grown on a petroleum substrate for animal feed.

We will see impressive new advances in the food field. AID, for example, is investing $40 million to improve selected grain legumes, which make up the major source of protein for half the world's poor. They are important because they supply the high quality protein lacking in most cereal grains. Recent breakthroughs in the science of biological nitrogen fixation offer considerable promise in increasing the capability of certain legumes such as cowpeas and soybeans to obtain most of their nitrogen fertilizer requirement through bacteria living on their root nodules.

We are working on new crop introductions, such as winged bean, a tropical legume whose protein content is about identical to that of the soybean. The National Academy of Sciences has identified dozens of little known or underutilized tropical crops that it believes have a great untapped potential for developing countries, many of which make less demand on soil, water and fertilizer than temperate crops previously introduced in these regions.

A really significant research breakthrough has been achieved in the physiological techniques required to introduce spawning in milkfish. This is the most widely cultured marine fish in Southeast Asia. Simply put, you mix a cocktail of valium and an antibiotic to calm the large adult fish and reduce infections, and then add a hormone injection, and presto -- you rear massive quantities of fish fry from eggs instead of netting fry in the open sea for stocking fish ponds. This means vast new quantities of milkfish can now be reared in millions of hectares of unused mangrove swamps and brackish coastal lowlands, providing a new source of protein at very low cost. For several million rural coast dwellers, typically the most poverty-stricken in these countries, it means higher incomes and more plentiful protein.

At Alabama's Auburn University, fish breeders have produced a new hybrid tilapia that has a significant worldwide potential. Brazilian scientists recently discovered a herbivorous Amazon fish that will convert agricultural and food industry wastes to edible protein.

An international network of 10 malaria research institutes recently demonstrated that a malaria vaccine is biologically feasible, with 100% success in immunizing monkeys against human malaria. This means that within the next decade we can probably provide a vaccine for field trials that will most likely become widespread in the last decade of this country.

The critical global shortage of firewood is being attacked from many angles. The lack of firewood means that women who might otherwise be attending to essential chores must often walk for hours to gather dwindling supplies of wood to cook the evening meal and keep the family warm. Not only is firewood critically short, but due to overcutting there is massive destruction of tropical forests in the greenbelt around the equator. One partial remedy promoted by AID's experts is introduction of an extremely fast-growing tree from the Philippines called the leucaena. It can grow 65 feet and 16 inches in diameter within five years.

There are other answers to the firewood shortage. The simple lorena stove developed in Latin America and made of clay and sand uses firewood much more efficiently. Biogas digesters are supplying methane for cooking in India and many other countries. Solar reflectors are being increasingly adopted for cooking in many tropical areas.

What is the ultimate answer in providing village energy needs in developing countries? I. H. Usmani, the United Nations Development Program energy adviser, is convinced that it lies in photovoltaic arrays that use small silicon solar cells to convert sunlight directly for storage in local village batteries. Last year the cost per peak watt was around $7, but this is expected to drop to about 50 cents by 1986. At this level it can compete economically with oil and possibly other fossil fuels. Right now one village in Upper Volta is using the photovoltaic source to pump water and grind grain with practically no upkeep cost. There are problems of course -- the source of raw materials is limited and battery storage often presents a problem. If Usmani's prediction is true that solar cells are the answer for remote villages, within the decade we will have the promise of extending new communications into remote areas to power radio and TV, provide refrigeration for drugs and vaccines and, most important, for the first time offer millions upon millions of remote villagers electric lights to study and read by.

The use of very small hydroelectric power is another very promising possibility for villages. For example, $15,000 and a lot of labor permitted several villages in a remote valley of Papua New Guinea to install a mini hydro-electric plant. For the first time they not only have light for their schools and a regular water supply for the village, but for the first time in their lives each villager can have at least one daily hot shower.

Rolls Royce has just developed an inexpensive small turbine

that can generate electric power for rural areas using a low-grade alcohol as fuel. The corporation is actively promoting distilleries to convert cassava, sugar cane and agricultural wastes to alcohol in developing countries that are suffering most from escalating oil prices. And in Brazil, 2 million cars are now fueled with gasohol, local alcohol distilled from sugar cane.

A host of other intriguing technology is coming, all of which hold a promising potential for the 80s. Some are even far out. For example, the scientist who developed the concept for towing icebergs from Antarctica to supply water to Saudi Arabia has a new scheme. It involves a giant sausage-like steel skin about two-thirds of a mile long, which is pulled underwater by an ocean-going tugboat that is also semi-submerged and has a propeller over 100 feet in diameter. The scientist feels the first high priority application of this new approach would be in Chile. There, the giant Kennecot and Anaconda copper companies are desperately short of fresh water near the Atacama desert, one of the driest places in the world. He proposes to tow fresh water from the abundant supplies in the southern Chile rivers to not only help the copper companies, but to make the Atacama bloom.

I have barely touched upon some of the exciting new technologies awaiting us in the next 20 years. In seeking to use these technologies, developing countries face a bitter choice. they can be integrated into the present international system, which will only increase their technological dependence. Or, on the other hand, they may disengage themselves from that international system and be forced to forego benefits of worldwide scientific and technological advance.

Discussion at the Vienna Conference on Science and Technology pointed to the likelihood that they'll opt for limited integration, but will insist that it be within the terms of a new economic order for which they are pushing strongly. Negotiations are underway in the UN Conference on Trade and Development -- UNCTAD -- regarding codes of conduct governing technology transfer and the behavior of multinational corporations in sharing technology assets. Elsewhere negotiations continue on revising the Paris Convention governing international patents and licensing agreements. This debate on mandatory vs. voluntary codes comes down to the question of the extent to which industrial countries are prepared to subsidize technology flows competitive with their own industries. In our own country, corporations and unions are joined in arguing vehemently that export of technology means export of jobs, pointing to our experience in recent years with Korea, Hong Kong, Taiwan, Singapore and other new industrial nations.

In any event, we must do everything possible to enhance the negotiating capability of developing countries in their effort to acquire foreign technology. One practical way has been suggested by the National Association of Licensing Executives, which has asked AID to help them conduct training courses for developing country executives responsible for negotiating foreign technology license agreements. This is a positive step.

It is essential that we strengthen indigenous capabilities in these countries to decide among technological alternatives and to help them carry them out. Special measures must be adopted to ease the international flow of technology for the benefit of least developed countries.

Are Science and Technology Leading to a New Pattern of Development?

ANDRE DANZIN*

The author reviews some of the reasons for adopting a new model for national development departing from the conventional type based on gross domestic product.

I. INTRODUCTION

The nineteenth century came to an end with the idea of progress triumphant. Science and technology held out the promise of a rosy future for the world. Man, the captain of his fate, could choose his goals and plan the route to success provided that a solution was found to certain problems inherent in society; and those could be identified through clear-head analysis and solved through firm political resolve.

We are, today, far removed from that feeling of triumph, although technological progress has made far greater strides than even such a fertile mind as Jules Verne's could have imagined. Man has walked on the moon, and space probes are exploring Mars and Venus; hygiene and preventive medicine have doubled life expectancy; gas and electricity are distributed over long distances to hundreds of millions of homes; the same is true of sound and pictures and will be true, tomorrow, of written or printed information. Nuclear energy could move mountains and open up valleys; and we are discovering not only the secrets, but also the wealth of the ocean depths. All the same, more than ever before, the future is full of unknown quantities and man is beginning to turn his attention to the shortage of certain natural resources

From **IMPACT OF SCIENCE ON SOCIETY**, 29, July 1979, (118-129), reprinted by permission of the publisher, UNESCO, Paris.

which he had always thought of as being inexhaustible.

The end of the century will be dominated by this threat of shortage -- shortage of available capital for development investments, job shortages leading to all the physical and psychological disorders which attend unemployment, and the threat of food shortages, but above all there will be the shortage of energy. Energy is both the keystone and the basis of industrial civilization.

Energy can be turned into food through fertilizers and mechanized agriculture, into work because of the need to operate the machinery, and into rare raw materials because everything exists on earth as scattered elements. A great deal of energy nevertheless still has to be spent on processing low-grade ore and producing chemical changes.

We are emerging from a fair-weather period in which the abundance of fossil fuels and the progress of technology made everything easy. It explains the prodigious population explosion and the great increase in man's personal power. We are entering a more difficult period in which mankind will again be faced with many of the limitations that have beset it throughout the course of history, but which will now have to be confronted in entirely new terms. First of all, it will have to bear the burden of a vastly increased population. Secondly, it will have to safeguard everything recently discovered by technology that is necessary to economic survival and readjustment. A broad section of public opinion still trusts scientists to help man overcome the obstacle. Has science reached saturation point; is it at a loss, on the wrong track, or does it still contain a fund of ideas and resources that will enable us to adjust smoothly to the new circumstances we are up against? These are the questions which every scientist should, in conscience, try to answer.

II. CAN ONE DESCRIBE LIKELY TRENDS FOR THE FUTURE ?

We cannot see clearly into the future, but it can, in a sense, be 'probed' using the mathematical models that have been prepared on the basis of the factors that will be essential in our future development. The Interfutures project, led by Jacques Lesourne for the Organization for Economic Co-operation and Development (CECD), has made it possible to compare these models and assess their merits. All these attempts to represent our future development have one fundamental point in common: they are all designed to avoid major catastrophes. They do not, therefore, make any allowance for major famines due to climatic accident, or widespread epidemics caused by an unknown virus, or world conflict between the super-powers; but accidents or local conflicts are not excluded. Indeed, it is the major evils which must be avoided through a judicious policy of adaptation.

Although the combined conclusions of the predictions made by the various available models are unlikely to be truer to fact than

TABLE 1. Projected growth between 1975 and 2000.[1]

Region	Population around 1975	Total GDP around 1975	Projected growth	Population around 2000	Total GDP around 2000
Industrialized countries (Canada, Japan, USSR, United States, Europe)	1.07×10^9	$\$4.8 \times 10^{12}$ (i.e. an average of \$4,500 per capita per year)	GDP rising between 2.0 and 2.5 per cent per year	1.2×10^9	$\$8 \times 10^{12}$ (i.e. an average of \$6,700 per capita per year)
Take-off countries or areas (Brazil, China, Mexico, South-East Asia, OPEC countries, etc.)	1.2×10^9	$\$6 \times 10^{11}$ (i.e. an average of \$500 per capita per year)	GDP increasing by about 7 per cent per year (i.e. about 5 per cent per capita)	1.8×10^9	$\$3 \times 10^{12}$ (i.e. an average of \$1,700 per capita per year)
Rest of the world (India, Pakistan, Africa, etc.)	1.73×10^9	$\$5 \times 10^{11}$ (i.e. an average of \$290 per capita per year)	Increase in GDP slightly higher than population growth	3.0×10^9	$\$1 \times 10^{12}$ (i.e. an average of \$330 per capita per year)
TOTAL	4×10^9	$\$5{,}900 \times 10^{12}$	—	6×10^9	$\$12 \times 10^{12}$

1. This table is not a forecast. Its sole object is to make it possible to assess what must be done if certain growth targets, which can be taken today as being half-way between the desirable and the possible, are adhered to.
Source: population—according to United Nations statistics and predictions; GDP—for 1975, taken from a report by Maurice Guernier, Club of Rome.

any one of the proposed hypotheses, we may accept that there will not be any significant divergence from this aggregate conclusion. Referring particularly to scenario X proposed by Wassily Leontief in his book The Future of World Economy, and simplifying his findings a great deal for the sake of easier discussion, we can divide the world into three distinct areas: (a) the industrialized countries; (b) the geographical areas in which the preconditions -- natural wealth or the skill of their peoples -- exist for economic 'take-off'; and (c) areas which are intrinsically too poor to participate in growth.

This classification is, of course, a gross simplification of the facts. There are virtually underdeveloped areas in the industrialized countries and pockets of wealth in the poorer countries but, when seen in the context of the overall resources shortages with which the world may be confronted, these exceptions -- true though they may be of certain individuals or groups -- can be ignored.

The projection for the year 2000 described in Table I calls for the following comments:

As regards population, the predictions contain a very slight error due to inertia factors inherent in population phenomena.

Regarding the growth of GDP, we have accepted the premise that the whole of the development process is dependent on the momentum provided by the industrialized countries. We shall come back to this question later. It is, however, not conceivable at the present time that this momentum can be maintained if the annual rate of GDP growth is lower than 2 or 2.5 per cent. This is why we have settled for this hypothesis to begin with.

The growth of 'economic take-off' regions has been taken to be in the vicinity of 7 per cent per annum. The projection is low in relation to the wishes of the peoples involved, the average starting level and the end result since at the end of the century, the per capita purchasing power in these countries will be on average hardly more than a third of the purchasing power of an average inhabitant of an industrialized country in 1975.

The poorer peoples, who by then will have reached the 3,000 million mark, will be more or less as deprived as they are now, taken individually, even though it is estimated that the global GDP of the areas they live in will have doubled in twenty-five years.

What Must Be Done?

This projection, which departs from Leontief's scenario X, is unsatisfactory as regards the correction of inequalities, but nevertheless enables the world GDP to be doubled in twenty-five years. This means that between 1975 and 2000 mankind will have succeeded in opening up as many new mines, in building as many new houses, in establishing as many schools and hospitals, and in setting up as many new factories as there were in 1975.

As regards energy, Robert Lattes[1] has calculated that an average annual growth rate of 3.5 per cent, which would approximately double the GDP in twenty-five years, would entail an overall consumption of energy of approximately 250,000 million TEP,[2] provided that we succeed in conserving 20 per cent of the energy hitherto needed to achieve past performance regarding the volume of GDP. We know that on average, between 1950 and 1975, energy consumption rose roughly on a par with GDP growth. Robert Lattes's theory is therefore both ambitious and optimistic since he assumes that there will be a 20 per cent improvement resulting from a successful voluntary energy-saving policy.

The projected need for energy growth cited for the period 1975 to 2000 should be related to the volume of consumption over the period 1950-75 which amounted to 100,000 million. If economic growth were to continue at the same rate between 2000 and 2025 -- which, with inevitable demographic expansion, it will be forced to do -- TEP consumption will have to amount to at least 400,000 million during the first quarter of the next century, assuming that further progress will have been made in the GDP/energy ratio. Here we reach the limits of physical potential, which betoken probable shortages of which we are beginning to feel the effects.

In the middle term, still according to Robert Lattes, 250,000 million TEP is a figure not completely out of reach. If we succeed very rapidly in developing atomic energy and coal, if we develop certain areas of the African, Asian and South American continents where there is still a vast unexploited potential of hydraulic energy, if we are able to improve our use of natural gas and take advantage of the additional limited though not insignificant potential of solar and geothermal energy, we could keep the shortage within almost acceptable limits at a figure 10 to 15 per cent below the quantity of petroleum required. Minor crises might result, but not destructive ones. This entails a vast programme of research, development and investment.

The risk of similar impasses would have been brought to light if we had looked at the three other most likely areas of shortage, namely labour supply in order to contain unemployment within acceptable limits, the need for capital for industrial investments and public utilities, or again food resources. Man has not yet reached the limits of growth, but there are not many years to go -- possibly twenty or thirty -- before he will have to 'find a new form of growth'. This is what the Club of Rome meant when they declared that we were entering a phase in which 'a solidarity of survival' would be the only means of avoiding acute crises.

III. WHAT CAN SCIENTIFIC AND TECHNOLOGICAL RESEARCH DO?

Before examining in what way scientific and technical research can be used as the therapeutics of development, it should be pointed out that science and technology today are the prerogatives of the industrialized countries. Ninety-six per cent of researchers and engineers are concentrated there, and the rest of the world's scientific personnel living in the developing countries, often remarkably able people, are a marginal fringe who look chiefly to their colleagues in the more advanced countries. Mobilizing science and technology therefore amount to asking the affluent countries to make the means available whereby the poorer countries can develop. This raises a number of problems, especially as regards genuine freedom for the less privileged nations and the criterion of reciprocity that should be established between the south and the north. Once again, it seems that the north will propose its solutions to the south, unilaterally, as it were.

Before we go on to discuss the foregoing comments, we should ask ourselves how research could be used as an aid to development in the countries at the economic take-off stage, and in other countries. Here it is important to make a clear distinction between three modes of action between which there is a fairly sharp divide: (a) the accumulation of knowledge with a view to understanding and explaining; (b) the mobilization of available knowledge with a view to securing new aids to development; (c) seeking to change existing economic and social conditions through technological innovation.

Science in the service of learning

One of the biggest hindrances to the study of the solutions required in the present world situation is the refusal to face the facts, the refusal to understand and, once one has understood, the refusal to accept the consequences of what has been understood. It interferes with the affluent nations' false sense of comforts, whilst the poorer countries prefer ideological explanations and the vain hopes engendered by them. Moreover, society is not constructed to take into account long-term concerns -- what is going to happen in five years or more is of minor interest to a member of parliament mindful of a forthcoming election, or a head of a firm concerned with his end-of-year balance sheet. It nevertheless takes more than fifteen years to develop a new source of energy on a broad scale, to irrigate and fertilize an arid tract of land or to train the personnel needed for the industrialization of a traditionally agrarian land.

The knowledge amassed through scientific and technological research, which could be used to describe the present situation of mankind and the deep-lying trends in its development, its 'prospects for the future', is the only instrument we have to bring home to people the fact that it is becoming necessary to build a solidarity of survival.

The principal tools are available, and attempts at a systemic analysis leading to the construction of socio-economic models should not be lightly brushed aside. It must, of course, be borne in mind that these models are not perfect, that they reflect the subjective views of those who devise them and that their apparently magical power of prediction must be offset by a sound critical sense. All the same, they are an excellent medium for asking questions. Today, indeed, it is more important to know how to pose the relevant problems at the right time than to know how to solve them. The difficulty is not so much to know what to do when a clearly identified obstacle arises but to identify the obstacles and classify them in a suitable order of priority.

The inadequacies of these models have been pointed out often enough for there to be no point in reiterating them. It should be said, however, that our main shortcoming is our failure to express most socio-economic situations in terms of quantified data,[3] and our failure to represent and describe certain stages of discontinuity or change (crises, invention of a new technology, etc.). What is more, we are still very short of indicators of the quality of life.

Multidisciplinary research, at the interfaces of mathematics, informatics and the so-called social and human sciences, here has a large number of objectives, and the means to reach those objectives ought to be provided without delay.

To quote an example of remarkable achievements in this field, we must come back to the OECD's Interfutures project, the setting up of the FAST (Forecasting and Assessment for Science and Technology) team by the Commission of the European Communities, and last but not least, the Unesco study group on Research and Human Needs.

IV. SCIENTIFIC KNOWLEDGE TO SUPPORT DEVELOPMENT

A considerable fund of scientific and technical knowledge has been accumulated over the past hundred years. What ought to be done is to use this knowledge to satisfy needs that have been identified as fundamental for man today. Without in any way claiming that it is the chief priority, the need to find alternative sources of energy is obviously one of the most important issues. Solar energy is a typical example of an application for which a deliberate, sustained and systematic effort in the fields of applied research and development, to the virtual exclusion of any other, would benefit a broad section of mankind, especially people living in tropical areas, which are otherwise so critically underprivileged.

Many other similar examples could be quoted in agronomy, the food industries, the development of ocean resources, and in the field of compound materials. Until now, applied research has taken relatively little account of the specific needs of the developing countries. It is a handicap but also an asset since a relatively unexplored technological speciality is usually rich in potential.

When one thinks that solutions must be found within fifteen to twenty years, the urgency of the situation means that there is no hope of the developing countries being able of their own accord to initiate applied research geared to their specific needs. Only if the affluent countries commit themselves to this task, in a spirit of pull together or perish, will progress therefore be made. This presupposes that the industrialized countries will be able to set aside the necessary financial and human resources from their own surpluses. Implicit in such a decision, running counter to man's natural selfishness, is the meeting of at least two conditions, namely that there should be a sufficient reserve of wealth for sharing to be acceptable; and that a broad section of public opinion should learn to appreciate the imperative need for sacrifices compelled by the present world situation.

However, we should note in passing that all our attempts at a deliberate application of available scientific and technical knowledge are inspired by our concern to follow the same path to development as in the past.

V. WHERE IS SCIENCE LEADING US?

In the two approaches described above we advocated the use of scientific and technical research as the medium for a policy based on the rational and aiming to control the future. The scientific approach is conceived of successively as a learning mechanism and then, through applied research, as an instrument for solving problems to do with needs. Basic research plays practically no role at all in these considerations since its results are taken for granted. With basic research the irrational comes into play, and we must ask ourselves whether that is not the most significant phenomenon.

Basic research is research that is carried out for the joy of knowing, to have done with the anguish of not understanding, to increase the power of imparting knowledge through teaching. It is the expression of that fundamental drive in man who will always try to surpass himself, exceed his own limits, explore the unknown. Basic research has no direct economic purpose. Even if it wished to it would be unable to identify its goals, for either the applications are too far off or the way to practical innovation is too complex. Nevertheless, the findings of basic research generate and sustain change. This is why it must be regarded as one of the principal instruments of social change, which today has taken over from genetic mutations in building our environment.

We can in fact extrapolate what has been observed about the history of the biosphere and agree with Jacques Ruffie that the distinctive feature marking the development of mankind has been the constant growth of complexity and the broadening of psychological experience. As Hugues de Jouvenel suggests, this trend persists but is now man's doing. But research alone would not suffice to give mutations a significant level of action. It provides a favourable environment and opens the way for the pooling of knowledge that will give rise to new products or concepts. But invention can only become innovation on a broad scale if there is a form of promotion that we call development, the most perceptible phase of which in industrialized society is market trial and selection. If the product or the new process passes this test, there will then be widespread applications of it which may have a modifying influence on economic and social conditions.

Over the last forty years, the arms policy and the conquest of outer space have been the mainstay of this process of trial and selection. Enormous sums of money have been poured into these prestige or power projects. Paradoxically, we owe to these efforts, made with what appeared to be very precise ends in view, the spectacular technological advances that encourage communication between men and systems of interdependence, e.g. aeronautics, electronics, data processing and telecommunications. Thus, without any conscious or even unconscious design on anybody's part, technological changes have continued to be conducive to greater complexity and psychological experience. A network of message transmission has forged links of solidarity between the different continents, giving each of their inhabitants as it were the gift of ubiquity. Economic and social interaction has become increasingly complex and the broadening of psychological experience is no longer an individual but a group phenomenon.

What form will innovation take?

A quotation from Andre Malraux springs to mind: 'The third millennium will be that of the mind or it will not be.' For how can one fail to see that the success of the pattern of increased material consumption, which one might think the industrialized countries to be set on, has coincided with the virtually spontaneous emergence of the instruments for a society based on communication, a civilization of the mind in which social values could be radically changed? And man, having played at being the sorcerer's apprentice, has the good fortune to find himself faced with a surprising choice: either the imminent shortage of resources and (probably) the atomic bomb; or the construction of a system of interdependence, the main aims of which would be linked to the development of culture and the mind.

Of course, the emergence of the mind will not come about without setbacks or surprises, but the foreseeable trend in technological innovation for the coming thirty years will probably be a further development in communication media and a growth of interdependence. There seems to be no way of stopping research and development efforts for military purposes, but the main trends in this research are known. Apart from the quest for 'nuclear miniaturization', it will continue to be focused chiefly on propulsion, optics, electronics, artificial intelligence, software -- all of which can serve to further progress in communication technology. As regards space research, its applications are also conducive to the perfection of complex observation systems (meteorology, remote sensing) and data transmission, broadcasting and television.

Nevertheless, space and armaments will not be the only areas of innovation. Economic warfare will perhaps carry on for a long time with undisciplined violence between the industrialized countries, the East and the West, and the North and the South. There seems to be every likelihood that the more affluent nations will seek to compensate for their demographic inferiority by continuing to impose their technological superiority, which implies constant progression. Competition will be particularly acute in the field of the international division of labour. For the United States, Europe and Japan, it will be a case of offsetting, by increased productivity, the handicap of their labour costs in relation to the costs and conditions of labour in the economic take-off areas. For Europe and Japan, particularly, it will be a matter of holding on to a commodity which they can trade with the countries in possession of the raw materials they lack.

It is therefore to be expected that there will be a considerable development of microprocessors and sensors in the applied fields of robot technology (industry) and business technology (services), and that the applications of remote mailing, tele-conferencing, remote reprography systems and computer networks will be brought into general use, as will the civilian applications of space communication. This new stimulus to productivity will make it necessary to shorten working hours, at least for mass production. This will make it possible to fulfil the second prerequisite for a 'civilization of the mind', namely that there should be enough spare time for culture, self-improvement and 'do-it-yourself' hobbies.

A possible answer

The knowledge acquired in recent years through basic scientific research presents other facts that hold out promise for the future. Their main characteristic is that most of them depart from the field of physics and enter that of alternative technology, namely biology, information theory, and multidisciplinary interaction between data processing and the human and social sciences. The application of discoveries in molecular biology and basic genetics will no doubt lead to changes in industrial activities and to new products, and will have social consequences comparable in importance to those brought about by quantum physics, by wave mechanics and Einstein's theory of relativity with the birth of the electronics and nuclear industries.[4]

There is strong evidence that the human sciences like philosophy, psychology, sociology, history and many other disciplines, which were at the forefront for centuries, will make a spectacular comeback, since their development responds today to new needs in human society. They will be increasingly called upon thanks to the new instruments of communication technology.

I therefore think that an affirmative reply can be given to the following question: is science leading us to an unknown form of development? Yes, science and the use that men will make of science in their struggle to conquer all forms of power, are leading us to a new pattern of development in which achievements will no longer be measured in terms of increased GDP or the consumption of energy, but by the quantity of information exchanged by men and by the quality of their communication. There are no physical limits to this type of growth. This is why I feel it is a fallacy to say that we can make no further progress and that we are doomed to zero growth. But what it entails is a fundamental shift in the emphasis of our endeavours. Power will no longer be seen so much in terms of material might and property, but rather in controlling the processing and transmission of information. In this way, man will come to realize that he is primarily 'made for the things of the mind'; and if this were so, the earth could probably feed, clothe, house and educate 10,000 million human beings.

On no account am I predicting a golden age. There is every likelihood that the struggle for the new form of power -- information -- will be as embittered as in the past but at least it will not lead us inexorably to major physical catastrophe. If one were optimistically inclined, one might even dream of a society of well-being and love, a civilization of amenities.

VI. IS THE CHANGE POSSIBLE?

Summing up what we have just said, several forces would appear to emerge over and above man's deliberate efforts to control his future: (a) the advent of shortages, especially of fossil energy reserves and probably, also, of the available capital needed to ensure the steady growth of consumption; these are negative forces; (b) the action of two groups of positive forces, one resulting from the explosion of information technology, the other arising from the free time now available as a result of vastly increased productivity -- which was already true in agriculture and is now affecting manufacture and services.

It is as though brake and accelerator were working in conjunction to lead us on to a civilization of complexity and communication providing a new pattern of development for mankind which could probably be extended to the whole world, whereas it is doubtful that the GDP civilization is universally applicable.

Today, however, we are in the confused transition period and nobody can imagine what the next stage of (relative) stability will be like. In these pages, we have not made any suggestion as to the poorer countries' ability to find their own answer to the problems they are faced with. It is not a matter that we lightly ignore, but we are forced to acknowledge objectively that the developing countries behave as though they were fascinated by the apparent success of Western civilization, and attempt to reproduce at home all the means of achieving it, especially wide-scale industrialization which they consider to be a token of accession to power. No force will be able to prevent those who think they are behind from imitating those whom they regard as being ahead of them.

The following remarks from a report to the Club of Rome by Jean Saint Geours, Michel Courcier and Maurice Guernier[5] should, I think, be regarded as highly significant. In it they stress that the industrialized countries, determined to conduct a long-term policy of adaptation, will make a contribution to the emergence of a new order by taking themselves as the starting-point and by concentrating their action on themselves. If this is so, the transition can probably take place without doing irreparable damage.

But will the transition come about within the extremely short time that still remains in order for acute crises to be avoided? It is a matter of learning how to go about it, and it depends on the determination of decision-makers, provided that public opinion is prepared to make the obvious, and necessary, sacrifices. Here we may quote Arnold Toynbee, who said that the situation could be seen as a race between education and catastrophe. And in this connection, the crises which disrupt our economies or our political equilibrium should not necessarily be thought of as negative: it is the price we have to pay if we are to achieve a 'solidarity of survival'.

VII. CONCLUSION (AND SOME WISHES)

To conclude, there are three points I would like to make concerning scientific and technological research programmes, and would like to express the wish:

1. That governments and international organizations should give sufficient attention to: (a) the actual situation in the world and to describing it by applying conclusions conducive to system analysis and mathematical modelling; and (b) a critical analysis of the results yielded by the models, made by study groups consisting of qualified persons from a variety of backgrounds.

2. That the industrialized countries' research and development programmes should take into account the specific needs of countries at present undergoing industrialization, and that development personnel should be trained for the economic take-off countries as part of these programmes.

3. That adequate funds should be allocated providing study groups with scope for imagination in proposing new patterns for

society compatible with a world population of 10,000 million -- with particular emphasis on changing the intellectual, spiritual and moral values of the more affluent countries.

The small Unesco study group on Research and Human Needs might constitute a nucleus to lead and co-ordinate world-wide endeavours to solve these fundamental problems.

NOTES

[1]Member of the Club of Rome.

[2]1 TEP = one tonne equivalent petroleum.

[3]A recent symposium was held in Paris on behalf of Unesco, organized by the Institut de Recherche d'Informatique et d'Automatique (IRIA) and the Institut Europeen d'Etudes Superieures et de Recherches en Management, Brussels. Some forty international specialists took part in the symposium, the theme of which was 'Problems Raised by Mathematical Modelling of Social Phenomena'.

[4]cf. Einstein's 'Unfinished Revolution', in the business columns of The Economist, Vol. 270, No. 7071, 10 March 1979.

[5]'Les Pays Industriels et le Nouvel Ordre Economique Mondial' (report), June 1978.

Technology Transfer: What Do the Developing Countries Want?

Miguel S. Wionczek

The fact that "The Economist" of London published recently a two pages long article about the UNCTAD discussions of an international code of conduct for transfer of technology suggests that the importance of that issue for the future of North-South economic relations has been finally recognized in industrial countries of the North Atlantic area.[1] The recognition comes after about three years of unsuccessful pleas on the part of the developing countries to take up at intergovernmental level for regulatory purposes the matter of international technology trade, the only part of world commerce left out of the scope of multinational arrangements. The developing countries pleaded for such action for the twofold purpose of establishing some mutually acceptable guidelines for technology trade and of linking it with their developmental needs.

Between 1970 and the summer of 1975 proposals for the regulation of international technology trade, made by a large group of the developing countries at UNCTAD and elsewhere, were meeting strong opposition of major technology exporting countries on a number of grounds. The developing countries were being told on every occasion that technology being a non-defined and complicated object of international transactions, its trade did not lend itself to any international regulation; that technology being mostly private property could not be subject of international regulation, and, finally, that any attempt to regulate international technology trade would affect negatively technology flows to the developing countries because any regulation would scare technology sellers from entering into contracts with restrictively-minded buyers in small, uncertain and underdeveloped markets.

From **INTERECONOMICS**, No. 3, 1976, (76-79), reprinted by permission of the publisher.

I. PROPOSALS OF THE GROUP OF 77 AND COUNTER-PROPOSALS

The almost theological discussions about the feasibility and possibility of international regulation of technology trade gave place to a more practical and pragmatic debate only when the developing countries as a group presented to Western industrial countries and the socialist block in May 1975 detailed proposals of a code of conduct on international transfer of technology.[2] The draft outline was elaborated by experts of the so-called Group of 77 participating in the UNCTAD Intergovernmental Group of Experts on a Code of Conduct on Transfer of Technology that met in Geneva twice in the spring and the fall of last year. It took the form of a draft of the international convention that covers the following fields: objective and principles, scope of application, national regulation on transfer of technology transactions, guarantees, special treatment for developing countries, international collaboration and applicable law and settlement of disputes. The draft of the Group of 77 has not been invented by the experts from developing countries. It represents an improved and refined version of proposals elaborated in Geneva in May 1974 by a private group of 15 technology experts from Western, socialist and the less developed countries, convoked under the auspices of Pugwash Movement on Science and World Affairs, an informal scientific organization which counts among its members a score of Nobel Prize winners.

In answer to the draft of the Group of 77 whose main purpose was to prove that international regulation of technology trade is both possible and feasible, governmental experts from the Western industrial countries drafted last fall a counterproposal of similar length and coverage. Both proposals were submitted in early December 1975 to the first session of the UNCTAD Commission on Transfer of Technology. They are going to become subject of international negotiations at the UNCTAD IV, scheduled for May 1976 at Nairobi, Kenya. The potential importance of the forthcoming negotiations can be understood only if one takes note that the Seventh Special Session of the UN General Assembly, in which Messrs. Kissinger and Genscher played such an important role, agreed by concensus that international code of conduct on technology transfer should be negotiated at the Nairobi Conference and thereafter so that it could become reality before the end of 1977.

II. DISAGREEMENT DUE TO MISCONCEPTIONS

After the meeting of the UNCTAD Commission on Transfer of Technology it is only fair to state that the gap between the respective positions on the code of the developing countries and industrial countries continues to be very large particularly in respect to the legal nature of the code. The fundamental disagreement is whether the code should be merely a set of voluntary guidelines or should be made binding in an international agreement and ultimately national legislations, as the developing countries propose at this state. This disagreement should not obscure, however, the degree of progress achieved between May and December 1975 by the developing and the industrial countries in respect to the general content of the code. Nor should the persistence of disagreement make anyone forget that socialist countries decided to participate in the exercise by defining their own detailed position on the major issues covered by the two above mentioned proposals. There are reasons to believe that socialist countries who participate heavily in international technology trade as importers from the West and exporters to developing countries, may bring their draft proposals to the UNCTAD IV.

A number of preliminary comments on the draft code, proposed by the Group of 77, has been made in recent months by such important bodies as the International Chamber of Commerce and the Licencing Executive Society and by important economic journals published in industrial countries.[3] While some parts of the proposal seem to be agreeable and fit for formal negotiations, others are being rejected. Such mixed reaction should not surprise anyone. It reflects the nature of informal prenegotiations on any internationally important subject. The progress of the code could, however, accelerate if the interested parties in technology exporting countries had the opportunity to understand better what the developing countries really propose in that respect. Judging by the first Western commentaries misconceptions continue to abound.

III. AGREED STATEMENT ON THE CODE OF CONDUCT

For the purpose of creating better conditions for a businesslike dialogue, the authors of the draft outline who represent, among other countries, Argentina, Brazil, Mexico, Peru, Venezuela, Irak, Egypt, India, the Philippines, Algeria, Nigeria and Ghana, elaborated during the UNCTAD Commission on Transfer of Technology meeting the agreed statement on the code of conduct which may be summarized in the following terms:

The important role of technology in the social and economic development of all countries, particularly the developing countries, has been universally recognized. Accelerating the rate of economic growth is not simply a matter of capital formation but, among other factors, of selecting the appropriate technology.

The relative ease with which the accumulated stock of technological knowledge can be transmitted across borders has rendered technology transfer from one country to another more immediately attractive than indigenous technological development.

In addition, the technological dependence of developing countries has been increasing, since they do not possess adequate research, engineering, and organisational capabilities to assimilate and adapt the imported technology to their own needs. Developing these capabilities in itself is an important aspect of the transfer process.

The need to accelerate the transfer of technology to developing countries has been constantly emphasized at the United Nations and in other international organizations including the World Intellectual Property Organization (WIPO). However, concern has been growing about the increasing number of obstacles to the effective and economical transfer of appropriate technology, which adversely affect the technological capabilities of developing countries and often tend to perpetuate technological dependence.

In several developing countries today, transfer of technology transactions, whether by public or private enterprises, is being regulated by government authorities. This regulation aims primarily at ensuring that the terms are consistent with the objectives of national development, including the development of national technological capabilities, as well as strengthening the bargaining power of the recipient enterprises. The experience of developing countries which have such regulations provides evidence of the prevalence of restrictive business practices, abuses of industrial property rights, the weak bargaining position of developing countries' enterprises, the overwhelming burden of the direct and indirect costs of transfer of technology in the balance of payments of recipient countries, and the various techniques by which transfers of technology are institutionally tied together with other aspects of trade and investment thus rendering it difficult to isolate or identify the technology components.

However, national regulations vary from country to country, both as regards their scope and application. In addition, such regulatory action represents a one-sided burden falling entirely on the countries importing technology. There is need therefore to restructure and improve existing relations between suppliers and recipients of technology so as to facilitate access to appropriate technology under equitable terms. It has become clear that present imperfections in the market for technology require the formulation and adoption of international regulations. A Code of Conduct agreed to both by technology supplying and receiving countries could set minimum binding standards based on an equitable balance of the various economic interests involved, while taking into account the particular needs of the developing countries. It is within this broad framework that the formulation of an international Code of Conduct on Transfer of Technology should be viewed.

IV. FUNDAMENTAL POSTULATES

The Code of Conduct for Transfer of Technology as proposed by the Group of 77 is based on certain fundamental postulates. The most important is that all countries have the right of access to technology in order to improve the standard of living of their peoples. Transfer of technology can become an effective instrument for the elimination of poverty and economic inequality among countries and for the establishment of a more just international economic order. An unrestricted flow of information on the availability of alternative technologies and for the selection of appropriate technologies is necessary in order to build up the technological capabilities of developing countries.

A major feature of the Code of Conduct as envisaged by the developing countries is its universality. The Code is intended to be applicable to all countries and to all enterprises, whether supplying or receiving technology. The universality of the Code will lead to a more equitable relationship between suppliers and recipients of technology transfer transactions benefiting all countries since almost every country is an importer of technology. One of the important purposes of the Code is to establish an appropriate set of guarantees to suppliers and recipients of technology alike, taking fully into account the weaker position of recipient parties in developing countries.

Another major feature of the Code is its flexibility. The Code explicitly recognizes the right of all countries to frame their own laws and regulations in accordance with their policies, plans, and priorities. The Code is intended to supplement and strengthen the national regulations, not to supplant them.

The Code of Conduct proposed by the Group of 77 also provides that technology transfer arrangements shall be governed, with regard to their validity, performance, effect and interpretation, by the law of the countries utilizing the technology in their economies. These countries shall exercise legal jurisdiction over the settlement of disputes pertaining to technology transfer transactions, except where arbitration is permitted by national regulations and agreed to by the parties concerned.

Finally, another major feature of the Code is its legal character. The Code of Conduct is intended by the Group of 77 to be an international legally binding instrument, necessary to ensure that its provisions are fully and universally implemented in all countries to regulate transfer of technology.

V. NEED FOR A CODE OF CONDUCT UNIVERSALLY ACCEPTED

By now the need for a code of conduct has been accepted by all groups of countries represented at the UN as can be judged by the following quote from the consensus declaration adopted by the

Seventh Special Session of the UN General Assembly, held in New York last September:

"All countries should cooperate in the elaboration of an international code of conduct for technology transfer, corresponding in particular to special needs of developing countries. The work on this code should thus continue within UNCTAD and be concluded so that decisions, including the decision on the code's legal nature, can be taken at UNCTAD IV, with the objective of adopting a code of conduct before the end of 1977."

Moreover, comments forthcoming from private parties in industrial countries on the proposal submitted by the Group of 77 do not question anymore whether the code on transfer of technology is possible or feasible. A statement by International Chamber of Commerce not only accepts its feasibility but declares that "the conditions for cooperation (in respect to the elaboration of the code) are propitious and the work now being undertaken should be capable of being brought to a successful fruition, provided all parties approach the issues with realism and understanding of the others' problems."

Those fully cognizant of the full text of the draft outline, prepared by the Group of 77 in UNCTAD, can hardly deny that while defending their interests the developing countries approach the issues with considerable degree of realism. Their proposals do not ask technology owners for anything that might be considered as confiscatory, unfair or retroactive. First, they do not want nor expect to receive any proprietary technology free of charge; secondly, their quest for some preferential treatment is only secondary to their request for elimination from technology trade of restrictive business practices that are illegal in most technology exporting countries; thirdly, they do not consider that their draft of a code involves the issue of retroactivity although it opens the door for the possibility of renegotiating existing technology contracts. Moreover, the draft of the Group of 77 proposes guarantees for *both* sellers and buyers of technology.

The main unresolved issue is that of the legal character of the code. Those who object to a legally binding instrument argue that most technology is produced and traded by private owners. The large majority of other goods and services are, however, also owned and traded privately. If the above mentioned objection had general validity, then it would not be possible to have any international agreement in respect to commodity trade or on regulation of service transactions. The existence of a large number of international regulatory agencies and international commodity agreements strongly suggests that a *legally binding* code of conduct for transfer of technology falls within the limits of the practices of international law as currently applied.

NOTES

[1]Twisting whose arms?, in: The Economist, November 29, 1975, pp. 79-80.

[2]UNCTAD, Report of the Intergovernmental Group of Experts on a Code of Conduct on Transfer of Technology, Annex I and III, TD/B/C, 6/1, Geneva, May 16, 1975.

[3]International Chamber of Commerce, Draft Code of Conduct on Transfer of Technology: Comments on the Report of the UNCTAD Group of Experts, Doc. No. 225-1/68, Paris, Nov. 14, 1975; Technology Transfer -- A Self-evident Truth, Intereconomics, No. 11, 1975; The Economist, op. cit.; and Karl Wolfgang Menck, Technology Transfer -- Problems of a Code of Conduct, in: Intereconomics, No. 10, 1975.

The Pressing Need for Alternative Technology

ROBIN CLARKE*

Whether big or small, rich or poor, societies faced with the harsh consequences of industrialization need to return to solutions never tried before. There are five general responses usually evoked by dilemmas born of scientific and technical progress and its impact on the human condition. Only one of these responses meets most fully and rationally the current social demand for succour. Alternative techniques are described which can help reverse the process of man's growing technico-economic frustration.

'Technology--Opium of the Intellectuals' was the title of a famous article in the New York Review of Books a few years ago. In it, the author argued that we in the industrialized nations had become enslaved and addicted to technology which, by providing material comforts, covered up the deeper and more important social, psychological and political shortcomings of present forms of society. This view of technology, while by no means a majority one in any part of the world, has recently grown in importance, particularly in the industrialized world and especially among the young. It has led to a view that it might in the future be a good idea to do away with technology altogether and return to forms of society in which human and social issues once again become the main concern.

To some extent, I believe this critique of technology to be justified. It seems almost wholly so in those cases where an improved technology is urged on people to cover up more fundamental problems, such as a lack of social justice. Thus the argument that new technology will promote economic growth so that a country's gross national product (GNP) becomes larger and everyone's slice of the economic cake will get bigger is often used as an excuse for not cutting that cake in a more equitable manner. At this level technology can indeed be used as a hard drug which promises nirvana

From **IMPACT OF SCIENCE ON SOCIETY**, Vol XXIII, No. 4, 1973, (257-271), reprinted by permission of the publisher, UNESCO, Paris.

but only at a huge and hidden social cost.

I shall deal mainly with a different but related problem. The view just outlined implicitly assumes that there is only one form of technology, and that that form is the existing type of technology we see today widely used in the developed countries and increasingly applied in the developing ones. This idea creates much confusion, for the shortcomings of contemporary technology then become the evils of all technology--and hence the rise of anti-technological schools of thought in the industrial civilizations.

The argument which I wish to advance here is that it is the form of contemporary technology which is primarily at fault, and not the existence of technology itself. I shall therefore first examine the nature of the technology we use today. Second, I will suggest alternative forms of technology which could be used or invented to replace current technology. And finally I will look briefly at the future relationships likely to evolve between alternative technology and developing and developed countries throughout the world.

I. THE NATURE OF CONTEMPORARY TECHNOLOGY

In the developed world, contemporary technology is almost universally regarded as polluting. Though this is by no means the most serious of the criticism which can be levelled at today's technology, we will deal with it first because it is by far the most common. And, of course, it is unquestionably correct. The technology we use is polluting in many different ways: factories discharge effluents, sometimes noxious and always offensive, into rivers, the sea and the atmosphere.

In several parts of the world the eating of shell-fish has become dangerous due to the high levels of heavy metal residue found in them. Nuclear devices, both military and peaceful, liberate unwanted and potentially harmful amounts of radiation into both water and air. Particulate matter accumulates in the atmosphere leading to smog. The air is so heavily dirt-infested in industrial areas that household cleaning becomes a twice-a-day routine. Dangerous chemicals accumulate in foodstuffs, giving them peculiar tastes and other undesirable properties. The discharge of waste heat from factories and power plants heats river and lake water to such a degree that eutrophication and subsequent death of aquatic life becomes a familiar problem. Agricultural soil is treated as though it were some kind of chemical blotting paper whose only function is to provide domestic plants with sufficient nitrogen, phosphorus and potassium. The soil structure deteriorates mechanically, and the highly complicated ecology of important soil organisms is irreversibly upset. According to one calculation, the United States has lost, since the time the prairies were first put under the plough, one-quarter of the topsoil available.

Such a list of the polluting effects of contemporary technology could be, and indeed has been many times in the past few years, greatly extended. To this problem there are now a number of standard responses. The first can be described as the 'price response': pollution, this riposte runs, is the price we pay for an advanced technology, and it is well worth the price; true, we have a pollution problem (though it is greatly exaggerated), but it is of minor importance in comparison to the real benefits technology produces. The price response is heard most often in the developed world but it is also found in developing countries in a slightly differing form: bring us your polluting factories and we will learn to live with the pollution that results, for it is a small price to pay for a means of escape from the grinding poverty in which we live.

The second rejoinder, and this is the one most widely found in scientific and technical circles, is the 'fix-it' response. Advocates of this position accept the seriousness of the pollution problem, or of much of it, and claim that serious and concerted action must be taken to restore the environment. This action, however, will involve more technology, not less, and the clever use of sophisticated devices to monitor and then lower pollution levels, if this is found necessary. Into this category of declamation fit advertisements for electricity boards urging users to take to 'clean fuel' and substantial international programmes, such as Unesco's own Man and Biosphere. The 'fix-it' response is primarily scientific and technical, and sometimes technocratic.

The next two possible responses are more radical. The first of them--the away-with-it response--has already been discussed. The argument used here is that the price we pay for advanced technology is far too heavy, and that we have to learn to live either without technology at all or at least with a great deal less than is now the case. This response is almost solely confined to the developed countries, and is remarkable in its absence in the developing countries where there may be a very minimum of technology in practice. Generally, it seems, people who are forced to live without technology quickly become unhappy with their situation when they see others benefiting from it.

II. THE ALTERNATIVE POSSIBILITY

Fourth, there is the 'alternative response'. In essence, this claims that the form of technology now in use is intrinsically polluting, and no amount of extra technical effort will ever change that situation. This response claims, however, that not <u>all</u> technologies are intrinsically polluting and that new forms of technology can and should be devised to remedy a deteriorating situation. Thus instead of burning fossil or nuclear fuels, with their particulate and thermal pollution, we should develop technologies such as the use of solar and wind power which are intrinsically non-polluting. The alternative response, with which we will mostly be concerned here, needs careful distinction from the 'fix-it' answer which sees nothing fundamentally wrong with the form of technology in current use. The alternative response sees current technology as fundamentally flawed and advocates radical alternatives. The alternative response is becoming increasingly common in the developed countries but is also found (though less commonly) in the developing countries.

These retorts to the most common criticism of contemporary technology--the pollution it produces--are all based to some extent on technical evaluations. There is, however, a fifth response which is not technical but political, and radical. It suggests that pollution is an invention of capitalist elites to disguise from the people their real political plight and the facts of their exploitation by profiteers. Pollution, it is argued, is not important except in the sense that it is a product and symptom of capitalist society, whether that society be the victim of either private capitalism or what is known as State capitalism.

Each of these five rejoinders has powerful advocates and, as we shall see, the choice between them is made usually on ethical and emotional grounds, rather than on logical ones. Indeed, it may be impossible to characterize any one as more logical than the other, or even as simply 'better'. It is largely a question of taste and philosophy, not subject to scientific analysis, and this makes the situation complex and difficult. I should stress, however, that each position demands serious consideration and the attempt to characterize them all in a pithy way is not meant to imply criticism of any one of them. Such a characterization is useful for the five responses are used not only to answer the critique of pollution by contemporary technology but the other criticisms which are now widely voiced. It is to those that we now turn.

Probably the most important feature and criticism of contemporary technology is economic. The type of technology we use in developed countries is extremely capital-intensive, so much so that it tends to become the prerogative of those countries which are richest, and of those groups within the countries which are the richest. What this means is vividly illustrated by a single statistic. In a labour-intensive economy, it takes perhaps the equivalent of six months' salary to buy the equipment needed to provide work for one man. In a capital-intensive, advanced-technology economy, the equivalent figure is 350 months' salary. It is thus easy to see why development using Western technology has been such a slow process.

However large figures for international aid from the rich to the poor countries may be, providing jobs in the developing world by using advanced technology is a very, very expensive business. At the same time, that very same technology is not designed to provide jobs as such; instead, very often it is designed to eliminate jobs, to replace them by automatic processes. It has been said, and with some justification, that our technologies are designed to eliminate the need for people and to maximize the need for capital. It should be noted that this is not a political criticism as such, for the economic problem is no less painful for non-capitalist countries. It is simply that the type of technology we use places great emphasis on the economy of large-scale operations and is often poorly adapted to decentralized, local situations. In this sense, contemporary technology is as badly suited to accelerating development as any that can be imagined.

III. RESOURCES ARE UNEVENLY SHARED

I have tried to summarize how this criticism is subject to the five responses, discussed above, in Table 1. For instance, the radical political response to this situation is that if resources were equally split both between and within countries, current forms of technology would be equally accessible to all. While this is undeniably true, it is a fact that neither social nor natural resources are evenly split in this way now; and that even if the Herculean task of international legislation improves the accessibility to resources, legislation will not affect the distribution of natural resources within national territories.

The third criticism most commonly made of contemporary technology concerns its use of natural resources. Essentially, our technology is in the sense of the industrialized world an exploitive one, wrenching from the earth mineral resources which have taken billions of years to accumulate and using them up within a few centuries. The arguments about how long our resources will last if used in this way are well known, of course, and can continue interminably. But it is obvious that we have a technology that uses resources such as metal and fossil fuel faster than they are created by natural processes. For this reason, there will come a time when scarcity becomes a serious problem.

In this context, as any competent economist will point out, the question of 'limits' to growth or consumption is not of central concern. What happens is that as a resource becomes scarcer, poorer quality reserves have to be used increasingly and their sources become ever more difficult to get at. Long before any resource runs out, then, an economic crisis is precipitated when the cost of obtaining a resource begins to equal the utility of getting it. If we were to continue burning fossil fuel for a few more centuries (at most), we would probably end up spending more energy obtaining the resource than is liberated by burning it. It should be noted that we have long since passed this energy break-even point in the field of agricultural products. In the developed countries far more calories are used in obtaining a food than are liberated by eating it. This has led the ecologist Howard T. Odum to claim that the potatoes we eat are "made partly from oil', referring to the petroleum products consumed by farm machinery. In a primitive agricultural tribe, by contrast, every calorie of energy used in farming produces the equivalent of about fifteen calories of food.

The fourth criticism made of technology today is that it is capable of wide-spread misuse. The technology of nuclear power, for example, is difficult to distinguish from the technology of nuclear warfare; the latest medical advances are apt to find themselves applied in centres developing biological weapons before they are in hospitals; and in the capitalist countries the pace and type of technical advances are very closely geared to the profit motive. The existence of this flaw in modern technology gave rise a few years ago to the whole 'social responsibility' movement in science in which it was argued that scientists are themselves responsible for the uses to which their work is put. Again, there is much argument over exactly what constitutes a misuse of science or technology and what a proper use. But clearly, just as modern technology has made contemporary man more secure from the whims and misfortunes of the environment in which he lives, so too has technology added a new and threatening dimension to life by making possible the annihilation of the human race.

IV. TECHNOLOGY AND SOCIAL VALUES

Many more criticisms can be made of technology today but, unlike the previous four, these are more social in nature and closely related to each other. Globally, the most important may be the destructive effect of our form of technology on local, developing-world cultures. Built into a technology one can always find the values and ideals of the society that invented it. So when we use contemporary technology in development programmes, we export a whole system of values which includes a certain attitude to nature, to society, to work and to efficiency. As yet no developing, local society has been able to withstand the effects of this onslaught, with the result that such a society always changes to meet the incessant demands of the new technology. The end of this process is a global uniformity of cultures, all perfectly adapted to high technology but everywhere the same.

Similarly, modern technology is highly complicated and requires a trained specialist elite to operate it. As a result, ordinary men and women are deprived of the ability they previously had to control their own environment. There exist opinions as to how unfortunate this is, but we should stress here the fact that it is so; and in any systematic account of the flaws and virtues of contemporary technology the fact must be recorded. Equally, the technology used today is based mainly on the virtues of highly centralized services. To be sure, centralization has many advantages but we should not ignore the disadvantages it brings with it. Technical innovation becomes very expensive, people become totally dependent on the existing system; the system itself, through centralization, becomes highly liable to both technical accidents and the activities of saboteurs. The last have only to remove a weak link in the chain to cause chaos over many interlinked systems covering hundreds or thousands of square kilometres. Centralization also precludes the use of diffuse energy sources, such as solar and wind power, which by their nature are extremely difficult to centralize.

I will make two further points in criticism of contemporary technology. The first is that technical knowledge today has become a separate part of all knowledge. By this, I mean that technical knowledge does not develop naturally out of local technologies but forms a distinct body of knowledge on its own, with almost no links with what preceded it. For this reason, the idea of craft activity--which of course involves its own technology--has become pitted against the demands of new technology. The choice that confronts us almost daily is whether a product can still be something made with skill by craftsmen in limited quantity, or whether that product must be mass produced in the latest way, by someone requiring a quick training programme only, in large quantities. This disadvantage of modern technology must be held responsible for the widespread alienation of workers in industrial society who are thus reduced to cogs in a machine and condemned to the performance of meaningless and repetitive manipulations as a means of earning their living.

To summarize, the principle criticisms of modern technology are thus: high pollution rate; high capital cost; exploitive use of natural resources; capacity for misuse; incompatibility with local cultures; dependence on a technical specialist elite; tendency to centralize; divorce from traditional forms of knowledge; and alienating effect on workers (see Table 1).

As the table shows, to all these points there are in essence five different types of response. And as I have already hinted, it seems very doubtful that there is any rational or logical way of characterizing any rational or logical way of characterizing any one of these responses as being 'better' than another. To do so means answering questions such as: 'What kind of world do we want to live in?' 'How highly do we value an equal technical chance for men all over the globe?' And 'Can men ever really get satisfaction from the activity we call work?' Each of us has his or her own answers to these questions and, consciously or unconsciously, personal views dictate the kind of response we choose to make to these technical dilemmas.

As this issue of *Impact* is mainly about the 'alternative response', obviously I shall evaluate only this particular solution. In the circumstances, however, this is probably justifiable because it is much the newest of the possible responses. Certainly, until the 1960s such an alternative had not been given any serious thought. Probably only now are we in a position to begin to outline some ideas for an alternative technology.

V. ALTERNATIVE FORMS OF TECHNOLOGY

To take the above criticisms seriously is to say that an alternative technology should be non-polluting, cheap and labour-intensive, non-exploitive of natural resources, incapable of

TABLE 1. Technical dilemmas and some social responses

Technical dilemma	Price response	'Fix-it' response
1. Pollution	Pollution inevitable and worth the benefit it brings	Solve pollution with pollution technology
2. Capital dependence	Technology will always cost money	Provide the capital; make technology cheaper
3. Exploitation of resources	Nothing lasts for ever	Use resources more cleverly
4. Liability to misuse	Inevitable, and worth it	Legislate against misuse
5. Incompatible with local cultures	Material advance is worth more than tradition	Make careful socio-logical studies before applying technology
6. Requires specialist technical élite	Undertake technical-training schemes	Improve scientific technical education at all levels
7. Dependent on centralization	So what?	No problem, given good management
8. Divorce from tradition	This is why technology is so powerful	Integrate tradition and technical know-how
9. Alienation	Workers are better fed and paid; what matters alienation?	More automation needed

'Away-with-it' response	Alternative response	Radical political response
Inevitable result of technology; use less technology	Invent non-polluting technologies	Pollution is a symptom of capitalism, not of poor technology
Costs of technology are always greater than its benefits; use less	Invent labour-intensive technologies	Capital is a problem only in capitalist society
Use natural not exploitable resources	Invent technologies that use only renewable resources	Wrong problem: exploitation of man by man is the real issue
Misuse so common and so dangerous, better not to use technology at all	Invent technologies that cannot be misused	Misuse is a socio-political problem, not a technical one
Local cultures better off without technology	Design new technologies which are compatible	Local culture will be disrupted by revolutionary change in any case
People should live without what they do not understand	Invent and use technologies that are understandable and controllable by all	Provide equal chance for everyone to become a technical specialist
Decentralize by rejecting technology	Concentrate on decentralized technologies	Centralization an advantage in just social systems
Tradition matters more than technical gadgets	Evolve technologies from existing ones	Traditions stand in the way of true progress
Avoid alienation by avoiding technology	Decentralize; retain mass production only in exceptional cases	Alienation has social, not technical, causes

being misused, compatible with local cultures, understandable by all, functional in a non-centralist context, richly connected with existing forms of knowledge and non-alienating. But immediately one is struck by the fact that the technology of, say, a primitive agricultural tribe in New Guinea or a hunter-gatherer society in the Mato Grosso of Brazil would probably fulfil all these boundary constraints. Yet this is not what we mean by an alternative technology. Primitive technology certainly has some links with alternative technology but is generally held to be a long way from it. Indeed, the evolution seen is that at some time in the past primitive technology led to industrialized technology, and that at some time in the future industrialized technology will lead to alternative technology.

The alternative, in other words, does not seek to jettison the scientific knowledge acquired over the past three centuries but instead to put it to use in a novel way. Space heating, in the primitive context, was achieved by an open wood fire. In the alternative context, it might still be achieved by burning timber--provided the over-all rate of use was lower than the rate of natural timber growth in the area concerned--but in a cheap and well-designed stove which optimizes useful heat output against the need for fuel. Or it might be provided by a cheap solar heating system, a small electrical generating windmill or simply by first-call insulation. This difference between primitive and alternative technology is important for it has in the past led to charges that the alternative is retrogressive, essentially primitive and ignores the utility of modern scientific knowledge. This is not the case.

The most compelling case for alternative technology can probably be made in the field of energy. In the developed world there is much controversy over the future of energy supplies. As our remaining fossil fuels are burnt up, a desperate struggle goes on to make nuclear energy both competitive in price and safe. Neither is easy. Even the future of enriched uranium looks far from being a long-term affair. Breeder reactors are generally held to be a neat solution to this problem, although the technical problems they pose are still far from solution.

There is the added danger that as such reactors breed plutonium, if they were to become widespread over the earth's surface, the possibility of plutonium falling into the 'wrong' hands is very real. Plutonium is not only a very toxic substance in its own right but it can, of course, be used in an atomic bomb. Estimates of the number of nuclear weapons that could be made--without the need for uranium enrichment plants--from the plutonium that will accumulate over the next two decades from nuclear fission are truly staggering. Add to this the problems of disposing of radioactive materials which are the by-products of the fission reaction (a problem still not solved, although the nuclear age is more than twenty years old) and those of preventing sabotage and accident in nuclear-power stations, and it is then clear that the path we follow is fraught with danger. The prospect that all these problems will be resolved by the development of safe, controlled nuclear-fusion reactors is still too distant to be realistic.

VI. THE FLAW OF THERMAL POLLUTION

In any case, all these energy technologies suffer from one fundamental flaw. Because they use up stored energy, they produce large quantities of thermal pollution. There is a real chance that if we continue to use such sources, and our energy demand mounts over the next 100 years as fast as it has in the past 100 years, we will heat up the earth to a point where noticeable and unwanted long-term changes in climate will ensue.

Is there any alternative? The alternative technology recipe for solving world-energy problems runs something like this. First, the developed countries must accept that there is a ceiling to the amount of energy they can use, and they must become more concerned with saving energy than with supplying it. Second, an intensive effort to make use of all those energy sources which are supplied to the earth in real time must be made: these include hydroelectric schemes, geothermal energy, tidal power, solar and wind energy, and timber as fuel. The first three of these are limited to particular regions but this is no reason why they should not be used to the fullest extent. Solar and wind energy are found more universally and, if coupled to the energy which could be obtained by burning timber, they form an interesting distribution pattern over the earth's surface. In almost any habitable place, energy is or could be available from the use of the sun or the wind or timber. In places where there is little sun, wind and wood are often common. And where timber and wind are rare, there is usually plenty of sun.

In the developed world, these sources have been largely neglected because no single one of them is capable of supplying all energy needs. In northern latitudes, for instance, it is difficult or impossible to heat a house sufficiently well with solar energy. But as experiments have recently shown, houses in northern France can be designed to gain two-thirds of their heat from a very simple and cheap installation known as a solar wall. If the remainder could be provided with a little wind power and timber burning, the problem is essentially solved at the level of the household. There is a very real chance that if we accepted multiple solutions to our energy problems we could solve them by what have been called biotechnic means: using energy sources at roughly the same rate as they are naturally generated on the earth, hence creating no problems of thermal pollution whatsoever.

The disposal of sewage is another area where the need for an alternative is compelling. The problems of the current system are classic: expensive sewage installations are needed, together with large volumes of scarce and purified water, to sweep our sewage into processing units which discharge into rivers and seas a rich effluent causing severe pollution problems. As sewage contains important quantities of organic materials, the land is consequently always in deficit, particularly where animal excreta are not returned to it. (In modern intensive factory farming, this is becoming more and more of a problem.) So sewage disposal causes water wastage, agricultural depletion and severe pollution.

VII. A SOLUTION WE HAVE MADE INTO A PROBLEM

In any rational scheme, we would have found ways of returning our sewage to the land where it belongs. To reduce expense, we would do this not with a centralized scheme but at the family or community level. And we would use our precious supply of purified water for more suitable purposes and tasks. In fact, all this is technically quite easy to achieve. In Scandinavia there is a device on the market which will compost family sewage and turn it over a period of about one year into a small quantity of extremely rich but sterile and odourless fertilizer which can be applied directly to the garden. The device uses no water and can digest kitchen scraps. Why this solution is not more common in the developed world is hard to understand.

It is nothing short of tragic, furthermore, to see developing countries investing huge amounts of hard earned foreign exchange into expensive sewage disposal schemes when this, altogether much more efficacious, solution is at hand. The irony of the situation is compounded when we realize that in some of the drier developing countries there simply will never be sufficient water available to provide a 'Western-type' sewage disposal scheme for everyone. In today's society sewage has become a problem: it should be, and could again become as indeed it once was, a solution.

How do these examples of alternatives in energy and sewage disposal measure up to the nine boundary constraints listed in Table 1? Clearly, they do well in terms of pollution (No.1), capital cost (2) and use of resources (3). Equally, they are essentially decentralized techniques (7) and their principles would be easily understood and controlled by anyone (6). Further, partly because they are decentralized, they would be difficult to misuse (4); indeed, a general principle for this constraint is that technical systems designed to operate optimally on the small or medium-small scale are usually difficult to misuse wherever that misuse involves a scaling up (as it usually does). Put another way, it is not easy to envisage what a solar bomb or a wind-powered missile would be like.

Certainly wind-power and composting are old and traditional technologies (8), found in many parts of the world. Neither has been much touched by scientific progress; it is not optimistic to assume that if our new knowledge were applied to either, we would find surely that traditional use had already discovered, perhaps intuitively, many of the important functional principles but that significant and perhaps radical improvements could now be made. The gearing and control mechanisms on a windmill, for example, can be much improved over what was possible in Holland three centuries ago. On this ground also, their development would be compatible with local culture in many areas of the world (5).

About alienation (9), little can be said, for alienation is usually produced primarily under conditions of mass production. True, there might be some mitigating effect through the introduction of technical substitutes, but it would not be a strong one. In general, an alternative technology can be designed to meet the nine boundary constraints listed, but in practice a substitute will always meet some conditions better than others. In a real world, this need not surprise us, nor need it be taken as proof of the impracticality of the idea. The important point is that by listing a series of goals for technology to meet, technology is lifted out of the moral vacuum in which it has existed for so long. It can thus, once again, become a moral activity, and like all human activities will probably always fall short of moral perfection in one of another respect.

VIII. NOVEL DESIGNS FOR DWELLINGS

There is not sufficient space to detail all the other possible alternatives to modern technology. Today a great deal of interest in construction is leading to some novel and satisfactory designs for dwellings made from cheap local materials, realized to a high degree of insulation, and with almost complete independence from external services. Designs have been made for dwellings which provide their own energy, process their own proper wastes and trap and purify their water supply. These designs usually fulfil all nine boundary conditions, although their weak points still tend to be that they are too complicated and costly to count yet as perfect examples of alternative technology. But real progress has been made.

Similar advances are now being tested in the field of food production. For example, one small-scale system in the United States produces high-quality fish protein at a truly enormous equivalent yield in relation to surface used, without relying on external sources other than the sun and human excrement. The fertile overflow from a domestic septic tank is led to a small pond over which a timber and glass structure has been built to capture the sun's energy. In the pond are grown insect larvae in great quantities, feeding on the rich nutrient in the pond and thriving in the hot, humid conditions. Once a week these larvae are removed and fed to *Tilapia* fish in another small pond contained in a plastic geodesic dome which acts as a hot house, heating the water in the pond to the 25-30 degrees C in which *Tilapia* thrive. In a single summer the fish grow to edible size, and the water is then used to fertilize the vegetable garden. This very ingenious, closed cycle system has much to recommend it; there are without doubt many possible variations applicable in many different parts of the world.

Similarly, much work is being done on the difficult question of protection of domestic crops from predators. Alternative technologists have to find a different solution to that of applying polluting, dangerous and expensive sprays. There are several possible approaches. Perhaps the most important lies in fostering highly diverse, ecological food-growing systems rather than the monoculture to which society is now so addicted. There is evidence that diverse-species food production can be more productive than that of single species. Ecologically, production of this kind clearly stimulates a healthy species balance, with less danger of the monumental and truly savage attacks made by predators and disease organisms where and when only one crop is grown.

Alternative techniques such as these will have to be complemented by the biological control of pests and systematic, companion planting programmes in which the beneficial effect some species of plants appear to have on other species is used to the full. Cheap and biologically degradable sprays might also be acceptable; both nicotine and garlic sprays have been shown to be effective against a wide range of pests. Alternative technology will have to find sound biological and ecological means of maintaining the altered states of nature which farming implies in order to replace those blunderbuss spray technologies which our current clumsy approach to things biological has deemed to be the most appropriate means.

IX. THE FUTURE OF ALTERNATIVE TECHNOLOGY

In the past three or four years the idea of alternative technology has blossomed in the developed world, particularly in the United States, United Kingdom, Sweden and France. Earlier two organizations, the Intermediate Technology Development Group in the United Kingdom and the Brace Research Institute in Canada, had been set up to design and stimulate the growth of an alternative economic technology which would be labour-intensive and use local materials--and hence be more accessible to the developing countries. Since then, many more, less formal institutes and organizations have appeared, proclaiming additional constraints on the technology they wish to develop, in some cases more than the nine listed earlier.

This year some of these institutes are carrying out their first research, and their membership is growing considerably. It must be stressed that not all of these are concerned with rural alternative technology; some are directing their attention to the urban situation, where the demands of an alternative technology may be different in kind but not different in principle. Considerable numbers of people, many of them young, are seeking life styles which can be supported by this type of alternative technology in preference to the 9 to 5 office or factory routine which conventional society and technology offer.

The change in attitude that has come about, therefore, is that the alternative which was first seen as a means of more rapid development for the Third World has become something of an obsession for the disenchanted in the so-called developed world. Recently, there has been less talk of the implications of alternative technology for development, and much more of the need for viable alternatives in countries which are usually considered to be developed. Whether this change is for the good is not clear, and at first glance it looks like a regression.

Those who urge labour-intensive, alternative technologies on developing countries place themselves in an exposed position. Countries without a real technological base tend to see alternatives as second-class options. After all (they contend), why should they accept forms of technology which the developed countries themselves do not normally use? The intermediate technologist have thus become, in many eyes, the 'new imperialists' trying to tell the developing world what is good for it. The story sounds all too familiar.

Yet the situation is more complicated than that. For one thing, considerable interest is to be found in the developing world for what is normally termed village technology or small-scale technology which can be operated at the village level and used to improve material conditions on the micro-scale. India, in particular, is a stronghold of such thought, but there are indications from other countries too that they find the idea of value.[1] And if one is discussing the people actually facing development problems in the Third World, they may often be more interested in making a simple pump from local materials than in their governments' far-reaching schemes for a nuclear power programme, or a green revolution which will help only the larger and richer farmers. What people in the developing world think about such things might then be imperfectly articulated by their governments.

The important moral is that what must happen is that the new alternatives be developed. If that is not done, the developing countries will have no choice to make about their own future. They can in effect only continue in their present state or adapt themselves to the existing technology of the developed world. That is a poor choice. Those of us who believe that the future could have more to offer than the technocratic nightmare are intent on widening the options available for ourselves and for future generations wherever they may be.

NOTES

[1]See J. Omo-Fadaka, 'The Tanzanian Way of Effective Development', Impact of Science on Society, Vol. XXIII, No. 2, April-June 1973.

TO DELVE MORE DEEPLY

BOOKCHIN, M. Ecology and revolutionary thought. New York, N.Y., Times Change Press, 1970.

CLARKE, J.; CLARKE, R. The biotechnic research community. Futures, June 1972.

CLARKE, R. Technology for an alternative society. New scientist, 11 January 1973.

ERIKSSON, B.; HARPER, P. Alternative technology guide. Undercurrents, no. 3, 1972.

HERBER, L. Towards a liberatory technology. Anarchy, vol. 7, no. 8, August 1967.

Further Information is Also Available From:

Brace Research Institute, McGill University, McDonald College, Ste Anne de Bellevue 800, Quebec (Canada).

Intermediate Technology Development Group, 9 King Street, London WC2E 8HN (United Kingdom).

New Alchemy Institute-East, Box 432, Woods Hole, Massachusetts 02543 (United States of America).

Strategy for the Technological Transformation of Developing Countries

UNCTAD SECRETARIAT*

To recapitulate the background to the formulation of a strategy, the last hundred years or so science and technology have spearheaded the transformation of agriculture and industry and the economies of the industrial parts of the world; but this momentous transformation has so far mainly bypassed the third world. The developing countries, spurred on by this example, see in science and technology a means of attaining their own technological transformation and of sharing prosperity with the industrial countries in a world of increasing interdependence. At the same time, they have understood that the pace and direction of technological change are an ineluctable consequence not of stages of growth but of deliberate social, economic and political choices. Their governments, already coping with a complexity of means to provide their peoples with a more secure and brighter future, are therefore devising strategies for the technological transformation of their societies as well.

The starting-point for any strategy for technological transformation has to be an identification of the main features of the present situation. The developing countries:

(a) Account for almost 75 per cent of the world's population but only 20 per cent of its income;

(b) Contribute 17 per cent to world industrial output (including mining, manufacturing, construction, electricity, gas and water supply), while the developed market-economy countries and the socialist countries of Eastern Europe account for 53 per cent and 30 per cent respectively;

(c) Derive 22.1 per cent of their income from agriculture, compared with 4.5 per cent for developed market-economy countries;

(d) Have generally an industrial structure in which consumer

From **PLANNING THE TECHNOLOGICAL TRANSFORMATION OF DEVELOPING COUNTRIES** (UNCTAD Secretariat), 1981, (59-63) reprinted by permission of the publisher, U.N. Publications, N.Y.

goods account for nearly half of output. Capital goods, in which advanced technology is embodied and which form the basis for raising productivity, provide only one quarter of industrial output, compared to nearly one half in developed market-economy and socialist countries;

(e) Have an overall literacy rate of less than 40 per cent, in contrast to about 95 per cent in the developed countries;

(f) Spend only about 0.3 per cent of their GNP on R and D, compared with some 4 per cent of NMP in the socialist countries of Eastern Europe and 2 per cent of GNP in developed market-economy countries. The corresponding numbers of scientists and engineers engaged in R and D activities are 1, 81 and 43 per 10,000 of total population;

(g) Hold only 1 per cent of the world total of patents. Six per cent of the 3.5 million patents currently in existence are granted by developing countries and about 84 per cent of these are owned by foreigners. Most of them are never used in the production process in developing countries;

(h) Disburse in direct payments for the use of patents, process know-how, trade marks and technical services annually about $10 billion. If indirect costs (for instance, overpricing of imports of intermediate products and equipment, profits on capitalization of know-how and price mark-ups, and other costs inherent in their technological dependence) are included, the total cost would probably be of the order of $30 billion to $50 billion;

(i) Face a situation in which about 90 per cent of world trade in technology takes place among developed countries, of which more than one half involves transactions within transnational corporations based in these countries.

However, since 1950 the national product of the developing countries has grown nearly 3.5 times, at about 5 per cent per year; in per capita terms it has nearly doubled. Industrial output has risen six to seven times; and a whole spectrum of new and modern enterprises has emerged. Gross capital formation has increased six times, rising from a little over one tenth to around one fifth of GDP. Enrollment in institutions of higher learning has increased tenfold. One out of four students in such institutes in the whole world now comes from a developing country. Development planning has been adopted as an active instrument of policy, with the public sector -- as a producer as well as a consumer and employer -- playing a strategic role in many countries. Though the developed countries are still far ahead, the dynamics of the technological transformation of the developing countries as a whole are not as forbidding as has often been suggested.

With the ground covered over the last 25 years, the foundations have therefore been laid for the technological transformation of the developing countries. The following outline suggests a strategy through which the next steps towards the attainment of this objective can be taken. Such a strategy is a prerequisite to setting targets and evolving an adequate policy and institutional framework. Achievement of the targets becomes the responsibility of policy-makers in developing and developed countries, through the formulation of precise measures for translating the strategy into reality.

I. OBJECTIVES OF A STRATEGY

A strategy for the technological transformation of developing countries should pursue two closely related and mutually reinforcing objectives, designed (i) to strengthen the domestic technological capacity of developing countries through the creation of a research and development infrastructure with appropriate and effective linkages to the production structure and thereby lessen their dependence on developed countries, including dependence on their transnational corporations; and (ii) to overcome the external constraints on the transfer of technology from developed to developing countries, opening up as widely and freely as possible access to the world's storehouse of technologies to enable their rapid diffusion, adaptation and assimilation in the production structures of developing countries.

II. NATIONAL ACTION TO STRENGTHEN TECHNOLOGICAL CAPACITY

1. Policies and Planning

The government of each developing country should formulate a comprehensive and coherent national policy for the strengthening of the country's technological capacity. Such a policy should create and reinforce in each country an autonomous decision-making capacity in technological matters in accordance with the realities of its political, economic and social situation and its development objectives.

The technology policy should be implemented through the adoption of technology plans as integral parts of national development plans. The technology plans should embrace essential responsibilities, such as budgeting, management, co-ordination, stimulation and execution or technological activities and cover specific requirements at the sectoral and intersectoral levels for the assessment, transfer, acquisition, adaptation and development of technology. They should reflect short-term, medium-term and long-term strategies, including determination of technological priorities, mobilization of natural resources, dissemination of the existing national stock of technology, identification of sectors in which imported technology would be required and determination of R and D priorities for the development and improvement of endogenous technologies. Particularly important in this connection is to initiate now the process of substitution of imported technologies by domestic ones.

Governments should take fully into consideration the interrelationship of technology with other social and economic variables in the process of the formulation of development policies and therefore, through their investment, taxation and other policies, ensure that the application of technology serves the interests of the majority of the population.

2. Technological Infrastructure

A prerequisite for the formulation and implementation of a national technology policy and its attendant technology plan is the creation of the necessary infrastructure, which should include the establishment of national technology centres at the highest possible political level. The centres should, inter alia, co-ordinate the activities of national technology institutions and ensure close linkages with the productive sector, for instance through the establishment of appropriate mechanisms such as technological extension services. They should also mobilize and direct technology towards sectors of critical importance, such as energy, capital goods, food processing and pharmaceuticals, and to areas where it is most needed, such as rural development, urban planning, and the ensuring of an effective role by women in the transfer and development of technology.

3. Financial Resources

An integral part of technology policy and planning should be the allocation of resources for technology, particularly for R and D. A technology budget as part of the national budget should ensure identifiable resources. By diverting a portion of the resources currently spent on importing technologies to measures designed to stimulate endogenous R and D and by other means, developing countries should increase their expenditure on research and development and technological innovation or adaptation from its current low level to 2 per cent of GNP in 1990 and 3 per cent in 2000.[1] Through their own national and collective efforts, as well as those of the international community as a whole, the developing countries should attain by the year 2000 the goal of carrying out 20 per cent of world R and D expenditures and having a larger share of R and D personnel.

III. CO-OPERATION AMONG DEVELOPING COUNTRIES

1. Policies, Programmes, And Projects

Parallel with their efforts to achieve national technological autonomy, developing countries should strive towards greater collective technological self-reliance. The newly established regional centres on technology and similar centres for specific sectors could serve as main instruments of policy and action in this area. Among other things, such efforts might include:

(a) Programmes and projects for joint acquisition and utilization of technology;

(b) Co-operative arrangements for consultancy, design and engineering services, for instance through the establishment of twinning arrangements between the specialized institutions of developing countries.

(c) Joint R and D, in particular in respect of programmes for which a minimal level of staffing, equipment and financing is required for viability;

(d) Information systems on all matters pertaining to acquisition (including terms and conditions), adaptation, development and utilization of technology, with a view, *inter alia*, to strengthening the individual and collective negotiating capacity of developing countries vis-a-vis suppliers of technology.

(e) The establishment of linkages between regional centres for the transfer and development of technology;

(f) The establishment of intergovernmental centres or similar institutions in sectors of critical importance so as to optimize common resources.

2. Preferential Arrangements

Developing countries should, as an integral part of their efforts to achieve collective self-reliance, pay particular attention to the special problems of the least developed, land-locked and island developing countries. More advanced developing countries should share technologies with such countries on preferential terms and provide facilities for training so as to facilitate access to technology already available in other developing countries. Any strategy for technology co-operation among developing countries should place special emphasis on the development of technologies and the solution of problems of particular significance to the least developed countries, such as control of drought and desertification, health, food production, housing construction and energy.

IV. CO-OPERATION WITH DEVELOPED COUNTRIES, INCLUDING THEIR TRANSNATIONAL CORPORATIONS

1. Decommercialization Of Technology Transfers

At present many of the technologies needed by most developing countries are either publicly owned or freely available in the developed market-economy countries. Yet the developing countries' access to them is severely constrained because the availability of technologies in the public domain is often made subject to private decisions through a complex process of packaging and control of know-how. In the socialist countries, technology is in the public domain or controlled by public enterprises which, even where they may be considered private legal entities, are not subject to the profit motive as the main determinant of their actions or international transactions. Determined action could be taken to facilitate developing countries' access to this type of technology. Therefore, as a significant contribution to the strategy for the technological transformation of developing countries, developed countries should make available to developing countries without charge, or for nominal payment technologies required by developing countries which are in the public domain or are freely available. When the transfer of such technologies to developing countries is subject to private decision, either by a process of packaging and control of know-how or by any other means, developed countries should progressively decommercialize such technologies, especially in those areas which cater to the satisfaction of critical needs of developing countries -- for example, pharmaceuticals, food and food processing, housing and building materials, public transport, telecommunications and energy supplies.

The developing countries' access to technology in the public domain is further constrained by lack of information on the nature of technologies available and lack of that minimum degree of organized ability to take advantage of them. Access to these technologies could be improved by means of the establishment in developed countries of institutions or mechanisms that would keep up-to-date registers of technology in the public domain, to which the developing countries could be given the freest and the fullest possible access. Even when a charge is made to cover costs, it could be reduced to a minimum. Specially favourable conditions, particularly in terms of cost and availability of information, could be granted to the least developed countries.

2. Transnational Corporations

Governments of developed countries should take effective

measures to ensure that transitional corporations whose headquarters are based in their countries suitably restructure and reorient their technology activities in developing countries to bring them into conformity with the host country's technological policies and, *inter alia*, ensure that they:

(a) Substantially increase R and D activities in their subsidiaries in a manner compatible with the research and development priorities of the host country;

(b) Contribute to the increase of R and D expenditures in developing countries, through full utilization of institutions in developing countries and by entering into R and D contracts and similar arrangements;

(c) Provide for greater research and development co-operation between their subsidiaries and the network of national research and development institutions;

(d) Do not enter into technology transfer arrangements that would hinder or limit the economic and technological development of developing countries;

(e) Make available to the governments of host countries the fullest information on the terms of technology transfer between them and their subsidiaries in an "unpackaged" form;

(f) Make available to the relevant comprehensive information systems of the United Nations full information on the terms of technology transfer to their subsidiaries and to local enterprises;

(g) Encourage their subsidiaries to transfer technologies to local enterprises in the host country on favourable terms and conditions;

(h) Fully utilize consultancy, design and engineering organizations of developing countries in the establishment and execution of projects in those countries;

(i) Co-operate with developing countries in their efforts to establish a "critical mass" of scientific, technological and managerial manpower, through the institution of in-plant training facilities, support to national training institutions and the like, and the establishment and development of consultancy services;

(j) Replace within an agreed time-frame scientists and technologists from developed countries employed in their subsidiaries by qualified personnel from the developing countries concerned, to the maximum extent possible.

3. Research And Development

There is a need for co-operative programmes designed to build or strengthen R and D capabilities in developing countries, such as:

(a) The establishment of joint arrangements between institutions of developed and developing countries;

(b) The monitoring of joint R and D projects oriented to the needs of developing countries and the close association of research workers of developing countries with such projects.

Developed countries should earmark a fixed percentage of their R and D expenditure specifically for the solution of the particular problems of developing countries.[3] An important aim of this form of co-operation should be to strengthen the R and D capabilities of developing countries by their active participation in such activities. In addition, it would make the R and D of developed countries more responsive to the needs and problems of developing countries.

V. ACTION BY THE INTERNATIONAL COMMUNITY

International action to support a strategy for the technological transformation of developing countries must be taken in a coherent and co-ordinated manner, in conformity with the principles and objectives of the new international economic order.

1. Restructuring The Legal Environment

All countries should accelerate their efforts towards restructuring the existing international legal framework and evolving a new one that guarantees a process of equitable transfer of knowledge and the strengthening of the science and technology capacities of developing countries. This should include, *inter alia*, the early adoption of:

(a) An international code of conduct on the transfer of technology;
(b) A code of conduct for transnational corporations;
(c) The revision of the Paris Convention for the Protection of Industrial Property;
(d) Arrangements for the provision of copyrights in the field of science and technology on more favourable terms to developing countries.

In the process of restructuring the existing international legal environment governing the transfer and development of technology, the international code of conduct on the transfer of technology at present under negotiation should play a dynamic role. All members of the international community should implement and apply the provisions of the code, once adopted. International action should be oriented particularly to facilitating an expanded international flow of technology on equitably and mutually agreed fair and reasonable terms and conditions in order to strengthen the technological capabilities of all countries, particularly the developing countries.

In pursuance of the objectives of the code of conduct,[5] developing countries should adopt and implement national policies, laws or regulations on transfer and development of technology and evolve policies and institutions for their collective self-reliance in this area. Developing countries should also consider structuring their national legal framework so as to promote the development of indigenous technology and thereby facilitate their technological transformation and increased participation in world production and trade.

After the current revision of the Paris Convention for the Protection of Industrial Property, all countries should continue to review the ways in which the industrial property system can become an effective instrument for the economic and technological development of developing countries. In particular, efforts should be made to promote the exploitation of patents in developing countries.

2. Technological Information System

An international technological information network should be established within the United Nations system, with maximum utilization of the existing institutions, including national and regional centres for transfer and development of technology. This should be designed to meet the urgent needs of the developing countries for information on, *inter alia*, alternative sources of technology supply, terms, conditions and costs of all major factors contributing to the use and application of technology, technology transfer transactions and their conditions, analytical and evaluative data on such transactions, including the operations of transnational corporations, and the technical data contained in patents.

3. Reverse Transfer Of Technology

A major problem faced by a number of developing countries in formulating a human resource development policy supportive of their technological transformation is the migration of skilled manpower to developed countries (reverse transfer of technology). Taking into consideration the recommendations made at the fifth session of the United Nations Conference on Trade and Development, concerted efforts should be made by the international community to formulate a comprehensive plan of action, and a strategy for its implementation, to ensure that skilled migration from developing to developed countries will constitute an exchange in which the interests of all parties concerned are adequately protected. In addition, intergovernmental negotiations should be initiated to establish an international labour compensatory facility, while developing countries should negotiate and put into effect schemes for regional and interregional co-operative exchange of skills.

VI. RESOURCES FOR THE IMPLEMENTATION OF THE STRATEGY

For the technological transformation of developing countries a vast increase in financial resources, as well as technical and operational assistance commensurate with the requirements of developing countries, will need to be provided. Hence every possible effort should be made by all countries, and particularly by developed countries, to increase financial and other contributions to programmes, projects and activities aimed at strengthening the technological capacity of developing countries and accelerating their technological transformation.

A key instrument in this respect will be the United Nations Financing System for Science and Technology, decided upon by the General Assembly in its resolution 34/218 of 19 December 1979. The system should, when it commences operations in January 1982, have the following characteristics;

(a) It should be used for the strengthening of the scientific and technological capacities of developing countries, including the acquisition of technology;

(b) It should mobilize and channel all types of financial resources, particularly from developed countries;

(c) Its resources should be substantial and supplementary to the resources that now exist, and should be furnished on a predictable, automatic and continuous basis;

(d) The volume of its financial resources should be sufficient to contribute effectively to the implementation of the strategy;

(e) It should be independent in its operation, and establish close co-ordination with the competent organs of the United Nations system;

(f) The developing countries should be enabled to participate fully in its operation.

NOTES

[1]On the assumption that R and D expenditure in the world will grow at an average rate of 5 per cent a year, it will rise from the present level of $130-$150 billion to about $350 billion in 2000. With the same rate of growth of GNP in developing countries, it will reach about $2,000 billion. With 3 per cent of their GNP spent on R and D, developing countries would account for some 20 per cent of world expenditure on R and D in 2000.

[2]Particularly relevant in this connection is the urgent implementation of the specific measures agreed to in chapter 6, "Special treatment for developing countries", of the draft international code of conduct on the transfer of technology now under negotiation (see TD/CODE/TOT/33) and certain provisions of Conference resolutions 39 (III), 87 (IV) and 112 (V).

[3]It may be difficult to make a distinction between problems that are more specific and problems that are less specific to developing countries. Various inquiries undertaken by international organizations indicate that current expenditure by developed countries on R and D for the benefit of developing countries ranges from 0.3 per cent to 2.7 per cent of their total expenditure on R and D (A/CONF.81/PC/42, para. 102).

[4]Particularly relevant in this connection is the urgent implementation of the specific measures agreed to in chapter 7, "International collaboration", of the draft international code of conduct on the transfer of technology now under negotiation (see TD/CODE/TOT/33).

[5]For the text on objectives and principles transmitted to the United Nations Conference on an International Code of Conduct on the Transfer of Technology by its President, see TD/CODE/TOT/33, chap. 2.

Guiding Technological Change

CHARLES WEISS, JR.*
MARIO KAMENETZKY*
ROBERT H. MAYBURY*

I. INTRODUCTION

The purpose of this paper is to promote discussion among decision-makers, from both public and private sectors of a developing country, on the topic of orienting technological change and innovation in support of development. Briefly, the argument is as follows:

(a) Technological change -- whether in production or in programs for social welfare -- cannot be left to chance; deliberate effort is needed to guide this change to support development.

(b) Guiding technological change requires effort on the part of many institutions in a country: government policy-making and regulatory agencies; enterprises and public bodies in the productive sector; financial institutions; and scientific, engineering, and technical organizations;

(c) To be effective in guiding technological change, these institutions must possess an ability to make technological decisions, referred to here as "technological capacity", this capacity inheres even in small-scale farmers and entrepreneurs -- perhaps the majority of a nation's producers.

Decisionmakers should explore the many dimensions of

Reprinted by permission of the authors. They are with the Science and Technology unit of the World Bank at Washington, D.C.

technological capacity and consider its status in the institutions of their own countries.

II. TECHNOLOGICAL CHANGE - THE RESPONSE TO RESOURCE SCARCITY

A fundamental first step towards making a clear case for the importance of technology in promoting economic growth and social welfare is to adopt an adequate concept of technology. One of the most satisfactory definitions is that given by the Economic Policy Council of the United Nations Association of the United States of America:

> Technology means human knowledge used to achieve ends. Thus it includes everything from the manufacture of computers to the marketing of breakfast cereals, and from the use of public health measures to prevent disease to the growing of more abundant and better crops. It also includes the knowledge of how to organize society in order to achieve desired ends (corporations, educational institutions, health care organizations) and how to cope with related consequences.

A similar definition was provided by the Argentine physicist Jorge Sabato. He said that, "Technology is empirical and scientific knowledge organized in sets for its use in the production and marketing of various goods and services."

The use of knowledge, organized as a technology to achieve a desired end, necessarily draws upon some set of resources -- natural, financial, and human. In any society, traditional or modern, the availability of resources constrains the effort to achieve an end. As history has shown, moreover, when a resource becomes scarce, there is a strong impetus to search for a better way to use the resource or even to discover new resource stocks. These improvements in technological capability have, in fact, been critical to the development of human societies.

Developing countries today face some of the severest scarcities of resources imaginable: inadequate supplies of human skills, lack of funds, increased cost of material inputs, and weak institutional capacity. These scarcities, coupled with a burgeoning population, constrain every effort these countries have made to life the burdens of poverty from the shoulders of their people. If progress is to be made, these countries must find more effective and efficient ways of using their scarce resources. These betters ways can only be found through innovation in existing production processes and social patterns.

III. UNDERSTANDING THE ROLE OF TECHNOLOGY IN THE PROCESS OF CHANGE

Although enterprises and other productive units of a society can be expected to introduce innovations in techniques, products, and the use of resources in times of difficulty or when change seems to be advantageous, neither the pace nor the pattern or direction of such change will always satisfy human needs or those of social development or meet the declared goals of a nation's economic development. Experience suggests that a nation should make deliberate efforts to orient technological change in such a way as to support these needs and goals. In short, technology must be mobilized for development. To be effective, these deliberate efforts should be based on an understanding of the process of technological change.

Technology is not only "hardware" -- the inputs, outputs, and equipment of productive activity -- it also includes the associated "software" components upon which the technicalities of production depend; the patterns of organization and management that provide the framework for productive activity and are therefore critical in determining its outcome.

The complex nature of technology introduces complexity into the process of technological change. Besides modifying the way in which a given productive activity is performed, a successful change requires that attention be paid to social, economic, cultural, and institutional adjustments throughout the system in which that particular activity is integrated. To illustrate, any change in agricultural practices the aim of which is to increase the productivity of the farming system requires simultaneous adjustments in the physical infrastructure, such as roads and communications, marketing, or processing facilities and in the institutional infrastructure within which rural activity takes place -- organizations for marketing outputs and buying inputs, financial and technical assistance, incentives and constraints on change, codes of behavior and administrative arrangements.

Technological change can be felt in many places throughout a society, often with attendant strain on people, the social and political structures, or the environment. Any negative effect must be compensated for or, if possible, avoided altogether. The Green Revolution is an example of a planned increase in physical production that helped consumers but had negative second-level effects, such as the skewing of the pattern of distribution of income, wealth, and power among farmers. Other examples of change are known to have brought about damage to the environment or increased hazards to human health. There may be a possibility of appraising the likely effects of a proposed technical change so that those planning the change can consider preferred alternative courses of action. This process is called technology assessment.

A further source of complexity in the process of technological change is the critical influence on choice of technology of the following two factors: the sources of technology available to a country and the behavior of those responsible for choosing technology.

Under normal conditions, a country should be able to obtain technology from both domestic and foreign sources. However, few of the developing countries have succeeded in building up domestic technological capacities, so their only practical source is foreign. This unbalanced situation creates two great problems; much of the technology available is inappropriate, since it has been designed to meet specific requirements in countries with very different levels of economic and industrial development, factor endowment, and socio-political and cultural conditions, and the ready availability of foreign technology greatly reduces the incentive to the development of an indigenous technological capacity. A distressing aspect of the latter problem is that a developing country should possess an indigenous technological capacity even to choose and apply technology from foreign sources in the ways that are most beneficial to its own development.

Those who choose technology, whether they are in production or in programs for social development, influence the course of technological change critically by their attitudes toward innovation and their responses to price signals and social indicators. Industrial firms and farmers rely on profit motives in making their choices of technology. They are, consequently, responsive to incentives granted by the government to increase those profits. Those incentives should direct the choice of technology toward the selection of techniques that will provide adequate profits and at the same time contribute to an increase in social welfare.

On the other hand, manager, purchasing officers, and others in government agencies are obliged to be sensitive to prices of resources and to social issues such as: generation of employment, distribution of income, balance in regional growth, and protection of the environment when selecting technologies for the operation of public enterprises or public service programs.

Both groups, but especially the small farmers and entrepreneurs, may be uninformed about alternative choices and may be intensively averse to the risk of failure that tends to accompany innovation. As a consequence, they may be guided in their choice by the first effective salesman who offers a technology already proven somewhere else. But this technology may be socially inappropriate even if it is financially attractive.

IV. INSTITUTIONS AND FUNCTIONS FOR GUIDING TECHNOLOGICAL CHANGE

A deliberate effort to guide the course of technological change requires action at the following three levels:

(a) the technical institutions that generate, assimilate, and disseminate technology -- public research laboratories and technological institutes, research and engineering departments of enterprises, consultancy firms, information and extension services;

(b) the uses of technology -- production units, agencies, and programs for meeting the objectives of social welfare;

(c) the institutional and legal framework, predominantly government policies, whose functions include the setting of priorities, the allocation of resources, and the regulation of, or direct participation in, the interaction between technological institutions and users of technology.

It is through the last level that patterns for the demand of technology will be shaped and the local supply side will be created and developed.

Thus, if economic policies prevent competition, firms will ignore technology that might improve quality. If financial authorities overvalue the local currency, firms will have an incentive to import equipment and raw materials in place of supplies that are available locally and will neglect local sources of technology. If government marketing boards pay too little for higher quality, processors of agricultural commodities will neglect their machines and will fail to take advantage of improved technology. If banks are unwilling to extend credit to small farmers and entrepreneurs, the latter will continue to face severely restricted technological choices. In this way, entrepreneurs may be led by market imperfections and government policies to "non-economic" behavior.

Government decisionmakers must cooperate closely with the technical specialists of a country to work along two lines. First, they should spell out the technological implications of national policies and plans: Are government tax credits, subsidies, loans, and direct grants promoting technological improvements? Are procurement and bidding practices encouraging local supplies of technology? What influence do tariff schedules and legislation to regulate foreign investment have on local value added? Do the project-appraisal criteria applied by financial institutions favor greater use of local factor endowments in investment projects? Do these institutions provide technical and managerial assistance for preinvestment work and implementation of projects?

Second, government officials should make an effort to introduce technology as a variable into policymaking, investment decisions, and planning. This involved not only attention to policy measures that affect labor and capital, but also, consideration of technology as a distinct variable factor of production which can be manipulated in order to accomodate the cost and availability of the other factors of production, the existing and desired level of local technological skills and capabilities and the required scale of production.

In working along these two lines, it is most important to adopt, insofar as possible, a holistic view of broadly defined elements in the economy, such as urban poverty or the production, storage, and marketing of food crops, in order to asses present and probable future needs for technology and to anticipate its effects. Efforts to alleviate urban poverty call for reorientation of part of the local technological community toward the development and design of low-cost hardware in close support of government or private efforts to provide the urban poor with basic services. Managers of public-sector programs for small-farmer agriculture should be aware of the need to develop and diffuse low-cost technologies for increasing agricultural productivity that are appropriate to local conditions. They should also improve the physical infrastructure of the rural areas they want to develop and promote community-based agro-industrial undertakings, health services and education programs.

Important as it is to ascertain the technological requirements of general socioeconomic goals, this is only one side of the effort required to guide technological change in support of development. This attention to what might be termed the "demand side" of a country's technological situation must be matched by an equal amount of attention to the "supply side" and its links with demand. The demand for technology -- that is, the technological requirements of development objectives -- must be satisfied by appropriate technological activities on the supply side. These activities, covering the broad areas of generation, import, and absorption of technology, include scientific research, technological research, engineering design, and a wide range of support activities, such as information systems, technical services, training, and high-level preparation of human resources.

A nation's decisionmakers face fundamental issues in their concern for this supply side: the problems in relying on foreign sources for the technology needed; the importance of developing local technological capability; and the approach to developing links between the supply of technology and the demand for it. The necessity of resolving these issues in full harmony with a country's socioeconomic development goals provides the rationale for establishing a national technology policy.

In addressing these fundamental issues concerning the supply of technology, a national technology policy must set priorities, allocate resources, and take other measure toward achievement of the following:

(a) The creation and strengthening of research laboratories, technological institutes, and information, quality-control and technical-assistance services;

(b) The development of local groups for providing pre-investment and project-implementation services;

(c) The awakening of awareness among financial institutions of the importance of technological choice in investment projects;

(d) The organization of an education system that will provide the technological skills required at the various levels of the economy.

No concern of technology policy is more urgent than that of fostering mechanisms for linking the supply of technology with the demand for it. In this effort, decisionmakers seek answers to such questions as: which are the best institutions and instruments for locally organizing technology in such a way as to facilitate its practical application? Who should provide incentives for designing and implementing investment projects in such a way as to make maximum use of the local supply of knowledge and of the services required for putting it to use? Who should study the demand for knowledge in the various sectors in order to give proper direction to efforts to supply it locally?

These questions call attention to the contribution of local technological institutions and the technical bodies of financial institutions to the building of bridges between the supply of technology and the demand for it. This contribution is made through investment projects, which are recognized as one of the principal channels for deliberate development action. By involving local institutions in the provision of preinvestment and project-implementation services, the investment project can powerfully reinforce the two-way flows between a country's technical institutions -- supply -- and its productions units and other users of technology -- demand.

Technology policy faces particular hazards in efforts to develop the supply side of technology. According to recent studies of such efforts in developing countries, research and technological institutes have frequently remained isolated from local users of technology. These studies advocate strict measure to gear the activities of technical institutes to the needs of local production and social action.

Another problem in developing the supply side of technology is the tendency to cater only to the technological needs of users in the modern sector -- large industrial enterprises and large-scale farmers.

The scientific and technical institutions often do not know how to establish contact with small farmers, tiny entrepreneurs, and artisans in order to assess their requirements for knowledge. The professionals and technicians who work at the technological institutes, research laboratories, scientific councils, technology registers, and technical information services are frequently too far from the poor people at the bottom of the economies both geographically and psychologically.

The main problem faced by development aid institutions and governments in their attempt to promote innovation among small farms and enterprises already in operation, is one of appropriate and timely financial and technical support. Still more difficult is the task of promoting social and technological change among people living under subsistence economies and struggling to realize some income. For them the question is that of promoting entrepreneurial initiatives by awakening these people to an awareness of the cause of their poverty and to their own potential for introducing progressive changes into their lives.

There is a need to explore possible new types of intermediary organizations for providing technical and financial assistance at these lowest levels of an economy. Intermediary organizations should work from within the communities they are trying to help and training talented young people from the region as field-based general practitioners to accomplish tasks in promotion, implementation, and trouble-shooting at the grass-roots level. These field-based practitioners should go well beyond mere extension activities to become village developers and community leaders. There is increasing evidence that non-governmental community-based organizations could reach high levels of efficiency when playing this intermediary role.

V. TECHNOLOGICAL CAPACITY: A DECISION-MAKING ABILITY NEEDED THROUGHOUT A SOCIETY

Guiding technological change to support development can now be seen to entail many technological decisions -- that influence the allocation of resources and the efficiency of productive units -- throughout the three sets of institutions involved in a country -- the technical institutions, the users of technology, and the government.

The ability to make these decisions is a critical factor in guiding technological change, one that may be called "technological capacity." This capacity inheres in the technical institutions of a country, in the technical staff of financial institutions and government agencies, and in productive enterprises. Moreover, since even village workshops and cooperatives, individual small farmers, and entrepreneurs make daily technological decisions, they too, possess at least a modest level of technological capacity.

In guiding technological change, a country requires technological capacity for a wide variety of tasks: identifying and defining technologically feasible development goals and strategies at the sector and project levels; shopping for, selecting, negotiating for, and acquiring technology and ensuring its adaptation to local problems, conditions, and resources; implementing, using and improving that technology in solving practical problems as they arise; evaluating the effect of technology and assessing its social and environmental impact; generating new technology; training qualified technologists and ensuring that they are used productively; keeping up to date on global technological developments and on technological elements of problems, such as food and energy, that require global management; and advising the government and the public at large on issues of national and international policy that require an understanding of technology.

VI. CONCLUSIONS - A CALL FOR PROMOTING STUDIES AND TRAINING ON TECHNOLOGY POLICY ISSUES

It is our intention in this discussion to increase the

awareness of decisionmakers of the critical need for technological capacity in the efforts of a country to orient technological change toward development. Having traced the locus of this capacity in sufficient detail, these decisionmakers may recognize the principal areas in their economies where this capacity may be deficient. It is likely that areas of deficiency exist across the entire range of sectors and institutions that make decisions affecting technological activities -- from the technical bodies of financial institutions through individual firms in private and public sectors to the small-scale producer in urban and rural areas.

We have given considerable thought to the question of deficiencies in technological capacity and to a possible program for overcoming them. Our experience indicates that there are two widespread fundamental deficiencies: first, a deficiency in knowledge of the principal technological features of a nation's economy, and second, a deficiency in the supply of individuals who possess the understanding and skill required in technological decisionmaking.

To overcome the deficiency in knowledge, a nation-wide study is needed in order to describe the existing scientific and technological institutions, technological behavior of entrepreneurs and managers and policy instruments which influence the direction of the technological change. The same study should then analyse the effectiveness of these institutions and instruments in accomplishing the socio-economic objectives established by the country for its development and present a program with the changes required in institutions and instruments in order to increase the technological capacity of the country in its pursuit of the development goals.

To overcome the deficiency in the supply of technologically minded decision makers, it is necessary to increase -- through appropriate training -- the awareness among decision-makers from all parts of the economy -- government agencies, financial institutions, technical institutions, enterprises and communities -- on the following issues.

--influence on the direction of technological change of economic and financial policies and legal instruments

--effects produced by different technological patterns of development on the use of resources and the environment

--management of technology as a distinct, variable factor of production which can be fitted to each investment project and sector strategy

--links between the scientific and technological infrastructure of a country and production enterprises

--means for improving the technological capacity at all levels of the society and the economy of a country

--management of the human factor in the design and implementation of technologies.

ACKNOWLEDGEMENTS

The authors express their indebtedness to the following sources for ideas and factual material that in some instances have been used in this text with little or no alteration.

Eckaus, R. "Criteria and Orientations for Science and Technology for Development," Technology in Society, vol. 1, (Summer 1979), pp. 123-135.

McInerney, J. The Technology of Rural Development Staff Working Paper no. 295 (Washington, D.C.: World Bank, October 1978).

Rothchild, D., and R. Curry. Scarcity, Choice and Public Policy in Middle Africa (Berkeley: University of California Press, 1978).

Sagasti, F. Science and Technology for Development, Main Comparative Report of the Science and Technology Policy Instrument Project (Ottawa: International Development Research Centre, 1978).

Technology Transfer Panel, Economic Policy Council of the United Nations Association of the United States, The Growth of the United States and World Economies through Technological Innovation and Transfer (August 1980).

Vaitsos, C. Technology Policy and Economic Development, A summary report undertaken by the Board of the Cartegena Agreement for the Andean Pact Integration Process (Ottawa: International Development Research Centre, 1976).

The Socio-Economic Iceberg and the Design of Policies for Scientific and Technological Development

MARIO KAMENETZKY*

I. THE SOCIO-ECONOMIC ICEBERG AND THE POLICY-MAKERS

The socio-economic structure of a country can be modelled after an iceberg. The small part of the iceberg that is above water represents those enterprises having technical structures and financial resources which make them visible to policy-makers. In developing countries the tip of the iceberg is formed of branches of multinational corporations and large local private and public firms. The part which is at the flotation line then represents small firms in the modern sector (many of which have at least one professional or technician among their staff). They are likely to experience the ebb and flow of the economic waves. The bottom submerged part of the iceberg represents the large majority of small farms, enterprises, miners, artisans, and cottage manufacturers and all the people living under subsistence economies.[1]

The makers of science and technology policy prefer large, modern enterprises as targets for the work of the technological structures they try to develop, despite the fact that large modern enterprises are those which need less attention from the local technological institutes or research laboratories. Large modern enterprises have their own internal technological capabilities to select the technologies best suited to their economic interests and to the psychological well-being of their owners and managers.[2] They are also able to put into use the technologies they select, using when necessary the most convenient local or foreign consulting and engineering groups to help them in the design and implementation work. Small enterprises would also benefit from building similar

---

Reprinted by permission of the author. He is with the Science and Technology unit of the WORLD BANK at Washington, D.C.

internal technological capacity. Their need, however, is seldom supported by fiscal and policy measures. Without tax and credit incentives, small enterprises do not invest in creating internal capacity nor ask for local technological services. They remain dependent on imported technologies which often do not make the best use of the country's resources and rely on imported technological services which often are overly expensive.

Even in the few instances where local sources of technical support and advice to production have been successfully mobilized in support of the small enterprises, they have not reached the potential and existing entrepreneurs who are at the bottom of the iceberg. The distribution of knowledge has not been based upon needs but upon economic power. Ability to pay for technical services and technologies is a function of current income, accumulated wealth, or both. Those who already possess technical assets are the ones who can increase their skills and talents further.

Those at the bottom of the iceberg rely on the few empirical skills that they have inherited or learned while struggling for their survival. This group has neither economic incentives nor psychological motivations to search for knowledge -- not even for free knowledge -- that might increase their productivity. They fear the risk associated with any technological change. Failure for them is neither a simple loss of benefits nor a decrease in capital; failure for them can mean starvation.

The science and technology institutions built with the enterprises at the top of the iceberg in mind do not know how to establish contact with those at the bottom in order to assess their requirements for knowledge. The professionals and technicians who work at the technological institutes, research laboratories, scientific councils, technology registers, and technical information services are frequently too far from the people at the bottom, both geographically and psychologically.

A physicist finds it much easier and more rewarding to work at the National Atomic Energy Commission along lines that allow him to exchange ideas with the most noted nuclear scientists of the day than to spend his time with the peasants in some unknown village trying to develop a solar crop dryer that is simple enough to be built at the nearest metal workshop with a few imported parts.

A chemical engineer is much happier collaborating in the adaptation of a process for a large corporation than trying to launch a cooperative dairy undertaking in some marginal agricultural region.

A psychologist prefers to counsel rich entrepreneurs or managers and their spouses in the large cities rather than help poor people overcome blocks, prejudices, and centuries-old feelings of frustration that inhibit the increase of their productivity, the enrichment of their daily lives, and the full development of their human potential.

The problem of working with present and potential entrepreneurs at the bottom of the iceberg is highly complex. It is so even in a country such as Denmark, which has an almost totally literate population, a tradition in the organization of agricultural extension work and farm cooperatives that dates from

1800, and an industrial tradition that is proud of having organized a technological institute as early as 1906. In 1978, the Danish Technical Information Center for Industry (DTO) was able to work only above the flotation line of the iceberg, and the lower limit of entrepreneurial size and structure among the clients of the DTO was represented by firms each of which had at least one engineer on its payroll.[3] Efforts were being made by the Danish technological institutes in Copenhagen and Jutland to create organizations and mechanisms capable of working much below the flotation line of the iceberg in providing technical assistance to enterprises that do not have internal technical structures.[4]

To overcome these weaknesses of previous approaches to the planning of S & T activities, a new approach will be proposed. It will be a systemic approach because it will emphasize the interconnections and relationships among S & T activities at all levels of the socio-economic iceberg. But as much use as possible will be made of the organizations that have emerged from earlier planning efforts.

II. THE PROPOSED MODEL: A SYSTEMIC APPROACH

In the systemic approach (see Figure 1), each field of S & T activity simultaneously rests upon and encompasses the lower ones. Each higher level receives signals and inputs from the lower ones and delivers outputs to them. But the use of knowledge in investment projects, in the operation of the resulting installations and in solving social problems is the primary activity in building the system. Without it the whole structure breaks down, and each level, especially the upper ones, begins working for its own objectives, a cloud floating on its own.

Instead, when all the levels are integrated into a production-based and socially-oriented system, self-reliance is increased. Local and foreign inputs and signals can then be received, processed at the various levels, and transformed into outputs and signals to be delivered either within the system or to foreign systems.

The development of the systemic model will take into account as agents of S & T activities at the various levels:

(i) present and potential small entrepreneurs;

(ii) regional technical-assistance groups and field-based practitioners;

(iii) technological institutes;

(iv) national groups that provide project design and implementation services;

(v) institutions for scientific research and training;

Figure 1: The Systemic Model for S&T Development Planning

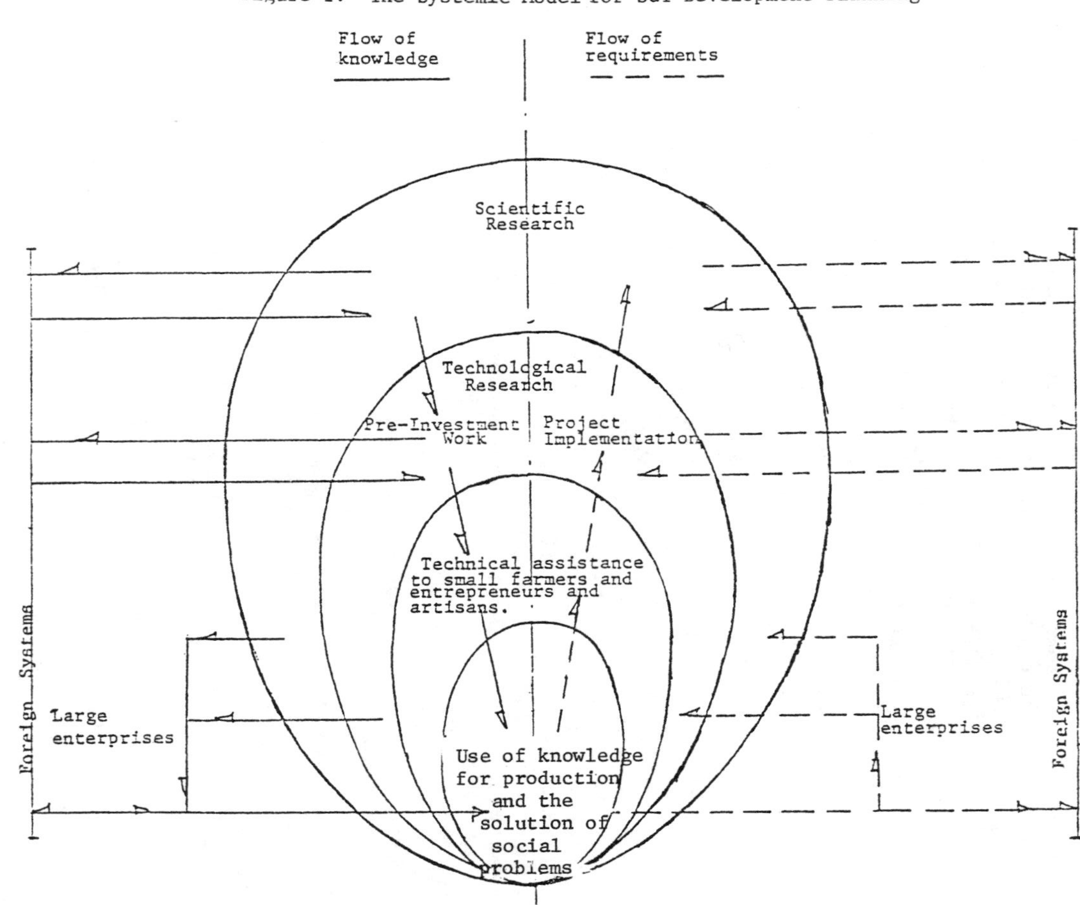

(vi) large enterprises;

(vii) financial institutions;

(viii) governments; and

(ix) science councils

I now consider each of these separately.

A. Present and Potential Small Entrepreneurs

The provision of technical assistance to present and potential small entrepreneurs who are living in villages and poor urban communities should have the highest priority among the S & T activities which aim at increasing the efficiency in the use of knowledge for production and the resolution of social problems.

For an appropriate design of the equipment and organizations required in order to build -- or in some cases rebuild -- the productive and social infrastructure in villages and poor urban communities it is not enough to assess their technological requirements and their financial and economic possibilities. It is also necessary to assess the system of values and beliefs that establish the frame of reference which orients people's behavior in their performance of production activities and in the use and consumption of the resulting products and services.

Moreover, the successful implementation of projects at this level of the economy requires that villagers and urban poor be able to analyze that frame of reference and make by themselves the necessary adjustments in their system of values and beliefs. Otherwise, their perceptions of the world (weltanschauung) and life styles will not be in accordance with the new technological patterns. Conflicts that often emerge from this situation may misdirect organizations and delay implementation schedules in projects which in other respects are well designed and engineered.

In successful grass-roots development programs, like the one by the Sarvodaya movement in more than 3,000 villages of Sri Lanka, financial and technical resources are channeled to the villagers only after they have been awakened to an awareness of the causes of their poverty and of the tremendous potential for taking their personal and social development into their own hands and brains.[5]

B. Technological Institutes

These include industrial and agricultural research and extension services, and the like. These institutions should work directly with the large and small enterprises of the modern sectors of the economy. Instead, their contribution to village and community development is better channelled through the field-based practitioners and regional technical-assistance groups.

Many times the amount or type of knowledge that field-based practitioners consider necessary for village or community development cannot be made available by regional technical assistance groups. In such cases, the technological requirements should be conveyed to the technological institutes and their advice communicated to villages and communities by the regional groups in a way that will be easy to understand for the recipients.

The institutes should have the highest degree of autonomy possible and should develop capacity to generate new technologies, adapt existing technologies to local conditions and provide operational support and advice to production units. They should also develop skills in marketing the technologies which they can offer and the technical services which they can provide. Entrepreneurs and managers from the modern sector and leaders of the regional technical groups should participate in the design of the objectives for the institutes and in the control of their accomplishments.

C. National Groups That Provide Project Design and Implementation Services

Consulting and engineering organizations are established with the aim of providing project design and implementation services to enterprises, especially those in the modern sectors of the economy. Regional technical-assistance teams should also turn to them when the tasks of regional planning and the design and implementation of projects at village and community levels exceed the expertise and engineering skills available at the regional and local level.

In providing project design services, these groups should examine each technological alternative to a given investment in the light of its opportunities, constraints and objectives.[6] The technological choice in an investment project is critical to the efficient allocation of a country's resources.

Project-implementation services include detailed engineering of the installations, procurement of the equipment and materials for the production facilities, supervision of the construction and installation of the production units, the training of personnel, and the initiation of operation of the production units. The quality of project-implementation services has a direct bearing on the efficiency of production units.

D. Institutions for Scientific Research and Training

A country needs to have at least one small unit in each scientific discipline to teach that subject and to develop an elementary research and scientific information capacity. The main task of these units is to stock and update existing knowledge. As a by-product, they may sometimes make a contribution to the production of new knowledge. Their activities should be supported through the recurrent budget for higher education.

In countries with limited financial and human resources, extraordinary resources may sometimes be allocated to a particular one of these basic research and teaching units. This may occur when its activity will provide inputs to industrial and agricultural research performed by the technological institutes.

On the other hand, countries with abundant resources and a fully-developed S & T infrastructure can invest in scientific research beyond merely updating the pool of knowledge for the education of new generations or feeding new knowledge to technological research for the productive sector. These better-off countries can also grant scientific research devoted primarily to expanding the frontiers of knowledge. In all cases, extraordinary resources for scientific research should be allocated after a careful assessment of the expected social usefulness of the proposed work and an evaluation of the efficiency that the unit may attain in carrying on the work.[7]

Occasionally, a highly gifted scientist comes along in a country with limited resources and an underdeveloped S & T infrastructure. He may want to pursue some field of science that is not directly related to local social objectives or problems. It is simply beyond the scope of this paper to propose a solution for such a situation, except to mention that work by such a gifted scientist does not always require large investments. He may value freedom to think and social esteem more than money. Moreover, he may best consider transferring to a laboratory outside his country where, with more adequate means he may obtain results that can be made universally available through the scientific literature. Such knowledge, unlike technology, is not generally appropriated by individuals or corporations.

E. Large Enterprises

More often than not, large enterprises do not need outside technical assistance for acquiring, adapting, or even creating knowledge required for designing, implementing, and operating their undertakings: When they do need the help of external institutions, they have the human resources required to establish direct relations with the technological institutes, scientific research laboratories and the groups that provide project design and implementation services.

F. Financial Institutions

The term "financial institutions" refers to commercial banks, development banks, venture capital companies and special institutions for the promotion and support of research, invention and innovation.[8]

Development banks and even commercial banks should become involved in the process of selecting the technology for the investments they support. They should also finance the costs of technological activities that are required by the enterprises in order to increase the efficiency of their production units.

Special institutions should be created to finance the risky investments that are associated with the development of new technologies and the adaptation of existing technologies to local conditions by enterprises at the top of the iceberg or its flotation line. The main instruments for financing the development of inventions through technological research are conditional loans, that is, loans which are totally or partially forgiven in case of failure of the research project.

Venture companies are needed to help launch newly developed products or processes onto the market. Equity participation is often the preferred way for financing such new undertakings.

The enterprises at the bottom of the iceberg can neither develop nor adapt technologies by themselves and they certainly cannot afford to pay technological institutes to do this for them. The only practical way to help these units is to provide grants to cover the costs of research work or technical services provided to them by the technological institutes at the request of the regional technical assistance group.

G. Governments

The role of the governments is to strengthen the supply of and to promote the demand for local S & T services. Government is able to increase the demand for local S & T services and products by taking, among others, the following measures:

(i) Granting income-tax rebates on expenditures of enterprises for locally performed scientific and technological research;

(ii) Setting procurement policies of the public sector that would favor efficient local production, especially by small entrepreneurs, and help decentralize the economy. For example, food for public school nutritional programs could be purchased by each school from small suppliers at the local level instead of being delivered by large enterprises to central warehouses for its distribution to the schools;[9] and

(iii) Setting procurement policies that would favor local consulting and engineering groups when project design and implementation services are required by development projects supported by public enterprises and government agencies.

Government may also strengthen the supply side through measures such as the following:

(i) providing grants for scientific research;

(ii) setting education policies that would make the students aware of their own inner potential for searching out and implementing solutions to local problems at all levels of the economy;

(iii) organizing technological institutes and institutions for the promotion and financing of inventions and innovations;

(iv) creating a legal and political environment that will allow the free exchange of ideas and the exploration of different solutions for social and economic problems; and

(v) providing grants to regional technical-assistance groups to pay for research, engineering and training work aimed to satisfy technological requirements that these groups and practitioners perceive as necessary for the development of a region, a village or an urban community.

H. Science Councils

Science councils should limit themselves to action only at the level of scientific research. They should coordinate research activities to avoid duplication and promote the formation of efficient research groups to work on research along lines of high social value, such as the following:

(i) classification, characterization, use, and industrialization of local species of plants and animals;

(ii) adaptation of foreign varieties of plants and animals and foreign agricultural practices to local conditions;

(iii) adaptation of mining and metallurgical processes to the characteristics of local minerals;

(iv) the scaling down of large-scale production processes to satisfy the demands of small markets at comparable costs; and

(v) solution of local social, cultural, and environmental problems, such as overpopulation, pollution, malnutrition, endemic diseases, violence, and natural disasters.

III. DUALITIES IN THE SYSTEMIC APPROACH

An important premise to bear in mind when applying the systemic approach to the design of S & T policies is that attributes which would normally appear to be opposites and hence contributing to an unproductive duality, are actually complementary and even synergistic, with all the benefits this carries. I now discuss a few of these pairs of attributes.

A. Hardware/Software

The technologies for equipment and installations, what is termed the hardware, should in no case ever be designed apart from its corresponding software -- technologies for the organization and management of the productive undertakings and their human environment. New hardware should be introduced at a rate that will allow societies and individuals to adjust to change. One way of promoting this adjustment is by reshaping organizational and legal patterns that too often are practices that have been established thousands of years ago for societies with primitive hardware.

B. Natural Science and Engineering/Social Sciences and the Humanities

Scientific and technological research should be oriented principally toward production problems. But efforts to solve production problems will also bring related social and human problems to the attention of the technological institutes and research laboratories, generating lines of research on education, health, social relations, life styles, art, and so on.

C. Small and Traditional/Large and Modern

The groups at the grass roots level that provide technical assistance to the traditional sector for village and community development should interact with technological institutes, research laboratories and project design and implementation groups that service farms and industries of the modern sector. At all levels technologies, small or large, old or modern, should be considered as potentially useful. The question is not the size or the age of the tools but rather their fitness to the job, the efficiency with which they use the resources, and their contribution to the full development of the human potential. Computers are used by the Sarvodaya Movement to keep control over stock in a cooperative network of several hundred community stores in Sri Lanka. On the other hand, even the most modern pharmaceutical industry may under certain conditions prefer to resort to hand-bottling and labeling instead of investing in costly and sophisticated high-speed automatic equipment.

D. Specialization/Multidisciplinarity

The degree to which the technological institutes must be specialized will depend on the type and size of the market for each of the services required and will vary from one country to another. In very small countries at initial stages of development a single multidisciplinary team of generalists can fulfill the needs of agricultural and industrial research, technical assistance and information services. In other countries the need for specialization may go far beyond the point at which separate institutes for agriculture and industry are required. It may be necessary to promote technical groups that will serve specific industrial products or agricultural crops. An institute for copper may be needed in Chile or Zambia, for example, or a research group for llamas and alpacas might be useful in Bolivia. even then, the question can be asked whether those groups should be organized as separate, independent institutes or as autonomous units within larger, multidisciplinary industrial or agricultural technological institutes. Economies of scale would be realized from the sharing of common administrative and ancillary services and synergism would result from cooperation among the different specialties. Even the research on human and social aspects mentioned earlier would benefit from integration within multidisciplinary technological institutes.

The benefits of multidisciplinarity are still greater in the work of the field-based practitioners. Otherwise the picture is the one seen repeatedly in many countries where several different extension workers, each of whom specializes in a given crop, visits the fields to teach the villagers appropriate practices and to provide them with seeds, fertilizers, and other inputs. In addition, villagers receive sporadic visits from public-health workers and social workers. None of them can establish deep human relationships with the villagers, for the villagers take them for what they really are -- external, occasional, sometimes timely advisers. Full development of the human potential at the bottom of the iceberg would require instead agents of change who were permanently engaged in development of the environment that they share with the villagers.

IV. POLITICAL ASPECTS IN THE DEVELOPMENT OF THE SYSTEMIC APPROACH

Any model for S & T decision-making and planning implies some particular ideology. The basic ideological premises of the proposal set forth in this paper are the following:

(i) A strong commitment to the preferential application of knowledge toward the development of the lowest levels of the economy through participative, self-reliant undertakings. As McNamara, former President of the World Bank, urged in 1972, this commitment is necessary if one really wants "to improve the welfare of the lower 40 percent of (the) populations (of the developing countries) which are living in conditions of the most abject poverty." [McNamara, R., 1972, p. 17] The Bank's new President, speaking on the same subject, stated that "experience demonstrates that development strategies that bypass a large segment of a society's people are not the most effective means to raise a nation's standard of living." [Clausen, A. W., 1982, p. 9]

(ii) A willingness to provide adequate local technological support for the preparation, implementation, and operation of development projects at all levels of the economy. The proposed strategy should operate at the two levels described by Mahbub ul-Haq: "On the one hand, a modern sector which grows fast and experiments with all kinds of price incentives and tolerates the prevalence of inequalities for some time. On the other, a large traditional sector where organization and institutional framework overcome the scarcity of capital and development is taken to marginal men through the organization of rural and urban works programs." [Ul-Haq, M., 1976, p. 42]

(iii) Acknowledgement of the fact that the state of the art of technological development allows psychological and socio-cultural needs -- participation, autonomy, education, recreation, communication -- to be satisfied simultaneously with the basic needs for survival -- food, shelter, clothes, and health. Attempts to satisfy the latter without the participation of the beneficiaries often end in the establishment of repressive controls or in harassment of the feeblest producers by big interests. In either instance, individuals at the lowest levels of the economy see their freedom for emotional, intellectual and physical expressions taken away, their initiative destroyed, and their psychological well-being impaired.

The interaction of this ideology with the prevailing political patterns will finally establish the extent to which the proposed model can be made a reality in each particular country. As Denis Goulet has written "Technology policy embraces a vast network of domains relating to a nation's scientific and technical pool, material and financial infrastructure, overall incentive system, attitude toward outside agents, degree of control over the direction and speed of planned social change, level of integration into global or regional economies, and relative priorities attaching to technological modernity itself." [Goulet, D., 1977, p. 145]

Taking this complexity and the political constraints into account, the designers of a national policy for S & T will need to choose from among the instruments and organizations that seem generally feasible, those that can be used under particular national conditions. Among other things the designers should:

(i) Analyze whether the existing economic policies will promote and support the building of the technological capacity needed to accomplish development objectives and guide technological change in the expected direction, at the necessary speed. The analysis should establish which policies have to be modified and propose the creation of new ones when necessary; and

(ii) Define the type, size, and scope of the scientific and technological organizations that would be required for support of the various social and economic activities.

A systemic approach for the design of a national policy for S & T can be seen as encompassing three processes:

(i) A process of finding out who is dependent on whom for the supply of knowledge and resources for adapting, creating, and using knowledge. By the same process it should also be possible to discover on what kind of knowledge the system depends;

(ii) A process of incorporating all participants in a system with their relevant motives, perceptions, and skills and of integrating the unique outputs of each level of activity into a synergistic whole; and

(iii) A process of co-learning and of using jointly both knowledge rising from the bottom and knowledge purposely sent down from the top of the socio-economic iceberg.

NOTES

[1]In the more developed Latin American countries, fewer than 1,000 enterprises are found at the top of the iceberg, and millions are found at the bottom.

[2]The word "technology," as it is used in this paper, refers to the organized set of scientific and empirical knowledge required to design, implement and operate a productive or a social system undertaking. Technology is put into use through technical services like engineering, quality control, trouble shooting, marketing, programming and budgeting, accounting and management.

[3]From a total universe of around 90,000 industrial firms only 3,000 were among the clients of the DTO.

[4]Data and comments on Danish technical tradition and the work of the DTO have been obtained from its Director, Mr. Kjeld Klintoe. Information on the efforts of the technological institutes has been provided by Mr. Jorgen Hammervig from the institute based in Copenhagen. In accordance with the Danish tradition of establishing close linkages between science and production, it is also worth quoting Mr. H. C. Orsted, the Danish scientist who discovered the electromagnetic effect. In 1824 he wrote, "The results arrived at by scientists must be made available to the layman in a popular form and in a language understandable to ordinary people. The manual worker, artisan and farmer must be educated to look at nature with understanding." Orsted had only contempt for those who "from made-up words build a castle of soap bubbles which looks very grand." He dedicated a paper in which he gave brief instructions as to how to boil saltpetre to "My dear fellow countrymen of the honourable farming class." The quotations have been taken from a special issue of _The Danish Journal_ published in 1977 to commemorate the 200th anniversary of Orsted's birth.

[5]See more details on the patterns of action of the Sarvodaya movement in [World Bank, 1982, Annex I]; [Ariyaratne, A.T., 1979] and [Goulet, D., 1981]

[6]See more on project design in [Kamenetzky, M., 1981] and on the role of consulting engineering design organizations in [Kamenetzky, M., 1979]

[7]See more on the allocation of resources for research work in [Araoz and Kamenetzky, 1975]

[8]See a description of some examples of institutions for the promotion of research, invention and innovation in [Rao, K.N. and Weiss, Ch., 1982]

[9]Example taken from an unpublished study of government procurement practices by Albert Araoz.

[10]In describing the three processes I have made free use of the pattern for design of organizations proposed in [Smith, W.E.,

Lethem, F.J., and Thoolen, B.A., 1980]

BIBLIOGRAPHY

Araoz, A. and Kamenetzky, M., 1975, Proyectos de Inversion en Ciencia y Technologia. Criterios para su formulacion y evaluacion en paises en desarrollo. El Pensamiento Latinoamericano en la Problematica Ciencia - Technologia - Desarrollo - Dependencia, J. Sabato, ed., Paidos, Buenos Aires.

Ariyaratne, A. T., 1979, Collected Works, Vol. I. The Netherlands.

Clausen, A. W., 1982, Address to the Board of Governors in Toronto (Canada), World Bank, Washington, D. C.

Goulet, D., 1977, The Uncertain Promise, IDOC/North America, New York.

Goulet, D., 1981, Survival with Integrity: Sarvodaya at the Crossroads, Lotus Press, Sri Lanka.

Kamenetzky, M., 1979, Preinvestment Work and Engineering as Links between Supply and Demand of Knowledge in Integration of Science and Technology with Development, Thomas, D. B. and Wionczek, M. S., ed., Pergamon Press, USA.

Kamenetzky, M., 1982, Choice and Design of Technologies for Investment Projects, World Bank, Science and Technology Unit, Washington, D. C.

McNamara, R., 1972, Address to the United Nations Conference on Trade and Development in Santiago (Chile), World Bank, Washington, D. C.

Rao, K. N. and Weiss, Ch., 1982, Government promotion of Industrial Innovation, World Bank, Science and Technology Unit, Washington, D. C.

Smith, W. E., Lethem, F. J., and Thoolen, B. A., 1980, The Design of Organizations for Rural Development Projects: A Progress Report, Staff Working Paper 375, World Bank, Washington, D. C.

Ul-Haq, M., 1976, Developing Country Perspective in Technology and Economics in International Development, report of a seminar sponsored by the U. S. Agency for International Development, Washington, D. C.

World Bank, 1982, The Industrial Sector: Problems and Progress. Mimeographed paper prepared by the Science and Technology Unit for the training program on scientific and technological aspects of development, Washington, D. C.

Technology Planning

UNCTAD SECRETARIAT*

I. THE NEED FOR TECHNOLOGY PLANNING

Since there are many worth while objectives for which special planning is not proposed, and yet technology planning is recommended, it may first be asked why this should be so. The purpose of this probe is not merely to show that technology planning is not a wasteful or unnecessary exercise but also to base the approach to technology planning on a precise reading of the need for that exercise. The nature of the need must govern the way in which the problem is tackled.

If either of the following statements were true, technology planning would be otiose: (i) the market mechanism functions properly to take care of the development and use of technology; (ii) even though the market mechanism may not take care of technology adequately, general development planning does, and a special category of technological planning is unnecessary. The need for technology planning can be established only by denying both (i) and (ii).

Statement (i) could be easily dismissed if technology development did not bring in any private profits, since if that were the case technology would not possess the signalling device through which the market mechanism operates. However, the development of technology as well as its proper use can be very profitable indeed. The question then arises whether the profits related to technology provide proper signals for the development and use of technology, or whether they are fundamentally defective. It would appear that the deficiencies of the guidance provided by profits are indeed fundamental. The defects relate partly to the

From **PLANNING THE TECHNOLOGICAL TRANSFORMATION OF THE DEVELOPING COUNTRIES**, (UNCTAD Secretariat), 1981, (42-50), reprinted by permission of the publisher, U.N. Publications, N.Y.

nature of technology as an influence on production but also to the dependence of technology on signals, related to other influences on production -- in particular, resource prices.

First of all, it is easy to see that many of the ways in which technology affects production have strong "externalities". For example, when a new method of production is invented, it helps not only the inventor but others who can use it. To avoid this problem, the new technique may be patented; but this leads to fresh difficulties, since the patent owner then has a monopoly power, which is made particularly restrictive by the fact that potential purchasers of the right to use that technique -- especially if they belong to a technologically backward economy -- may not even know precisely what it is that is being sold. The combination of externality and ignorance -- both typically present in dealings involving technology -- make the market mechanism especially unsuited for the development and use of technology. The conflict between incentives to develop better techniques and incentives to use the ones that have been developed is fundamental in a market system, the former calling for stricter patent laws and the latter calling for exactly the opposite.

Earlier, the role of technology in enhancing production possibilities was shown to affect a number of distinct though interrelated problems, viz., the methods problem, the products problem, the skills problem and the tools problem. It is important to note that the fundamental conflict between development and use exists in relation to each of these problems. For example, skill formation is helped by the production experience of workers, but this does not form part of the profits of the enterprise in which the workers are employed. To "internalize" that externality, the movement of labour might be restricted, so that the enterprise could benefit more from the skills developed. But that restriction -- apart from being a limitation of freedom -- would also, by preventing the movement of skilled workers to production units which have greater use of their skill, act against better use of the skills that had been developed. Technology is not a factor of production that can be freely bought and sold in the market in the shape of a fully "private good", and the failure of the market mechanism to "deliver" has much to do with this basic character of technology.

A second difficulty arises from the fact that the impact of the development of technology may be felt only in the rather long run, very often after many years. In view of the strict financial discipline of private firms and the high rates of private discount, these long-run benefits may be systematically undervalued. Obviously, the social valuation will be different only if the social rates of discount differ from the private rates, but it has been argued that there are good reasons for taking the social rates to be below the corresponding private ones.[1] If the society can take a view less geared against valuing the future, it may see the market decisions as systematically inoptimal from its point of view.

A third problem concerns the importance of interdependence in the development of technology, even when the interdependences work through the market. Externalities are interdependences that work outside the market, but there are market-based interdependences which the market mechanism may also fail to deal with appropriately because of lack of co-ordination. Development of technology in field A may be held up by lack of development in field B, or vice versa, and though co-ordinated development of the two fields could sustain both of them, the market may not succeed in arranging that co-ordination. Though it is not difficult to imagine a Walrasian "auctioneer" who irons out these interrelations -- a hypothesis on which so much of traditional neo-classical analysis rests, often implicitly -- this piece of fantasy is rarely of use in understanding what really happens in markets.[2] The problem is particularly important in the development of new technology, which requires co-ordination between many different sectors over a long period of time if it is to be profitable.

Fourth, significant uncertainties are unavoidable in the field of technology, since it is a characteristic of expanding knowledge that the result of the effort cannot be known beforehand. The market mechanism has to operate through this haze of uncertainty, and though under certain assumptions this would create no serious problem for the market, the reality tends to be rather different. Uncertain investments -- even though they may be very profitable on average -- are often neglected by the market mechanism, especially when the markets are themselves very primitive -- as they frequently are in developing economies.

All these problems limit the operation of the market mechanism because of certain characteristics of technology itself. In addition, problems arise from the limited capacity of the market mechanism to deal with related factors; for instance, the imperfection of capital and labour markets may misdirect technological change. This problem has been discussed in the context of the use of appropriate techniques of production, but the question of misdirection also arises in connection with the development and acquisition of technology. Indeed, there is much evidence that technological changes in developing countries are often ill adapted to the resources available in those countries; sometimes the changes are better suited to the resource positions of the developed countries from which the techniques come than to those of the developing countries to which they are sold.

It is thus clear that the market mechanism may fail in many different ways to provide a framework within which technological development can flourish. In so far as these deficiencies can be remedied by planning, it may be asked whether they will not automatically be taken care of by overall development planning, and why there should be a special category of planning activities for technology if development planning can include technology planning as well. Technology is an integral part of the economy, and every action of production, consumption, investment, etc., is connected with the underlying technological base of the economy. It might, thus, seem natural that technology planning should be subsumed within overall development planning rather than treated as something on its own, and that the failure of the market mechanism to deal with the development of technology adequately is not a sufficient reason for establishing a separate mechanism specifically for technology planning.

This view of technology planning would have been quite appropriate if the overall development planning traditions did not exhibit biases that work against giving adequate attention to the development and use of technology. Unfortunately, however, such biases do exist, and -- what is more important -- they arise not for ad hoc reasons but from systematic deficiences in traditional development planning. If the market mechanism has its systematic failings, so has overall development planning. Briefly, the biases can be placed in three distinct categories, to wit: tangibility bias, simplicity bias, and articulation bias.

The presence of a bias towards tangibility is distressingly common in development planning. Indeed, the standard models of growth, which focus on capital accumulation, with technical progress exogenously given (or worse, simply absent), have provided the theoretical underpinning of development planning for a long time. This is true not only of aggregative growth models of the traditional type but also of typical multisector models of input-output, activity analysis, and linear -- and non-linear -- programming of disaggregated optimal growth. Though growth models of the more theoretical type have sometimes involved endogenous technical progress, or "learning by doing", in the models that have seen application these influences have been typically absent. Greater concern with tangible resources has much to do with this neglect of the more subtle, less quantifiable aspects of technology. The obviously physical nature of an irrigation dam, a steel mill, heavy motor traffic, or a building boom, gives more tangible evidence of economic development, and the development planning models have -- perhaps understandably -- tried to emphasize the elements in their strategy.

The simplicity bias shows in the concentration on those variables which can be dealt with more easily in a planning exercise: easier quantifiability; simplicity of the assumed relations, e.g., between inputs and outputs (x amount of iron ore of the content needed per unit of steel, etc.). Technology is inherently complex as an entity and as an influence; the development planner has frequently responded to Rudyard Kipling's plea: "Teach us delight in simple things".

Aside from tangibility and simplicity, there is also the issue of articulation, for planning is notoriously biased towards the articulate. This applies to the relative amount of attention paid to one group of people -- the rich, the urban, or the educated, who are less mute than others -- than to another, but it also applies to the question of taking note of different types of influences with which planning is concerned. Technology has often been given inadequate attention or even ignored, largely because of failure to articulate the need to take it seriously. Silence may breed respect, but more often it leads to neglect. One of the main arguments for technology planning is to provide appropriate attention to technology in development through reasoned articulation. The point is not so much that overall development planning cannot take full care of technology planning, but that it does not. Hence the need for special attention.

TABLE 1

Indicative dynamics of technological transformation in developing countries

Sector	Ratio of per capita availability of goods and services in developed market-economy countries to availability in developing countries, 1975 [a]	Annual percentage rates of per capita growth required for reaching 1975 levels of developed market-economy countries		
		In 50 years (by 2025)	In 35 years (by 2010)	In 25 years (by 2000)
Agriculture	2.0	1.4	2.0	2.8
Mining	3.0	2.2	3.3	4.5
Manufacturing	16.0	5.7	8.2	11.7
Consumer goods	11.0	4.8	7.0	10.1
Intermediate goods	13.0	5.3	7.5	10.8
Capital goods	28.0	7.0	10.0	14.3
Services	14.0	5.4	7.8	11.1
Total	11.0	4.8	7.0	10.1

Sources: UNCTAD, *Handbook of International Trade and Development Statistics, Supplement 1977* (United Nations publication, Sales No. E/F.78.II.D.1), and United Nations, *Statistical Yearbook 1977* (United Nations publication, Sales No. E/F.78.XVII.1).

[a] Rounded figures. Developing countries exclude China and other socialist countries of Asia.

II. FROM NEED TO OBJECTIVES

An understanding of the need for technology planning leads to an assessment of the objectives of that exercise. If the need arises partly from the failure of the market mechanism, then the objectives of technology planning must include undoing the biases of the market allocation. The market may leave the technological base severely deficient, and technology planning has to be geared to changing that situation. Technology planning is concerned with the acquisition, adaptation and development of technology, and the focus clearly has to be on those elements thereof which the market would tend to neglect. If a transnational firm decides to set up a branch in a developing country using the technology employed elsewhere, the market will see to it that that technology is used; there is no need for technology planning to achieve that result. But the transnational firm may not pay adequate attention to the long-run effects on the economy of the developing country of trying out other types of technology that are less certain, long in gestation, and full of externalities or of market interdependences. In the search for suitable technology, which is not necessarily what the market delivers, such potentially neglected possibilities deserve particular attention.

The biases of technological choice are not confined to transnational enterprises or even to relatively large concerns. The neglect of externalities, interdependences, long-run gains and uncertain benefits is a general feature of market allocation and constrains the choices of firms of all types and sizes. The first objective, therefore, is the acquisition and adaptation of technology from all available sources, domestic and foreign, on the best possible terms.

But the ability to acquire and adapt is itself an art that requires cultivation; as noted earlier, this leads to the objective of non-dependence in the long run. This objective belongs to a somewhat deeper level of technology planning, going beyond the pursuit of the technology that serves best the immediate economic need. Certainly, one aspect of technology is that it enhances the ability to develop further technology. There is plenty of evidence that technological capability requires more internal generation and greater independence from external sources. This is not, it should be stressed, the same thing as technological autarchy; there is no reason why each country should use only techniques developed within its own frontiers. If there is technological give and take, then each country develops capabilities of its own; what is to be avoided is the tremendous asymetry that characterizes the relation between developed and developing countries in the modern world. So "self-reliance" has to be understood as a denial of this asymmetry rather than a demand that each country rely only on its own homespun technology.

Both the "acquisition and adaptation" objective and the "self-reliance" objective fit well into the aims of general development planning. Nevertheless, the biases of development planning -- concern with tangibility, simplicity and articulation -- often leave these objectives under-recognized. Even when recognition exists on paper, the pursuit of these objectives may be artificially limited. The objectives under discussion may appear to be too obvious to need stating; however, the point of clear and assertive articulation of these objectives is to make them operational and indeed actually used. This is only the beginning of the exercise; the next steps are taken up below.

III. FROM OBJECTIVES TO CRITERIA

Objectives have to be translated into more precise criteria before they can be used. It is useful to distinguish between two types of criteria: (i) those that serve as rules of thumb, and (ii) those that are explicitly related to attempts at optimum choice. The former aim at transforming the basic objectives into some crude indicators in terms of which choices can be made, and the latter translate the objectives into an objective function which is maximized subject to the identified constraints (economic as well as social and political ones). It is reasonable to argue that the latter method is better if the objectives translate precisely, if all the relevant factual data are available, and if detailed investigation has negligible cost (in terms both of resource use and of the delay in taking decisions). However, it is also reasonable to argue that none of these conditions is easily satisfied, so that, though exercises based on type (ii) criteria may be ultimately more attractive, type (i) criteria cannot be easily dismissed.

It is sometimes thought that rules of thumb are so arbitrary that it is difficult to discuss them rationally. This is not so, since the rationale of the different rules of thumb may be eminently suitable for examination, and two rules of thumb may be easily compared. For example, one for the allocation of resources for the development of technology may be to divide them between the sectors according to the gross value of the output of that sector. More is apparently "at stake" in a sector with a higher value of gross output. In assessing this rule of thumb within its chosen rationale, it is useful to compare it with another -- that the allocation between sectors be made according to net value added. There is no doubt that the latter is a better indicator of what is "at stake"; the gross output of a sector with a high value of material inputs exaggerates what the economy gets out of that sector, since material inputs from elsewhere are counted in the gross value. Hence within the chosen perspective the rule of thumb of pursuing gross value may be rejected in favour of the rule of thumb of espousing value added. In economies in which labour is fully employed, even the net value added may exaggerate the contribution of that sector since labour has to be drawn from elsewhere for that sector to produce the value added. In such a context, it may be appropriate to focus on the net value added minus the value of wages. This calculation of "profits" will require correction if externalities are strong, and then the chosen focus may more sensibly be the externality-corrected value of the net contribution of the sector (net of material costs and net of labour costs). When full employment does not obtain, the deduction of labour costs may be eschewed. These alternative rules of thumb may be compared within the chosen perspective -- a very limited one -- of what is "at stake" in the sector as a whole. Such comparisons may be worth doing despite the limitations of the perspective, since the complexity of technology planning makes some limitations inevitable, and perfectionism is not a very helpful motto for the technology planner.

However, the perspective within which these approaches are to be compared would need to be examined on its own. The important point is not so much what is "at stake" in the sector in question but the difference that can be brought about in that sector by technology planning. If a sector has a high total contribution but is known to be quite unresponsive to additional resource investment in technological development, then it would seem pointless to put any more resources for technological development into that sector. This way of viewing the problem leads naturally to considerations of cost-benefit analysis, an assessment of what is obtained by putting more resources into that sector, i.e. the benefit compared with the cost. Cost-benefit analysis, in its turn, leads to optimization exercises with objective functions and constraints.[3] The argument has thus moved slowly from type (i) criteria to type (ii) criteria. It is worth noting that the difference between the two types of criteria is not really one of kind but one of degree, since the rationale of rules of thumb carries with it -- usually implicitly -- some background cost-benefit picture, and since optimization approaches make use -- usually implicitly -- of some rules of thumb to simplify calculation and diagnosis.

This is not the occasion to discuss the details of cost-benefit analysis, which have been extensively explored elsewhere.[4] It is, however, useful to examine the relationship between the objectives identified in the present exercise and the criteria that cost-benefit analysis may use. There may be a conflict here. It is often overlooked that cost-benefit analysis has no given set of goals, and that the methods it uses deal with the relation between goals -- whatever they may be -- and policy decisions at the micro-level. It is important to recognize this, since the goals of technology planning can also be incorporated in the framework of cost-benefit analysis. This does not imply a recommendation that cost-benefit analysis should be carried out in terms of goals that are in sharp conflict with the usual objectives of pursuing some conception of national welfare based on assessment of overall prosperity and its distribution. For the objectives of technology planning, as identified here, are not really at variance with these broad-based and widely supported goals -- indeed, they are derived from these more basic goals. Underlying the assertion that one of the objectives is to "acquire and adapt" technology "from all available sources -- domestic and foreign -- on the best possible terms" is the implication that there are some criteria for judging terms -- and that leads back to the more usual objectives of investment planning: more consumption, better distribution, better quality of life, etc. The objectives of technology planning are ultimately "intermediate" to these welfare objectives.

The issue of "self-reliance", however, raises an important problem of evaluation. If technological self-reliance -- or, more appropriately, "non-dependence" -- is to be pursued for the sake of developing future technological capability (rather than for, say, national pride), then "self-reliance" as an objective is two stages removed from the welfare goals. Self-reliance is pursued for greater technological capability, and greater technological capability is sought for considerations of national welfare. This two-stage remoteness is inherent in the nature of technological planning, since the policy strategy has to take note also of the effects, and not only of the results of the first stage.

It is important to realize that this two-stage remoteness is a common problem of investment planning; though some of the remoter effects are more difficult to assess in the case of technology planning, nothing new in principle is involved. It is also worth noting that even when there is a move away from the more speculative effects of "self-reliance" and a concentration solely on the effects of technology on economic production, remoter effects will still be involved. This is because technological investment today will bear fruit over a long period, and the exercise of technology planning -- even at this more direct level -- must include the assessment of what is likely to occur in the distant future. Investment in technology -- like other investment -- involves forgoing some immediate benefits (e.g., consumption) for the sake of some future benefits (e.g., future consumption), and this type of intertemporal choice is a common element of both general investment planning and technology planning.

The criteria to be used in technology planning, then, are of two types. Following the type (ii) approach which is ultimately more satisfactory, the criteria for technology planning may be linked with general cost-benefit analysis, the latter method being used in choosing technology variables. The ultimate objective remains the pursuit of social welfare, and technology planning extends that pursuit to the specific sphere of programming technological decisions. The type (i) criteria take more _ad hoc_ targets -- for example, raising the proportion of investment in technology by a certain percentage, or increasing the allocation of investment in sectors with high net value added or a high likelihood of technological response. As has already been argued, the type (i) approach can be seen as a rough way of doing what type (ii) would do, using some basic logic and applying it in a simple fashion. In practice, the reliance on the type (i) approach may be considerable, but since there is no basic conflict between this and the more sophisticated type (ii) approach, there is some advantage in bringing in cost-benefit analysis when this can be done sensibly.

IV. FROM CRITERIA TO PLANNING

Planning is based on objectives and criteria. But it also depends on choosing fields of action, identifying control variables, and selecting instruments. To reiterate, the technology problem has been split roughly into: the methods problem; the products problem; the tools problem; and the skills problem. This classification may serve as the starting-point for a discussion of fields of action. Technology planning will be concerned with developing better methods, a better range of products, better tools, and better skills. Obviously, nothing much of interest can be done in the general area of technology planning without a clear understanding of the state of technology in these different respects. It is, therefore, essential to begin in investigating the technology profiles of the various sectors of the economy, paying particular attention to potentially important sectors.

The notion of "potential importance", though a difficult one to define with precision, is a necessary consideration since it is clearly impossible for reasons of time and resources, to construct a technology profile of every part of the economy with equal care. This is where some rough use of type (i) criteria will be sensible. Obviously, other things being equal, a sector with a large net value added is potentially more important to look at. Similarly, a sector that is known to have undergone technological change, either in the domestic economy or abroad, may be potentially more important -- in other words, world-wide technology trends should be taken into consideration.[5] On the basis of such broad criteria, some sectors may be chosen for more detailed investigation than others. The general strategy at this stage is to construct as useful a profile of the technological situation as possible for systematic evaluation of planning options.

Even for a given sector, or for that matter even for a given industry, the construction of a technology profile is a far from trivial exercise. It is an act of description, and description always involves choice -- to wit, choosing, from the large range of possibilities, a particular subset of relevant statements. The notion of relevance relates, again, to potential for technological change and the importance of such change. There is thus an implicit evaluation exercise in the homely task of constructing a technology profile. The important point to bear in mind in all these exercises is the eventual interest that motivates the description. With this focus on potential relevance, a technology profile has to go into such details as processes and methods used, product types covered, skills employed, tools and machinery used (types as well as vintages). In assessing the possible fruits of planned change in each sector, such a profile will be an essential first step. Since profile-making is expensive and time-consuming, it is, at the same time, important to avoid going into unnecessary detail; what is needed is functional efficiency.

In evaluational planning, it is necessary to examine the alternatives that exist and can be chosen. Part of this exercise may take the direct form of considering alternatives that are known to exist -- ones that are already in use elsewhere or ones that are believed to be technologically evolved even if they have not been tried out. In assessing such alternatives, even the rough use of cost-benefit analysis may be important. The costs may not be known precisely, nor may the exact benefits to be expected, but it could still make sense to guess the possible results and see what the net benefit looks like from the point of view of changing technology. If there are major gaps in the understanding of the probable costs and benefits, it will be necessary to supplement the first rough estimates by "sensitivity analysis", checking how the results change as the assumptions are varied.[6] It may also be useful to apply the "range method", by which the variables are given not point values but ranges of values and it is then checked whether the proposal is a good one under the most unfavourable assumptions (in which case it should be accepted), or a bad one under the most favourable assumptions (in which case it should be rejected). There would, of course, remain some cases in which a decision was not clearly indicated, and for these a narrowing of the range of values would be called for; however, the straightforward decisions could be separated out first and the more controversial decisions taken up progressively.[7]

Such direct exercises can be performed for all the particular problems listed, viz., the methods problem, the products problem, the skills problem, and the tools problem. The costs and benefits of each proposed change -- with an estimate of "best guess values and of ranges of possible values -- can be systematically examined with the general objectives of development planning and the criteria corresponding to them. Sometimes the effects may be particularly difficult to estimate, e.g., the effect of using newer methods on skill formation, or of technological experience on future capability in evolving better techniques. In such cases, where "best guess" values mean nothing and ranges are themselves figments of the imagination, it is clearly useless to try cost-benefit analysis. In these cases, such sophistication should obviously be abandoned, since technology planning is concerned with concrete results. The point at which abandonment is sensible is clearly a matter of judgement, and no mechanical formula can be applied.

In addition to direct exercises, there are indirect methods of technology planning. They are concerned not with selecting particular proposals for change in methods, products, skills or tools, but with providing institutional facilities that encourage positive development in the field of technology. For example, extension services can be expanded, consultancy facilities offered or the incentive structure changed through fiscal reform (see chap. VI). In this indirect approach, the control variables are not technological changes themselves but institutions that govern the selection, use and development of technology. Such decisions are particularly difficult to assess by the formal methods of cost-benefit analysis, since benefits in particular tend to be hazy and imprecise and may not be easy to marshal within the format of that analysis. Here, type (i) criteria would be much more likely to provide scope for sensible discussion and debate.

A combination of direct and indirect methods may be the best way of undertaking technology planning in most developing countries today. Some particular fields may offer scope for direct choice and even for sophisticated use of cost-benefit assessment, for example: specific techniques of production, specific proposals of product type variation, or the expansion of skill formation in a particular field through schooling. On the other hand, there are likely to remain quite a few areas in which technology planning can be handled only indirectly and rules of thumb must be used. The uniform use of a single, comprehensive technique of technology planning to be applied to every sector and every problem is difficult to envisage.

V. FROM PLANNING TO PROGRAMMES

The general features of technology planning have to be translated into specific programmes, which together make up the plan. The strategy outlined so far suggests how this might be done. At the risk of over-simplification, it may be useful to split up the task of technology planning into a number of components, reflecting specific exercises:

(1) Specification of objectives;
(2) Profile making;
(3) Direct/indirect division;
(4) Importation and adaptation of techniques;
(5) Evolution of domestic technology;
(6) Manpower training programmes;
(7) Technological incentive structures;
(8) Sectoral plans;
(9) Integration.

1. Specification Of Objectives

This part of the exercise has already been discussed at several levels. To sum up, there is the need to be clear about the objectives of development planning in general, and here the focus has to be on the particular concept of social welfare that may be used to specify the objective function of planning. The level of national consumption, its distribution, the removal of poverty, the fulfilment of certain rights-based objectives, etc., can be considered in line with the social situation and political values. At a more immediate level, the objectives of technology planning relate to the best use of available technology (covering both the choice of particular techniques and the terms and conditions governing their use, e.g. royalties) and to the development of technological capability and self-reliance (in the particular sense already analysed). The general lines are clear enough, but the details of the objectives can be decided upon only in the context of the exact political, social and economic circumstances obtaining in the country in question.

2. Profile Making

This consists in making a general profile of the technological situation as a background to the exercise of technology planning. There is a need here for selection in terms of priorities, as has already been discussed, and rules of thumb, reflecting a rough-and-ready rationale, have to be used.

3. Direct/Indirect Division

Another early exercise involves determining the respective domains of direct and indirect planning. This depends partly on the nature of the existing institutional mechanism and on whether the government is confident of its own ability to do direct cost-benefit calculation. But it is also a question of political strategy, viz., to what extent the government wants to operate directly and to what extent it would rather leave it to other organizations, working in accordance with rules, incentives and guidance provided by the government. Strong direct participation will undoubtedly be necessary in some fields, e.g., particular education programmes for specific skill formation, while a decentralized approach will certainly be more convenient in others, e.g., evolving the details of on-the-job training. But there is much scope for variation depending on the nature and commitments of the government.

4. Importation And Adaptation Of Techniques

One of the areas of great importance for technology planning is the use of technology developed abroad. This raises issues of adoption and of adaptation. In so far as this exercise is done directly, cost-benefit analysis will clearly be worth considering, possibly with range values and with sensitivity analysis. Even if rules of thumb are used, they have to be kept under constant scrutiny since their rationale calls for such close examination, as has been noted above. If the procedure is indirect, then the focus has to be on the indirect instruments, such as incentives to adapt technology to local conditions and encouragement -- regulational or financial -- to train local personnel in the use and modification of the techniques employed. Even under the indirect approach, it will be the responsibility of the government to explore the types of adaptation it wants, the nature of the modifications it demands, etc. Some of these aims will have a longer-run perspective than others, since adaptation and modification take time. But it is most important for the government to view these questions in an active rather than a passive way and to seek changes in particular directions over time. Though the directions may be determined by rules of thumb (e.g., a greater use of aluminium rather than copper if the domestic supply situation so requires, or the use of more semi-skilled labour if the domestic training programmes can cope with that more easily), they are subject to scrutiny and to rough cost-benefit examination in so far as the government is able to carry it out. Among the effects of importation and adaptation to be considered are the indirect results of skill formation and, as discussed above, the even more elusive goal of developing the technological capacity of the country.

5. Evolution Of Domestic Technology

Though adapting imported techniques rather than using them without modification helps to develop domestic ability, the long-run objective of self-reliance requires deliberate fostering. Traditions are hard to develop, but there are many recent examples of the remarkable success of some erstwhile backward economies in seizing the technological initiative. This is partly an institutional question, and the government can help by setting up facilities for innovative activities, providing financial incentives and support, and encouraging recognition and the use of the results of innovation. It is most unlikely that cost-benefit analysis can give any very precise guidance on this, other than ruling out some outrageously expensive schemes or proposals that have little chance of success. The main reliance has to be on rules of thumb, and they can be formed not as optimum rules but as appropriate directions of change, moving away from both the biases of the market mechanism and the biases of the typical overall development planning. The factors involved in each of these biases were discussed in section A of this chapter; they have to be examined to decide how vigorous a correction is needed.

6. Manpower Training Programmes

Skill formation is, of course, a central feature of technology planning. Targets for training must take into account the need for trained personnel arising from the importation, adaptation and evolution of new techniques, including both new methods of making old things and methods of making new things. In deciding on the priorities for training, for example, as between agronomists and electrical engineers, note will have to be taken of the manpower requirements of planned technological changes. This is obvious enough and hardly needs stressing. Less obvious, however, is the question of technical innovation, which is also related to decisions on training and on facilities for teaching and research. Decisions may have to be taken on whether to give priority to -- using the same example -- agronomic innovation or electrical engineering innovation, and the policy regarding manpower training in these fields will also reflect these innovational priorities rather than merely taking into account the manpower requirements of the techniques already evolved.

7. Technological Incentive Structures

The importance of incentives has already been emphasized in the context of indirect ways of carrying out technological planning. Given the central importance of this issue, it is useful to treat it as a component of technology planning in its own right. The incentives may include taxation measures, subsidies, prizes, publicity, information exchange, etc., and they affect virtually every aspect of technology planning. Even in the case of a very active government that performs many direct exercises, the need for a proper incentive structure can hardly be over-emphasized, since the government has to rely on the initiative of people outside government departments contributing in the acquisition, adaptation, modification, development and use of technology. Once again, formal cost-benefit analysis is not likely to provide much precise guidance, and rules of thumb derived from overall objectives have to be used. There is also much to be learned from actual experience of the operation of incentive systems, and the incentive structure can be adapted according to what is learned this way. The art of arranging incentives for technology is in itself a subject of some importance.

8. Sectoral Plans

The various components of technology planning outlined above must cover sectoral technology plans as well. The plans for importing or adapting technology, evolving new techniques, providing manpower and organizing incentives will naturally relate to particular sectors in very specific ways. It may nevertheless be useful to go through the technology over again sector by sector, in order to check whether anything important has been left out. It is also useful from the point of view of implementation to spell out the implications of general technology programmes for particular sectors. Thus, without treating them as in any way independent, sectoral plans can be separately specified within the general plan. Sometimes this will also permit the specification of a technology package for a particular sector, including the unpackaging of imported technology and its gradual adaptation and modification, the evolution of new departures, the absorption of specially trained manpower, etc. (if such a package fits into the different layers of the technology plan considered under different headings). It is, in general, useful to view the same set of policies from more than one perspective: e.g., in a problem-oriented way (products, methods, tools, skills), in a task-based way (adaptation, importation, evolution), and in a sectoral way (each sector taken separately).

9. Integration

It has been emphasized throughout this chapter that the exercise of technology planning is not to be regarded as independent of development planning in general. Therefore, any technology plan must give some consideration to the question of integration. It was pointed out that overall development planning often fails to attach sufficient importance to the requirements of technological development because of the existence of a number of biases; the tangibility bias, the simplicity bias and the articulation bias were all mentioned in this context. In providing correctives, technology planning has to be geared to counteracting these biases. The bias from inadequate articulation is perhaps the simplest to deal with, though arguably the most important in terms of its effects. In the last few decades, articulation and pressure in favour of the planning of education and health have led to remarkable achievements in these fields in many countries. The same can happen in the case of technology if technology planning succeeds in shifting sufficient attention in that direction. The technology planner must not shy away from stating what may appear to him to be obvious, since one reason for the neglect of technology is the absence of articulation. But the tangibility bias will militate against the pursuit of technological results that look elusive, and the complexity of cost-benefit analysis when applied to this field and the difficulty of applying other traditional methods of planning will continue to make the espousal of technology planning rather an uphill task. The cards are thus somewhat stacked against technology planning when it comes to integration, but integration nevertheless has to be sought to make sense of the technology programmes. The spelling out of rationale, explicit specification and reasoned examination of rules of thumb, together with limited use of cost-benefit yardsticks, can go a long way towards achieving greater acceptability and integration of technology planning. Integration is obviously essential, since the need for better technology and the rewards it brings relate ultimately to the general goals of development planning. As has already been emphasized, the objectives of technology planning are, in an important sense, "intermediate" to the general objectives of planned development. This should be borne in mind when preparing a technology plan.

NOTES

[1]See, for example, A.K. Sen, "Isolation, assurance and the social rate of discount", The Quarterly Journal of Economics (Cambridge, Mass.), vol. 81, 1967.

[2]Some of the more important limitations have been admirably discussed in K.J. Arrow and F.H. Hahn, General Competitive Analysis (Edinburgh: Oliver & Boyd; republished, Amsterdam: North-Holland, 1979).

[3]The relationship between shadow-price-based cost-benefit analysis and objective-constraint-based optimization has been discussed in A. Sen, Employment, Technology and Development (Oxford, Clarendon Press, 1975), chaps. 11-13.

[4]See, for instance, Guidelines for Project Evaluation (United Nations publication, Sales No. E.72.II.B.11); and I.M.D. Little and J.A. Mirrlees, Project Appraisal and Planning for Developing Countries (London, Heinemann, Educational Books, 1974). See also UNCTAD, Handbook on the Acquisition of Technology by Developing Countries, op. cit., annex.

[5]This is a major element in science and technology planning in the USSR. See "Experience of the USSR in building up technological capacity", study prepared by G. E. Skorov at the request of the UNCTAD secretariat (TD/B/C.6/52), September 1980; and "The planning of science and technology development in the USSR" by B. Milner and G. Smirnov (to be published by UNCTAD).

[6]See L. Squire and H. G. Van der Tak, Economic Analysis of Projects (Baltimore, Johns Hopkins University Press, 1975) chap. five, pp. 45-46.

[7]See A. Sen, Employment, Technology and Development, op. cit., pp. 112-114; and also his "Labour and technology", in J. Cody, H. Hughes and D. Wall, eds., Policies for Industrial Progress in Developing Countries (New York, Oxford University Press, 1980), pp. 126-128.

Strategic Choices in the Commercialization of Technology: The Point of View of Developing Countries

CONSTANTINE V. VAITSOS

This article refers to some of the policy considerations arising from the actual mechnisms of technology purchase by developing countries within or outside the framework of direct foreign investment[2]. Given the complexity and extent of the subject matter we will not deal with issues related to the selection of appropriate technologies nor with the existence (or not) of domestic scientific and technological activities which are complemented or substituted by know-how flows from abroad. Furthermore, we will not deal with broader issues referring to the re-enforcing interrelationship between income distribution, consumption structures and technology utilization. Rather we will limit ourselves to the evaluation of the process by which production know-how becomes a negotiable unit (a merchandise) which is traded among transnational enterprises and developing countries. We will thus refer to the market within which technology is being commercialized. In the latter part of the article we will draw parallel conclusions between the experience of developing countries on concession agreements in the extractive industries in the first part of this century and their negotiations in technology licensing agreements in the 1960s. The empirical evidence provided comes from research undertaken in the Andean Pact countries.

I. THE MARKET OF TECHNOLOGY: PROPERTIES AND CHARACTERISTICS

We can distinguish four basic properties which have specific repercussions on the market for technology.

From **INTERNATIONAL SOCIAL SCIENCE JOURNAL**, Vol XXV, No. 3, 1973, (370-386), reprinted by permission of the publisher.

1. In the process of its commercialization technology is usually embodied in intermediate products, machinery and equipment, skills, whole systems of production (like turnkey plants), even systems of distribution or marketing (like cryogenic technology in ships that transport liquid gas), etc. Thus, know-how represents a part integrated in a larger whole. As a result, the market for the former is not independent but constitutes part of the market for the latter. This market integration of various inputs creates non-competitive conditions for each one of them since they are sold in a package form.
2. In the formulation of the demand for information, as in all other markets, a prospective buyer needs information about the properties of the item he intends to purchase so as to be able to make appropriate decisions. Yet, in the case of technology, what is needed is information about information which in many cases could effectively be one and the same thing. Thus, the prospective buyer is confronted with a structural weakness intrinsic in his position as purchaser with resulting imperfections in the corresponding market operations.
3. The use of information or technology by a company or person does not in itself reduce its availability, present or future. Thus, the incremental cost in the use or sale of an already developed technology is close to zero for someone who already has access to that technology. In cases of minor adaptations (due to scale, taste, local condition differences, etc.) the firm incurs certain costs that can be estimated and usually do not exceed a figure in the tens of thousands of dollars. From the point of view of the prospective purchaser, though, the incremental cost for developing an alternative technology with his own technical capacity might amount to millions of dollars. Given market availabilities, the price between zero or tens of thousands of dollars, on the one hand, and millions of dollars, on the other, is determined solely on the basis of crude bargaining power. As in the case of bilateral monopolies the price settled within the range is _a priori_ indeterminate.
4. Technology or knowledge in general are not only irreducible over time while they are being used by a given firm or an individual; their availability is also not diminished for those who possess them if others use them too. This creates what economists call a public good or property. Yet, appropriation for the commercial use of knowledge is sought through restrictive methods such as legal systems (patents), contractual obligations in licensing agreements, acquisition of ownership of firms, etc.

Despite the potentially multiple sources of supply of technology for developing countries (given the relatively early vintage of such know-how[3] empirical evidence indicates high market concentrations of diverse types. Such concentrations imply sequential monopoly characteristics in the case of the market for technology with corresponding price, income and other effects. We

TABLE 1. Percentages of total outlays by sector paid to countries by Chilean licensees for royalties, profit remissions, intermediates, etc.

Sector	Countries	Percentages of total payments by the whole sector going to the countries that appear in the previous column
Food and beverages	Switzerland and United States	96.6
Tobacco	United Kingdom	100
Industrial chemicals	Federal Republic of Germany and Switzerland	96.6
Other chemicals	United States, Federal Republic of Germany and Switzerland	92
Petroleum and coal products	United States and United Kingdom	100
Rubber products	United States	99.9
Non-metallic minerals	United States	97
Metallic products (except equipment)	United States	94
Non-electric machinery	United States	98.7
Electric equipment	Netherlands, United States, Spain	92
Transport equipment	France, Switzerland	89

See: Gastón Oxman, in *La Balanza de Pagos Tecnológicos de Chile*, September 1971 (mimeo.).

TABLE 2. Countries with the highest number of technology contracts, the larger direct investments, the larger credits extended by private firms and the larger receipts from the sale of international intermediates and capital goods in Chile

Countries	Number of licences	Total volume of foreign investments 1964–68	Total volume of foreign private loans 1964–68	Total receipts from intermediate and capital goods, royalties and profits in 1969 derived from 399 technology contracts
		$	$	$
United States	178	43,103,000	120,299,000	16,849,000
Federal Republic of Germany	46	14,517,000	28,181,000	4,238,000
Switzerland	35	2,941,000	18,250,000	3,949,000
United Kingdom	30	2,264,000	8,121,000	3,896,000
France	17		6,051,000	2,606,000
Italy	12			
Canada		25,181,000	4,789,000	
Netherlands	10			2,575,000

Sources: ODEPLAN, *El Capital Privado Extranjero en Chile en el Período 1964-1968 a nivel Global y Sectorial*, Santiago, August 1970; and G. Oxman, op. cit., p. 35.

will be using data from research undertaken in Chile to describe three types of concentrations.

First, there is a concentration in the total payments involved by sector with respect to the country of destination of such payments. On the bases of 399 contracts analysed, Chilean licensees (national and foreign owned) paid for royalties, profit remissions, intermediates, etc., the percentages of total outlays by sector to countries as shown in Table 1.

This type of very high country concentration of destination of payments from the various sectors (which in turn is the mirror image of the concentration of origin of resources) basically depicts two interrelated causal factors. On the one hand it indicates the lack of diversification or lack of attempts to diversity potential sources of supply on the part of the purchaser. Quite often he prefers to receive resources in package form from the same origin since an alternative strategy of diversification would have implied costs of obtaining information, friction on initiative, etc. The second causal factor stems from the fact that the country concentration, expressed above, often reflects a company concentration. Arrangements of patent cross-licensing among transnational corporations, cartel agreements, tacit segmentation of markets (particularly of developing countries whose size prompts such arrangements) often constitute common behaviour rather than the exception.

A second type of concentration involved reflects the joint existence of technology contracts, of foreign investments (direct as well as loans) and of the purchasing of intermediate and capital goods. For example, the following table to Chile lists by order of importance the countries with the highest number of technology contracts, the larger direct foreign investments in Chile, the larger credits extended by private foreign firms and the larger receipts from the sale of intermediates and capital goods to their Chilean licensees from whom payment of royalties or receipts of dividends are also obtained.

Table 2 indicates an almost complete one-to-one correspondence in the order of countries appearing in each of the four columns. Since the listing of countries reflects in practice the firms involved, the table indicates once more the existence of a collective exchange of factors of production and intermediates in a package form. Direct foreign investment implies the concomitant 'sale' of technology from parent to subsidiary. Also, the propensity to use technology commercially, prompts direct foreign investment. Furthermore, the sale of technology and capital generate the sale of products embodying the former or tied to it. This concentration of resources in a package creates special monopolistic conditions due to the absence of competitive forces for each of the inputs involved which are exchanged as a collective unit.

The third form of concentration refers to the market structure of the recipient countries. In a sample of foreign-owned subsidiaries in Chile, 50 per cent of them had a monopoly or duopoly position in the host market. Another 36.4 per cent were operating in an oligopoly market where they occupied leading positions. Only 13.6 per cent of the foreign subsidiaries in the sample controlled less than 25 per cent of the local market[4]. Similar concentration indices were noted in Colombia. Thus, foreign resource suppliers operating within high protective tariff walls[5] are able to pass to the final consumer, through market domination, monopoly rents that are related to the other two types of concentration examined above.

Thus, the three kinds of concentration are intimately connected. Market concentration and control in the host country, coupled with high tariff and other protections, bring high effective returns in such markets. These returns are then passed on through tied arrangements of inputs to foreign suppliers of collective units, often resulting in domestic tax avoidance (as distinct from tax evasion). Furthermore, country or firm concentration avoids competition even among alternative packages of inputs. Thus, the market of technology commerce and of direct foreign investments, due to their high and compounded imperfections through various forms of concentration, call for particular counter policies by the host governments so as to protect national interests.

II. CONCESSION AGREEMENTS IN THE EXTRACTIVE INDUSTRIES AND LICENSING CONTRACTS ON TECHNOLOGY COMMERCE

Descriptive models of concession agreements usually include, among others, the following general considerations applying to the initial conclusion of agreements in the first part of this century[6]:

1. The host country negotiating from weakness, 'excessive permissiveness' in initial agreements.
2. Lack of knowledge of other concession agreements.
3. Competence of negotiating government officials.
4. Hedging on the part of the concessionnaire for pressures during subsequent periods by arranging the initial agreement according to the tactics of a 'defensive negotiator'.

The Host Country Negotiating From Weakness, 'Excessive Permissiveness' in Initial Agreements

As in the case of concession agreements the developing countries approach the initial negotiations for technology purchase from a position of considerable weakness. In many instances the scarcity of domestic capital or foreign exchange reserves needed for an investment can hardly be compared with the domestic scarcity of the technical know-how required. The main way by which a plant could be made technologically viable was for it to be designed on the basis of processes supplied from abroad; the only legal way by which a product could be manufactured was through the licensing of the patent that covers that product or its process. The key factor in the competitive viability of a firm could thus be the necessary foreign technical assistance.

Developing countries lacking the basic technological infrastructure must purchase know-how from the technologically endowed world. Until they have mastered an already supplied technology or are able to copy foreign techniques they are dependent and hence in a very weak initial negotiating position. (Of course, in the case of direct foreign investment, unless the host government 'intervenes', there is not even a minimum negotiating position since, presumably, the subsidiary serves the total interests of the parent corporation.)

The history of concession agreements teaches us that during the initial periods of these agreements, governments 'have been known to be incredibly permissive, at least when judged by hindsight . . .In Latin America until World War I, concessionnaires could usually count on nominal income tax rates, waivers of import duties and insignificant obligations . . .Nominal as these commitments were, concessionnaires were known to bargain hard in order to get them lower still'[7]. In reviewing technology commercialization contracts at the end of the 1960s one cannot help being struck by the incredibly permissive attitude of the governments of developing countries in a way that parallels their permissiveness in the same area in the 1910s.

An area where such permissiveness is most evident is that of the tied sale of intermediate products and capital goods together with technology and/or direct investment flows. Such tied product sales have critical income and balance of payments effects, as we will see below.

In a sample of more than 150 technology licensing contracts studied in diverse sectors in Bolivia, Colombia, Ecuador and Peru more than two-thirds explicitly required the purchase of intermediates from the technology supplier or his affiliates. Even in the absence of such explicit terms, control through owner-ship or technological requirements and specifications stemming from the nature of the know-how sold could determine quite strictly the source of intermediate products. Thus, benefits for the supplier and costs for the purchaser are not limited only to explicit payments such as royalties or dividends. They also include implicit charges in various forms of margins or in the concomitant or tied sale of other goods and services.

Such an outcome is of particular importance since developing countries are becoming increasingly dependent on the import of intermediates and capital goods as they advance in their industrialization. For example, in Colombia two-thirds of the total import bill in 1968 covered materials, machinery and equipment for the industrial sector while the other third included final products for consumption and intermediate goods for the agricultural sector[8]. Similar dependence and structure of imports is to be expected for countries like Chile and Peru and other nations of comparable industrial development. For the whole of Latin America it has been estimated that during the period 1960-65 about $1,870 million were spent annually for the import of machinery and equipment. These amounted to 31 per cent of the total import bill of the area. They also accounted for about 45 per cent of the total amount spent by Latin America on capital goods during the same period. For individual countries this relationship worked out at 28 per cent for Argentina, 35 per cent for Brazil, 61 per cent for Colombia and 80 per cent for Chile[9]. As far as intermediates are concerned, industry samples in Colombia have indicated that imported materials represented in 1968 52-80 per cent of total materials used by firms in parts of the chemical industry. In rubber products the corresponding ratio was 57.5 per cent and in pharmaceuticals 76.7 per cent. It was only in textiles that the ratio of imported intermediates to total materials used fell, to 2.5 per cent[10]. Similar figures were reported for Chile. For example, imported intermediate products amounted to more than 80 per cent of total materials used in the pharmaceutical industry and between 35 and 50 per cent of total sales of the firms involved. This heavy dependence on imports of intermediates and capital goods has important repercussions on the recipient countries if one considers the fact that the majority of such imports are either exchanged between affiliated firms or tied to the purchase of technology. For example, it has been estimated that about one-third of the total imports of machinery and equipment in Latin America are realized by foreign subsidiaries[11]. The prices of those imports traded among affiliates correspond more to the global strategy of the transnational firm given related government policies, rather than to any adherence to prices, competitive or otherwise, established in markets characterized by arms-length relationships among firms.

Studies undertaken in the Andean countries in various sectors on these issues revealed the very critical importance of transfer pricing as a mechanism of charging purchasing countries for the factors of production (technology and/or capital) bought. Defining 'overpricing' as the following ratio:

$$100 \times \frac{\text{FOB prices on imports in Andean countries} - \text{FOB prices in different world markets}}{\text{FOB prices in different world markets}}$$

the country results presented the following characteristics.

In the Colombian pharmaceutical industry the weighted average overpricing of products imported by foreign-owned subsidiaries amounted to 155 per cent while that of national firms was 19 per cent. The absolute overpricing for the foreign firms studied amounted to six times the royalties and twenty-four times the declared profits. For national firms the absolute overpricing did not exceed one-fifth of the declared profits[12]. The sample evaluated included 25 per cent of the imports of foreign firms controlling about half the Colombian market, and 15 per cent of the imports of the most important nationally owned firms.

In Chile, of 50 products for which international prices were available corresponding to the imports of 39 firms: 11 were not overpricing; 9 were overpriced between 1 and 30 per cent; 14 were overpriced between 31 and 100 per cent; 12 were overpriced between 101 and 500 percent; 2 were overpriced above 500 per cent. In

TABLE 3. Percentage overpricing in imports corresponding to 22 firms in Peru

Percentage overpricing	No. of nationally owned firms involved	No. of foreign-owned firms involved	Percentage overpricing	No. of nationally owned firms involved	No. of foreign-owned firms involved
0-20	4	3	100-200	1	2
20-50	1	5	200-300	0	2
50-100	1	2	300 +	0	1

TABLE 4. Contracts in the Andean Pact countries containing clauses on exports

Country	Total number of contracts	Total prohibition of exports	Exports permitted only in certain areas	Exports permitted to the rest of the world
Bolivia	35	27	2	6
Colombia	117	90	2	25
Ecuador	12	9	—	3
Peru	83	74	8	1
TOTAL	247	200	12	35

terms of ownership structure of the importing firms: 0-30 per cent overpricing was reported in 13 nationally owned firms and in 6 foreign owned; 31-100 per cent overpricing was reported in 5 nationally owned firms and in 3 foreign owned; 100 per cent + overpricing was reported in 2 nationally owned firms and in 10 foreign owned.

In Peru imports corresponding to 22 firms were studied with the results as shown in Table 3.

In all three countries studied overpricing registered on imports of foreign-owned subsidiaries was on the average consistently and significantly higher than that of nationally owned firms. Foreign subsidiaries apparently use transfer pricing of products as a mechanism of income remission thus significantly underdeclaring their true profitability.

In the electronics industry in Colombia comprehensive samples corresponding to firms that controlled about 90 per cent of the market showed overpricing ranging between 6 and 69 per cent. In the Ecuatorian electronics industry twenty-nine imported products evaluated in relation to the Colombian registered prices gave the following results: sixteen of them were imported at prices comparable to the Colombian ones, seven were overpriced at up to 75 per cent and six at about 200 per cent.

Earlier studies undertaken only in Colombia presented a weighted average of 400 per cent overpricing in the imports by foreign-owned subsidiaries in the rubber industry and zero overpricing for nationally owned firms. Also small samples in the Colombian chemical industry indicated weighted average overpricing ranging between 20 and 25 per cent.

A significant point needs, furthermore, to be added. The above investigations and their results were based on comparisons of 'overpricing' or discriminatory pricing, which in turn imply the comparison between two different prices. Yet, income flows occur on the basis of <u>pricing</u> and not just of 'overpricing'. The former implies the comparison between price and costs while the latter that between prices.

The permissiveness and often complete ignorance in several developing countries of the extent and implications of these issues are not restricted to critical income and balance of payments costs. Such pricing policies also have a significant effect on the type and magnitude of tariff and non-tariff protections offered to cover not only production inefficiencies but also accounting allocations of implicit charges. Furthermore, if questions were raised with respect to the deteriorating terms of trade of developing countries, what would be the result of import substitution of finished industrial products by imports of intermediates and capital goods whose markets are even more monopolistic?

Another area of permissiveness in negotiating contracts of technology imports is that of export opportunities for goods produced by such technology. One of the most frequent clauses encountered in contracts of know-how commercialization is that of export prohibition. Such restrictive practices generally limit the production and sale of goods utilizing foreign technology solely within the boundaries of the receiving country. Some allow exports only to specific neighbouring countries. Of the total of 451 contracts evaluated in the Andean Pact countries, 409 contained clauses on exports which can be tabulated as in Table 4.

In Chile from 162 contracts with information on the matter 117 totally prohibited any form of exportation. Of the remaining forty-five the majority limited export permission to certain countries. The exact number of these partial export permits could not be estimated from data offered by Chile. Thus, in the four countries where precise figures were available about 81 per cent of the contracts prohibited exports totally and 86 per cent had some restrictive clause on exports. In Chile more than 72 per cent of the contracts prohibited exports totally.

In terms of ownership structure the following percentages were noted in the Andean countries on the various forms of export restrictions with respect to the total number of contracts providing relevant information: wholly foreign owned subsidiaries, 79 per cent; nationally owned firms, 92 per cent.

The lower percentage figure noted for wholly foreign owned subsidiaries is of limited significance since control through ownership can dictate export possibilities.

The figure with great significance is the one referring to nationally owned firms. In our sample it indicated that 92 per cent of the contracts of Andean-owned firms prohibited the export of goods produced with foreign technology. And this occurred at the time when, with the establishment of their common market, the Andean nations were trying to integrate their economies by, among other things, increasing inter-country trade. Agreements reached between governments are therefore greatly conditioned by the terms reached among private firms whose relative bargaining power is totally unequal. Also, efforts by UNCTAD and individual governments to achieve preferential treatments for the exports of manufacturing goods from developing countries have to be considered within a market structure which does not permit such exports through explicit restrictive clauses. Technology, an indispensable input in industrial development, through the process of its present form of commercialization becomes a major limitation on development.

To understand the meaning and repercussions of a contract one must evaluate it in its totality. Often terms that are defined in one clause are conditioned or modified in another. Also, without explicitly stating something so as not to violate local legislation one can achieve desired ends through indirect, legally accepted means. For example, one can indirectly affect the volume of production or control the sources of intermediates through certain quality clauses. Or through control on the volume of production (which is permitted under certain patent legislation) one can control the volume of exports (which is not permitted by the same patent legislations).

Restrictive clauses in contracts of technology commercialization are of various types. For example, in Bolivia out of 35 contracts analysed (and in addition to the export restrictions and tie-in clauses on intermediates cited above) the following terms were included: 24 contracts tied technical assistance to the usage of patents or trade-marks and vice versa;

22 tied additional know-how needed to the present contracts; 3 fixed prices of final goods; 11 prohibited production or sale of similar products; 19 required secrecy on know-how during the contract and 16 after the end of the contract; 5 specified that any controversy or arbitrage should be settled in the courts of the country of the licenser. Also, 28 out of the 35 cases contractually set quality control under the licenser. Similarly in Chile out of 175 contracts 98 had quality control clauses under the licenser, 45 controlled the volume of sales and 27 the volume of production. In Peru, of 89 contracts 66 controlled the volume of sales of the licensee. Some clauses prohibited the sale of similar or the same products after the end of the contract. Others tied the sale of technology to the appointment of key personnel by the licenser.

The list of clauses included in contracts of technology commercialization and the impact they have on business decisions prompt the question as to what crucial policies are actually left under the control of the ownership or management of the recipient firm. If the volume, markets, prices and quality of what a firm sells; its sources, prices and the quality of its intermediate and capital goods; its key personnel to be hired and the type of technology used, are all under the control of the licenser then the only basic decision left to the licensee is whether or not to enter into an agreement of technology purchase. Technology, through the present process of its commercialization, thus becomes a mechanism of control of the recipient firms. Such control supersedes, complements or substitutes that which results from ownership of the capital of a firm. Political and economic preoccupations voiced in Latin America concerning the high degree of foreign control of industry can properly be evaluated not only within the direct foreign investment model but also within the mechanism of technology commercialization. For this reason the term 'technology transfer' is here considered as inappropriately representing the phenomena involved and their implications.

An additional issue needs to be mentioned. The type of clauses encountered in contracts of technology commercialization violate basic anti-monopoly or anti-trust legislations in the home countries of the licensers. Since the extra-territoriality of laws is in general not applicable (at least operationally) or not acceptable, it befits the technology-receiving countries to legislate and regulate accordingly so as to protect the interests of the purchasing firms. Industrialized countries have in the last half century or even earlier undertaken to define in one way or another in their legal structure the extent to which private contracting and the exercise of business power can operate within a market mechanism[13]. Developing countries have still to show an adequate understanding of the issues involved in their commercial laws, especially those that regulate the application of industrial property matters, etc.

As in the case of concession agreements, 'nominal as the commitments of the foreign investor are, licensors are known to bargain hard in order to get them lower still'. It is not uncommon in the experience of licensing agreements in Latin America for the licenser--after, for example, he has been guaranteed an implicit (but very real) return of 15 per cent on sales through intermediate product overpricing--to bargain very hard to increase his royalty receipts from 3 per cent to 3 1/2 per cent on sales. Tax and accounting experts are specially flown to the developing country to bargain hard for the 1/2 per cent in royalty increase even if multiple returns have been secured through other arrangements.

It appears that sellers of technology bargain very hard on small differences of royalty rates as a strategic choice that leaves other much more important negotiable aspects outside of the negotiating process. The government or company negotiator concentrates attention and negotiable 'trade-offs' on elements perhaps completely marginal. By defining royalties as the only cost of technology purchase (since they are explicit) one overlooks the more important implicit costs, such as import overpricing.

Lack of Knowledge of Other Agreements

Acceptable competitive market conditions assume *a priori* sufficient and equitably available information. Yet, in a bargaining framework information is an instrument upon which the whole system of relative power is based since the latter (i.e. bargaining power), is, among other things, a function of the knowledge of what the counterpart is gaining from different configurations of policies and situations. Furthermore, acquisition of information implies certain costs which need to be evaluated in relation to the benefits to be received. Thus, availability of information cannot be assumed as given but needs to be introduced as one of the policy variables in a country's confrontation with foreign investors. The economics of knowledge in the foreign investment model, an area quite unexplored, could show the possibilities and limits of exercising certain bargaining pressures.

The history of concession agreements teaches us that in the early stages of such agreements it was, quite often, practically impossible to obtain copies of the actual documents of the terms of the concession. Several countries considered concession agreements as secret documents and hence non-available to anyone except the negotiators. In other countries concession agreements were defined as, theoretically at least, in the public domain, but only a very small number of copies were reproduced and hence, knowledge of them was very restricted. If the countries themselves did not make available information to their public and to each other, clearly the foreign companies would not go against their self-interest and publish the terms of agreements. Quite the contrary; 'only a great deal of detective work--combing legal libraries of universities in developed countries, scanning trade journals for clues, trading information with other governments, etc.--could yield much information'[14]. It took some time to understand the necessity of exchanging information as an explicit policy by governments of developing countries. This, together with other factors, led to the establishment of such institutions as the Organization of Petroleum Exporting Countries (OPEC), which has as one of its main functions the dissemination of information to the definite interest of the member countries. The result of this enhanced availability of information led, together with market and risk factors, to the signing of 'model contracts' in the post-war period.

Reflecting on the process and procedures, with respect to information handling of technology commercialization during the 1960s, one encounters cases which are similar or parallel to those in the initial concession contracts. Due to mis-specified concepts of confidentiality and secrecy, contracts of technology purchase are kept completely secret. In countries that do not apply an exchange control mechanism in contractual agreements, information is restricted to the two contracting parties. In countries where government regulating bodies intervene in the contractual processes between private parties, inadequately functioning administrative procedures limit the degree of knowledge of contractual terms. The members of royalty committees know general industrial terms usually in an intuitive sense only, and from memory. Of course, no explicit mechanism exists for inter-country comparisons. Governments very scrupulously guard the contractual agreements of their neighbouring countries, in that way believing they preserve the national interests. Effectively, what they jointly achieve is a reduction of their knowledge and bargaining power by segmenting the market of information and accentuating problems of relative ignorance. Government agencies in groups could certainly proceed to inform themselves about market conditions in technology commercialization around the world as well as to inform each other about it and about terms of domestic agreements. The benefits derived from such a policy could certainly outweigh the actual or imaginary costs of secrecy among nations about their contracts with foreign technology suppliers.

In addition countries could introduce the use of the 'most favoured nation' principle. (This principle has been in use in international trade arrangements, like the GATT, and lately in the concessions area[15]). A number of countries can defend their interests if lower terms in one of them would also apply to others for contracts of 'similar' technology. An explicit introduction of such a clause could open the way to smooth renegotiations of contracts without the conflicts resulting from forced renegotiations. The whole system of most favoured nation could, actually, be an indirect way of achieving a common front for purchasing technology in an environment where negotiations by nature or by intent differ in time.

Competence of Negotiating Government Officials and Lack of Institutional Infrastructure

One of the factors that contributed to the relative weakness of government officials in their negotiations of initial concession agreements was their sheerly inadequate understanding of the complex financial bookkeeping practices of the large multinational companies. Analyses of concession agreements demonstrate that tax authorities were quite unprepared to deal with the issues of transfer pricing between parent and subsidiaries as a means of shifting untaxable profits from one country to another. 'The terms of the early income tax agreements in some countries may appear very strange to those used to sophisticated tax systems. Government officials sometimes agreed to both depreciation and tax deductions of reserves for replacement of the same asset, tax liability to be determined by foreign accounting firms whose accounting principles were those of the shareholder and not those of a tax authority, allowance of deduction for interest designed to permit tax-free shifting of profits outside of the host country, and so on and so on. Government negotiators found terms which they did not understand, or, for a fee (perhaps in the form of loans for shares or well paying positions on boards of directors) terms which they were fairly certain their superiors or political opposition would not understand'[16].

The above passage is quoted because, if one substitutes the words 'concession agreements' for 'licenses of industrial know-how', it approximately describes the reality in the 1960s in technology commercialization. Quite often the complexity of evaluating modern technology further aggravates any already existing inadequacies in financial analysis which translates technical coefficients into economic units of measurement. In the case of transfer pricing of intermediate (technology-embodying) products questions arise as to whether government officials are competent or not to handle this matter, as well as to whether an institutional framework exists which could attempt to scrutinize transfer prices. Also, as was indicated earlier, wholly foreign-owned subsidiaries capitalize know-how for reasons not related to control. The companies pay royalties, levy depreciation charges on intangible assets and reduce their excess profit tax base through the capitalization of the same know-how.

In the process of evaluating and negotiating the purchase of foreign technology government officials are called upon to perform a dual task. First, they try or should try to scrutinize technology commercialization within a complex of other resources that are being jointly sold, like intermediate products or machinery. (If more sophistication is used to determine opportunity costs then they should evaluate, in addition, the complex of other resources or inputs whose use is being foregone, or, in various cases displaced.) Furthermore, a given technology is tied in with the transfer of capital, the extent of market opportunities (i.e. export restrictions) as well as the ability or inability to use other forms of complementary or substitute technology. Even more, technology commercialization is related to the tax system that regulates the distribution of net benefits, the tariff policy that determines the extent of effective protection, etc. Evaluation of the purchase of a given technology should thus try to scrutinize the whole package of interrelated effects of various inputs as well as policies and their implications rather than limit itself to specific elements such as royalties, direct employment effects, direct balance of payments effects, etc. Social scientists, seeking rapid recommendations, have been often led to an excessive use of ceteris paribus assumptions. This, in turn, has probably misdirected attention to very marginal evaluation techniques. What is needed is not an analysis in ceteris paribus terms but a conceptualization of 'the problem as a whole to identify the ceteris'[17].

Second, technology (or its process of commercialization) appears to be the least identified and understood factor of production. Its exchange takes place in the vaguest manner to the minds, at least, of the purchasers. Countries have developed specific definitions as well as elaborate systems (which still leave much to be desired) for the classification and evaluation of the transfer of other resources. One need not spend very much time in a central bank or a customs office to notice the elaborate mechanisms of registration, classification, etc., for the transfer of financial capital or goods among countries. Generally, know-how is still commercialized under the broad, vague, and in many ways economically ill-understood term of 'technology'. Tautologically, we define the import of technology as the import of know-how. But the question arises as to what actually is, at least operationally, the technology a country is importing for a given industry, or process, or product.

An attempt to distinguish the different components brought together under the broad term of 'technology' for negotiating purposes was initiated by the Colombian Comite de Regalias in 1970 and was formalized by Decision 24 of the Andean Pact. This included first, a separation of elements of industrial property (e.g. patents and trade-marks) and strictly technological components of a contract. With respect to the latter, an effort was made, where possible, to distinguish between production manuals supplied only once, at the beginning of a licensing agreement, continuous technical assistance over time, technology embodied in imported intermediates and capital goods, etc. Since technology lacks units of measurement, instead of negotiating payments of these disaggregated components of know-how as a percentage of sales, Decision 24 established that negotiations should be undertaken with respect to the contribution of foreign technology to the profitability of recipient firms. For example, an imported cost-saving technology is to be negotiated with respect to its effects on increasing net returns rather than as a percentage of total sales. Effects on profitability for the firm, and on value added for the host country, determine, in market prices, the imputed value of technology. This, in turn, determines the maximum permissible price to be paid for that technology. In addition to this imputed value, for the commercialization of know-how, one needs to determine the price equivalent of comparable technologies in different markets around the world. Not uncommonly, the same type of product can be produced by more than one process. Furthermore, and of great importance to developing countries, there are many different sources of supply, with different prices, offering a given type of technology for the industrial needs of these countries. For developing countries, the opportunity cost of technology in the process of its commercialization (not its imputed value), can be determined only through knowledge of available alternative sources of supply and their respective prices. Provisions for such action, through explicit search in the international market, were included in Decision 24 of the Andean Pact.

Defensive Negotiations and the Element of Time in Progressive Bargaining Over Different Periods

In the history of concession agreements, the initial demands of the concessionary have been described as being based on the 'exaggerated emphasis of the defensive negotiator'[18]. Once an agreement is reached and capital is invested the bargaining power of the foreign concessionary clearly diminishes. He thus attempts to use his bargaining power right at the outset, when it is strongest. Furthermore, consistent with the tactics of the 'defensive negotiator', initial terms are expected to be higher than average during the life of the contract. The concessionaire _expects_ his initial terms to be reduced.

In principle, technology bought at a given period and sunk investments are quite similar notions if they are evaluated over a time range. Both of them are irreversible in time. The use of information during one period certainly does not diminish its availability in the future. On the contrary, its 'availability' is enhanced as it is being mastered, and once mastered, it cannot be lost. Thus, reacquisition of the same information in some future period involves, intrinsically, no additional cost, since this information is already embodied in machines, processes and skills used in the past.

This property of decreasing imputed costs over time generates conflicting interests and varying degrees of dependence between supplier and recipient of information since its value depends strictly on the point of time being evaluated. If history then repeats itself, the licenser probably is setting his initial terms 'on the exaggerated emphasis of the defensive negotiator' and expects his terms to be lowered, but not without hard bargaining. The renegotiations undertaken by the Colombian Comite de Regalias and the notes included in the minutes of this committee gave support to this conclusion.

A more explicit introduction of the time factor in the strategy of the government, or company negotiator of the developing country, has definite implications on his expected behaviour. First, he has explicitly to define his negotiating horizon over a relevant period where subsequent renegotiations will take into account the continuously shifting power or dependence relationship. What should be maximized is not the use of negotiating power at the initial negotiation, but the use of changing negotiating power over time. Second, the negotiator from the developing country should preplan the institutional means by which renegotiations can be opened so as to avoid alienating potential investors or technology suppliers by the negative effects of forced renegotiations. Clauses in the initial contract should exist which will smooth the road for the reopening of negotiations. The 'most favoured contract' clause has already been indicated as an example.

In addition, with respect to technology purchases, he should very carefully tie payments to received benefits from the know-how bought. As discussed earlier, quite often technology contracts are signed without any differentiation between payments of patents, technical assistance, plant designs, etc. Clearly, as suggested earlier, each one of these has a different incremental contribution to the buyer at different periods. For example, in the pharmaceutical industry, technical assistance is quite often limited to a production manual which can be learned very quickly, while the key to the technology buyer's dependence on the seller is the patent covering the product or its production process. By differentiating the types of know-how received, allocating payments for each one of them separately and indicating different periods of duration for each, the negotiator not only rationalizes his procedure of technology purchase, but also smoothes the way for renegotiations at different intervals.

NOTES

[1]An earlier version of this article was presented in 1970 to the Junta del Acuerdo de Cartagena as reference material for the preparation of the proposal on treatment of foreign direct investments, licensing agreements, patents and trade-marks. A Spanish version was published by Comercio Exterior, September 1971.

[2]For a more analytical treatment of the subject see C. V. Vaitsos, Income Generation and Income Distribution in the Foreign Direct Investment Model, Oxford Press University (forthcoming).

[3]Research undertaken in the petrochemicals industry indicated that during the time when technology was most likely to be sold to developing countries the originators of such know-how accounted for only 1 per cent of total licensing agreements. The rest was divided between commercial imitations and followers (52 per cent) and equipment manufacturers (47 per cent). See R. Stobaugh, Utilizing Technical Know-how in a Foreign Investment and Licensing Program.

[4]See Corfo, Comportamiento de las Principales Empresas Industriales Extranjeras Acogidas al D.F.L. 258, p. 16, Santiago, Chile (Publication No. 9-A/70).

[5]The infant industry argument and tariff protection certainly need a re-evaluation if 'infancy' is ascribed to companies like General Motors, ICI, Philips Int., Mitsubishi, etc., whose subsidiaries dominate the market of key industrial sectors in developing countries.

[6]See for example L. T. Wells, Jr., in 'The Evolution of Concession Agreements', paper presented at the Harvard Development Advisory Service Conference, Sorrento, Italy, 1968 and of R. Vernon, 'Long Run Trends in Concession Contracts', American Journal of International Law, Vol. 61, 1967.

[7]R. Vernon, op. cit., p. 83.

[8]See data from Banco de la Republica, tabulated by INCOMEX, Clasificacion Economica de las Improtaciones, 1969.

[9]Preliminary estimates by CEPAL presented by F. Fanjzilber, Elementos para la Formulacion de Estrategias de Exportacion d Manufacturas, Santiago, Chile, July 1971, p. 91-4 (ST/ECLA/Conf. 3/L.21).

[10]See C. V. Vaitsos, Transfer of Resources and Preservation of Monopoly Rent, p. 48-54, Harvard University, Center of International Affairs (Economic Development Report, No. 168).

[11]Estimates were based on figures of OECD and the survey of current business as analysed by F. Fanjzilber, op. cit., p. 94-5.

[12]For a full analysis of the above see C. V. Vaitsos, Transfer of Resources and Preservation of Monopoly Rent, op. cit.

[13]For tie-in restrictions see Section 1 of the Sherman Act and Section 3 of the Clayton Act in the United States. On similar and related issues (such as export restrictions) see Article 85 (1) of the Rome Treaty establishing the European Economic Community, Article 37 of the 1945 Price Ordinance of France, the Economic Competition Act of 1958 of the Netherlands, the Antimonopoly Law of Japan, etc.

[14]L. T. Wells, Jr., op. cit., p. 6.

[15]See actions of the Federal Government of Nigeria in 1967 based on the 'most favoured African nation clause', OPEC, Collective Influence in the Recent Trends Toward Stabilization of International Crude and Product Prices.

[16]L. T. Wells, Jr., op. cit., p. 9-10. Wells makes specific reference to such 'very strange' terms that exist in clauses of contracts between the Republic of Liberia and the Bethlehem Steel Corporation, the Liberial Mining Company, Ltd and the Liberian American Swedish Minerals Company.

[17]See Y. Aharoni.

[18]R. Vernon, op. cit., p. 84.

Progressive Technologies for Developing Countries

KEITH MARSDEN[1]

This paper will discuss the choice of technologies for developing countries. The word choice is used advisedly. The theme will be that technology is an important variable in development strategy, not an immutable force requiring adjustments in other factors to make way for it. Indeed, technology may have greater inherent flexibility in the short and medium terms than the human factors (skills, attitudes, behaviour, propensities and motivation) which co-operate with it in production. It will be suggested that government has a vital role to play in influencing investment decisions, directly and indirectly, so that they result in the optimum utilisation of available resources.

This approach is not exactly new. The concepts involved have been examined in the professional economic journals over a number of years.[2] But it is only recently that a growing body of opinion within the developing countries has begun to question whether the form and timing of technical change have always been in their best interests.[3] The present paper will take this critical examination further. The analysis will be illustrated by case study examples drawn from investigations in the field.[4]

The first part will show that the technology transferred directly from the highly industrialised States has not always been appropriate for the developing countries. It should be made clear that rejection of advanced technologies is not advocated. Selected judiciously, they undoubtedly have an important contribution to make in certain areas and circumstances. The purpose of this part is to demonstrate the need for greater discrimination in their application than has been shown hitherto. Technological choice is defined in its widest sense. It embraces not only different ways of making an identical article (where the choice may be strictly limited) but also choice of product mix, i.e. between competing or substitute articles in the same broad product group (e.g. plastic

From **INTERNATIONAL LABOUR REVIEW**, Vol 101, May 1970, (475-502), reprinted by permission of the publisher, International Labour Office, Geneva.

shoes or leather shoes, bricks or cement, cotton or synthetics).

The second part will suggest various criteria against which to measure the suitability of alternative technologies. The term "progressive technologies" has been coined to convey the characteristics required of technical innovations in the developing countries. They should stimulate economic progress by making optimum use of available resources. They should be conducive to social progress by enabling the mass of the population to share the benefits and not just a privileged few. They should represent technical progress, measured by improvements over existing methods and not by reference to external standards which may be irrelevant. And they should be progressive in a temporal sense, i.e. their characteristics will change over time in response to the society's ability to pay for them and capacity to employ them effectively. In other words, the concept is dynamic, not static.

The third part will consider the possible types and sources of progressive technologies and the action that could be taken at the national and international levels to increase their availability.

Finally, the fourth part will examine the policy measures open to governments to ensure that progressive technologies are actually selected and installed in their countries, and used efficiently.

I. SOME REASONS WHY THE DIRECT TRANSFER OF TECHNOLOGIES FROM INDUSTRIALIZED TO DEVELOPING COUNTRIES MAY BE INAPPROPRIATE

Capital is Dearer and Labour Cheaper in the Developing Countries

In the advanced economies labour is scarce and dear, and capital relatively plentiful and cheap. The opposite is true of most developing countries. It follows that a machine that is economic with wage rates of $2 an hour and interest charges of 4 per cent per annum may be uneconomic when interest charges are 15 per cent (or should be) and wages $0.10 an hour.

Case Study Illustrations

The total costs of alternative methods of making wooden window frames have been compared in the two kinds of countries described above. It was found that in the labour-expensive country two special-purpose machines, one performing four-sided planning and moulding, and the other a doubled-ended tenoner, were the cheapest combination for output capacities in excess of 50,000 units per annum. In the cheap-labour/dear-capital country, however, they would be uneconomic unless the capacity exceeded 450,000 units per annum. Up to an output of 64,000, which encompasses the great majority of carpentry workshops, the lowest unit costs would be achieved by using single-purpose planing and thicknessing machines and single-ended tenoners, with a considerably lower investment per worker.[5]

The importance of the current wage level in determining the viability of new agricultural technology can be illustrated by the history of the mechanical reaper in American agriculture. The first McCormick mechanical reaper was patented in 1834. Only 793 machines had been sold by 1846. By 1858 the total had reached 73,200. The sudden upsurge in the 1950s can be explained by the rising cost of farm labour, due to the relatively inelastic supply and the increased demand for labour coming from railroad building and urban development. Higher wages lowered the farm-size threshold below which mechanical reaping was uneconomic. Thus, in 1849 a McCormick reaper cost the average farmer in the mid-west the equivalent of employing an agricultural labourer for 97.6 days at the current wage rate. During the period 1854-57, it had fallen to the equivalent of 73.8 man-days. Hence, the threshold point dropped from over 46 to roughly 35 acres, bringing it within the range of many more farms in Illinois.[6]

Large-scale Production May be Inefficient in the Conditions Prevailing in Some Developing Countries

Much modern technology has been designed for use in large-scale plants. Certain prerequisites of large-scale operations have therefore to be fulfilled before it can be employed economically. These are not always present. Large-scale capital-intensive production is not efficient if markets are small, scattered, highly seasonal or fragmented; if distribution channels are not well organised; if workers are not used to factory discipline; if management does not know the necessary managerial techniques or, though aware of them, cannot implement them because they conflict too strongly with accepted customs, beliefs, systems of authority, etc., of the employees; or if there are no service engineers who can get the complicated machinery going again when it breaks down. Scarce capital is wasted if it is invested under such circumstances. And these are factors that cannot be changed overnight by a simple planning decision.

Case Study Illustrations

The large public-sector shoe factory which operated at 20 per cent of capacity because it had no means of reaching the small private shoe retailers who handled 90 per cent of the shoe trade.

The battery plant which could satisfy a month's demand in five days.

The woollen-textile factory which had a 10 per cent material wastage figure (costing precious foreign exchange) because its management did not know how to set and control material usage standards.

The $2 million date-processing plant which had been out of action for two years, ever since a blow-out in the cleaning and destoning unit, because there were no service engineers who knew how to repair it.

The confectionery plant which was inactive for most of the year because 80 per cent of sales were made during the month of a religious festival.

The ceramic factory whose quality was poor as it had to use up two years' stocks of the wrong glaze, imported in error, because the general manager had too much work to attend to all details satisfactorily. This was attributable to a lack of middle management and supervisory personnel willing to accept responsibility, combined with a reluctance to delegate authority on his part--both reflections of prevailing social attitudes.

The radio assembly factory whose production line broke down repeatedly because of the high rate of absenteeism among key workers.

These economic and social realities explain why large-scale factories in some developing countries achieve a much lower level of labour productivity and capital productivity (which is more important) than do identical plants in the long-industrialised countries. The relevance or importance of these factors obviously varies from country to country. An assessment needs to be made by those who frame policy. Economies of scale are not just the product of technical coefficients. They cannot be realised unless the co-operating economic and social factors are favourable. This does not imply any inherent inferiority or inability, but merely the lack of opportunity in many countries in the past to acquire the particular skills and habits required in all strata of society or to build up the infrastructure which is essential for smooth business operation.

Advanced Technologies May Reduce both Employment and Real Income in Certain Circumstances

Even when run below capacity, advanced machinery can often make products cheaper than is possible by traditional methods. As a result a large number of indigenous craftsmen may be put out of business even when the innovation is a product not previously produced within the country. In the mass market there are very few new items which do not have their equivalents among traditional products and, where incomes are static, the traditional product will often be replaced by the new one.

The redundant workers may not be absorbed into the new

factories because machinery is being employed instead of labour and the difference in productivity is so great. Some consumers may receive some benefit through lower prices, but this may be offset by a decline in average real income in the community at large. This will occur if--

(a) the new substitute product has a higher proportion of imported materials and components than the old; and

(b) resources cannot be easily transferred to satisfy higher momentary demand for other products, because the surplus capital is tied up in specialised equipment and the unemployed labour lacks the educational background or the social mobility needed in the new occupations.

The fact that many countries are experiencing growing unemployment, a rising import bill and domestic inflation all at the same time would tend to confirm this explanation. In this case the persons who gain most from the technical change are the inhabitants of the already rich countries exporting the machinery, the materials and components required by the new technologies. In other words, the main benefits "leak" abroad. So in these cases there is little opportunity to compensate the unemployed by transfer payments from those with higher real incomes elsewhere, or to wait until a long-term growth in employment opportunities absorbs them.

This is the crucial difference between the industrialised nations, which have initiated technological change themselves, and the developing countries, which acquire new methods only from abroad. In the first group, technology "producers" and technology "users" have established close interrelationships which facilitate the flow of knowledge between them. The sources of innovation are widely diffused throughout the economy. Each change is relatively small in relation to what has gone before.[7] Innovations represent not just advances in scientific knowledge in the abstract but economic solutions to specific problems. The demand for new products springs out of larger incomes. Even when substitution occurs, the decline in the sector employing an obsolete technology is more than offset by the expansion in those using and making the new. Resources are progressively transferred from one to the other, although the geographical concentration of declining industries still presents difficulties (depressed areas).

In the developing countries, however, the innovation may take the form of an alien "transplant" that kills off competing activities in the traditional sectors but has to be fed through external linkages established with suppliers abroad to be kept alive itself.[8] This is because the indigenous industry lacks the specific skills, materials and equipment to satisfy its immediate requirements, and time and money are too short for it to make the necessary adjustments. Decades or even centuries of development cannot be compressed into the couple of years it takes to build a plant and get it running. Thus the innovation may make little contribution to the spread of employment or to raising the skills upon which self-sustained growth depends. The backwash effects can exceed the spread effects[9], and the technological and income gap becomes wider.

Case Study Illustration

One country imported two plastic injection-moulding machines costing $100,000 with moulds. Working three shifts and with a total labour force of forty workers they produced 1.5 million pairs of plastic sandals and shoes a year. At $2 a pair these were better value (longer life) than cheap leather footwear at the same price. Thus, 5,000 artisan shoemakers lost their livelihood; this, in turn, reduced the markets for the suppliers and makers of leather, hand tools, cotton thread, tacks, glues, wax and polish, eyelets, fabric linings, laces, wooden lasts and carton boxes, none of which was required for plastic footwear. As all the machinery and the material (PVC) for the plastic footwear had to be imported, while the leather footwear was based largely on indigenous materials and industries, the net result was a decline in both employment and real income within the country.

Policy Guidelines for the Selection of Progressive Technologies

It is clear from the above case studies that the best machinery for one country is not always the best for another. What kinds of technology would be progressive in the senses that we have defined, i.e. will induce balanced economic and social development in which all members of the community will find opportunities for increasingly productive employment and rising living standards?

Obviously, this paper cannot provide an inventory of appropriate techniques. These will vary from country to country, and from one period to the next, according to circumstances. The list of products, each requiring specific technical knowledge, is far too long.

Nor can the problem be reduced to a single economic criterion. Such criteria have been offered in the past but none has proved to cover all situations or offered practical guidance to the industrialist or policy maker.[10]

This section will therefore run through some broad economic and social considerations which might be borne in mind when economic policy is being framed. It has become evident that the investment decisions of individual entrepreneurs or enterprises do not necessarily achieve maximum benefit to the society at large. There are three main reasons for this. The first is that money costs may not reflect real or social costs, so a choice which would maximise private entrepreneurial (or public-enterprise) profits will be different from the one which would achieve optimum economic efficiency or social welfare from a national point of view. The second is that non-rational factors may influence decisions unduly, e.g. prestige considerations, assiduous salesmanship on the part of machinery manufacturers, the confusion of technical efficiency with economic efficiency, the belief that the latest must be the best irrespective of circumstances, etc. The third reason is that the individual entrepreneur is often unaware of the full range of technical alternatives that exist, or may not have ready access to them.

We shall discuss the measures that can be taken to correct these biases later in the paper. Before doing so, it is necessary to indicate the broad relationships that should normally be observed if technical progress is to result in balanced economic and social development. They are expressed in the form of general guidelines. These are not precise formulae but rather rules of thumb or check-lists for planners. Again, specific case studies will be used to illustrate the principles involved.

Wide Disparities in Capital Intensity Per Worker in Different Sectors Should be Avoided

Recently, more attention has been focused on the problem of the widening gap between rich and poor, both between nations and between different segments of the population in the developing countries themselves. The dual structure of many of these economies appears to be hardening. The Director-General of the ILO has pointed out that "within developing countries, there is a gap between the privileged high-income minority and the poverty-stricken, illiterate mass of the population; there is the gap between the workers in the modern industrial sector and the unemployed peasants; there is the gap between the inhabitants of areas of booming economic activity and those sections of the population who live in the most backward and stagnant areas".[11] He has urged that the closing of the gap "is a job to which governments, workers and employers must direct their attention in the years immediately ahead". How can this be done? Several lines of attack are necessary, including a wider spread of educational and training facilities among the people. Investment policy, however, must play a central role. Recent studies have shown a close correlation between capital intensity and labour productivity.[12] The relationship between productivity and earning capacity is obvious. Thus, if the gap is to be narrowed, it would seem necessary to raise the productivity of all those in the traditional and subsistence sectors by a substantial injection of capital in these occupations. This does not mean setting up a few isolated "model" schemes in these areas. The demonstration effect of these model units is lost if the masses do not have the capital resources or training to follow suit. A country with an average income per head of $200 (75 per cent of the population of the developing world as a whole have less than this) cannot afford to provide the same equipment on the average to its workers as a nation twenty times as rich.[13] This implies that modernisation must be seen as a continuous process, in which there is a widespread improvement in methods, which use more equipment or more expensive materials than has been traditional, but far less than is currently the practice in the advanced countries. As we shall see, the few spectacular capital-intensive projects may be an essential part of development, but they cannot be the main source of progress if this gap is to be closed.[14]

The short-term social advantages of reducing the disparities between the modern and traditional sectors are clear. But there have been fears that the price to be paid might be a retarded rate of economic growth in the long run.[15] The evidence on this is not conclusive. In theory, the rate of capital accumulation ought to be higher in large-scale, capital-intensive industry. This does not always work out in practice, perhaps because of the economic and social factors already mentioned.[16] On the other side, it can be said that the concentration of resources in the modern sector, which is a feature of many development programmes, has not been conspicuously successful in achieving a satisfactory rate of growth. A stagnant agricultural sector can slow down expansion elsewhere.[17] And the lack of viable investment opportunities for small family savings may lower significantly the community's propensity to save. Perhaps even more important, this bottleneck can prevent or deter those rare individuals with a high achievement motivation to channel their entrepreneurial energies into industrial pursuits.[18] For all these reasons, a more balanced distribution of capital investment might prove to be economically as well as socially desirable.

It is significant that, during the historical growth of the now advanced countries, a close harmony was maintained in the development of one sector and another, and between capital intensity and income levels. In the United States, for example, average capital intensity per worker in industry was equivalent to only 1.7 times the average net output of the entire labour force (all sectors) in 1880. In 1948 it was 1.8 times.[19] Variations from one industry to another were relatively small (mostly within a range of 3:1).[20] In other words, industrial investment never "ran ahead" of society's ability to save out of past and current incomes. If a developing country were to reproduce the current American capital intensity in industry (around $20,000 per worker)[21] by a direct application of the latest techniques, this would be equivalent to up to fifty times the average net product or national income per worker. Obviously, only a few workers can benefit from greater mechanisation in this way. The present lopsided economies and huge disparities in productivity and incomes reflect such attempts in the past.[22]

Case Study Illustration

The economic and social advantages of a balanced industrial development strategy can be illustrated by the case of a typical traditional bakery in West Africa.[23] This is a family enterprise employing perhaps one or two hired hands. The premises consist of a single room with a brick-lined oven built on to the back. Firing is by charcoal. The dough is mixed and kneaded by hand in bowls and on wooden tables. Proving takes place in wicker baskets. The single-plate oven is loaded and unloaded with a wooden peel, and the loaves are left to cool on racks before being sold direct from the bakehouse or delivered to retail outlets by handcart or bicycle. Productivity per worker averages eighty 14-ounce loaves per day, selling at 6d. each; capital investment per workers if 100 pounds and wages are 10s. per day.[24]

Taking a shadow interest rate on capital of 15 per cent per annum, the cost breakdown per loaf (in pence) would be as follows:

Materials	3.4
Labour	1.5
Fuel and power	0.3
Depreciation and interest	0.2
Total cost	5.4
Profits	0.6
Selling price per loaf	6.0

Consumption of bread per head is the equivalent of half a loaf a day (i.e. 49 ounces per week compared with 40.6 ounces per week in the United Kingdom in 1955 and 55.1 ounces in 1965). Thus, a town of 100,000 would consume 50,000 loaves daily, requiring a labour force of 625 spread among, say, 125 bakeries (five workers each).

How best should such an industry be developed? One could introduce the latest processing equipment including pneumatic handling of flour, continuous mixing and dividing, proving on overhead conveyors, baking in a travelling turbo-radiant oven and feeding the loaves direct from a cooling chain to a battery of slicing, wrapping and palletising machines. A single-plant bakery of this type would produce 50,000 loaves daily. In order to deliver the bread to the customers in time (they could no longer collect it themselves from the neighbourhood baker), a fleet of delivery vans and a chain of distribution outlets and stores would be required. The total investment in land, buildings, machinery, vehicles and working capital would be of the order of 600,000 pounds. Labour productivity would, of course, be much higher than with the traditional techniques. An estimated sixty employees would be needed. As the skills required would be higher (managerial, supervision, clerical, administrative and servicing of complex equipment), an average wage of 1 pound per day might be necessary (in the United Kingdom, average earnings in the bread industry are nearly 4 pounds per day). Additional materials (the wrapping paper, office records) and fuel (petrol for delivery vans) would raise the production cost, and some of them would have to be imported, requiring foreign exchange. The resulting breakdown of cost per load (in pence) would therefore be as follows on next page:

Materials	4.0
Labour	0.3
Fuel and power	0.5
Depreciation and interest	2.0
Total cost	6.8
Loss	0.8
Selling Price per loaf	6.0

In other words, such a plant would not be economic in these circumstances. There are other factors which would also militate against a project of this nature:

(a) The existing bakers could not accumulate the capital required. If the project were to be executed, it would probably need government finance and administration, or government financial participation. In case of a government venture, the difficulties of finding officials with the right kind of experience and motivations to provide the standards of efficiency, hygiene, quality and spread of services demanded by the consumer in this industry are well known; such managerial experience in the private sector is scarce.

(b) Five hundred and sixty-five bakers would be thrown on the labour market. Retraining would be required before they could find other jobs, imposing an additional cost on the community. The chances of finding alternative employment would be limited as the consumer would be receiving no benefits from the technical innovations through lower prices. So demand for other products would not increase.

(c) The livelihood of other trademen would also be affected adversely, e.g. brick-makers, and producers of wicker baskets and charcoal, as their products would be replaced by imported machinery and materials (e.g. steel bread tins).

(d) Advanced European bread-making plant has been designed to handle high-grade white flour, which is in general demand there. These white flours and the usual additions of vitamins, eggs, milk, etc., are more expensive than the locally milled brown wheat and maize mixtures containing a higher portion of the husks. As the choice of materials is limited by the different mixing and baking characteristics of the various flours, the use of advanced plant would, in fact, raise the material costs substantially above the figure indicated in the cost analysis above (in the United Kingdom a 14-ounce white loaf costs 11d. today, which reflects both high material and labour costs). There is also a question whether countries which have a chronic shortage of home-grown cereals, and where the average citizen suffers from an unbalanced diet, can afford to waste some of the more nutritious parts of the grain by using white flours.

(e) Technological factors would also have an impact on consumer tastes and preference in respect of the shape, colour and texture of the loaf. In these basic foodstuffs one encounters a pronounced resistance to change among consumers (still one of the principal obstacles to productivity growth in the French baking industry, for example). If such a plant were erected, there would be some risk that it would work at well below capacity because of the continued preference of most customers for their traditional loaf.

It is clear that the latest western bakery technology would not be the most rational choice in the socio-economic climate of the country in question. If the whole package of advanced technology is not suitable, it may be that certain elements could play a part. To start with, modern technical innovations could be introduced at two levels, first in the dough-kneading operations, which are very tiring and time-consuming if performed by hand (and unhygienic), and second in the firing system for the oven. A Vienna-style, reciprocating T-arm mixer with revolving bowl would perform the work of one man in a five-man enterprise (25 per cent increase in productivity). Cheap second-hand models in good condition are readily available, as these batch mixers are being replaced in Europe by continuous mixers.

On the baking side, an oil-fired fuel system would enable more constant temperatures to be maintained, increasing fuel efficiency and reducing wastage through over-baking and burning. A reduction in the wastage rate from 6 per cent to 1 per cent would result in savings in all the cost items. Thus, the potential savings of materials, labour and fuel of the two innovations could amount to 0.7d. per loaf, offset by an increase in 0.1d, in capital costs. The net savings of 0.6d. per loaf could be shared between higher wages per man (from 10s. to 12s. per day), lower prices to the consumer (6d. to 5.75d. per loaf) and higher profits for the entrepreneur (0.6d. to 0.7d. per loaf).

The above technical changes would be economic only for the larger baking establishments (five to ten workers), These would then expand their share of the trade, absorbing the workers from the smaller, less competitive enterprises in the process. But the rate of productivity growth in the industry as a whole need not exceed the growth of the market, which is dependent on both the price elasticity and income elasticity of demand as well as on income changes in the community as a whole. Assuming a growth of real income per head of 3 per cent per annum (which would require complementary changes in other sectors similar to what is being advocated in the bakery industry), the end result of a phased process (spread over ten years) of modernisation of this type should be higher wages, lower prices (assuming no inflation),larger profits, and fewer bakery firms but a constant labour force. Higher earnings and profits in the trade would result in increased demand for the products of other industries, hence raising employment levels there.

A Technology Should Suit the Economic and Social Environment in Which it is Employed

The scale of organisation and the kind of technology which goes with it should match the economic and social characteristics of a country. Artisan workshops and small factories using simple equipment may be the most suitable first stage of industrialisation in a subsistence economy with low purchasing power and little monetary exchange, with bad communications and a predominantly peasant population where authority is vested in tribal chiefs or large landowners. In more advanced societies with a developed exchange economy, a more homogeneous, mobile population and a basic infrastructure, more highly mechanised small and medium factories will take the lead. In the process of time, some of these will grow into large-scale enterprises as they acquire experience, as markets expand, as the level of education and scientific skills rises and as a professional managerial cadre can be developed.

In some industries where little technical choice exists, large-scale capital-intensive plants may have to be set up at an early stage, e.g. in heavy chemicals, oil refining, and iron and steel, Large organisations (not necessarily capital-intensive) may also have an important role to play as catalysts for activity in ancillary trades and industries with which complementary relationships are forged (as in Japan).[25] In countries where land and other natural resources are underexploited there may be some inducement to adopt a more capital-intensive approach than would be desirable in countries experiencing heavy population pressure on the land. The political system and its economic instruments represent another variable in determining the appropriate scale and capital intensity.[26] It is salutary to remember, however, that when income had reached $500 per head in France, 70 per cent of the French labour force in the manufacturing, mining and construction industries was still employed in establishments with less than fifty workers.[27]

Small firms using labour-intensive methods have several advantages for developing countries. They can be managed successfully by the personal supervision of the owner, without the need for sophisticated control procedures or a complex hierarchy of authority. Hence they economise scarce managerial skills which cannot be built up in a country overnight. At the same time they act as a seed-bed for management personnel for the future. They can provide immediate executive experience for indigenous entrepreneurs, and so reduce the dependence on expatriate managers and technicians, who can be a considerable drain on foreign currency reserves without passing on much of their expertise to subordinates. Direct personal contact can be maintained with suppliers and customers. The hours of work can be more flexible and the inter-personal relationships less formal. Thus the transition from an individualistic peasant life, where the rhythm of activity is ruled by the seasons, is made easier. Small firms are often more flexible in the sense that they can more easily move into new lines or drop old ones, particularly if they use simple, standard equipment rather than high-speed, special-purpose machines. They can save on transport costs by on-the-spot processing of widely dispersed materials. Overhead costs are low. By product specialisation they can often achieve high economies of scale. And unlike large-scale industry, which usually needs to be in close proximity to the markets and services found only in big cities, small units can be viable in the provincial towns and villages, thus leading to the harmonious development of town and country.[28]

Case Study Illustrations

In Japan, which has achieved the most rapid sustained economic growth of any country in this century, small and medium firms (under 300 employees) have taken the lead in many industries and as late as 1954 employed 59 per cent of the total labour force in industry. In Japan small firms make 100 per cent of the toys, 87 per cent of household utensils, 81 per cent of leather shoes, 78 per cent of printing ink, 71 per cent of farm utensils and 70 per cent of cotton textiles, for example. Much of their technology would be classed as obsolete in the United States.[29]

In the United Arab Republic small firms (less than 100 workers) achieved a higher productivity of capital than did large firms in fifteen out of the twenty major industrial classifications in the 1960 census. Yet their capital intensity was generally much lower. A similar picture is found in data for India, Taiwan, Japan, Chile and Ecuador.[30]

New Technologies Should Stimulate Output in Indigenous Industries and Be Capable of Being Reproduced Locally

A technology is more likely to be progressive if it can be reproduced within the local engineering and material manufacturing industries within a short time and if it processes indigenous raw materials. The immediate benefit will be the saving of foreign currency, which is a serious limiting factor in most countries' expansion plans. Just as important in the long term is the creation of indigenous industries for machine tools, machine building and repair, and material and component manufacture. The experience of North America, west and east Europe and Japan has shown that these industries are indispensable features of a balanced industrial structure. They provide employment and incomes for large numbers of the people. They accelerate the process of technical change through their intimate contacts with the technology users. Their familiarity with the environment ensures that technical developments are practical and economical in the local setting. And developments in this sector tend to have a beneficial multiplying effect on other areas of the economy. Such indigenous industries would lessen dependence on intellectual imports.[31]

To build up such industries in the developing countries would mean starting on the ground floor and at a relatively low technical level in many cases. The technologies chosen must therefore be within their capabilities. By a combination of imaginative improvisation and adaptation, the absorption of scientific knowledge from abroad, an emphasis on technical training, tolerance of initial imperfections by the customers, and accumulated experience and confidence coming from self-achievement, each country could establish a strong and healthy technology-producing sector over a period of years. The Soviet Union and Japan have both demonstrated in this century that the transition from a large imitative to an innovatory role can be accomplished in this way.[32]

Complete technological self-sufficiency is not being proposed. Most developing countries are too small for that, and it would involve quite wasteful duplication. But if these industries are to get off the ground, some loosening of the present technological ties with the advanced nations would seem to be required. Customs unions between countries at a similar stage of technical and economic development might be one answer, with specialisation according to the complementarity of resources and skills. This appears to hold out better prospects than the present policy of importing advanced technologies on the grounds that this is the only way of reaching the quality standards needed to sell manufactured goods in the world markets. This policy has not been too successful in achieving its objectives so far.[33] Export performance is not simply a coefficient of a certain capital intensity. Human skills and motivations, particularly managerial, are usually more important. And exports are possible without complex machinery even in technical fields, as the second case study below will show.

Case Study Illustrations

A ceramic factory making floor and wall tiles formerly imported its hand-operated presses. As a result of close co-operation with small engineering workshops in its locality it was able to have replacement presses made locally, using castings moulded from scrap metal in small foundries and machined on general-purpose lathes and drilling machines. The tiles themselves were made from indigenous clay deposits and fired in kilns composed mostly of local refractory bricks. Thus output, income and employment were stimulated in a number of other industries and trades, e.g. scrap metal, foundry, carbon, refractory, engineering, mining. This multiplier effect was just beginning to make itself felt when it was decided to build a modern large-scale ceramic plant in place of the existing one, with high-speed fully automatic presses, continuous tunnel kilns, etc. This equipment required special steels and engineering skills, refractories with a high aluminum oxide content and technical know-how, which were not available locally (and were not likely to be for many years). Therefore they had to be imported. Also, because of the high speed of operation, very malleable clays were required and these, too, had to be imported. In the end the consumer got a poorer-quality, dearer product because the breakage rate was higher due to inadequate temperature control in the tunnel kilns (technological inexperience) and clumsy handling during glazing operations (inadequate supervision in the new factory). Employment and net output declined in the ceramic and allied industries listed above and the country's trading deficit widened.

The second case study illustration is more encouraging. An Asian country which had formerly imported its sewing machines decided to promote its own machine-building industry. A nucleus already existed in the small workshops manufacturing replacement parts for imported models. Profiting from the temporary protection afforded by import restrictions, local entrepreneurial initiative quickly appeared to co-ordinate and expand the activities of these specialised workshops and to set up assembly units. In a few years the sewing machine industry, equipped with general-purpose lathes and drills (rather than multi-spindle boring machines and special jigs) was turning out models at 60 per cent of the price of previous imports. The local sewing machines had a more limited range of operations and were less accurate, but because of their lower price they had opened up a new market among small-scale clothing and footwear establishments, thus increasing their efficiency. By 1966 import restrictions could be relaxed and the industry was strong enough to have established a thriving export trade to neighbouring countries.

The Productivity of Capital Should be Maximised and the Social Cost of Production Minimised

In countries where capital is scarce and labour unemployed or underemployed (i.e. most developing countries) the emphasis should be put on maximising capital productivity rather than labour productivity. Where there are a number of alternative methods of making a given product, the one which achieves the highest output for a given capital cost should be selected (other things being equal). When setting up new industries, priority should be given to those products and processes which need less capital for the same value of output than would competing substitutes, e.g. cotton ginning and spinning rather than synthetic fibre production.

The idea is basically that you can make the largest "cake" to share among the population by using your scarcest ingredient (capital) most economically, even if this may mean using more of another, more plentiful, ingredient (labour) in the "factor mix" of any production process. It should be pointed out that there are certain kinds of labour--particularly those with entrepreneurial, managerial and advanced technological skills--which are also scarce and need to be husbanded too. And there are many ways by which the utilisation of existing plants can be raised without further investment. The whole range of management techniques, from production planning to improved marketing, can be applied to raise the productivity of existing capital as well as of labour.[34]

What about price and production costs? Should not the technique be chosen which results in the lowest unit costs? This is probably the most important consideration in private investment decisions. It would also be desirable from a national point of view if money costs reflected the real costs to society of using the factors of production. Unfortunately, this is not always true. Official interest rates paid on capital made available to large enterprises (public and private) are often fixed artificially low in the developing countries. The "usurious" rates of moneylenders may, in fact, be a truer representation of the supply price of capital, the level of risks involved and its marginal productivity in use. Similarly, current wage rates, low though they may be, can be said to overprice labour if there are large numbers of unemployed seeking work. The real cost to society of providing productive jobs for them (which would increase the size of the national "cake") would certainly be less, if not zero (e.g. in the case of cottage industry).[35]

Another cause of distortion is the practice among some developing countries of overvaluing their currencies and rationing foreign exchange, thus tending to underrate the real costs of projects making heavy demands on foreign exchange.

Imperfect mobility of labour between occupations and differentials in the capital market are further reasons why money costs are not always the best indicator of economic efficiency or social welfare. It has therefore been proposed that planners should use "shadow prices", which would attempt to measure the real cost to the community of the factors of production when making investment decisions. Similarly, direct government intervention on the wage/interest rate structure may be desirable to ensure a closer identity between private economic advantage and the broader national interest than exists at present.

Case Study Illustrations

A tanning industry project in the Middle East envisaged building a small model tannery to act as a training centre and to demonstrate new techniques, together with a number of new buildings to rehouse existing tanneries, thus improving working conditions and separating the industry (with its obnoxious smells) from the living quarters. The total capital costs were projected as $2.5 million for an output of $15 million per annum (a high capital productivity). The buildings and some of the machinery could be made locally, so the import content was small. Demand for leather was growing at 5 per cent per annum and labour productivity was expected to rise at this rate due to improved methods and conditions; thus the total labour force in the industry of 3,000 would remain the same.

This project was rejected, however, on the grounds of not being modern enough. In its place was substituted a scheme for a large government-owned tannery estate, costing $15 million, equipped with the latest imported machinery and with a total capacity 50 per cent in excess of the existing firms. Labour productivity would be doubled but the savings in wages would be more than offset by higher capital costs (interest and depreciation) if a shadow interest rate were used. The productivity of the capital employed would be only 25 per cent of the anticipated level in the first project proposed. Employment in the industry would be halved, the existing equipment made obsolete and the import bill increased by more than $8 million. The present firms would be broken up and experienced owners made redundant. Little improvement in quality could be expected because further foreign exchange to buy better hides and tanning materials (which, together with technical know-how, were the primary determinants of quality) could not be afforded. And in international terms they would not end up with the most up-to-date process but with an expensive "white elephant", because heavy sole leather and even some upper leathers were being replaced rapidly in world markets by synthetic material (e.g. vulcanised rubber compounds for soles, corfamporomeric material and PVC for uppers).

Thus the more modest scheme was not only more appropriate for the particular internal circumstances of this developing country but also gave it greater flexibility to take advantage of world technological developments when it had the necessary resources (i.e. a petrochemical industry).

On the other hand, there are undoubtedly examples of advanced capital-intensive technology satisfying these criteria best. A case in point is a fibreboard plant in an African country. This cost $2 million and employed only 120 workers directly, because the higher pressures and great bulk involved required very heavy machinery. However, it processed the residue of sugar-cane and maize stalks which would otherwise have gone to waste. Thus the value added during the process was high, and it provided additional incomes to the farmers. The finished product was a good, cheap substitute for certain kinds of wood for furniture and housing. This wood had previously been imported, so foreign currency was also saved. This project therefore served the national interest in several respects.

II. SOURCES OF PROGRESSIVE TECHNOLOGY AND WAYS OF INCREASING ITS AVAILABILITY

I have indicated various economic and social objectives of technical progress and suggested some general guidelines which might be borne in mind by those who frame economic policy affecting public and private investment decisions. The problem does not cease there. The optimum choice can only be made if a full range of alternatives is available. Unsuitable techniques are often applied because there is nothing else on the market except machinery which has been designed to meet other needs. The full spectrum of scientific and technical knowledge must be brought to bear. The brand new, the present day and the past are all potential sources which should be tapped. Let us examine the sources in more detail and put forward some ideas of how their yield can be increased by international action.

Specially designed technologies

The most effective means of overcoming economic backwardness would be to apply accumulated scientific knowledge to the solution of the particular problems of the developing countries. There is undoubtedly a great need for new technologies which will incorporate recent inventions but at the same time take account of the scarcity of capital and of certain managerial and operative skills in the developing world. Innovation is required too so that local raw material can be substituted in certain processes for the different types which are imported at present. Varying climatic conditions may demand new solutions to familiar problems. Working parties have been formed in India and Britain to undertake research into these questions and the United Nations Advisory Committee on the Application of Science and Technology to Development is keenly interested in the problems of adaptation of designs and methods. Much valuable pioneer work has been done by specialist institutes like the Tropical Products Research Institute in England.

Technological research institutes are now being set up in some countries under United Nations Special Fund auspices. But only the surface of the problem is being scratched. If just a small proportion of the ingenuity and creativity which goes into satisfying the exotic demands of an affluent society could be diverted to this end, a very real contribution to overcoming world-wide poverty could be made. Perhaps each of the advanced countries could earmark a certain percentage of its aid funds to sponsor research of this kind, preferably within the developing countries themselves so that it is based upon first-hand knowledge of the local situation.[36]

Case study illustrations

Unconventional sources of energy, such as solar heat to distill fresh water from salt water, and as an energy source for industrial purposes.

"Hover" (air-cushioned) vehicles for transport over swampy ground.

Self-maintaining machinery.

The processing of date palm fibres to replace wool and hair in upholstery stuffing.

The extraction of creosote and charcoal from the husks and shells of coconuts, and biological insecticides from coconut oil.

The derivation of food proteins from oil.

The "spray" process, which may reduce the capital cost of making steel considerably.

Modern technologies

As emphasised previously, modern up-to-date technology from the most industrialised countries can play a part. To deprive oneself completely of these techniques would be just as wasteful as their indiscriminate application in the past. What types are likely to pass through the screen that has been proposed? One can distinguish four main groups.

The first consists of technical know-how with little or no capital element. Improved ways of making or growing things as a result of a deeper understanding of the chemical, physical and biological properties of products and materials fall into this category. The quicker this knowledge is incorporated into current practice the better, and extension services and demonstration units have vital roles in this dissemination. There would appear to be no major economic obstacles, though social resistance may be encountered.

The second group consists of technologies where the tool element can be easily separated from the labour-saving element. One particular process in a series of operations may have to be performed by a particular machine if consistent quality and precision in the final product are to be maintained. The ancillary operations, particularly materials handling, can be carried out by hand methods if labour is abundant and cheap.

The third category covers machines which replace non-existent human skills, or skills which would be more expensive to train in terms of educational facilities.

The fourth embraces all those modern technologies that may be the only effective means of exploiting a country's physical resources, which would otherwise lie idle, and which could form the basis of other indigenous industries.

Case study illustrations

Colour charts, penetrometers and triaxial compression testing machines for measuring the properties of soil and clays--leading to improved crop selection and rotation, higher land yields, cheaper, more durable roads, and improved ceramic products.

A modern gas or oil-fired fuel system for bread ovens giving more precise temperature control, and resulting in even quality and reduced wastage.

Diagnostic machines for locating and identifying defects in automobiles.

Numerically controlled machine tools which economise on skilled labour.

Infra-red scanners and tungsten-tipped drills for discovering and tapping hidden water and oil resources.

Long-established designs

Classic designs which have proved themselves at a similar stage of development in the now advanced economies could still be relevant to the developing countries. They may not be included in the current machinery catalogues because they have been superseded in the advanced countries by more expensive, labour-saving devices. They have to be dug out of the archives of patent offices and long-established machinery manufacturers. Trade associations could carry out such sifting and collating, sponsored by multilateral or bilateral aid funds. They, in turn, would send the designs to the research institutes in the developing countries, which would disseminate technical specifications, blueprints, drawings, etc., to the engineering workshops and manufacturing firms.

The major international companies which set up subsidiaries in the developing countries could contribute considerably in this. The Philips electrical concern has given a lead by establishing a pilot radio assembly plant in Holland where simple, commonly available tools are tried out in conditions which simulate those encountered in its overseas operations.

The examples given below are just indications of what might be the appropriate "new" techniques to replace the existing ones in some countries as a beginning. These should be regarded as steps and not platforms. In each industry improved techniques need to be introduced successively over the years so that productivity is raised progressively.

Case study illustrations

Bakery industry. Steam pipe ovens which ensure an even dispersion of heat by means of coiled steam-pipes; draw-plate ovens in which loading and unloading are speeded up by putting the plate of the oven on wheels and rollers; Vienna, T-arm kneaders in which a single reciprocating arm kneads the dough in a rotating mixing-bowl. This equipment is more advanced and efficient than brick-lined, open-flame ovens and hand-mixing, but is much less capital-intensive than turbo-radiant travelling ovens or continuous mixers.

Ceramic industry. Oil or coal-fired Hoffman kilns; hand-operated jiggers for forming plates, semi-automatic presses for tiles, gravity-fed extruders for pipes. These are all superior to traditional methods but less expensive than tunnel kilns and fully automatic equipment.

Shoe industry. Simple Blake sewing machines (first introduced in 1859) for stitching the sole to the upper and insole. This is quicker than hand-stitching but may be more appropriate than vulcanising or injection-moulding equipment for soling in these countries.

III. POLICY MEASURES TO FACILITATE THE IMPLEMENTATION OF PROGRESSIVE TECHNOLOGY

The need for discrimination has been demonstrated, various yardsticks proposed and possible sources of supply of progressive technologies indicated. International action to increase the availability has been outlined. What remains to be examined is the most crucial aspect of all--how to ensure that these ideas are turned into cold metal and bricks and mortar with the desired characteristics in the countless investment decisions being made throughout the developing world.

This is no easy task. A general awareness of the issues involved is insufficient. When the investment is in the public sector, the officials responsible may take a broad national view. But they may lack a first-hand acquaintance with the down-to-earth, practical operating details which have a bearing on the decision. Private industrialists will naturally tend to seek private advantage, which may not be identical with social welfare, as we have seen. All in all, it sounds like a counsel of perfection to expect the individual investor to be, on the one hand, omniscient, i.e. familiar with all the alternatives available and able to make a quantitative judgment on each and, on the other hand, dedicated to the common good, i.e. to reject the pursuit of private gain whenever it conflicts with the interests of the whole community (even if he could perceive or acknowledge it).

In other words, it is unrealistic to think in terms of providing each potential investor with a check-list of do's and don'ts, together with a full inventory of alternative techniques, and just to wait until common sense and social justice prevail. This would be administratively unworkable and psychologically unsound. The only practical approach is for governments to create a legislative framework and a general economic climate in which individual investment decision, public and private, tend to coincide rather than conflict with the national interest. What is advantageous for the enterprise must also be advantageous from the point of view of the national economy. Only if there is such a coincidence of these interests can decision-making be decentralised and diffused and the price mechanism left to determine the distribution of resources between competing ends. This would not prevent special cases from being examined individually by the direct application of the investment criteria I have indicated.

What specific measures could be taken by governments to create the economic and legal environment most conducive to optimum choice and, once selected and installed, to ensure that progressive technology is utilised efficiently? Some possible steps are listed below.

(1) The formation of customs unions with other States at a similar stage of development and with complementary resources. These would encourage a new international division of labour and a competitive stimulus for efficiency, while avoiding head-on, heavily one-sided encounters between rich and poor nations in the international trade and technology fields.[37]

(2) Higher official interest rates to raise the price of capital vis-a-vis labour costs. This would tend to bring more labour into productive employment as well as increase the propensity to save. It should be made clear that this is not to suggest that governments should pay a higher interest rate on what they borrow from advanced countries or international agencies.

(3) Providing indigenous industries with ample scope to expand, develop and diversify over time without bumping their heads against competing industries which are technically more advanced because greater resources (uneconomically priced) have been placed at their disposal. Giving a clear run ahead to indigenous entrepreneurs is likely to be more conducive to growth and development than imposing protective subsidies and quotas in an attempt to have the best of both worlds.[38]

(4) Tax concessions and political guarantees to attract foreign capital and know-how, accompanied by legislation requiring all companies to buy a certain proportion of raw materials, components and replacement machinery locally within a fixed time period (as in Mexico).

(5) The setting up of documentation and information centres to keep track of past and current technological developments throughout the world. These would establish close liaison with international and other national advisory services for the selection of equipment.[39]

(6) The provision of widespread primary and technical education facilities at the apprentice level, combined with night-school tuition and upgrading courses for practising

operatives, supervisors and managers. Vocational, instructor and management training institutes sponsored by the ILO already function in many countries, while UNESCO programmes cover school, college and university education.

(7) Training courses for managers and planners in feasibility study and cost/benefit analysis techniques to increase the "rationality" of investment decisions and in the use of other management tools (e.g. work study) which will increase the efficiency of existing manufacturing methods. The ILO and UNIDO are operating here.

(8) The encouragement, by state subsidies, grants, etc., of trade and research associations for each industry, sponsored and run by the members themselves. Special budgets could be allocated for importing standard machines to be stripped down, adapted and eventually reproduced locally.[40]

(9) The institution of incentive rewards schemes for inventions, as well as patent protection for local adaptations of foreign designs.

(10) The formation of common facility co-operatives and joint production workshops to raise the productivity of artisan and handicraft industries.

(11) The provision of extension services for small-scale entrepreneurs, providing advice on product and process development, technical skill formation and the selection and use of appropriate technologies. Again, the ILO is active in this work through small-industry institutes and experts on individual assignment.

(12) Long-term planning of manpower and skill requirements in the various sectors of the economy, closely related to the foreseen rate and character of technical change.

(13) The adoption of factory legislative and safety regulations which provide adequate working conditions and safeguards for all groups of workers but do not create dual standards (i.e. for those within and those outside the practical jurisdiction of the laws) and act as barriers to expansion for the smaller enterprises.

(14) The creation of central quality-control and inspection schemes to ensure that products destined for export meet external quality standards, but without imposing unrealistically high standards on total production within the country.

(15) Priority in the allocation of import licences for machinery and materials to organisations that have already demonstrated the aptitudes, skills and motivations required for success in the export markets.

(16) Systematic market research surveys abroad to identify precise consumer needs (and appropriate distribution channels) which might be satisfied by the use of relatively labour-intensive techniques. UNCTAD and GATT have already sponsored and carried out such investigations on behalf of member governments.

(17) The establishment of special small-business development banks to reduce the differentials in capital accessibility between the traditional and the modern sectors. The World Bank is giving technical and financial assistance in this area.

(18) The planned distribution of industry to backward areas to provide more employment opportunities outside the major cities and

to reduce income inequalities between regions. Processing of agricultural and other land-based products are obvious choices (the FAO has substantial interests in this field).

(19) Financial incentives (e.g. tax rebates on training costs) to international companies to set up apprentice training schools, management development programmes and planned succession to management positions for indigenous staff. This would reduce the foreign exchange costs of expatriate staff, while ensuring that their essential expertise in operating, servicing and managing more advanced technology is passed on to local personnel.

(20) Devaluation of currencies to ensure that the importer has to pay the real cost of foreign machinery and materials, and that a proper evaluation is made in initial feasibility studies.

(21) State-financed hire-purchase and rental schemes with lower interest rates for imported second-hand machinery and locally made equipment.

(22) Subsidised factory premises in provincial towns and villages to slow down the population drift to the cities. The subsidies could be equivalent to the cost of housing and other facilities which would otherwise have to be provided in the cities.

(23) "Tax holidays" for foreign machinery and component manufacturers who set up local design and production plants to develop indigenous technologies.

(24) Public information campaigns to increase the prestige and consumer acceptance of indigenous technologies and products.

Where such policies and programmes have been introduced, the results are promising. They appear to open up new avenues for a dynamic attack on poverty in the developing countries in which the progressive and widespread introduction of new methods (new, that is to say, compared with <u>their</u> traditional ones) could lead to a better use of their current resources and achieve a rapid and sustained growth shared by the whole people.

NOTES

[1] International Labour Office.

[2] See, for example, A. E. Khan: "Investment criteria in development programmes", in <u>Quarterly Journal of Economics</u> (Cambridge (Massachusetts)), Feb. 1951; Hollis B. Chenery: "The application of investment criteria", ibid., Feb. 1953; W. Galenson and H. Leibenstein: "Investment criteria, productivity and economic development", ibid., Aug. 1955; O. Eckstein: "Investment criteria for economic development and the theory of inter-temporal welfare economics", ibid., Feb. 1957; A. K. Sen: "Some notes on the choice of capital intensity in development planning", ibid., Nov. 1957; "Some aspects of investment policy

in underdeveloped countries", in International Labour Review, Vol. LXXVII, No. 5, May 1958; A. K. Sen: "On optimising the rate of saving", in Economic Journal (London), Sep. 1961; and S. A. Marglin: "The opportunity costs of public investment", in Quarterly Journal of Economics, op. cit., Feb. 1963.

[3]For instance, the Indian Planning Commission, in a recent letter to Secretaries of Union Ministries and Secretaries of States and Union Territories, has suggested that in the formulation of projects for the Fourth Five-Year Plan there should be a conscious attempt to depend more and more on indigenous, labour-intensive technology where it is economically sound, rather than on imported technologies which have a labour-saving bias. Reported in Times of India (Bombay), 4 Aug. 1969.

The United Nations Economic Commission for Asia and the Far East (ECAFE) has stated: "The emphasis given to advanced technology as a strategy for rapid industrialisation in developing countries has resulted in the accentuation of a dualistic economy. It has often given rise to social destruction and aggravation of unemployment without producing much change in the standard of living of the majority." See Appropriate technology for small manufacturing plants (Bangkok, ECAFE, 21 May 1969).

Similarly, the United Nations suggests that "the accelerating pace of technical change renders investment choices that much more difficult and, given the scarcity of capital in relation to the supply of labour in the developing countries, it is far from certain that the wisest decisions were always made". United Nations: World Economic Survey, 1967 (New York, 1968), Introduction, p. 27.

[4]Some of this case material has been published previously in ILO: Human resources for industrial development, Studies and Reports, New Series, No. 71 (Geneva, 1967). Certain aspects have also been examined at the macro and micro levels in other papers by the present writer, e.g. K. Marsden: Appropriate technologies for developing countries (Geneva, ILO, 1966) (mimeographed); Technological change in the handicraft sector of developing countries (Geneva, ILO, 1968) (mimeographed); Factors affecting the management of small enterprises (Geneva, ILO, 1968) (mimeographed); "Integrated regional development: a quantitative approach", in International Labour Review, Vol. 99, No. 6, June 1969; and "Towards a synthesis of economic growth and social justice", in ibid., Vol. 100, No. 5, Nov. 1969.

[5]G. K. Boon: "Choice of industrial technology: the case of woodworking", in Industrialization and Productivity (United Nations), No. 3, 1961. Other case studies of alternative techniques are given in the same author's Economic choice of human and physical factors in production (Amsterdam, North Holland Publishing Company, 1964).

[6]See Paul A. David: "The mechanisation of reaping in the

ante-bellum mid-west", in H. Rosovsky (ed.): Industrialisation in two systems (New York, John Wiley, 1966). A similar trend has been traced in Britain, though the introduction of mechanical reapers was more delayed and the widespread diffusion more prolonged because of the more plentiful supply of labour and smaller size of farm. See E. J. T. Collins: Sickle to combine: a review of harvest techniques from 1800 to the present day (Reading, University of Reading, 1968).

[7]Professor W. Lockwood has pointed out: "If Japan's experience teaches any single lesson regarding the process of economic development in Asia, it is the cumulative importance of myriads of relatively simple improvements in technology which do not depart radically from tradition or require large units of new investment." See W. Lockwood: The economic development of Japan: growth and structural change 1868-1938 (Princeton (New Jersey), Princeton University Press, 1954), p. 198.

[8]Professor Hirschman introduced the concepts of backward and forward linkages to illustrate how the benefits resulting from an innovation in one area might spread to other sectors. But in the absence of existing indigenous know-how, these new linkages may be more readily established with foreign suppliers whose experience can be tapped immediately. Hirschman also underestimates the disruptive effects of new products and technologies on the existing network of interrelationships built around the substituted products. See A. O. Hirschman: The strategy of economic development (New Haven, Yale University Press, 1958).

[9]See Gunnar Myrdal: Rich lands and poor: the road to world prosperity (New York, Harper & Row, 1957).

[10]Criteria proposed in the past include the following: equalising the marginal productivity of each factor of production, A. P. Lerner: "On the marginal productivity of capital and the marginal efficiency of investment", in Journal of Political Economy (Chicago), Feb. 1953; maximising the surplus to turnover or capital ratio, Galenson and Leibenstein, op. cit., and A. K. Sen: Choice of techniques (Oxford, Basil Blackwell, 1968); and maximising the present value of the future income stream, Eckstein, op. cit.

The difficulties experienced in translating these criteria into practical policy guidelines are several; market prices are often an unreliable guide to social costs and values, owing to the existence of monopolies, external economies and diseconomies, imperfect factor mobility and market imperfections; the problem of determining, in a vacuum, the values which should be given to the "shadow prices" of labour, capital and foreign exchange which would measure their social opportunity costs; the fact that profitability is the result of the interaction of a host of economic, social and technological factors which will vary from country to country and industry to industry and which cannot always be quantified in advance; the realisation that certain ex-ante assumptions, such as that wage earners do not save, that wages will not rise as fast as productivity or that fiscal measures cannot be used to influence the rate of saving, are not very realistic assumptions in the real world; the difficulty of fixing the rate of discount of future income expectations, when this will vary with each individual in the community according to his time horizon and personal preferences; and the problem of deciding upon the weight to be given to the social and political costs of unemployment and the greater income inequality which might result from the application of growth-maximising criteria.

[11]David A. Morse: "Narrowing the gap", in ILO Panorama (Geneva), Sep.-Oct. 1966.

[12]See United Nations Centre for Industrial Development: Criteria for the development of manufacturing industries in developing countries (New York, 1966) (E/C.5/111/Add.1). Of course, the growth of labour productivity is not simply the result of larger capital inputs. E. F. Denison has pointed out the importance of the so-called "residual factors" (technical progress, higher skills, better management, etc.) in recent American economic growth. See E. F. Denison: The sources of economic growth in the United States and the alternatives before us (New York, Committee for Economic Development, 1966). This fact has sometimes been taken to imply that such "windfall gains" would accrue automatically to the developing countries if they selected the latest techniques, incorporating the full quotient of the technical progress which had occurred historically in each field. But in practice, technical change cannot be easily separated from its socio-economic environment. Technological innovations which are not accompanied by appropriate changes in the skills, education, attitudes and behaviour patterns of management, workers and consumers may result in a lowering of economic efficiency, i.e. a decline in the output/total factor input ratio.

[13]The stock of total reproducible capital averaged $25,860 per worker in the United States in 1966; see United States Department of Commerce, Bureau of the Census: Statistical abstract of the United States, 1967 (Washington, DC, 1968), tables 492 and 494. The relatively small variations in investment per worker from sector to sector (including agriculture) have been noted by Simon Kuznets in Modern economic growth: rate, structure and spread (New Haven, Yale University Press, 1966).

[14]This point has been made by Professor E. Hagen in Development of the emerging nations: an agenda for research (Washington, DC, The Brookings Institution, 1962).

[15]See Hirschman, op. cit., Galenson and Leibenstein, op. cit., Sen: Choice of techniques, op. cit., and A. Gerschenkron: Economic development in historical perspective (Cambridge (Massachusetts), Harvard University Press, 1962).

[16]Data for India and the United Arab Republic--two developing countries with long-established industrial sectors--show that small firms tend to realise a higher surplus/capital ratio than do large enterprises. See Census of Industrial Production, 1960 (Cairo, Department of Statistics, 1962) and Capital (Calcutta), 22 June 1968. The explanations are manifold: use of cheap, second-hand equipment, greater experience and involvement of owner-managers, achievement motivation of the entrepreneurs compared with the desire to maximise lifetime earnings of salaried managers, greater flexibility in resonding to market demand, low overhead costs, etc. These factors are considered at greater length in Marsden: "Towards a synthesis of economic growth and social justice", op. cit.

[17]See Food and Agriculture Organisation: The state of food and agriculture 1966 ome, 1966), Chapter III, for a good analysis of the interdependence of agriculture and industrial development.

[18]The role of achievement motivation as a driving force behind economic development is described in D.C. McClelland: The achieving society (Princeton (New Jersey), D. Van Nostrand, 1961).

[19]Derived from D. Creamer: Capital and output trends in manufacturing industries, 1880-1948, Occasional Paper No. 41 (New York, National Bureau of Economic Research), tables 1 and 2; and S. Kuznets: Six lectures on economic growth (Glencoe (Illinois), The Free Press, 1959), table 6.

[20]See Kuzents: Modern economic growth: rate, structure and spread, op. cit., tables 3 and 6.

[21]The net value of structure, equipment and inventories per worker in American manufacturing industry amounted to $9,300 in 1965 (at 1958 prices). However, the gross value of this reproducible capital stock, which measures the stock's capacity for current production, is a more meaningful figure for our purpose than the depreciated book value. R. W. Goldsmith estimates that the gross value of reproducible assets used for production was 61 per cent higher than the net value in 1958, and that the gross capital stock increased during the post-war period at an average rate of 8.6 per cent per annum in current prices. Thus an estimate of $20,000 for gross reproducible capital per worker in American manufacturing industry in 1969 at current prices is not likely to be far from the mark. See Statistical abstract of the United States, 1967, op. cit., tables 1114 and 1126; and R. W. Goldsmith: The national wealth of the United States in the post-war period (Princeton (New Jersey), Princeton University Press, 1962), tables 10 and 14.

[22]The extreme case may be Venezuela, where labour productivity in the petroleum industry, which employed 1.5 per cent of the working population in 1963, was sixty times the average productivity in agriculture, which was responsible for 28 per cent of employment. See Plan de la Nacion, 1963-66 (Caracas, 1963).

[23]An interesting analysis and description of the Nigerian bakery industry is given in P. Kilby: African entrepreneurship: the case of the Nigerian bread industry (Hoover Institution Study, Stanford University Pres, 1965).

[24]1 pound = US $2.40; 1s. = $0.12; 1d. - $0.01.

[25]See T. Ando: "Interrelations between large and small industrial enterprises in Japan", in Industrialization and Productivity, op. cit., Bulletin No. 2, 1959.

[26]In a centrally directed economy, where all means of production are under state control and where output is determined by the plan and not by consumer demand working through the market, wide variations in capital intensity from one sector or enterprise to another will not produce the same effects as in a market economy. Prices, wages and income distribution are not determined by costs and productivity. An enterprise does not go out of business if its costs are higher than the prices fixed by the central authorities. The scale of operation is less influenced by the structure of the market, and not at all by the ability of individual entrepreneurs to accumulate capital. Employment can be maximised by central allocation and decree. Thus, in the Soviet Union, more capital-intensive technology was introduced in the 1920s and 1930s than might have been appropriate elsewhere. Nevertheless, it has been shown that in the metalworking industry, for example, Soviet planners did adapt Western technology to their own factor proportions by substituting labour for capital wherever possible. Even the Gorky automobile plant, which was built with the assistance of Ford engineers, was redesigned in significant ways in the light of special Soviet conditions. And in many plants it was common to find basic processes highly mechanised, whereas ancillary operations such as materials handling were carried out with large quantities of labour. See David Granick: "Economic development and productivity analysis: the case of Soviet metalworking", in Quarterly Journal of Economics, op. cit., May 1857, and idem: "Organization and technology in Soviet metalworking: some conditioning factors", in American Economic Review (Evanston (Illinois)), May 1957.

[27]Figures quoted by B. F. Hoselitz: "The entrepreneurial element in economic development", in United Nations: Report on the UN Conference on the Application of Science and Technology for the Benefit of the Less Developed Areas (New York, 1962).

[28]For a fuller analysis of the role of small-scale enterprises in development, see H. W. Singer: International development: growth and change (New York McGraw-Hill, 1964); E. Staley and R. Morse: Modern small industry for developing countries (New York, McGraw-Hill, 1967); Economic Bulletin for Africa (United Nations Economic Commission for Africa), June 1962; and K. Marsden: "The role of small enterprises in the industrialisation of the developing countries", in ILO: Report on the ILO Inter-Regional Seminar on Programmes and Policies for Small-Scale Industry within the Framework of Over-all Economic Development Planning (Geneva, 1968) (Mimeographed).

[29]Ministry of International Trade and Industry: Vital statistics of production (Tokyo, 1955), quoted in Ando, op. cit.

[30]See L. K. Mitra: Employment and output in small enterprises in India (New Delhi Bookland Private Limited, 1967); ILO: The development of small enterprises in Taiwan, Republic of China (Geneva, 1967) (Mimeographed); K. Miyazawa: "The dual structure of the Japanese economy", in Developing Economies (Tokyo), June 1964; Consejeria Nacional de Promocion Popular: Pequena industria y artesania en Chile (Santiago, 1968); The artisan community in Ecuador's modernising economy (Menlo Park (California), Stanford Research Institute, 1963) (mimeographed); and K. Marsden: The role of small-scale industry in development, with special reference to Egypt (Cairo, Institute of Small Industries, 1964) (mimeographed).

[31]The Director-General of UNESCO, in a speech to the United Nations Economic and Social Council in July 1966, put forward the view that the resort to foreign "magic" was the characteristic of underdevelopment, which would persist until science and technology became part of the indigenous culture. Similarly, the Director-General of the World Health Organisation has pointed out: "Technology may be international in substance, but its method of application must be adapted to the situation in which it is to be applied." See Dr. M. G. Candau: "Knowledge, the bridge to achievement", in WHO Chronicle (Geneva), Vol. 21, No. 12, Dec. 1967, p. 518.

[32]See A. Maddison: Economic growth in Japan and the USSR (London, George Allen & Unwin, 1969). Although many technological innovations in both countries had Western origins, they were mostly adapted so that they could be reproduced by the indigenous machine-building industry. Thus, already in the first decade of this century, 60 per cent of total capital investment in durable equipment (excluding military) was satisfied by local production in Japan. In the 1930s net imports of durable equipment had fallen to less than 4 per cent of total domestic investment. See. H. Rosovsky: Capital formation in Japan 1868-1940 (Glencoe (Illinois), The Free Press, 1961). This machinery was manufactured mostly in small establishments. Even in 1965, 47 per cent of the labour force engaged in the manufacture of non-electrical machinery in Japan was still in establishments

with fewer than a hundred workers. Japan statistical year book, 1967 (Tokyo, Bureau of Statistics, 1968), table 119.

[33]Exports of manufactured goods from developing countries reached only $6.87 thousand million in 1966, or 19 per cent of the total exports. Imports of machinery, chemicals and other manufactures were running at more than three times this level. See United Nations: Statistical yearbook, 1967 (New York, 1968), table 14. What proportion of these imports could be saved, and what benefits would result in the form of more jobs and higher incomes among the producers of indigenous substitutes, if less effort were devoted to chasing this will o' the wisp objective of higher exports via advanced technology, and instead more attention were paid to making optimum use of domestic resources to satisfy known markets at home and abroad?

[34]Utilisation of the other imputs--land and materials--needs to be taken into account. Advanced technologies can sometimes offer better material utilisation, but this may not be realised if the requisite labour and managerial skills are lacking (cf. the ceramic factory case study). It is sometimes argued that where managerial and operative skills are scarce, it would be wiser to encourage large-scale enterprises to invest heavily in automatic processes so that relatively few unskilled workers have to be controlled, rather than to foster labour-intensive factories which need many more workers and higher craft skills, thus complicating management and increasing training costs. Put like this, the argument is persuasive, But are these the only alternatives? By product and process specialisation, a number of small enterprises can produce the same output as one large factory. And managerial efficiency in small enterprises depends more upon the innate entrepreneurial characteristics and motivations of the owner than upon formalised control procedures requiring higher education and institutional training. Similarly, unlike high-level mechanical and electronic engineering know-how, craft skills can be acquired on the job, so that the wage and supervisory costs of the apprentice need not exceed his marginal product, and material wastage is minimised. New techniques can be injected by machinery and material suppliers as well as by government advisory services. This is how the Japanese made full use of their labour force, at a low capital and training cost, while husbanding their managerial resources, and still gaining the advantages of a high division of labour.

[35]The distortion of the price structure resulting from the use of selective government controls is discussed in Gunnar Myrdal: Asian drama. An inquiry into the poverty of nations (New York, Pantheon, 1968), Chapter 19.

[36]The United Nations Advisory Committee on the Application of Science and Technology to Development recommended, at its 13th Session in April 1970, that the targets for the Second United Nations Development Decade should include the allocation of 5 per

cent of the non-military research and development expenditure of the developed countries to specific problems of developing countries.

[37]So far such agreements have been confined to small groupings of neighbouring countries, e.g. the East African and the Central American common markets. But it may be that Tanzania and Pakistan, say, are more complementary than Tanzania and Kenya, for example. The real crux is whether the developing countries are willing to accept lower-quality products from their trading partners in place of imports from the advanced countries, as the short-term price to be paid for a long-term mutual self-help programme. There is also the question of the product mix, which influences technological choice and may determine factor proportions within narrow limits. Will they be prepared to exchange cotton textiles for leather and leather products which can still be made by labour-intensive, capital-saving equipment, locally produced, rather than insist upon importing capital-intensive machinery and material inputs from the advanced countries in order to manufacture the most modern versions of these products (clothing and footwear made from nylon, terylene, corfam, PVC, etc.), and where little technological flexibility is possible? Such alternatives are available for a wide spectrum of consumer products and services. A high proportion of the products of the latest technology did not exist at the turn of the century. Yet income per head in the United States in 1900 was well over $1,000 at today's prices. High income growth could be attained by choosing the products which can be produced in ways that are most appropriate for their current resource endowment. The opportunity to make choices of this kind throws a heavy burden on the planners and political leaders of the less developed countries, particularly as they are members of the relatively high-income minority in which consumer demand is concentrated and as they have generally been most exposed to current Western consumption patterns. But the key to harmonious economic and technological change probably lies here.

[38]The problems encountered in India as a result of a co-existence of the most advanced and the most primitive technologies in the same industry, e.g. textiles, and from the attempts to cushion the effects of such confrontations by the use of selective controls and subsidies, are described in Ministry of Industry: Development of small-scale industry in India: prospects, problems and policies. Report of the International Perspective Planning Team sponsored by the Ford Foundation (Delhi, 1963).

[39]Scientific and technological information centres have played an important part in Japanese and Soviet economic development. In 1963 about 210,000 abstracts of foreign scientific papers were made by the Japan Information Centre for Science and Technology. See Japan Council for Science and Technology: Science and technology in Japan (Tokyo, 1964), p. 15. In the Soviet Union the All-Union Institute of Scientific and Technical Information has an abstracting service which digests 400,000 scientific

papers a year. See Maddison, op. cit.

[40]The need for indigenous design and development organisations is recognised in India and Tanzania. The Indian draft Fourth Five-Year Plan, 1969-74, states: "Self-reliance in the technological sense implies the existence and effective functioning of indigenous organisations for design, construction and engineering of projects as well as capability for design and development of machinery, equipment and instruments indigenously manufactured. At present there is unwholesome dependence on foreign agencies for these services. As long as this deficiency remains, local talent will not have scope to develop, and excessive dependence on foreign help will be prolonged . . . It is only by participating actively and in positions of responsibility that such skill and confidence are generated and scarce high talent human resource is developed. It is therefore of vital importance for the future development of the country that urgent attention is given to promoting and encouraging health development of adequate design and engineering organisations, staffed by highly qualified personnel and working under proper leadership." (Planning Commission, Government of India: _Fourth Five-Year Plan, 1969-74_ (New Delhi, 1969), p. 44).

In the Tanzanian Second Five-Year Plan the view is expressed that "the tendency in the past to think of mechanisation primarily in the context of tractorisation has led to the neglect of opportunities to improve hand- and animal-drawn equipment, the need for simple processing equipment, water-lifting and reticulation devices, and on-farm transport such as ox carts and trailers, as well as feeder transport vehicles". The Plan provides for the expansion of the Tanganyika Agricultural Machinery Testing Unit to engage in designing simple and inexpensive farm implements which could easily be made from available raw materials. Particular emphasis would be placed on intermediate technology, including the construction of ox carts, ox-driven water-pumps, hand-operated water-pumps for irrigation, ox-drawn earth scoops, ox-drawn tool bars and crop-driers. See United Republic of Tanzania: _Second Five-Year Plan for Economic and Social Development 1969-1974,_ Vol I: _General analysis_ (Dar-es-Salaam, 1969), pp. 37 and 38.

Choice of Technology and Industrial Transformation: The Case of the United Republic of Tanzania

DAVID A. PHILLIPS*

The United Nations Second Development Decade witnessed a considerable amount of promotional effort in the field of labour-intensive (appropriate) industrial technology. This work arose from the growing realization that employment opportunities in developing countries were not being increased adequately to absorb the labour supply. Some of the pioneering work was included in studies by the International Labour Organisation of Colombia, Kenya, Sri Lanka and others [1]. Much academic effort has gone into identifying appropriate technologies, evaluating their efficiency, both in static and dynamic terms, identifying the determinants of technological choice, and the policies and strategies suitable for their promotion.[1]

This paper is concerned with the relationship between choice of technology and industrial strategy. The ultimate objective of industrialization, it may be safely assumed, is not merely to provide current employment or more manufactured goods, but to contribute to and if possible accelerate the long-term growth and enrichment of the economy (however measured). Consequently, it is also reasonable to assume that the objective of labour-intensive industrialization, although it may to some extent substitute current for future employment, cannot be divorced from the objective of economic growth. Society will trade off long-term growth for current employment, if necessary, but at a low and decreasing rate.[2] If long-term growth is a key objective, then industrial strategy may be regarded as of primary importance.

The contention of this paper is that the choice of appropriate industrial strategy, in terms of pattern of growth and composition of output of industrial goods, should precede the selection of techniques of production. The reasons for this are, first, that this approach is more likely to ensure the long-term growth of

From **INDUSTRY AND DEVELOPMENT**, No. 5, 1980, (85-106), reprinted by permission of the publisher, U.N. Publications, N.Y.

industry in terms of output and employment; and, second, that, as will be argued, choice of technology has been and is inevitably determined to some extent by the composition of output. In such a case it would be impractical to arrive at an appropriate industrial strategy by basing it simply on labour-intensive projects without any wider concerns.

The composition of industrial output is an important parameter not only because of its effects on choice of technology, but also because it is a measure of the "transformation" of industrial production. Transformation is taken here as meaning the process by which industry is restructured from the type of structure associated with a colonial-type primary producer economy to that of an economy producing a balance of manufactured and primary goods. This transition involves characteristic shifts in the pattern of industrial output.

General political and economic forces have influenced the choice of techniques both directly and through their effects on the composition of output; consequently it is impractical to promote labour-intensive technology without taking account of a country's political and economic environment. This is not a new idea, but in view of the widespread efforts that have taken place to promote labour-intensive technology in isolation from the political and economic factors that affect the characteristics of industrial production, it is worth restating.

To provide an illustrative background, we examine the case of the industrial sector in the United Republic of Tanzania, a country whose allegiance to objectives of employment generation, via labour-intensive decentralized production, particularly over the period 1967-1977, is well documented.

I. TECHNOLOGY AND INDUSTRIALIZATION IN THE UNITED REPUBLIC OF TANZANIA

As was the case with other developing economies, the structure of Tanzanian industrial development was laid during the period in which it was under foreign control. This political inheritance affected, first, the composition of output, consumption and trade. Secondly, it affected the specification of products, scale and location of production, and technology. These effects were of the type commonly associated with dependence and widely documented.[3]

The country's colonial experience was to some extent atypical. Before the First World War it was a German protectorate. After 1918 it became a League of Nations mandate under British administration rather than a colony. After the Second World War the British continued to administer the country as a United Nations trust territory. In 1961 it became independent and in 1964 joined with Zanzibar to form the United Republic of Tanzania.

Thus colonial penetration of the economy was limited in comparison, for example, with that of Kenya. Nevertheless, the pattern of economic development does not seem to have been

Table 1. Industrial capital intensity, 1966-1976

Item	*1966*	*1967*	*1968*	*1969*	*1970*	*1971*	*1972*	*1973*	*1974*	*1975*	*1976*
Electricity use per worker (thousand kWh)	1.52	1.85	1.62	2.16	2.05	2.37	2.61	2.45	2.34	2.35	2.5
Depreciation per worker (thousand TSh)[a]	15.3	18.0	15.0	20.4	20.1	18.7	20.5	23.6	22.5		

Source: International Bank for Reconstruction and Development and *Tanzania Economic Survey 1977-8.*

[a]Constant prices.

Table 2. Plant size and industrial concentration, 1966-1974

Item	*1966*	*1968*	*1970*	*1971*	*1973*	*1974*
Number of registered enterprises	434	496	482	468	503	499
Enterprises employing 10-100 workers	346	408	340	348	362	350
Per cent of gross output	36	44	36	29	27	27
Per cent of employment	34	31	24	23	20.5	19
Enterprises employing 100-500 workers	83	74	98	105	120	127
Per cent of gross output	46	36	42	46	38	35
Per cent of employment	49	39	39	40	40	39
Enterprises employing 500 and more workers	9	12	14	15	21	22
Per cent of gross output	18	19	23	25	34	38
Per cent of employment	16	30	37	38	47	43

Source: Based on surveys of industrial production, 1965-1974 (Government of the United Republic of Tanzania).

fundamentally different from that of other developing countries.

With regard to the composition of trade, by 1911 the export sector had been developed, characteristically, around a range of primary products supplying German markets. These were principally (80 per cent), sisal, rubber, hides and skins, copra, coffee, cotton and gold [15]. Simultaneously with the emergence of production of cash crops for export and the shift of rural labour to the plantations, imported mass-produced manufactured goods started to erode the market for domestic manufactures.

The orientation of the economy towards primary exports and manufactured imports has remained substantially intact. At independence, in 1961, the composition of exports was dominated by unprocessed or partly processed primary products and it still is. In 1961, 50 per cent of wage employment was in the plantation sector producing sisal, coffee and cotton. Unprocessed raw materials and crops accounted for over 90 per cent of exports. Currently the export composition is the same, except that cloves from Zanzibar are included, and refined petroleum has been re-exported (to Zambia) since 1968. On the import side, manufactured consumer, intermediate and capital goods predominate, plus petroleum, which now contributes about 15 per cent to the total import bill. The composition of export trade is related to the composition of domestic output in the monetary sector, and to some extent also in the subsistence sector. The principal shift in emphasis, from sisal to coffee, has not changed the primary-product orientation of exports.

The erosion of markets for domestic cottage industry manufactures began around 1870. Research has produced evidence of flourishing spinning and weaving of locally grown cotton in various parts of Tanganyika [16]. Germans travelling through the country in the nineteenth century reported iron smelting and blacksmithy in various "industrial centres" that demonstrated skills comparable with those of pre-industrial Germany. Iron tools, chains, wire and weapons were produced with locally made equipment. Iron ore was mined in a number of places, and rudimentary blast-furnaces using wood, charcoal and limestone flux were functioning. By 1900, domestic production of cotton cloth and iron tools was all but extinguished because of inability to compete with imports. The quality of domestically produced cloth was poor, but iron hoes had been produced of reportedly superior quality to the imported products available around 1900. From then on Tanganyikan cotton was exported after ginning, to be reimported as finished cloth.

Spinning and weaving were not reintroduced into the country until 1960, when a 10-million-metre-capacity mill, financed by a consortium of local and foreign interests, was set up. Iron smelting has not as yet been reintroduced into the country; but local manufacture of iron hoes began again in 1970, when a farm implements plant was set up with Chinese aid. Ironically, the output of this plant over the first few years (up to 1 million units) was possibly lower than *per capita* output in 1880, when it was estimated that 150,000 hoes were passing through one market, Tabora, in the west [16].

The eradication of the cottage textile industry was complete, unlike that of India, for example, which survived severe pressure

Table 3. International Comparison of Industrial Structure, 1965

(Percentage of total output)

Industry	*Typical small primary producing country*	*Typical small industrialized country*
Food, beverages, tobacco	45.0	15.0
Textiles	13.0	8.0
Clothing, footwear	10.0	6.0
Wood products	5.0	5.5
Leather and leather products	0.2	0.8
Non-metallic mineral products	6.0	5.5
Rubber products	1.0	1.5
Paper and paper products	—	6.0
Printing, publishing	4.0	4.5
Chemicals, petroleum and coal products	7.0	9.3
Basic metals	—	6.0
Metal products and engineering	11.0	31.0
	100.0	100.0
Overall manufacturing share in GDP	10.2	32.2

Source: H. Chenery and L. Taylor, *Review of Economics and Statistics,* No. 50, November 1968.

Table 4. Structure of industry in developing countries, 1960, 1970 and in the United Republic of Tanzania, 1975

(Percentage of total output)

	United Nations study		*United Republic of Tanzania*
Industry	*1960*	*1970*	*1975*
Food, beverages, tobacco	28.5	24.8	29.4
Textiles, allied	21.4	18.4	21.2
Wood, paper, printing	8.3	8.4	6.9
Chemicals, plastics, rubber	14.6	16.3	18.0
Non-metallic mineral products	5.2	5.7	4.2
Basic metals, engineering	19.4	24.1	17.8
Other	2.8	2.2	2.8
Total	100.0	100.0	100.0

Source: Based on *Industrial Development Survey* (United Nations publication, Sales No. 73.II.B.9); *Tanzania Economic Survey, 1977-8.*

from British exports [17]. The Indian hand-loom sector currently accounts for about 30 per cent of total textile output, while attempts to revive the industry in the United Republic of Tanzania have been limited to the establishment of training centres. (Proposals have included, ironically, the introduction of the broadloom from the United Kingdom of Great Britain and Northern Ireland.)

The combined force of the expansion of plantation crops and the contraction of cottage industry, under pressure of the foreign trading system, must be conceived as one of the foundations of the subsequent evolution of economic structure and composition of output. This is reflected in the available data on the structure of GDP. The contribution of agriculture to GDP rose to something over 50 per cent, and in 1961 stood at 48 per cent. Subsequently the share declined to below 40 per cent, which was primarily due to the collapse of the sisal industry, which over three years caused the loss of about 60,000 plantation jobs (50 per cent of sisal employment). (In 1961, Tanganyika produced one third of world sisal output, and the industry as a whole accounted for 30 per cent of total wage employment.) The colonial powers who brought into being the export-dependent plantation economy also developed the cheaper substitute materials that undermined it. The collapse of the sisal industry resulted in almost zero wage employment growth in the country between 1962 (397,000 workers) and 1972 (403,000 workers).

Apart from agriculture, the other major sector was, characteristically, services. In 1961, at independence, finance, administration, hotels and distributive trades accounted for 34 per cent of GDP (42 per cent if transport is included). Therefore, 90 per cent of GDP was accounted for by agriculture, services and transport [18]. This reflected, first, the orientation towards export cash crops, and, secondly, the inflated banking, trade and administrative system required to feed the foreign trade sector and colonial administrative apparatus. In 1977, these sectors, however, still accounted for 86 per cent of GNP [19], while manufacturing, crafts, power, mining and construction accounted for 14 per cent.

On the wage employment side, heavy orientation remains towards these two sectors, with transport and services accounting in 1977 for about 44 per cent, and public administration (community services) alone, recently the fastest growing sector, accounting for 25 per cent [19]. The continued emphasis on public administration has its origin in the colonial system, subsequently reinforced by more recent developments in the socialized public sector.

Returning to the foreign trade sector, we find that the import ratio changed from 25 per cent in 1966 to 34 per cent in 1975, and 27 per cent in 1976, a high level by international comparison [20], but typical of an open and dependent economy exporting primary products in return for manufactured goods. The export ratio, which was around 25 per cent throughout the 1960s, fell back to 20 per cent in 1976. The country's position as a producer of primary goods and importer of manufactured goods showed itself particularly vulnerable when domestic drought coincided with severe price inflation of imports, resulting in a trade deficit of 31 per cent of GDP in 1975.

The emergence of the modern industrial sector after 1946 followed a pattern of investment largely predetermined by the colonial economic system as regards composition of output, based on the one hand on local processing of exportable primary products, and on the other hand substitution for previously imported consumer goods. This pattern reflected, first, the existing nature of the external trade sector, and, secondly, the emerging pattern of consumption oriented towards the wealthier urban population (foreign and indigenous). In either case a high level of dependence on exports or imports was required, which reinforced the divergence of domestic resource use and domestic demand. To the extent that new factories substituted for existing cottage industry based on local resources (e.g., in furniture, shoes, garments and beverages), dependence on imports increased.[4]

In 1949, the composition of organized industrial output was largely determined by the predominance of cash crops and urban consumer goods. The first industrial survey of 1957 [21,22] listed sisal processing; cotton ginning; sawmilling; vegetable oil extraction; tobacco curing; and manufacture of soap, leather, garments and furniture. Alongside these enterprises a depleted craft sector remained. Notable here was the increasing use of scrap material, based on discarded imported tires, tin cans, vehicle parts etc. This 'degenerate' form of craft industry was involved in manufacturing sandals, lamps, spray guns and domestic utensils. Imported synthetics were increasingly used in footwear and garment manufacture. Thus the material base of the cottage sector was transformed in that it also depended on imports.

Between 1949 and 1961, the commodity composition of industry remained substantially unchanged; only the scale, location and pattern of ownership and control changed owing to the internationalization of investment. Apart from one large-scale, plantation-based sugar factory and a brewery, which had existed before 1949, the period saw international interests setting up plants for the production of cola beverages, milled flour, canned fruit, dairy products, canned meat, paint, insecticide (mixing) and tin cans. All these plants directly supplied the export processing or consumer markets.

The economic and political forces that determined the composition of new industrial output also had considerable impact on technology. This was effected through their impact on location and scale of production, and also more directly through techniques introduced by capital-rich international companies. In the 1950s, the average manufacturing enterprise employed fewer than 40 workers. In 1961, despite the emergence of international investment, the industrial census still listed only 8 out of 700 manufacturing plants (as distinct from agro-processing) as employing more than 50 persons. However, development was confined to an enclave of exporting and importing largely at two ports, Dar es Salaam and Tanga. Four regions out of 18 accounted for 70 per cent of manufactured output. This process was typified by the location of the meat cannery at Dar es Salaam, some 400 miles from the cattle-raising areas, a location unsuited to the domestic

Table 5. Trends in Industrial Structure of the United Republic of Tanzania

(Percentage of total output)[a]

Type of goods	*1965*	*1969*	*1973*
Consumer	71.4 (56.2)	63.3 (59.2)	57.2 (58.9)
Intermediate	23.5 (39.6)	25.7 (30)	32.7 (33.1)
Capital	1.3 (2.7)	9.2 (10.1)	8.8 (6.9)

Source: Annual surveys of industrial production (Government of the United Republic of Tanzania).

[a] Value-added ratio in parentheses.

Table 6. Capital intensity and capital coefficients by size of manufacturing firm (Japan, 1957)

Size of unit (employees)	*Number of firms*	*Capital/labour (thousand yens)*	*Capital/value added (thousand yens)*
1-10	300 374	69	0.371
10-50	90 766	85 (group average)	0.265 (group average)
50-100	8 460	120	0.285
100-500	4 772	228 (group average)	0.384 (group average)
500-1 000	441	408	0.523
1 000-1 999	222	589	0.64

Source: B. F. Hoselitz, ed., *The Role of Small Scale Industry in the Process of Economic Growth* (The Hague, Mouton, 1968).

market. With independence, the characteristic features of neo-colonial industrialization were reinforced with regard to scale, location and external dependence of production. From 1961 on, a clear dichotomy began to develop between older, small-scale processing plants and larger-scale plants set up by international investment.

Transnational investment reinforced the suppression of domestic linkages and promoted external dependency because this was consistent with the trade-expansion objectives of the transnational corporations.[5] Investment was made in a range of import substitution or export processing plants during the 1960s, including coffee, cigarettes, textiles, sisal products, truck assembly and radio assembly [24]. Some diversification into production of intermediate goods occurred with cement and petroleum refining. Foreign backers were from the Federal Republic of Germany, Italy, Japan, Netherlands, Switzerland and the United Kingdom. East African companies (e.g., Chandaria) established production of glass containers, aluminium products and matches.

In 1964, manufacturing still represented only about 4 per cent of GNP [25], somewhat less than in neighbouring Kenya. Of this, 80 per cent of output was in the export processing and consumer goods sectors. In 1965, the international average manufacturing ratio for small primary producer economies was 10 per cent, with 70 per cent of output in export processing and consumer goods [26].

A watershed in policy came in 1967 with the Arusha Declaration, which called for the nationalization of external trade, banking and several major industries, and a major drive towards rural collectivization. This policy, with its emphasis on self-reliance, socialization and public control at the national and local levels, had implications of both the composition of output and technology. A further declaration in 1973 formally confirmed the previously stated objective of switching towards labour-intensive, small-scale industry. Meanwhile, however, public-sector control had reinforced the tendency to engage in large-scale, capital-intensive production.

Between 1967 and 1975, the composition of industrial gross output was altered with the establishment in the public sector of a range of intermediate goods industries such as tires, steel products, chemical fertilizer and farm implements. Between 1964 and 1975, the share of consumer goods and export processing fell from 80 per cent to 70 per cent of industrial output, and the overall manufacturing ratio rose to 10 per cent. Industrial output growth rates exceeded 10 per cent per annum [27]. By 1977, industry was a significant employer of labour in the wage sector, with 17 per cent of the total wage labour force. Import substitution had reduced imports of finished goods as a percentage of total supply of industrial products from 68 per cent in 1961 to 55 per cent in 1973. In the consumer goods sector, 30 per cent of finished goods was being imported in 1973, according to World Bank estimates.

The apparent advances in terms of structure and growth over the period 1961-1975 are, however, subject to several qualifications. While the intervention of the public sector resulted in control over the majority of investment in industry and a decisive shift to intermediate goods, these goods were highly dependent on imports. The ratio of imported inputs to industrial output grew steadily, particularly in the petroleum, steel, aluminium and metal products, tires and chemical fertilizer sectors. Industrial value added as a proportion of gross output declined from 32 per cent to 28 per cent, reinforcing the dependence on imports. At world prices it appeared at one stage that local value added in steel rolling was close to zero. The corollary was that domestic interindustry linkages were not established to any significant extent. One study [28] concluded that the intermediate goods industries were themselves export-import dependent, since 60 per cent of their output was purchased by the export-processing and import-substitution consumer goods sector. (The latter category included beer, soft drinks, furniture, radios and jewellery.)

The lack of interindustry linkages has meant that some of the production anomalies characteristic of underdevelopment and dependency have been perpetuated [29]. For example, the map of the mineral resources of the country was based on piecemeal surveys carried out largely by foreign firms looking for exportable minerals rather than domestically usable industrial materials. This is related to the fact that coal and iron ore, which were used in the nineteenth century, have since then been largely unexploited, although plans for their exploitation are now in hand. During the first phase of industrialization, industries such as cement were based on fuel oil. Urban and to some extent rural construction was developed on the basis of imported steel, aluminium, iron sheets, prefabricated concrete and glass. Bricks and tiles were hardly used except in isolated towns, villages and a few missions. Yet the raw materials for these products were widely available. The fertilizer plant was based on imported raw materials despite the known existence of potash and phosphate deposits. Ceramic products were entirely imported despite local deposits of kaolin, feldspar and other inputs.

Local production of aluminium and plastic utensils (import-based) has to some extent pre-empted exploitation of local materials. A glass-container plant, while using local beach sands, also imports soda ash, while a large soda ash mining project has been planned exclusively for export (to Japan). Until about 1975, a match factory imported wood splints, although located in a well-forested area. Pyrethrum has been exported and reimported as insecticide for local final processing. The drain through export of domestic raw materials from potential domestic processors has been especially obvious in the textiles (mentioned earlier) and leather industries.

This lack of integration has also resulted in waste. By-products such as molasses, rice and maize bran, cashew-nut-shell liquid, cotton and wood waste, coconut husks and scrap metal from steel rolling and fabrication plants have been lost.

Other technology-related variables in the pattern of industrial development also show characteristic behaviour. The geographical distribution of industry has, if anything, become more skewed over time. In 1975, industrial production remained concentrated in Dar es Salaam and four other towns. Dar es Salaam, with 4 per cent of the country's population, accounted for 60 per cent of industrial output, and out of 20 regions with 38 per cent of the population accounted for 91 per cent of output [18, 28, 30]. This centralization has occurred not only because of the trend towards larger-scale plants and centralization, but also because of stagnation of small supply-based enterprises in the private sector such as soap, jaggery, sawmilling and sisal processing.

Increasing centralization has occurred not only in higher-technology industries, which might have been more susceptible to scale economies, but also in leather, shoes, food processing and sawmilling. A particular example is the establishment in 1975 of a $2.5 million semi-automatic bakery in Dar es Salaam (in competition with existing small bakeries), with the high capital cost per worker (2 shifts) of $40,000 in 1975.

The capital-intensity of production increased steadily over the period 1966-1976. However, the growth of industrial employment apparently kept pace with industrial output, rising from 42,780 in 1968 to 75,350 in 1976, because since 1969, both labour productivity and capital productivity dropped, and thus the incremental capital-output ratio in industry rose rapidly, especially in the public sector. Employment kept pace with output growth only because of unsatisfactory output performance. The trend towards increasing capital intensity can be seen more clearly from estimates of investment and electricity consumption per worker as shown in table 1.

Between 1966 and 1974 the scale of production in industry also increased steadily as can be seen from table 2.

The contribution of the largest enterprises to both output and employment has risen steadily and rapidly while that of the smallest has declined correspondingly. Of the largest 21 enterprises in 1973, 12 were in Dar es Salaam, and these 12 alone accounted for 25 per cent of all manufacturing wage employment. The most highly concentrated industries were meat processing, beverages, tobacco, textiles, footwear, tires, cement, fertilizer and electrical appliances. Average employment per unit in the organized sector rose steadily from 86 in 1968 to 145 in 1976.

Outside the organized sector is a fairly extensive, if depleted, cottage sector. Data for this sector up to 1975 have never been satisfactorily collected. Efforts to collect such data in 1975 yielded a rather approximate estimate of 30 per cent of total industrial output and 84 per cent of industrial employment (including seasonal labour) [29]. On that basis one could perhaps surmise that a polarization was developing between traditional craft industry at the one extreme and large-scale production at the other. Such a phenomenon has been recorded elsewhere, in, for example, the Philippines [31].

The decline of small factories was speeded up by the exit of sections of the Asian community that had previously owned such enterprises. The real growth rate of small-factory value added was about 3 per cent per annum between 1964 and 1975, compared with 13 per cent for large-scale industry.

The next problem confronting Tanzanian industrialization has been the perennial excess capacity also associated with large-scale production and plants dependent on imports and/or exports. Excess capacity has occurred particularly in agro-based industries such as meat and fruit canning, kenaf, sawmilling and vegetable oil, but also in industries dependent on imports and export markets such as steel products and fertilizer. Cement production has also suffered considerable excess capacity. The principal reasons have been inadequate infrastructure and materials (for collection and transport), delay in receiving imports and market constraints. Steel products have been dependent on imported billets of a non-standard specification, and the 22,000-ton rolling mill has operated at below 50 per cent capacity because of both import supply bottle-necks and domestic market constraints. Fertilizer, cement and textiles have operated at below capacity at various times because of lack of spare parts, transport breakdowns and delays and shortages of materials, especially imported materials [28-32]. In 1974/75, some fruit canning plants were either closed or operating at something below 25 per cent capacity. In 1975, the meat cannery lost its British export market because of a minor change in processing regulations for slaughter-houses that the enterprise could not meet in the short term. This added to its location and supply problems and resulted in operation at well below 50 per cent of capacity.

A final problem of particular relevance to this discussion is the drain of profits through foreign investment, also directly linked to the external dependence of industries. If repatriation of profits is restricted and perceived investment risks are high, alternative channels of foreign payment are activated, particularly technology payments with over-invoicing or under-invoicing, or transfer pricing [33]. Both of these practices have been prevalent in the United Republic of Tanzania according to some fairly recent research [24, 34], owing to external dependence and the widespread existence of joint venture and/or management agreements with transnationals. Such agreements applied in 1974 to textiles, diamonds, tires, coffee, radios, cigarettes, brewery, tin cans, cola beverages, fertilizer, cement, cashew nuts, leather tanning, bicycles and vehicle assembly. In all cases they were minority or majority foreign holdings or technical services agreements. Specific instances of transfer pricing were unearthed in the course of government monitoring of operations, in, for example, leather tanning and tin-can manufacture.

The above discussion has highlighted (a) trends in the composition of output and (b) factors directly affecting technology choice and the performance of large vis-a-vis small enterprises. In the light of the discussion some final points are relevant. The composition of industrial output has been dictated by forces that have prevented the formation of domestic linkages and encouraged export processing and import substituting consumer goods. This is particularly reflected in the weakness of engineering production. In 1974, the engineering and fabrication sector comprised about 9 per cent of industrial net output (including one farm implements factory, truck assembly and light engineering and repairs.) Metal processing accounted for a further 3 per cent, consisting of a steel rolling mill, aluminum rolling and a handful of captive foundries attached to workshops. The combined contribution (bearing in mind the low local real value added in metals) was 12 per cent compared with an average figure of up to 20 per cent to other primary producer economies [26]. (See also tables 3 and 4.) The capital goods component of industrial production was inadequate as a base for indigenous technical development.

The goal of the third five-year plan was to effect a significant structural change in industry, based on large-scale public-sector investment and continuing centralization and increasing capital intensity of output [29]. At 1975 prices, the average (unweighted) capital cost per worker in industry was $18,000; and at the top end, paper milling, it was $100,000 per worker. These figures are to be compared with a probable total net domestic surplus (undiscounted) per economically active member of the population of a maximum of $1,000 over 1975-1995 in (1975 prices).[6] Even given the very high investment ratios (20-22 per cent), achieved, under such circumstances a capital constraint is likely to be operative, and employment growth in industry over 1975-1977 started to drop well below historical rates of 10-12 per cent. This has been reinforced by an external payments deficit since 1974. Urban unemployment, which was estimated at 10 per cent in 1971 [35], as a result shows considerable likelihood of reaching significant proportions, with urban population growth of about 8 per cent. The overall manufacturing ratio between 1975 and 1977 levelled out at under 10 per cent, rather lower than average for primary producing countries and signifying that industry is no longer acting as an accelerator to economic growth. The third plan projected an 8 per cent annual growth rate up to 1995, but current indications are that it will not be achieved.

This discussion of Tanzanian industrialization has focused on (a) the commodity composition and (b) the scale and technology of industry. Eradication of domestic linkages and introduction of capital-intensive techniques were two principal features of the industrialization process. The Tanzanian economy in the nineteenth century was able to produce, albeit at a low technical level, cotton cloth and iron tools, the tools and equipment to make these products, the equipment to make the tools and the basic material to make the equipment. In the twentieth century, by contrast, industry became oriented (albeit at higher technical level) either towards the first stages of production (mining and primary processing) or to the final stages (assembly and finishing operations) and was increasingly concentrated at external trade access points. The dominance of the public sector since 1967 seems to have aggravated certain tendencies, especially increased scale and capital intensity, while only altering superficially up to now the composition of output. Future plans for restructuring production appear likely further to increase industrial concentration, even though the political system was nominally decentralized to prevent that by providing a power structure giving weight to regional economic interests.

The lack of domestic linkages is particularly noticeable in the weakness of capital goods production, characteristic of most dependent developing countries; external dependence for all types of equipment carries with it technological dependency problems distinct from the others discussed.

In the absence of a viable capital goods sector, local

technological research and development have not showed significant progress beyond sporadic efforts (e.g., village mechanization).[7] The phenomenon of marginalization of scientific research and development, noted by Cooper [36], is evidently applicable in the United Republic or Tanzania, and the transnational corporations that participate in the ownership and control of several State enterprises include some whose world monopolistic position in the technological field is well known.

The pattern of industrial development has been determined largely by changes in patterns of production, consumption and trade. The technology of production has in turn been closely influenced by these changes plus the extra factor of changing location and centralization of production, and orientation towards imported rather than domestically available raw materials and fuel. The political system underpinning this pattern was initially colonial; later on it was aided by transnational private investment; and, since 1967, it has become bureaucratic centralist (or perhaps State capitalist), during which time a large public sector has emerged.

II. TOWARDS A STRATEGY FOR TECHNOLOGY AND INDUSTRIAL TRANSFORMATION

From the foregoing analysis two conclusions may be suggested: (a) choice of technology has been bound up with forces influencing macroeconomic development, particularly through the composition of industrial output; (b) optimal technology choice for the long term would have to attach priorities to particular products or groups of products and to allocation of inputs if it is to conform to long-term strategy priorities. We have tried to argue these points from the macro-level, focusing on the historical breakdown of domestic interindustry linkages that affected output composition and input patterns (and scale and location of production).

The transformation of industrial structure (composition of output) is central, either as cause or effect, to the development of the sector. This may be ascertained from Chenery and Taylor's study of development patterns in rich and poor countries [26] (see table 3).

The United Nations study cited in Table 4 has shown that developing countries as a whole are currently undergoing structural shifts of this type. Table 4 shows this trend, with the Republic of Tanzania's current position by comparison.

Table 5 shows the broad trends of the industrial structure of the United Republic of Tanzania.

The data are inconsistent and do not agree with some cited earlier, but nevertheless demonstrate the comparative smallness of the capital goods sector and the growth of intermediates at the expense of consumer goods.

The estimates in all show the shift from primary processing and light industry to heavy industry at higher stages of industrialization. This shift has been established from time series as well as cross-section data. In 1975, Tanzanian industry, despite the structural change described, was clearly in category A. At the opposite extreme one may cite the contribution of Japan's metals and engineering sector to total output, about 52 per cent in 1975, while its food, drink and tobacco sector contributed 10 per cent [37]. It has been shown [38] that such structural changes have historically coincided with an increase in interindustry transactions. The highest measured interindustry linkages backwards and forwards occur in the following industries: basic metals, textiles, leather, paper, wood products, chemicals, petroleum, and food processing [24, 39]. Within a closed (dynamic) input-output matrix, engineering would also have high backward and forward linkages.

A pattern of investment to achieve a high level of domestic linkages would necessarily be planned around the above-mentioned group of basic commodities, including machinery production. Such a strategy was that prescribed for the Union of Soviet Socialist Republics (the so-called Feldman model) and for India (the Mahalanobis model). Its application to small primary producer economies such as the United Republic of Tanzania has been advocated more recently by Thomas [40]. Such strategies have their intellectual basis partly in the Marxist distinction between department I and department II commodities (capital goods and consumer goods).

The justification for singling out capital goods may depend on certain rather restrictive conditions prevailing in the economy [6]. The principal condition is that the capacity of production of the capital goods sector is the binding constraint on investment and growth, rather than the savings rate or absorptive capacity of the economy. This is only likely to be the case with severe limitations on trade (or very low foreign trade elasticities) such as in the Soviet Union in the 1920s and 1930s.

However, spare parts shortages are a frequent constraint in many countries that prevent the full utilization of capacity. If capital goods production is also regarded as embodying technical progress, than emphasis on the capital goods sector may yield special benefits of adaptive technical advance, which are available to capital goods producers only. The peculiar benefits to the economy arising from indigenous production of machinery arise from its high interdependence with other industries and apparent inducement and innovation effects. Strassman [41, 42] has analysed innovation potential as a function of the proportion of its input or output that a purchasing or supplying industry buys from or sells to an innovating industry. High linkages mean a high coefficient of technological transmission. As an economy develops domestic linkages a given increase in final demand generates higher levels of interindustry demand and higher innovative effects. Technical progress embodied or disembodied has been credited with a major contribution to the growth of output per head [42].

Thomas [40] proposes a basic-industry strategy to integrate use of domestic resources with domestic demand as a principal element in the strategy for transforming small primary producer economies. The principal reason for this is to prevent surplus drains arising from an unequal exchange in trade and low income elasticities of demand for primary commodities; a classic example is the case of the Tanzanian sisal industry. The complementary factor is the drain of the surplus through technology payments and transfer pricing discussed above. Under these conditions it would be expected that a basic-industry strategy would yield a higher rate of domestically retained surplus than a strategy based on primary processing and permit faster growth. This is particularly the case if simultaneous planned development of the sector occurred that permitted the most rapid possible development of intermediate goods markets and most rapid possible attainment of scale economies. Similarly, generation of local technical progress may be expected to lead either to high rates of surplus or to lower capital-output ratios in industry, permitting faster growth.

Such a strategy would imply both a quantitative and qualitative shift in development strategy as a whole. The required pattern of investment may be determined by reference to the type of information contained in table 3. Even assuming that these data are not necessarily prescriptive, they suggest the desirable direction of change. Investment reallocation within industry would be designed to achieve a balance of capital, intermediate and consumer goods, with a medium-term to long-term target for production of capital goods as a proportion of industrial output. In the long-term strategy of the United Republic of Tanzania, metal and engineering industries are expected to account for 30 per cent of output by 1995.

To determine a desirable overall pattern of investment is one of the two objectives of our strategy. The second is to determine the optimal set of production techniques. It would follow from the conclusion of the above discussion that the question is not whether industrialization should be capital intensive or labour intensive because the efficient use of resource inputs is only one criterion for identifying appropriate technology. The other criterion is appropriate composition of output. Maximization of long-run output and employment would be a function of the structure of production as well as the choice of inputs in industry. If a balanced industrial strategy is the best approach to long-term growth, then the key issue for technology policy is the choice of appropriate technology in the basic and capital goods industries.

A basic-industry strategy for small economies carries with it, however, the serious problem that such industries are particularly susceptible to economies of large-scale production and are highly capital intensive. This applies particularly to petroleum and coal products, non-ferrous metals, iron and steel, non-metallic mineral products and chemicals [24, 44-46]. Furthermore, the "minimum economic scale" in many industries increases rapidly with innovation, and these industries are among those in which research and development are concentrated. Pratten estimates that oil-refining cost per ton falls by 60 per cent between 1 million and 20 million tons annual capacity. In synthetic fibres, costs at 40,000 tons capacity are 20 per cent below those at 4,000 tons. Steel production costs at 6.5 million tons are 10-18 per cent below those at 2.2 million tons. Pack [47] estimates that capital per man employed in textiles increased between 1950 and 1968 by a factor of 15-20.

Table 6, based on a Japanese study [48] shows the relationship between scale and capital intensity.

The data show a direct relationship between scale and capital intensity from the 10-worker level upwards. The data on capital coefficients, however, show that the smallest enterprises are the least efficient users of capital. This finding has been confirmed elsewhere; for example, in the ILO study on the Philippines [31].

Given the apparent need to build large-scale plants and the scarcity of savings relative to the investment required in such industries (see above on the Tanzanian third five-year plan) and the foreign-exchange constraint, the scope for a basic-industry strategy would therefore appear limited.[8] Thomas [40] argues that a "minimum optimum scale" would be a realistic objective for small economies, based on standard technology. However, the recognized minimum economic scale in, for example, steel production (using an electric-arc furnace) is about 250,000 tons per annum, which would cost upwards of $300 million in capital investment. A current proposal to establish a mini steel plant of 50,000 tons capacity in Nepal is expected to cost $95 million, including infrastructure, and its feasibility is doubtful. Such sums are not readily available to small primary producer economies. A recent study [49] estimated that in the United Republic of Tanzania the rapid sequential establishment of basic industries would in all cases except food processing result in a short-term reduction in the rate of growth of output (and employment), which would aggravate the shortage of savings and foreign exchange; consequently, in a basic-industry strategy, economic-efficiency objectives might have to be ruled out. The small domestic market would also act as a constraint on establishing efficient units in the absence of export opportunities and simultaneously expanding the production of interrelated intermediate goods. The imposition of one or other of these constraints--savings, foreign exchange or market--has in practice provided the justification for concentrating on the development of light industry in most developing countries.

Despite these constraints, there are several reasons why scale economies based on engineering estimates may not be readily attainable in developing countries. First, factory prices even in distorted markets are generally more favourable to small-scale, labour-intensive techniques. Secondly, scale diseconomies may operate, some of which have been cited in relation to the United Republic of Tanzania. For example, infrastructural gaps are frequent, e.g., lack of capacity of power transport and port facilities to handle lumpy increases in investment. (The Tanzania fertilizer factory, which cost $25 million in 1971, required a purpose-built jetty, which increased basic cost by 25 per cent. Subsequently, capacity utilization was reduced by continuous shortages of rolling stock, water and materials.) Further instances of failure in infrastructure, supplies and maintenance were mentioned previously. Thirdly, large-scale plants, because of the higher likelihood that finance, management and materials will be imported may be more prone to the various surplus drains, so that rates of profit are reduced. The classic example of surplus drain via technology payments and transfer pricing is in the pharmaceutical industry, studied by Vaitsos [33].

Fourthly, small-scale plants enjoy certain advantages. These include reduction in transport costs where the geographical spread of the market is restricted. This is especially true for heavy goods of low value, goods such as bricks, cement and timber, or perishables such as fruits and vegetables. Small plants may also be able to derive an advantage from using scattered, small deposits of raw materials unworkable economically on a large scale. Scrap materials in developing countries are probably more easily recycled through small plants. Many small chemical plants were reportedly constructed in China in 1958/59 with the use of scrap material [50]. Small-scale production is also appropriate for specialized custom-made products and services, as evidenced by the continuing 20-30 per cent share of small enterprises in output in Japan, United Kingdom and United States of America.

Small-plant economics are affected by the general location of supplies and markets. A decentralized integrated economy (with a degree of regional self-sufficiency) such as that of China, where provincial and country-level production has been oriented primarily to local needs [51], is likely to permit viable small plants to be established. If such a socio-economic structure is accompanied by a redistribution of income towards the poor, with an expansion in demand for lower quality or less highly refined goods, including consumption goods and rural construction materials, then smaller-scale, less capital-intensive techniques are also likely to be appropriate [12, 52].

Another way of introducing small-scale plants, related to the previous one, is through flexibility in the use of materials. Some materials requiring low capital-intensity production techniques and a small scale of production may act as substitutes especially where investment and consumption are oriented towards unsophisticated products. This applies, for example, particularly to building materials (e.g., bricks and lime can be substituted for cement, steel) and in fuels (coal, charcoal can be substituted for diesel).

In rural areas demand for clothing and footwear made of synthetics is likely to be limited. In addition, ceramic and earthenware may be economically substituted for plastic products (pipes, containers, domestic-ware etc.).

These points suggest that a development strategy designed to promote domestic integration by forging interindustry linkages, accompanied by corresponding changes in the social structure, may itself create an environment in which small plants are economic, at least over a transitional period.

In the United Republic of Tanzania, as can be inferred from the data of table 2, despite its stagnation over the period 1966-1974, the small-industry sector showed consistently higher gross output per head than the largest enterprises. This is partly because the small- and large-industry sectors included different industries and were thus not strictly comparable; in particular, textile spinning and weaving fell exclusively into the large-scale category. Large-scale industry over this period was also undergoing teething troubles. Nevertheless, assuming that the ratio of value added to gross output was similar for small- and large-scale production, labour productivity in small plants was higher. Since capital productivity was probably also higher, small plants maintained a competitive position. Statistics from other countries do not suggest such a clean-cut conclusion, but it is a general finding that for small enterprises employing fewer than 10 workers, this is not so, however (see table 4);[9] as a result, craft industry is thought to have limited prospects unless it becomes mechanized. (This conclusion applies to handicraft production for domestic consumption, not to production for export and tourist markets.)

Basic industry, particularly in China and India, has been established on a small-scale basis. It has included chemicals, fertilizer, cement, paper and paper products, a range of food processing industries, textiles, leather, engineering and metal products. Experiments at the Regional Research Laboratory at Jorhat, India, on small-scale cement production showed promising results, and more than 100 mini cement plants are being set up using vertical-kiln technology. In the Indian textile sector handloom or small-scale power loom weaving has accounted for more than 50 per cent of output. The Chinese have promoted the small-scale sector as a means of recentralizing the economy and achieving local self-sufficiency [51, 54]. Over 50 per cent of the nitrate and phosphate fertilizer produced has been produced in plants of under 10,000 tons capacity (compared with a standard assumption of economic scale of 200,000 tons or more). Fifty per cent of the cement produced is produced in plants of under 30,000 tons annual capacity (compared with a standard economic scale of 300,000 tons or more) [54, 55]. However, in 1973, China ordered several large ammonia-urea plants, and there are indications of a shift in policy away from the initial self-reliance approach of the 1960s. It is, however, not possible as yet to assess the prospects for mini basic industries.

Despite such cases, in certain heavy industries the threshold of capital cost for developing countries is formidable. Among these are petroleum and coal products, steel and non-ferrous metal processing, and petrochemicals. Motor vehicles are also subject to high economies of scale. However, assembly plants have in fact been established throughout the developing world, usually on a highly inefficient basis. A greater variety of choice of technology is possible in light industries than in heavy industries. In a ranking of industries according to their capital intensity, light industry would generally appear in the lower half of the range [24, 56]. Leather tanning and wood products industries are at the least capital-intensive end, while some food processing (e.g., grain milling) appears half way up. Considerable progress has, however, been made in India in developing small-scale production for sugar, vegetable oil, fruit and vegetable processing and other food industries. Mini maize mills are extensively used in Africa and mini rice mills in Asia.

Indian statistics [57] show that the share of output of registered small industries is highest in industries such as fruit and vegetable canning (60 per cent). The small-industry share of output is also high in textiles (30 per cent in handloom and 20 per cent in small power loom plants) [58], garments (over 70 per cent), textile goods (reserved for the small-scale sector only), knitwear

(95 per cent), tanning (79 per cent), leather products (83 per cent), sawmilling (75 per cent). Small-scale furniture and shoe production account for 49 per cent and 43 per cent, respectively, a share that would no doubt be much higher if unregistered industries were taken into account. The average share of the registered small-scale sector is about 33 per cent. The lowest small-sector shares are, predictably, in chemicals, steel, cement, non-electrical machinery and vehicles (all below 20 per cent). Particular importance attaches to the engineering industry. The economies of the industry suggest that large-scale production is not necessary for viability. In terms of physical capital intensity, machinery manufacture is not in the upper end of the range. The specialized nature of production does not lend itself to long runs except for simple components and some other mass-produced goods such as containers, machinery casings and vehicle bodies. The principal condition according to Rosenberg [42] is a market large enough to permit specialization. In small primary producer economies, the domestic market is limited both by small existing capital stock and few possibilities for investment. In such countries specialization in a limited range of standardized equipment seems desirable.

In this sector the Chinese experience is once again instructive. Reportedly engineering is decentralized, with heavy machinery the responsibility of the State and repair shops the responsibility of brigades. Small diesel engines and pumps are manufactured at the county level (current average population 400,000). Tractor maintenance and repair, light engineering and fabrication are carried out in local factories in communes (population 10,000-20,000). Innovation efforts are actively encouraged locally.

In India, light engineering forms a main element in the ancillary industry programme, based on a network of over 400 industrial estates. Industrial statistics [57] show that the small-scale sector accounts for a high share of production in the following industries (per cent): metal products, 50 or more; vehicle repairs, 91, and electrical machinery, 33. By comparison, the engineering industry in the United Kingdom, according to Pratten [44], in 1971 consisted predominantly of medium-scale plants (average 400 workers).

In Japan, transport equipment and fabricated metal products are among the most highly subcontracted industries. The history of sewing machine manufacture is also a good example of specialization through subcontracting to small enterprises [58]. In this industry, between 1950 and 1969, production increased from 0.25 million units to 4.8 million units through speicalization and standardization of components. The average number of parts produced by each enterprise fell between 1941 and 1963 from 60 to 3.

From the above evidence it seems fair to conclude that apart from heavy machinery production, involving, for example, the machining of heavy castings, there appears to be considerable scope for developing engineering in small developing economies such as the United Republic of Tanzania in both component and subassembly manufacture, light-to-medium engineering and assembly work, provided that careful standardization for the local market is adhered to. Consequently, technical innovation could take place indigenously, given an industrial structure with internal linkages. Thus, a basic-industry strategy may be suitable for an economy such as that of the United Republic of Tanzania. Heavy industries (petroleum, steel, chemicals, vehicles, heavy engineering) would be centralized to attain whatever economies of scale were feasible. For a second important group of industries a wider choice of technology and scale is possible. These industries could be decentralized. This second group could include heavy industries such as cement, chemical fertilizer, certain types of paper, and a wide range of light industries and engineering production. In this way a dual programme comprising a medium-to-large-scale heavy-industry sector and a small-to-medium-scale relatively labour-intensive industry sector could be established. The first sector, while possibly inefficient in terms of current factor endowment, would be justified because it would transform the composition of output. The second sector would be relatively resource-efficient in static terms, achieving objectives of allocative efficiency and others such as redistribution of income and regional balance. Both sectors combined would be designed to transform output and form domestic linkages and to raise economic efficiency.

The actual allocation of investment and production among sectors and between regions and centre, industry by industry, requires detailed investigation. The decentralized "efficiency" programme would incorporate cottage and small enterprises in "traditional" light industry and some basic-industry production, possibly in cement, fertilizer, engineering and paper. The centralized "transformation" programme would incorporate heavy industry and certain light industries where economies of scale are attainable.

III. CHOICE OF TECHNOLOGY IN A STRATEGY FOR INDUSTRIAL TRANSFORMATION—CONCLUSIONS

In the first part of this paper, the discussion of the industrial development of the United Republic of Tanzania illustrated the interrelationship between macro political and economic forces, composition of industrial output and choice of technology. In the process the discussion focused on the pattern of ownership and control of production, direction and composition of trade, and changing composition of output resulting from changes in political control. Technology was affected by composition of output (e.g., the advent of the import substitution regime resulted in the expansion of large-scale production), and also by the location of production, which in many cases dictated the size of accessible markets and feasible scale of production. In the second part of the paper some of the details of a technology and industrial-transformation strategy were sketched, with emphasis on basic industries and the contribution that appropriate technology might be expected to make in the basic-industry sector, particularly engineering.

All the various parameters of economic development are interrelated; it is not necessary to assume that choice of technology is dictated by the composition of output and that the composition of output is dictated by the aggregate political and economic system. The point is simply that by identifying such relationships between micro-choice and the aggregate system certain insights can be gained. One of these may well be that standard methods of evaluating techniques at the plant or industry level are deficient because they cannot in practice incorporate or quantify external effects that would justify giving priority to certain products over others in a manner consistent with the requirements of national long-term strategy. This may be particularly the case with the products of the engineering industry whose priority within the strategy arises largely from its external (linkage) effects.

The choice of appropriate technology, e.g., labour-intensive technology, cannot be divorced from choice of an appropriate industrial strategy (planned composition of output) because, first, the objectives to be achieved by choosing appropriate technology cannot be divorced from those of industrialization as a whole, i.e., long-term economic growth, and, secondly, because choice of technology is in any case bound up historically with the composition of output and with aggregate economic forces dictating the actual composition of output. Therefore, promotion of appropriate techniques in isolation from macro-strategy may be frustrated. The key issue appears to be the decision on strategy. A strategy of industrial transformation involves the development of a group of basic industries. Consequently, the critical issue is whether appropriate (labour-intensive and small-scale) techniques are available within this particular group of industries.

Since, 1950, developing countries have diversified their industrial sectors by establishing basic industry and moving away from colonial-type export processing or neo-colonial import substitution. This diversification has, however, been carried out, in the United Republic of Tanzania and elsewhere, largely by establishing large-scale plants of very high capital intensity. Furthermore, some of this diversification has been illusory because intermediate and capital goods projects depend heavily on imports largely because of their high level of mechanization and highly specified raw material requirements--both functions of capital-intensive technology. Many private companies in industrialized countries still view international investment primarily as a means of establishing protected overseas markets and not as a means of developing the recipient economy's indigenous resources for local industry. In such circumstances simpler labour-intensive technology that could be developed and supervised locally would be of prime importance, particularly in the basic-industry group.

NOTES

[1]See, for example, Bhalla [2], especially studies of textiles, sugar, cement blocks, can making and metalworking; Jequier [3], including studies of sugar, ceramics and footwear; Pickett and others [4], including leather, iron foundry products, maize milling, brewing, fertilizer, footwear, nuts and bolts. Also individual studies by Pickett and Robson [5]. A recent theoretical overview of the whole question is found in Stewart [6]. Further useful case material has inter alia come from Marsden [7] and Timmer [8]. An older, pioneering work on the subject is Boon [9]. Also see Sen [10], Stewart [11] and Morawetz [12].

[2]Stewart and Streeten [13] discuss exhaustively the relationship between current and future employment.

[3]For a recent example, see Leys [14].

[4]Marsden has noted this phenomenon when bakeries and plastic footwear have been introduced [7].

[5]A relevant discussion of multinational investment objectives appears in Kilby [23].

[6]Calculation based on third-five-year-plan projections (1975).

[7]For example, by the Tanzanian Agricultural Machinery Testing Unit at Arusha.

[8]The study Small-scale Industry in Latin America (United Nations publication, Sales No. 69.II.B.37) specifically excludes cement, fertilizer, paper and glass from small-industry programmes.

[9]See Sharing in Development [31] where this finding is supported. Dhar and Lydall [53] come to a different conclusion, however.

References

1. ILO, Employment, Incomes and Equality; A strategy for Increasing Employment in Kenya (Geneva, 1972).

2. A.S. Bhalla, ed., Technology and Employment in Industry (Geneva, ILO, 1975).

3. N. Jequier, ed., Appropriate Technology: Promises and Prospects (Paris, OECD, 1976).

4. J. Pickett and others, "Choice of technology in developing countries", World Development, vol. 5, No. 9/10 (September 1977).

5. J. Pickett and R. Robson, "Technology and employment in the production of cotton cloth" (Glasgow, David Livingstone

Institute, Strathclyde University, 1974).

6. F. Stewart, Technology and Underdevelopment (London, Macmillan, 1977).

7. K. Marsden, "Progressive technologies for developing countries", International Labour Review, January-June 1970.

8. P.C. Timmer, ed., Choice of Technology in Developing Countries: Some Cautionary Tales (Cambridge, Mass., Harvard University, 1975).

9. G.K. Boon, Economic Choice of Human and Physical Factors of Production (Amsterdam, North-Holland Publishing, 1964).

10. A.K. Sen, Employment, Technology and Development (London, Clarendon Press, 1975), chap. 5.

11. F. Stewart, "Technology and employment in LDCs", World Development, March 1974.

12. D. Morawetz, "The employment implications of industrialisation strategy", Economic Journal, December 1974.

13. F. Stewart and P. Streeten, "Conflicts between employment and output in LDCs", Oxford Economic Papers, vol. 23, July 1971.

14. C. Leys, Underdevelopment in Kenya--The Political Economy of Neocolonialism (Berkeley, University of California, 1975).

15. M. Yaffey, International Transactions Before and During the German Period (University of Dar es Salaam, Economic Research Bureau, 1969).

16. Helge Kjekshus, "Pre-colonial industries" (University of Dar es Salaam, December 1974).

17. A.K. Bagchi, "The de-industrialisation of India in the 19th century", Journal of Development Studies, vol. 12, No. 2 (January 1976).

18. Tanzania Economic Survey, various years (Dar es Salaam, Government of the United Republic of Tanzania).

19. Tanzania Economic Survey 1977-8 (Dar es Salaam, Government of the United Republic of Tanzania).

20. H. Chenery and M. Syrquin, Patterns of Development (London, Oxford University Press, 1975), p. 20.

21. Statistics on Industrial Production (Dar es Salaam, Government of Tanganyika, 1957).

22. K. Schadler, Small Scale and Craft Industry in Tanzania

(1967).

23. P. Kilby, Industrialisation in an Open Economy; Nigeria 1945-66 (Cambridge University Press, 1969).

24. J.F. Rweyemamu, Underdevelopment and Industrialisation in Tanzania; A Study of Perverse Capitalistic Industrial Development (London, Oxford University Press, 1973), pp. 150-159.

25. Ann Seidman, "Comparative industrial strategies in East Africa" (University of Dar es Salaam, Economic Research Bureau, 1969).

26. H. Chenery and L. Taylor, "Development patterns among countries and over time", Review of Economics and Statistics, No. 50, November 1968.

27. M. Roemer and G. Tidrick, The State of Industry in Tanzania 1973 (Dar es Salaam, Ministry of Economic Affairs and Development Planning, 1975).

28. N. Bhagavan and others, "Industrial production and transfer of technology in Tanzania" (University of Dar es Salaam, 1975).

29. D. Phillips, Industrialisation in Tanzania. Small scale industry, intermediate technology and a multi-technology program for industrial development" (University of Dar es Salaam, Economic Research Bureau, 1976).

30. Census of Industrial Production (Dar es Salaam, Government of Tanganyika, 1961).

31. ILO, Sharing in Development (Geneva, 1974).

32. G. Wangwe, "Factors influencing capacity utilisation in Tanzania manufacturing", International Labour Reivew, January 1977.

33. V.C. Vaitsos, "Strategic choices in the commercialisation of technology", International Social Science Journal, vol. 25, 1973.

34. Le Van Hall, "Transfer pricing; the issue for Tanzania" (Dar es Salaam, Institute of Finance Management, 1975).

35. R. Sabot, "Open unemployment and the unemployed compound of urban surplus labour" (University of Dar es Salaam, Economic Research Bureau, 1974).

36. C. Cooper and A. Herrera, in Science, Technology and Development (London, Cass, 1973).

37. Statistics on Japanese Industries, 1976 (Tokyo, M.I.T.I.).

38. M.Zaidi and S. Makhopadhyay, "Economic development, structural change and employment potential", Journal of Development Studies, vol. II, No. 2 (1975).

39. P. Yotopoulos and J. Nugent, The Economics of Development; Empirical Investigations (New York, Harper and Row, 1976).

40. Clive Y. Thomas, Dependence and Transformation; The Economics of the Transition to Socialism (New York, Monthly Review Press, 1974) chap. 6.

41. W. P. Strassman, "Interrelated industries and the rate of technological change", Review of Economic Studies, No. 27, October 1959.

42. N. Rosenberg, "Capital goods, technology and economic growth", Oxford Economic Papers, No. 15, November 1963.

43. R. Solow, "Technical change and the aggregate production function", Review of Economics and Statistics, August 1957.

44. C.F. Pratten, Economics of Large-scale Production in British Industry (Cambridge University Press, 1971).

45. S. Teitel, "Economies of scale and size of plant", Journal of Common Market Studies, vol. 1-2, No. 13 (1975).

46. Yearbook of Industrial Statistics, 1974 (United Nations publication, Sales No. 76.XVII.3).

47. H. Pack, "The choice of technique and employment in the textile industry", in Technology and Employment in Industry (Geneva, ILO, 1975).

48. B. Hoselitz, ed., The Role of Small Scale Industry in the Process of Economic Growth (The Hague, Mouton, 1968).

49. Kwan S. Kim, "The linkage effect of basic industries in Tanzania; policy issues and suggestion" (University of Dar es Salaam, Economic Research Bureau, 1976).

50. E. Wheelwright and B. Macfarlane, The Chinese Road to Socialism (Harmondsworth, Penguin, 1973).

51. C. Riskin, "Local industry and the choice of techniques in planning of industrial development in mainland China", in Planning for Advanced Skills and Technology, Industrial Planning and Programming Series (United Nations publication, Sales No. 69.II.B.8).

52. C. Blitzer and others, eds., Economy Wide Models and Development Planning (London, Oxford University Press, 1975).

53. E. Dhar and H. Lydall, The Role of Small Enterprises in Indian Economic Development (Bombay, Asia Publishing House, 1962).

54. J. Sigurdson, "Technology and employment in China", World Development, March 1974.

55. S. Lateef, "China and India, Economic Prospects and Performance" (Sussex, Institute of Development Studies, 1976).

56. K. Arrow and others, "Capital-labour substitution and economic efficiency", Review of Economics and Statistics, No. 43, August 1961.

57. Handbook of Statistics (New Delhi, Development Commission for Small Scale Industries, 1973).

58. S. Watanabe, "Subcontracting, industrialisation and employment creation", International Labour Review, No. 104, July-August 1971.

Technological Development in Latin America and the Caribbean

JORGE A. SABATO*

I. INTRODUCTION

The scope and limitations of this document are as follows:

(i) It refers to technology in its broadest sense, that is, as the organized body of all knowledge used in the production, distribution (by trade or any other means) and use of goods and services. Accordingly, it encompasses not only the scientific and technical knowledge generated by research and development (R&D), but also that which results from empiricism, tradition, manual skills, intuition, imitation, adaptation and so on.

(ii) Secondly, it recognizes that this broad scope means that technology is an essential component of economic, educational, cultural and political systems, and as a result has an influence throughout society. However, the document restricts itself to analysing technology as a component of economic and social development and it therefore presents an analysis of the interface between the structure of production and technology without dealing with the interfaces between science and technology, culture and technology or education and technology, except to the extent necessary for better understanding of the central theme.

(iii) Another important limitation of this document is that the analysis covers most, but not all, of the structure of production. It includes manufacturing, with its subsectors of consumer goods, as well as the basic infrastructure of energy, transport, communications and so on and the so-called "high-technology" industries (electronic, computers and data processing, nuclear power and the aeronautical industry), but it excludes agriculture, forestry, fisheries, and the finance and

From **CEPAL REVIEW**, 10, April 1980, (81-94), reprinted by permission of the publisher, U.N. Publications, N.Y.

insurance sectors, which are outside the author's province. It is possible that much of what is asserted for the other sectors is valid for these sectors, but this will have to be evaluated by competent experts.

In preparing this document the author has endeavoured to bear constantly in mind that:

(a) Latin America and the Caribbean (hereinafter referred to as LAC) is not a unit, but a collection of nations which are at very different stages of development and which have governments of different kinds, are carrying out development plans with different aims based on very different economic policies, are implementing subregional and bilateral agreements of various types, possess substantial areas of competition, and so on.

(b) The United States, though it has maintained a sort of special relationship with LAC, has political, economic, cultural, scientific and military interests which extend far beyond the context of LAC and its institutions, while at the same time it is the political, administrative, technical and financial headquarters of most of the transnational corporations operating in LAC.

As a result, the policies, strategies and actions recommended in this document are not those which might be valid for a more homogeneous group, but those most capable of being undertaken in such a heterogeneous framework, through existing or similar bodies.

II. TECHNOLOGY IN LATIN AMERICAN COUNTRIES: A BALANCE SHEET

It is necessary to draw up a balance-sheet of the situation in order to ascertain what has been done and where matters now stand. To begin with, it is important to mention that, in the two decades following the end of the Second World War, an intensive effort was made in Latin America to establish a scientific and technical infrastructure and create public awareness of the importance of science and the urgent need to develop this subject in our countries: a campaign which was crowned with success in the establishment of faculties of science in many Latin American universities, research institutes and centres, and national councils for scientific and technical research. However, technology was not very important in this effort, because it was assumed that once the capacity to produce science was acquired, it would flow in a continuous manner and become quite smoothly integrated within the structure of production, which was anxiously awaiting it. It is only in the past decade that attention has focussed on problems such as: When, why and how is demand for science created in specific circumstances? What are the relations between science and technology? Is technology merely "applied science"? How do the flows of supply and demand for technology move through the various socioeconomic circuits? Who benefits from the results of scientific and technological research? How and why are the structure of production and the scientific and technical infrastructure not properly interconnected? What relations are there between technology and foreign investment? What is technological dependence?--and so on and so forth. These and other similar questions have been studied with thoroughness and originality, leading to significant progress both in the academic field--in other words the field of studies and research on the group of problems involved in science, technology, development and dependence--and in the political field, which covers the action taken to use science and technology to attain specific objective of economic and social development, as summarized below:

Recognition of the existence of structural obstacles to scientific and technical progress. Study of these obstacles made it possible to draw a distinction between explicit and implicit scientific policy; to understand the causes of the usual attitudes of the governing elites to science and technology (hostility, lukewarm support or indifference); to explain apparent contradictions, such as the relative advance of certain branches of science (such as biology) and the backwardness of others (such as geology); to discover the impact of the import-substituting model of development on the assimilation of technology; to create awareness of the existence of a new international division of labour, centred on the production and consumption of technology; and so on. The practical result was the explicit formulation of science and technology policies and the establishment of appropriate bodies (ministries and the like) to implement them. Outstanding examples are Decisions 84 and 85 of the Andean Pact, the Science and Technology Plan prepared by CONACYT in Mexico (1976), the organization of the CNP (Conselho Nacional de Pesquisas) in Brazil, etc.

Recognition of the importance of technology as a carrier of values, so that importing technology means importing not only an organized body of knowledge, but also the production relations which gave rise to it, the sociocultural characteristics of the market for which it was originally devised, and so forth. Technology transmits the value system for which it was designed, just as if it bore a "genetic code" within its structure. This means that the scope of technological dependence goes far beyond merely economic considerations.

Comprehensive study of trade in technology on the basis of recognition that technology is a valuable commodity in the system of production, and that most movements of technology occur through trade and not through free transfers. This study was also accompanied by an examination of the technology market, highlighting its imperfections, criticizing its worst distortions and unfair practices, penetrating into the sanctified area of industrial property and discovering the importance of "unpackaging" in the importation of technology. The result has been the introduction measures to analyse and control the flow of imported technology (for example, the establishment of registers of technology), to govern relations with foreign investment (Decision 24 of the Andean Pact), to revise legislation on industrial property (as in Brazil and Mexico), etc.

Verification that most imports of technology have occurred through direct foreign investment, with recognition of the growing role of the transnational corporations in producing and marketing technology, the growing importance of movements of technology between the head offices of such corporations and their subsidiaries, and recognition that the concept of industrial property has broadened to include property which is not legally protected and is known as quasi-property (know-how, engineering services, trade names, place in the market and so on), which accounts for an increasing volume of commercial technological operations.

Recognition that the process of industrialization is leading to a growing "technologization" of LAC, measured in terms of the larger number of persons from various levels of society who have acquired scientific abilities or technical skills, giving rise to an important qualitative change in the structure of employment. The local output of technology is small compared with the flow of imported technology, but some encouraging successes have been recorded (PEMEX in Mexico, agricultural machinery in Argentina, machine tools in Brazil, and so on), as well as progress in "unpackaging" technology (Atucha nuclear power station in Argentina; the Brazilian iron and steel plan; petrochemicals in the Andean Pact, and so on) and growing activity in the field of adaptation of imported technology to local conditions, which means that the flow of internal innovatory activity is by no means nonexistent. The first cases of substantial exports of packaged and unpackaged technology are being recorded, and measures are being introduced to support and encourage them (preferential credits, tax relief, favourable exchange rates). Exports of capital and technology with the region, especially from the three largest countries, are beginning to acquire importance. Brazil, for example, exported unpackaged technology worth US$ 135 million in 1975, against a total of only US$ 3 million in 1967.

Critical analyses of multilateral and bilateral co-operation and assistance in the field of science and technology, and of the bodies and executing agencies involved. These have led to corresponding stimulation of a new strategy for co-operation and negotiation at the regional level (OAS and SELA), the subregional level (Andean Pact) and the international level (United Nations agencies), and adoption of a new attitude to the United States (declaration by the CACTAL conference, position in UNIDO and UNCTAD and so forth.

Increases in local consultancy and engineering capacity, in some cases to international levels of quality and quantity, thus making it possible to extend such services, in open competition, outside national frontiers and even outside LAC.

Significant increases in scientific and technical exchanges among the countries of the region and with the rest of the world.

So far, we have indicated the most significant advances. In order to complete the balance-sheet it is now necessary to mention the areas where there has been no progress, and possibly even retrogression. Perhaps the most important of all these areas is the limited impact, in the field of technology, of the science and technology development plans implemented in various countries, and the failure to link the structure of production and the scientific and technological infrastructure. In contrast to science which can develop in the isolated environment of a university, academy, institute or laboratory, technology operates in a much larger area of society, that of the units of the structure of production, with a wide range of active participants. In particular, entrepreneurs and managers in the industrial sector and farmers in the agricultural sector are of fundamental importance in introducing technology in their activities. These activities, however, have been and remain totally isolated from the policies, strategies, plans, agencies and actions related to technological development, which, as a result, have remained as if floating in a socioeconomic void, without real links with reality. In short, technology has so far been handled more as an item of data than as an operational variable to which the tools of economic policy must be applied if it is truly wished to achieve some impact on reality.

The importation of technology, whether by subsidiaries of the transnational corporations, private national enterprises or by State enterprises, is effected first and foremost in the light of the micro-economic interests of such enterprises, regardless of the ecological, socioeconomic and cultural consequences. There is implicit or explicit acceptance that certain assumptions are firm truths: (a) that technology from the central countries is the only, the best, and the most suitable technology; (b) that technology is neutral, in other words, value-free; (c) that any "modern" technology will, by definition, be of the greatest use for development; (d) that this technology is sufficiently well tested, and that its introduction therefore poses no risks. It is forgotten that such technologies are designed in the light of the availability of factors and resources in the country where they were created; that for that very reason they are capital-intensive and energy-intensive; that they are aimed fundamentally at meeting the needs of sectors of the population of those countries which, because of their incomes, stand far above the mass sectors of the importing country, so that a technology which, in a central country, meets the needs of a large number of consumers can, in a peripheral country, be of use only for the elites, etc.

Local production of technology has not been properly promoted: it has not been given the protection essential to permit competition with imported technology, nor has it been possible to introduce efficient means for its production.

Studies on technology in the fields of food, housing and health fall below those carried out for the industrial sector in terms of quality and quantity, so that they have received little attention, while the importation of technology for use in the production of goods and services for the privileged sectors has continued to increase.

There is still no sound theory on the role which the State should play as a producer and as an owner of units (industries, banks, business, insurance, and so on) which are major consumers of technology and which frequently behave *vis-a-vis* science and technology as regressively as the private sector, or even more so, thus giving the lie to the belief that nationalizing a unit production or bringing it into the hands of the State is sufficient to put an end to its technological dependence.

The brain drain has continued, and in a number of countries has increased, basically because of political persecution and ideological discrimination.

Demand for local technology remains weak, since under the prevailing rationale of the structure of production it continues to be more convenient to import technology than to produce it or purchase it locally.

Regional and subregional co-operation, which are essential for achieving the "critical mass" and thus jointly tackling the multitude of problems which need to be solved, are making slow progress, and in particular the ability to fulfill the formal undertakings entered into is very inadequate. No machinery has been established for trade co-operation in the field of technology.

Technological dependence and technological dualism have been denounced vigorously, but not thoroughly studied, and there is still no proper strategy to overcome them.

No country of the area, with the possible exception of Brazil, has yet passed from a defensive strategy, consisting of such actions as strengthening the infrastructure, operating registers of technology, and so on, to an attacking strategy, with emphasis on the production of technology and on aggressive negotiations with the external suppliers of technology. It is urgently necessary to recognize that the defensive strategy has a structural and operational upper limit, and that this limit can only be passed by going over to an attacking strategy.

The scientific and technical infrastructure is not linked either with the structure of production or with its own "owner", the State, thus showing that institutional obstacles of a socio-political and cultural nature can be as important as strictly economic obstacles.

Local efforts at scientific and technological development continue to be weak, and only Brazil has planned a significant change, through its Second Plan for Scientific and Technological Development, which provided for investments of the order of US$ 2,700 million for the three years 1975 to 1977. This situation is all the more serious since economic, material and human resources continue to be used very inefficiently, Skilled personnel still do not receive proper social and political recognition.

In these efforts there is a clear absence of projections and decisions regarding the relation between technology and the quality of life in the broadest sense. If this situation is not rapidly corrected, the consequences will be serious.

III. OBJECTIVES AND STRATEGIES

A Possible Common Objective

The above outline of the present situation defines the frame of reference within which it will be necessary to specify objectives and strategies for the better use of technology in the socioeconomic development of LAC. Three conclusions should be emphasized:

(a) There is now clear awareness that the problems are highly complex: much more so than was naively thought in previous decades. As Maximo Halty clearly put it:
"The first step towards solving a problem is to know that the problem exists. This step has been taken. Simplistic solutions have gradually been put aside: the problem is not solved merely by training skilled technical staff and increasing research funds. Neither is the evaluation and control of the importation of technology, with all its strategic importance, a complete solution on its own and in itself. The two are necessary but not sufficient conditions . . ."

(b) The LAC countries are basically consumers of technology but poor producers. As a result, they are spectators and not actors: passive recipients of what others do in the light of their own needs and interests, and they thus inexorably tend to adopt the general outlook of the suppliers, against which mere rhetorical protest is of no use. This leads to one of two equally harmful positions: to the worst kind of technolatry--slavish copying or imitation--or to furious denunciations of technology, which are completely sterile if no viable alternatives are proposed.

(c) International co-operation has taken place particularly, in respect of the science/technology interface, and the greatest efforts have been applied to creating and strengthening the scientific and technical infrastructure (training of staff, exchange of scientific and technical personnel, equipping of laboratories and pilot plants, establishment of institutions, technical service centres, etc.), and carrying out academic and field research on the many aspects of the problems of science, technology and development. The programmes applied to the interface between technology and the structure of production have been few in number, and so far very limited in scope and resources.

There is no doubt, in these circumstances, that the next stage should focus on objectives which are directly related to technology as an operational variable _in_ and _for_ the system of production, and should proceed on the basis of overall _attacking_ strategies which are consistent with the objectives and strategies of socioeconomic development. A fundamental preliminary question immediately arises: in view of the large number of nations which make up the continent, each with its own interests and its own conception of economic and social development, will it be feasible to define objectives and strategies which are useful for all in such a way as to make firmly based and continuing _co-operation_ possible and desireable? We are not referring, of course, to co-operation to strengthen the scientific and technical infrastructure, which can and should continue, since a firm basis exists for it, but to the area of socioeconomic development, because co-operation in the exchange of scholarship-holders and teachers, the organization of courses and seminars, the equipping of libraries and laboratories, and so on, is one matter, while tackling problems so bound up with power and interests as the regulation of imports of technology, reform of industrial property laws, and the evaluation of technology in relation to income distribution, for example, is very lifferent. What is convenient and desirable for country A may not be so for country B: thus, A may pursue socioeconomic development through extreme liberalization vis-a-vis foreign investment and technology, while B, in contrast, may pursue the same goal by restricting and controlling such investment; country C may hope to improve its non-traditional exports through the massive assimilation of imported technology, while D may give priority to meeting the basic needs of its rural population (food, health and housing), sharply reducing imports of technology for use in the production of luxury items and applying its greatest efforts to the development of indigenous technology, and so forth.

This is a consequence of the very nature of technology and its full participation in the process of production, which means that everything related to it is necessarily linked to conflicts of interests between classes, groups, countries, and so on. Any decision on technology benefits some and harms others, just as occurs with other variables of socioeconomic progress, such as wages, rents, interest, etc. In itself, there is nothing bad about this, since it is a natural consequence of the prevailing rules of the society; what is really important is to be aware that the situation is thus, and this is often forgotten or ignored, perhaps because technology is often confused with science. In scientific matters conflicts are usually academic, while in the field of technology they are political. "The capacity of technology to transform the nature and orientation of development is such that whoever controls technology controls development. Thus this is a fundamentally political issue" (Dag Hammarskjold Foundation).

The argument of this document is that it is possible to define at least one objective which, since it is shared by each of the countries, makes co-operation among all of them possible and desirable; this objective is that each sovereign nation should, by definition, attempt to achieve an autonomous capacity to handle technology, so as to be able to direct it and use it in the way most convenient and appropriate to its interests and objectives. However opposed the interests of nations A and B, or C and D, each and every one of them needs to know how to handle technology, just as it needs to know how to handle taxes, currency, income distribution and external trade.

Only insofar as a nation acquires this capacity for handling technology will it be able to achieve the desired objective of converting technology into a special tool for its own development, an operational variable in the system of production, subject to its own decisions and not to those of others. In this complicated game there is a crucial dilemma: either one manages technology, or one ends up by being managed by it. Whatever each nation does with the technology once it has learned to manage it will be exclusively a matter for its own policy, and its decisions in this regard will be taken in the light of its own plans and programmes, its specific characteristics and its degree of interdependence with other nations.

Why does the pursuit of this objective make co-operation between nations desirable? In the first place, because such co-operation is an essential element in achieving a local capacity to produce technology, the effort to secure which will, because of its magnitude in terms of human and material resources, necessarily demand all the co-operation which can possibly be obtained. Moreover, such joint action also offers the participating countries greater latitude and makes it possible to achieve reasonable scales of operation. Secondly, because concerted actions will help to develop the capacity to identify and formulate specific technological requests, a capacity which is notably absent at the moment. Finally, because it will make it possible to negotiate with the United States and other nations which provide technology to the region from a position of greater strength.

Considered as a political and social process, knowing how to manage technology also means knowing how to define it in the terms which are most convenient and appropriate to the objectives proposed, knowing how to produce it using one's own resources, knowing how to select it from the existing local or foreign stock, and finally knowing how to use it in the existing socioeconomic circumstances. Two areas must be distinguished in all this:

Area I, that of the structure of production of goods and services, where technology behaves as a commodity and the problem consists in the smooth and reliable supply, in quality and quantity, of the technology needed for the area's proper operation, in keeping with the inherent rationale of this structure of production and using the machinery and channels which normally operate within it;

Area II, that of the "global problems", where managing technology means knowing how to use it effectively to solve problems which, by their very nature, extend beyond the framework of the structure of production, such as weather control, the development of hydrographical basins, forest or desert management, the occurrence and control of natural disasters, urban planning, control of the environment, health protection, and so on.

In both areas the management of technology comprises various stages, as summarized in the following diagram:

It is clear that the rules are different in the two areas, as are the principal participants. In area I the economic system rules, and the principal protagonists are the entrepreneurs (private and public, national and foreign, industrial, commercial and agricultural), because they are directly responsible for producing the goods and services and therefore for taking the final decisions regarding the technology to be used. The State participates in its dual function as regulator and controller of the structure of production, and as owner of productive enterprises. In area II, on the other hand, it is fundamentally the system of penalties and rewards that governs the behaviour of the State as the entity responsible for managing the territory of the country and its resources, the health and protection of its inhabitants, the provision of urban infrastructure, defence against natural and social disasters, and so on.

In both areas I and II local production of technology is an essential element for achieving the desired autonomous capacity, not least because it will permit better management of the importation of technology and of technical assistance, for a sound local productive capacity substantially strengthens a country's

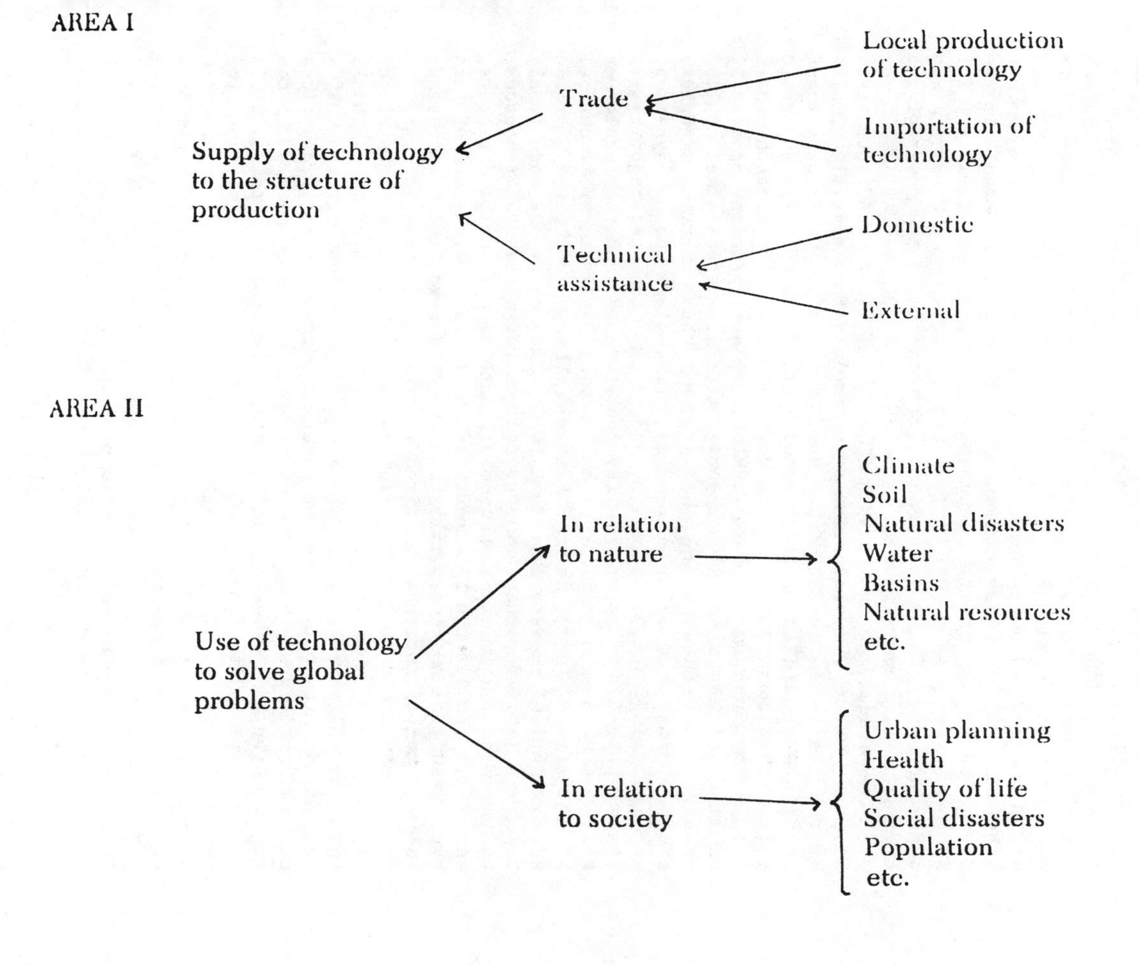
AREA I
Local production of technology
Importation of technology
Trade
Domestic
External
Technical assistance
Supply of technology to the structure of production
AREA II
Use of technology to solve global problems
In relation to nature
Climate
Soil
Natural disasters
Water
Basins
Natural resources
etc.
In relation to society
Urban planning
Health
Quality of life
Social disasters
Population
etc.

bargaining position.

While for thousands of years man produced technology in a spontaneous, unsystematic and almost amateurish way (i.e., in the manner of the craftsmen of those days), in recent decades this mode of production has sharply changed, and has been converted into a specific, organized, differentiated and continuous activity with its own identity, its own legitimacy and its own economic characteristics. Just as everyday goods are produced in establishments broadly referred to as factories, the same occurs with technology, with the difference that the technology factories or firms are given such names as "research and development laboratories", "R&D departments", and so on. In area I in the developed countries the technology factories and firms are the largest and most efficient units producing technology, and represent one of the fundamental components of the power of the transnational corporations.

The situation is different in area I of the LAC countries, however, where the production of technology fundamentally remains at the craftsman level: while such production is important and therefore should be suitably encouraged, it is far from enough to enable these countries to become efficient producers of technology.

The technology used to tackle the global problems of area II usually--although not always--originates from specialized applied research institutes along the lines of those dealing with lake studies, integrated basin studies, seismology and oil dynamics, urban planning, infant nutrition, and so on. The vast majority are State, para-State or university institutions, and the results of their work are distributed much more by means of technical assistance than by means of trade. These institutes provide services which are very important for the community and undoubtedly provide a basis for the knowledge and control of natural resources, the natural and human environment, natural and social disasters, and so on. In all the LAC countries there are institutions of this type, and in many cases they have achieved significant successes: indeed, some of them have acquired international prestige.

Only in the last few years have the principal aspects of the importation of technology into LAC begun to be understood, thanks to the establishment of National Registers of Technology.

Since this procedure has been introduced not only in various countries in LAC, but in many others in the Third World, the importation of technology is becoming more open, and this will in due course make it possible to bring into effect appropriate regulating legislation, like that which exists for the importation of other commodities. It will then be possible to control imports of technology in terms of cost use (reducing or eliminating provisions restricting its use) and content. Among the most important consequences of this process are the negotiations at present under way within UNCTAD to formulate a Code of Conduct for the Transfer of Technology.

In order to ensure that the supply of technology in area I is properly managed, the LAC countries should concern themselves first and foremost with promoting the local production of technology and controlling its importation. These two processes should be carried out simultaneously, since one is supported by the other, because imports cannot be replaced if there is no local replacement, and it is not possible to produce locally without a certain degree of protectionism against external production.

The promotion of production requires action both in the field of demand and in the field of supply. The promotion of demand for local technology can only be successful if the prevailing rationale of the structure of production is adapted to this end, which means using fiscal, economic and financial incentives to increase the consumption of local technology and introducing penalties of the same types for the unnecessary use of imported technology. A certain degree of technological protectionism will necessarily have to be established, not by means of "ad valorem" tariffs, but through preferences of a qualitative type, since in general technology is purchased on the basis of its "quality" and the trust inspired but its supplier, rather than its price. One possible mechanism for qualitative technological protectionism would be the giving of preference to local consulting and engineering firms in feasibility studies and design work on investment projects in the public sector, as in the Argentine legislation on "national contracts": the purchase of locally produced capital goods, the engagement of skilled personnel of local origin, and so on.

In order to improve supply, it will be necessary to strengthen the scientific and technological infrastructure and encourage the establishment and operation of enterprises producing technology, consultant services, engineering and design services, and auxiliary technical services. This encouragement should also make use of the machinery and procedures which are accepted and used in the structure of production: bank credit, reduction of taxes, other fiscal benefits, etc. The quasi-artisanal production which takes place within enterprises in the structure of production should also be suitably encouraged, for example by permitting tax deductions in respect of expenditure on the production of technology, with "soft" credits for the development of prototypes and the establishment and operation of pilot plants, and so forth.

The encouragement of production should be complemented with vigorous promotion of exports of technology, which have already been successfully started in various LAC countries and which are likely to experience explosive growth in the next few decades, especially towards the Third World, where countries which are still at an earlier stage of development find that technologies from LAC are more suitable than those from the central countries. In particular, the export of technological services is of the greatest importance, since it prepares the ground for the subsequent export of technological items, capital goods, etc.

As regards the control of imports of technology, the principal aim here is not to reduce the volume but to improve the quality and importance of what is imported, to bring it into line with local needs and resources, and to improve the terms on which the imports occur. In fact imports of technology are not only not going to decline, but are very likely to increase, insofar as the development of LAC becomes more thoroughgoing and extensive; what is needed is to avoid superfluous imports, replacing them by imports which are really essential and are negotiated on the basis of fair and non-restrictive conditions.

It is clear from the above that policies to promote the production of technology and control its importation must be mutually consistent, so as to ensure a smooth flow of technology which is suited to the structure of production.

Choice Of Technology

In recent years there has been increasing awareness of the need that all technology, before being supplied and used, must be selected from among various alternatives so as to ensure that the most appropriate technology has been chosen. This process of elimination should be applied not only to imported technologies, but also to those produced locally, since these are in many cases copies and adaptations of foreign technologies and as a result transmit the value system carried by their "genetic code", also sometimes called their "ideological content".

Of course, each country will make this selection on the basis of its own criteria and for its own purposes, but co-operation among various countries will be very useful in defining more precisely the questions of: (a) Why to select; (b) What to select; and (c) How to select.

Some Limitations

At this point in the document it is necessary to realize that technology, even when selected with the greatest care and supplied to areas I and II with maximum efficiency, is <u>not</u> a "magic wand" which solves everything, a "cure" for <u>all</u> the evils of underdevelopment, a "key" to open all the doors to happiness. It is a <u>necessary,</u> but not sufficient, condition to enable the LAC countries to pass beyond their present stage of development and succeed in reducing poverty, backwardness, malnutrition, disease and so on; to that end technology should be used in accordance with appropriate overall policies and with plans and programmes for socioeconomic development which are designed to solve these problems. Technology is sometimes expected to eliminate or reduce unemployment, protect ecosystems, uphold cultural independence, and the like, at the same time as it is being used in a context which ignores such objectives or--worse still--implicitly defends what it rhetorically purports to be combating. These are complex demands which cannot be satisfied merely by using appropriate technologies, although it is true that this can help, sometimes to a significant extent. There is a danger of proposing solutions which are unattainable, or which are not solutions at all but merely desires expressed with attractive rhetoric but little profundity. Thus, for example, the demand is often heard in LAC that modern capital-intensive and energy-intensive technologies should not be introduced into rural areas, but that the traditional technologies, appropriately improved and adapted, should be maintained. This position, which has a certain validity and enjoys a degree of recognition in mainland China, India, Indonesia, and so on, would seem to be difficult to apply in our continent. In the first place LAC is no longer rural, although it has rural sectors which have very rapidly been "opened up to the world", with the help of geographical mobility (in contrast to Asia, there are no large isolated plateaux). Moreover, it has a single language which unifies and promotes integration, widespread urban life patterns, growing penetration by industrial goods, a touching 19th century faith in Progress, and great admiration and respect for Technology with its magic products (radio, the cinema, TV, the telephone, antibiotics, electricity, agricultural machinery, and so on).

Finally, this position is difficult to uphold because the greater productivity of modern technologies makes their introduction almost inevitable. This does not mean that these technologies are the only possible ones, or that they must necessarily be capital-intensive and energy-intensive, for other solutions, better suited to the availability of local resources and factors, might be imagined and possibly developed, but this will only be possible by means of an intensive research and development effort and not simply through a sort of romantic "return to Nature". If modern technology is not appropriate and convenient, the only acceptable response is to produce even more up-to-the-minute technology, which is appropriate and convenient.

Furthermore, some criticisms of modern technologies, both for the agricultural and for the manufacturing sector, unfairly forget that the only way of breaking away from the old international division of labour has involved the use of technologies which, because of the increased productivity they brought to the economy, have enabled some underdeveloped countries to begin to produce industrial goods on acceptable terms. Without the help of those technologies, these countries would probably have continued to be mere producers of raw materials.

In short, this complex subject requires much more research before the present stage can be passed.

Some Important Obstacles

In order to attain the desired objective it will be necessary to overcome a group of obstacles of varying importance, including:

(a) The groups of interests which benefit from technological dependence and will not remain indifferent to a vigorous programme aiming at technological independence.

(b) The weakness of the State, which must play a leading role, and its meagre capacity to implement and ensure the implementation of decisions of a technological nature.

(c) The intellectual alienation of those groups in the ruling class which hold that nothing can change because "we are not capable", and others which hold that nothing can change because "they won't let us".

(d) The existing rationale, according to which it is better business to import technology than to produce it locally.

(e) Cultural dependence, holding that "any foreign technology is better . . . because it is foreign".

(f) The prevailing system of values, which gives action to provide for superfluous consumption by the elites priority over providing for essential consumption by the majority of the population.

(g) The slavish imitation practised by the periphery, whereby even the worst products and processes from the centre are copied.

(h) Local financial machinery, which fails to provide risk capital for the production of technology, will provide backing for any "prestigious" imports of technology.

(i) The poor links between the principal participants in the process: State officials, entrepreneurs and managers, scientists and technologists.

IV. POSSIBLE STRATEGIES

Two fundamental strategies are proposed: a strategy of co-operation, to help to make technological autonomy in LAC viable, and a strategy of negotiation, to help to make equitable interdependence between LAC and the United States on a basis of solidarity possible.

The strategy of co-operation should aim at the same general ends in both areas I and II: to support the national plans and programmes for technological development, to carry out a coherent set of actions to strengthen national efforts and expand their scope of operation, and to co-operate with the programmes of other subregional, regional and international bodies. However, the development of such a strategy cannot be the same in areas I and II: it will take place in each of them in conformity with their particular characteristics. Equally, it will not take place with the same intensity throughout the whole wide range of possibilities, but must select some priority directions where efforts will be particularly concentrated so as to use the scarce resources available with maximum efficiency.

Bearing in mind the particular characteristics of area I, it is proposed that the co-operation should focus on the two following lines:

(a) Promotion of production of technology, through action aimed at greatly strengthening the capacity to produce technology and promoting the creation of bilateral, multilateral and subregional capabilities which, at an appropriate time, could come to be articulated in a genuine regional technological capability.

The promotion of production should include both industrial and quasi-artisanal production, as well as the protection of them from foreign technology. The extension of production and protectionism to the whole region might be achieved through such appropriate instruments as the operation of Latin American technology enterprises, agreements for technological complementarity, regional preferences, etc.

(b) Promotion of trade in technology among the LAC countries, so as to increase the present meagre flow significantly, control the importation of foreign technology by seeking a reduction in the redundant imports which occur when various countries import the same technology, and develop adequate economic room to permit autonomous technological development.

Trade promotion should encompass both technology embodied in goods and in human knowledge and technology which is not embodied in technological goods and services, and as a result should cover capital goods, the emigration and immigration of skilled personnel, consulting services, design and engineering services, and so on. The instruments used should naturally be compatible with those used in trade within Latin America in general.

Just as in area I the strategy of co-operation proposed highlights the production of and trade in technology, for area II it is proposed that the co-operation should develop on the basis of "joint projects for production and technical assistance". It has already been noted that, because of the very nature of the global problems of area II, technical assistance is more important than trade; thus, for example, if country A has developed a given technology for flood control, while country B needs it for application on its own territory, it is very unusual for the technology to pass from A to B by means of trade: it is most likely to be transferred under an agreement between A and B, which might possibly include some payment, but not in terms of an actual price for the technology.

An important aspect of the global problems is that, by their nature, although one of them may be specific to one country or a sub-group of countries (for example, the protein deficit), there will certainly be another which is specific to another sub-group (for example, the ecosystem of desert zones) and so on and so forth, so that the set of problems as a whole will in fact be of equal interest to all the countries and, as a result, co-operation will prove to be of genuine mutual interest. There are also problems of regional--and even worldwide--scope, such as general weather control, earthquake prediction and control, exploitation of the seabed, mass hunger and poverty, the exhaustion of natural resources, waste management (industrial, mineral, nuclear and so on), urban marginality, world population growth, and so on, which will be impossible to study and solve without co-operation by all, which is therefore an ineluctable imperative.

The principal characteristics of the "joint projects" would be as follows:

(a) They would be defined and organized around problems to be solved, and not around disciplines. For example, there might be a joint project for the development and use of tropical forests, but not a silviculture project; for the replacement of animal protein in mass diets, but not a protein chemistry project; for the ecological control of marginal problems, but not an ecology project; for the use of nonbauxitic ores for aluminium productions, but not a mineralogy project, and so on.

(b) Thus defined, the projects will necessarily be multidisciplinary, and although some will be linked more to the "hard" sciences such as physics and chemistry and others to the "soft" sciences such as sociology and anthropology, the solution of the problems will require knowledge from various sources and disciplines. Moreover, although applied research will be one of their fundamental tools, the projects will not make use of "applied science", but of "technology", because their final objective is not to inquire into the problems, but to propose solutions, and for that purpose they will be able to use knowledge of any origin and nature, provided that it is useful for that purpose.

(c) The projects will aim to develop technological solutions which are feasible, viable, appropriate and convenient, on the basis of definitions and criteria explicitly set out in the programmes themselves.

(d) The joint projects will encompass all the stages, ranging from prefeasibility and feasibility studies to the production of technology--which will contain an appropriate mix of local and foreign technology--and its application and full use by society. The countries participating in each project will endeavour to participate fully in all these stages, so as to transform the static transfer from the owner to the recipient into a dynamic transfer in which all the participants give and receive.

(e) The joint action will mean that the countries which wish to embark on a specific project aimed at solving a problem of common interest will define the nature and structure of the project, the terms of their participation and the adaptation and use of the technological solution achieved and other technologies which it may have been possible to generate in the process.

(f) The projects will be implemented by suitable institutions such as research institutes, technology enterprises, university laboratories, research centres and so on, which will coordinate the organization, administration, control and implementation of each project.

(g) It will be essential, for each project, to ensure that it is related as closely as possible to the circumstances which it is supposed to improve; otherwise there will be a risk of producing solutions which might be satisfactory to their authors but impossible to apply. For this reason, the various interest groups linked with the problem which is to be solved should be properly represented in the management of the project. Thus, for example, in an endeavour to solve the problem of improving the everyday diet of the mass of the people through the addition of proteins to bread, the body responsible for managing the joint project should contain representatives not only of the scientists and technologists participating in the project, but also of the bakers who may possibly produce and market the new type of bread, the manufacturers of bread-making equipment, the producers of the raw materials used in producing bread, and so on.

(h) The final form of applying the solution reached will depend on the specific circumstances of each project. Although in general this will be done by means of technical assistance, there will undoubtedly be cases where it is most desirable to do so by means of trade. Thus, for example, in the case of flood control, the technological solution found will no doubt be transferred in the form of technical assistance; but in the case of the new type of bread, it is likely that marketing of the solution through the specific channels of the sector (for example, the manufacturers of bread-making equipment) will be not only the most rapid and efficient way of achieving the goal sought (to improve the diet of the masses), but perhaps the only feasibly way, despite the difficulties inherent in it.

(i) In the global projects which are regional or worldwide in scope, where because of their magnitude and complexity scientific, technical and economic leadership will in fact be in the hands of the developed countries, LAC will nevertheless demand genuine co-operation and, as a result, full participation in decision-taking and in the benefits.

The strategy of negotiation in the field of technology between LAC and the United States must naturally be part of the strategy of general negotiation between the two sides which is carried out in various forums. In the specific case of technological negotiations, the central concern should be to ensure that the United States recognizes that it is in its own political interest that the LAC countries should pass beyond their present stage of technological dependence, and that it should undertake to co-operate actively to ensure that this occurs in the shortest possible time and at the lowest possible cost.

The technological negotiations should encompass at least the following subjects:

(a) Regulation of trade in technology between LAC and the United States, under a "Code of Conduct for the Transfer of Technology", identical or similar to that currently being drawn up within the United Nations;

(b) Regulation of the behaviour of the United States-based transnational corporations, as far as technology is concerned, by means of provisions identical or similar to those which will appear in the United Nations "Code of Conduct for Transnational Corporations" which is being prepared;

(c) Regulation of the behaviour of the United States consulting and engineering firms which sell services in LAC in order to prevent restrictive practices or trade abuses arising from their power;

(d) Sustained support for the development of a local capacity to produce technology, and particularly for the establishment and operation of technology enterprises;

(e) Support for the LAC countries in their negotiations with the World Bank, IDB and other international financial institutions with a view to eliminating the conditions which "tie" their credits to the supply of foreign technology, stimulating full use of local technological goods and services (especially in consulting and engineering) and allocating "risk capital" for the establishment and operation of local technology enterprises;

(f) Active co-operation in "joint projects" among LAC countries and support to ensure that they achieve full participation in "joint projects" at the global level;

(g) Development of technological transactions with medium-sized and small industry in the United States, and increased use of the programmes of the Small Business Administration.

The negotiation strategy aims to ensure not only that a harmful confrontation with the United States is transformed into active co-operation, but also that the LAC countries engage in an active learning process on some central problems connected with the interface between the structure of production and technology and become aware of their principal shortcomings in this field. The United States knows how to manage technology as an operational variable and has an "implementation system" for decisions in this field, which functions in response to explicit orders from the political authorities. This is precisely an area where LAC falls seriously short, all the more so because there is not even full awareness of its necessity and importance. The negotiations will bring this out very clearly, while at the same time clarifying relationships and mechanisms which are still hidden behind an ideological tangle of absolute pseudo-truths. For this reason the negotiations must constitute a continuous and ongoing process.

Problems of Scientific and Technological Development in Black Africa

LANDING SAVANE*

Scientific and technological progress has promoted decisive improvements in man's living conditions in recent centuries. The twentieth century in particular testifies to the triumph of science and technology, but technical progress has not been equal in all the regions of the world. Nearly all the countries of Africa, Asia and Latin America have been by-passed by scientific and technological progress, which is concentrated elsewhere, and reduced to the humble role of technology consumers. The gap has grown steadily wider, but the situation can be remedied.

Technological backwardness is one, and not the least, of the factors in underdevelopment, and the concern of the United Nations about it is understandable. Ten years ago, in fact, the United Nations Conference on the Application of Science and Technology for the Benefit of the Less Developed Areas was held in Geneva. This meeting was followed in the same year (1963) by the establishment of the United Nations Advisory Committee on the Application of Science and Technology to Development. One of the activities of this committee was the World Plan of Action for the Application of Science and Technology to development, to which approval was given in 1971.

But the problems facing humanity as a result of uncontrolled technical progress were already receiving attention though a general awareness of the need to safeguard the environment; this was reflected in the Conference on the Human Environment held in Stockholm last June. Now the stage has been reached, mainly in the Western countries, where science is being called into question. We therefore feel it worth while to explore the underlying significance of the growing role of science and technology in Black Africa with a view to the establishment of guidelines for constructive action in this field.

From **IMPACT OF SCIENCE ON SOCIETY**, Vol XXIII, No. 2, 1973, (85-93), reprinted by permission of the publisher, UNESCO, Paris.

I. FUNDAMENTAL CHARACTERISTICS

It is generally accepted that the role of scientific and technological research in economic and social development has been and will continue to be large. Research of this kind gives men access to new areas of knowledge which they can use to better their welfare. National resources can be developed far more easily where technologies that respond to genuine needs have been developed as a result of a proper research policy. A coherent relation must therefore be established between research aims and development objectives.

Experience shows, however, that things are not always quite so well arranged. Despite innumerable declarations of intent and the steps taken to promote development-oriented research in the various countries of Black Africa, we cannot fail to see that no African country has really succeeded in mobilizing its full scientific and technological potential for constructive economic development. In our opinion this situation is essentially due to three sets of factors, which we shall now discuss and which seem to represent the essence of scientific and technological research in Africa.

1. All too often, research is still the work of foreign institutions. The interests of a foreign institution may conflict with those of its country of establishment, or else have no direct relation to that country's priorities. National authorities may readily accept and even encourage this situation of dependence on foreigners, taking a false pride in the number of research bodies which have been set up in their country. They may not realize that situations of that kind do nothing to enhance national scientific independence if the fruits of the research remain the sole property of the foreign firm which had executed it, like any discoveries which the firm makes overseas.

The very nature of African economies, whose activities are basically directed towards satisfying world market requirements rather than domestic needs ('extraverted' economies), facilitates the establishment of foreign institutions, particularly those of the principal customer country. These institutions can use that advantage to direct the host country's economic activity into channels which best serve their own interests. It is therefore an illusion to think that the interests of the host country are furthered by the installation of bodies of this kind. Many studies by present-day economists [1][1] have demonstrated that the extraversion of developing country economies is a fundamental cause of their underdevelopment and that these countries should start exerting every effort to make their development self-sustaining.

2. Often the scope of research is still narrow. Even at university level, research work is isolated from the day-to-day life of the surrounding population. In most countries the universities are fairly independent in their scientific and technological research. The reason why they often embark on research projects devoid of any practical significance for their county is that many of the university teaching staff are still foreigners. For the same reason, there is a tendency to

concentrate on secondary lines of inquiry at the expense of fundamental research aims of national importance.
institutions.

We must not forget that university research includes various kinds of doctoral work as well as more specific research topics. It may well be asked whether, if the universities adopt too nationalistic an approach, they may not develop in watertight compartments to the detriment of their international reputation. What is more, this kind of research looks very like a direct continuation to the university stage of primary and secondary education programmes. It is essential to stress the close relation between the educational systems of the former colonial powers and those of the newly independent African countries, and the decisive influence of the former on the latter.

The socio-economic and socio-cultural development aims of the African countries, however, seem to call for a fundamental reform of African universities towards a systematic response to national realities. The African nations could surely make a significant contribution to world research if they boldly followed their own bent and refused to tread the same path as the developed countries. Surely Africans are best able to revive their cultural values and their history, and set about developing technologies appropriate to their particular physical and human conditions.

This clearly brings us to the problem of decolonizing, not only the universities but African education as a whole. Without doubt this process should extend beyond the content of education to include its structure, the composition of its teaching profession (in which the majority must be Africans), and its mentality and attitudes (a necessary cultural decolonization). It is not surprising that under present conditions many universities constitute a permanent source of tension in different countries and that the problem of fundamental educational reform is central to student demands.

3. Research finds no widespread application anywhere. This is true even in the national research institutes; yet the only possible justification for scientific research is the improvement of human living conditions. In fact, in the rare cases where research at universities or State institutes produces interesting and potentially useful results, serious difficulties persist. In some cases, the State may decide to oppose, or at any rate minimize, the scope of discoveries which if publicized might conflict with the interests of highly influential business circles.

There can be no other reason why so little is made in Africa of its traditional pharmacy, the benefits of which have been abundantly demonstrated not only through research in certain universities but also through centuries of practice among groups of inhabitants, unfortunately all too few. In other cases, the State, whose role in industry is purely marginal, has no means of undertaking the additional studies which could turn an experimental discovery into an industrial process.

Moreover inventions, even when their industrial use becomes feasible, are of no interest to their potential users--the foreign companies which normally dominate the market. Since they are public property they are not protected by patents and so could not bring their users any profit not available to possible competitors; in addition, their use would call for new investment.

Another point is that, because no constructive policy exists for training national staff, national institutes are very often directed by foreign technical-assistance personnel who usually have very little knowledge of the beneficiary country's problems. National research workers, where they exist, furthermore, have very poor equipment and human resources; they are often in a precarious material situation themselves. A working background of this kind is clearly inconducive to any substantial or widespread progress in the fields of research falling within the scope of national institutes.

Nearly all research units in Africa are either foreign establishments financed or administered by a foreign institution, or else national, that is State-controlled, bodies. Taking into account the three sets of factors which I have described, scarcely a single African country could be said to have a 'science policy'. For such a policy to exist, national research programmes would have to be settled by a central authority, so that they could be co-ordinated and the various stages of their execution supervised as far as the mass application phase.

It is true that some countries have bodies which supervise the development of scientific and technological research (e.g. Senegal, Zaire, Ghana, Tanzania and Zambia); but these bodies seem to have had a secretarial function and to have been capable at most of assisting a particular line of research which has been considered worth while rather than playing a central decision-making and planning role for research programmes forming part of a systematic scheme directed towards national development.

II. UTILIZATION OF RESEARCH

I shall now attempt to identify the main problems and possibilities which arise from the utilization of science and technology in Black Africa.

Most African countries have embarked on a period of economic development characterized by the use of modern methods. This would not have been possible if they had not imported technology by attracting foreign capital and foreign industrial processes. It may well be wondered whether such development is in the long-term national interests of these countries. With this question in mind, after looking at the real significance of the transfer of technology now taking place between developed and developing countries we shall discuss the relations which exist in Black Africa between economic and industrial development and the protection of the environment. We shall then consider the links between population growth and scientific and technological development and conclude with a survey of prospects for the use of science and technology in Africa.

Transfer Of Technology In Black Africa

In the 1960s many African States became independent and their governments had quickly to formulate development strategies to further national economic and social progress. Very often, however, that development was regarded as a linear operation which consisted in catching up with the advanced countries, not in the economic and social reorganization of a whole society to enable it to stand on its own feet in essential matters. It is not surprising, therefore, that mass importation of technology was a current practice among African countries. Yet experience now shows that the transfer of technology often threatens the political and economic independence of the beneficiary country.

As one observer has put it: 'The problem for firms in developing countries is not simply to obtain advice and acquire expertise . . . It is also--what is often more important--to put that advice and expertise to economically profitable use . . . Ability of this kind is precisely what companies in developing countries frequently lack. Because of that, many beneficiary firms have good reason to rely on the intermediary role of production undertakings in developed countries which manifestly possess such ability [2].' This also explains why 'Assignment of patents and trademarks is generally associated with contracts for technical assistance [3].'

In addition, the corporation supplying the technology generally has for that very reason a technological monopoly in the beneficiary country. It is often even better off, for it can also obtain a legal monopoly, since African countries' domestic markets are so small that they are forced to accept conditions of that kind or else foreign industrialists would not invest in them.

Because of conditions which create a monopoly situation for the supplier firm, the beneficiary country is deprived of access to other suppliers' products which are available on the world market even on highly favourable terms. There are also usually stipulations obliging the beneficiary enterprise to provide the processing equipment. Hence, in the long run, contrary to appearances, it is not the beneficiary but the supplier country which profits from these transfers: in addition to the substantial amount of foreign currency it acquires from the sale of its technology, it derives considerable long-term benefits by operating the exported processes.

Clearly the access to existing expertise and technology which the beneficiary country thus acquires does nothing to strengthen its technological independence. On the contrary, there is a dangerous tendency to rely on foreign sources of technology and to adopt indiscriminately any methods or criteria coming from industrialized countries. It is important to remember, however, that any scientific or technological invention is related to the socio-economic and socio-cultural conditions in which it is born. Generally speaking, the technology of advanced countries tends to make an undertaking as economically profitable as possible, in a setting in which capital is often available and labour scarce.

Conditions are quite different in Black African countries, where, on the contrary, there is a surplus of labour and capital is rare. For the various reasons I have mentioned, the possibility of procuring the transfer of technology should never be preferred to a conscious effort by a society to create the technology appropriate to its specific developmental conditions. What is more, transfer of technology is meaningless unless it helps the country effectively towards real technological independence. Unfortunately, this very seldom happens in the present system of relations between developed and developing countries, particularly those in Black Africa.

Economic And Industrial Development And The Protection Of The Environment

We have seen that all too often the countries of Black Africa are still happy to import foreign technologies, and particularly those of advanced capitalist countries. Yet in recent years uncontrolled economic and industrial development in those countries has considerably aggravated the harm man does to his environment.

We may therefore inquire how the African countries react to this negative aspect of development in industrialized nations, and what place protection of the environment occupies in their development strategies. At this point we must re-emphasize that, for us, the environment means the total reality of the human world, including physical as well as social, economic and political elements; it cannot be limited to ecology alone.

Detailed studies have shown that the African economies are really an entity in which every sector conditions the others. Basically, the dependent capitalist structure of these economies is the origin of a 'modern' import-substitution industrial sector which exploits the agricultural sector and simultaneously prevents it from advancing, the driving economic force behind the process being the profit of the industrial sector.

In these circumstances present-day Africa is undergoing a process of economic and industrial development that is broadly marked by a serious degradation of the environment due to the frequent use by African farmers of methods which, under the conditions of the precolonial continent, did not dangerously threaten the environment. In the Africa of today, however, these methods irreversibly destroy not only the forests, which are far fewer than they were, but even the soil, in particular through the misuse of the pesticides and chemical fertilizers or through harmful cropping.

A start has been made towards creating some awareness of these problems. In October 1970, for example, the first conference on the rational utilization and conservation of nature took place at Tananarive. It adopted a series of pertinent resolutions which stressed the need for adequate protection of the priceless heritage represented by fauna and flora in Malagasy. In its resolutions it laid special emphasis on the need to promote intensive agriculture, rational animal husbandry and a vigorous forest policy.

But economic and industrial development raises other kinds of environmental problems. Certain large structures (particularly dams), for instance, have been built without any previous thorough study of the various consequences they might have for the environment. In Ghana, for example, the River Volta dam has created the largest artificial lake in the world, and this has affected the environment. Waterborne disease such as bilharziasis have returned and certain climatic changes have been noticed in the area, resulting in fresh problems for the indigenous population.

The fact is that a preliminary multidisciplinary study of the project would definitely have made it possible to avoid many of these disadvantages. Fortunately, in Ghana they are greatly outweighed by the clear benefits which the inhabitants have derived from the dam, particularly since the Ghanaian Government is giving them some help to deal with the situation. The same cannot be said of the Cabora Bassa dam project now under construction in Mozambique, which is still under Portuguese domination. In pursuing this huge scheme (to be the fourth largest dam in the world), the Portuguese authorities and their allies do not seem to have taken account of anything except their own selfish interests, or to have set any store by the health and welfare of the indigenous population.

Taking a broader view, we must repeat what we indicated about the transfer of technology; every imported or new technology should be matched to its proposed environment and introduced into this as gently as possible. This means that scientific research and decision-making about problems of this kind must be handled on a national and multidisciplinary scale. In our view, now that the environmental crisis is assuming alarming world proportions, the developing countries, particularly in Africa, have an exceptional opportunity to avoid a blind commitment to the paths of development of the former colonial nations. We shall thus be able to make a fundamental contribution to the protection of the human environment in this field also.

Population Growth, Scientific And Technological Development

It has become the custom to relate environmental problems to population growth, which is regarded as a negative factor for the environment. The growing pressure exercised by a constantly rising population is held responsible for the increasing quantitative and qualitative deterioration of fauna, flora and soils. This neo-malthusian approach, when taken a stage further, leads to the view that rapid population growth hampers a nation's economic and social expansion, and in particular its scientific and technological development. We must now consider the specific significance of this approach for Black Africa.

The neo-malthusian doctrines are familiar territory. Their central theme is that the countries of the third world are distinguished by a lack of capital for essential investment in various fields of industry, agriculture, manpower training, education and health. At the same time, because of their rapid population growth, their governments are faced with added social liabilities (in education, health and employment) and so are forced to spend on these sectors capital which could otherwise help to provide industrial and agricultural equipment and thus further the expansion of their national production capacity. Furthermore, steep population rise tends to lessen the relative significance of advances by overtaking the benefits which they promote.

Many factors seem to confirm the neo-malthusian hypotheses when applied to scientific and technological development. The inability of local authorities to meet situations such as the rapid rise in the numbers of schoolchildren and university students leads to a striking decline in educational standards. This is because, of course, of a lack of properly qualified staff and, above all, the steadily deepening crisis now affecting African educational systems. As I pointed out earlier, these systems have not been fully adapted to the new conditions which arose after independence.

In addition, at the universities, many of which are important research centres in Africa, the staff (mainly Africans) are in such great demand for teaching duties that they have scarcely any time left for research. Nearly all lecturers in medicine and pharmacology, for example, are practitioners as well as teachers and, because of the shortage of medical personnel in Africa, cannot easily take up research.

In the final analysis, then, the main obstacle to scientific and technological development in Africa south of the Sahara is not population growth but the failure to establish any scientific and technological policy which could abolish dependence in this field. To achieve this, governments must have the proper means of controlling and directing research institutes and a thoroughly effective policy for training and motivating research workers. In view of the actual conditions in African countries, political and moral stimuli would doubtless be more productive than material inducements. But these--indeed, the entire policy of reorientation which we are advocating--must await far-reaching political, economic and social changes.

Scientific and industrial development, however, is only one aspect of the general problem of economic and social development. The development strategies employed in Africa have seldom led to anything more than an illusion of development--the conspicuous affluence of a very small minority of the population based on exploitation of the vast majority and their continued subjection to extremely precarious living conditions. Consequently, as a matter of urgency, the African countries must abandon these patterns of growth without development of their 'extraverted economies' and turn to strategies of 'self-sustaining development', directed towards the welfare of those classes of society which have hitherto been barred from development (mainly the peasants) [4].

III. CONCLUSION: PROSPECTS FOR THE FUTURE

It is therefore clear in my mind that certain political, economic and social conditions must be created if a true scientific and technological development is to be possible in Africa. It would be foolish to underestimate the difficulties which are bound to exist even after those conditions have been established.

At the national level the major difficulty is the scarcity of human resources (senior scientific personnel) and of finance (investment capital). African countries must narrow and concentrate their ambitions, accept the sacrifices necessary for progressive research, and give applied research the highest priority. It must be clear that they cannot possibly adopt inappropriate international standards governing, for example, the correct proportion of the gross national product to invest in research.

At the company level, research will have to be directed towards immediate profitability. In a self-sustaining economy this is what is needed for national development.

Internationally, assistance from foreign sources must be properly geared to national research programmes. The partners in aid must therefore co-operate fully and honestly in applying science and technology to development. There is no doubt that this condition generally remains unfulfilled.

African States are definitely starting on the path to progress, and are certain to benefit from the establishment of that kind of co-operation among themselves. It is bound to further their technological liberation, seeing how truly essential it is that the narrow frontiers of African micro-States should disappear for this as for other purposes.

In our view, Africa has a pressing duty to protect its exceptional all-round potential against the deadly attack of the advanced civilizations. Its task is to find a way of its own which will permit a development commensurate with present scientific achievements and yet enable Africa to be a civilization in which the principal aim of the development process shall remain the fulfilment of every member of society.

PART II

STATISTICAL INFORMATION AND SOURCES

Main purpose of this section is to provide a current bibliography of data sources related to technology policy and development from a Third world perspective. However, this bibliography should be supplemented with the complete directory of United Nations Information Sources provided in Part IV of this volume.

I. BIBLIOGRAPHY OF INFORMATION SOURCES

AFRICAN STATISTICAL YEARBOOK, UN

Presents data arranged on a country basis for 44 African countries for the years 1965-1978. Available statistics for each country are presented in 48 tables: population; national accounts; agriculture, forestry, and fishing industry; transport and communications; foreign trade; prices; finance; and social statistics: education and medical facilities.

ASIAN INDUSTRIAL DEVELOPMENT NEWS, UN, Sales no. E.74.II.F.16

In four parts: (a) brief reports on the ninth session of the Asian Industrial Development Council and twenty-sixth session of the Committee on Industry and Natural Resources; (b) articles on multinationals and the transfer of know-how, acquisition of technology for manufacturing agro-equipment, fuller utilization of industrial capacity; (c) report of Asian Plan of Action on the Human Environment; and (d) statistical information on plywood, transformers, and transmission cables.

Banks, Arthur S., et al., eds.
ECONOMIC HANDBOOK OF THE WORLD: 1981. New York; London; Sydney and Tokyo: McGraw-Hill Books for State University of New York at Binghamton, Center for Social Analysis, 1981.

Descriptions, in alphabetical order, of all the world's independent states and a small number of non-independent but economically significant areas (such as Hong Kong). Data are current as of 1 July 1980 whenever possible. Summary statistics for each country include: area, population, monetary unit, Gross National Product per capita, international reserves (1979 year end), external public debt, exports, imports, government revenue, government expenditure, and consumer prices. Principal economic institutions, financial institutions, and international memberships are listed at the end of each description.

BULLETIN OF LABOUR STATISTICS. Quarterly, with supplement 8 times per year. Approx. 150 p.

Quarterly report, with supplements in intervening months, on employment, unemployment, hours of work, wages, and consumer prices, for 130-150 countries and territories. Covers total, nonagricultural, and manufacturing employment; total unemployment and rate; average nonagricultural and manufacturing hours of work per week, and earnings per hour, day, week, or month; and food and aggregate consumer price indexes.

COMPENDIUM OF SOCIAL STATISTICS, 1977. 1980, UN, Sales No. E-F. 80.XVII.6.

Contains a collection of statistical and other data aimed at describing social conditions and social change in the world. In four parts. Part 1 includes estimates and projections for the world, macroregions, and regions. Part 2 comprises data for countries or areas that represent key series describing social conditions and social change. Part 3 consists of general statistical series for countries or areas. Part 4 is devoted to information for cities or urban agglomerations. Includes a total of 151 tables, covering population, health, nutrition, education, conditions of work, housing and environmental conditions, etc. Provides an overall view of the world social situation and future trends.

DEMOGRAPHIC YEARBOOK, 1978. (ST-ESA-STAT-SER.R-7) 1979, UN, Sales No. E-F.79.XIII.I.

--Vol. 1. viii, 463 p. This volume contains the general tables giving a world summary of basic demographic statistics, followed by tables presenting statistics on the size distribution and trends in population, natality, fetal mortality, infant and maternal mortality, general mortality, nuptiality, and divorce. Data are also shown by urban/rural residence in many of the tables.
--Vol. 2: Historical supplement.

DEVELOPMENT FORUM BUSINESS EDITION. DESI/DOP, UN, Palais des Nations, CH-1211 Geneva 10, Switzerland. 24 times a yr. 16 p.

A tabloid-size paper, published jointly by the United Nations Department of Information's Divison for Economic and Social Information and the World Bank. Presents articles on all aspects of the development work of the United Nations, with emphasis on specific development problems encountered by the business community. Contains notices referring to goods and works to be procured through international competitive bidding for projects assisted by the World Bank and the International Development Association (IDA). It also includes a Supplement of the World Bank, entitled "Monthly Operational Summary", and a similar supplement of the Inter-American Development Bank (IDB), once a month, which provide information about projects contemplated for financing by the World Bank and IDB, respectively.

DEVELOPMENT FORUM GENERAL EDITION

A tabloid-size paper, published jointly by the United Nations Department of Public Information's Division for Economic and Social Information and the World Bank, having as objective the effective mobilization of public opinion in support of a number of major causes to which the United Nations is committed. Presents articles reporting on the activities of various UN agencies concerned with development and social issues (health, education, nutrition, women in development). Includes a forum for nongovernmental organizations (NGO's) and book reviews.

DEVELOPMENT AND INTERNATIONAL ECONOMIC CO-OPERATION: LONG-TERM TRENDS IN ECONOMIC DEVELOPMENT. Report of the Secretary-General. Monograph. May 26, 1982.

Report analyzing world economic development trends, 1960's-81, with projections to 2000 based on the UN 1980 International Development Strategy, and on alternative low and medium economic growth assumptions. Presents data on GDP, foreign trade, investment, savings, income, population and labor force, housing, education, food and energy supply/demand, and other economic and social indicators.

DIRECTORY OF INTERNATIONAL STATISTICS: VOLUME 1. 1982 Series. Sales No. E.81.XVII.6

Vol. 1 of a 2-volume directory of international statistical time series compiled by 18 UN agencies and selected other IGO's. Lists statistical publications, and machine-readable data bases of economic and social statistics, by organization and detailed subject category. Also includes bibliography and descriptions of recurring publications, and technical descriptions of economic/social data bases.

VOLUME 2: INTERNATIONAL TABLES. Sales No. E.82.XVII.6, Vol. II

Presents analytical summary of major income and product accounts for approximately 160 countries, by country and world region.

ECONOMIC AND SOCIAL PROGRESS IN LATIN AMERICA: 1980-81 REPORT. 1981, IDB.

Provides a comprehensive survey of the Latin America economy since 1970, with particular emphasis on 1980 and 1981. Part One is a regional analysis of general economic trends, the external sector, the financing of development from internal and external sources, regional economic integration, and social development trends (women in the economic development of Latin America). Part Two contains country summaries of socioeconomic trends for 24 States members of IDB. Statistical appendix includes data on population, national accounts, public finance, balance of payments, primary commodity exports, external public debt, and hydrocarbons.

ECONOMIC AND SOCIAL SURVEY OF ASIA AND THE PACIFIC, 1977. The International Economic Crises and Developing Asia and the Pacific. 1978, UN, Sales No. E.78.II.F.1.

In two parts: (a) review of recent economic developments and emerging policy issues in the ESCAP region, 1976-1977; and (b) the impacts of the international economic crises of the first half of the 1970's upon selected developing economies in the ESCAP region and the market and policy response thereto. Topics discussed include: the food crisis; the breakdown of the international monetary system; fluctuations in the international market economy comprising the primary commodities export boom, the associated inflation and the subsequent recession, and, finally the sharp rise in the price of petroleum.

ECONOMIC AND SOCIAL SURVEY OF ASIA AND THE PACIFIC, 1979. Regional Development Strategy for the 1980's. 1981, UN, Sales No. E.80.II.F.1.

Analyzes recent economic and social development in the UN ESCAP region, as well as related international developments. Focuses on economic and social policy issues and broad development strategies. In two parts: (a) recent economic developments, 1978-1979, covering the second oil price shock economic performance of the developing countries of the ESCAP region, inflation, and external trade and payments; and (b) findings of a two-year study dealing with regional developmental strategies, covering economic growth, policies for full employment and equity, energy, technology, implementation systems, international trade, shipping, international resource transfers, and intraregional cooperation.

ECONOMIC SURVEY OF ASIA AND THE FAR EAST, 1973. 234 p. (also issued as Economic Bulletin for Asia and the Far East, vol. 24, no. 4), 1974, UN, Sales No. E.74.II.F.1.

Contains a general summary followed by Part One, which covers: education and employment--the nature of the problem; population, labor force and structure of employment and underemployment in the ECAFE [ESCAP] region; the role of location--assumptions underlying the education policies of developing countries in the ECAFE region; momentum and direction of expansion of education; structuring the flow of workers into the modern science of education for self employment--the traditional and informal sectors; and the search for new policies--a review of current thinking. Part Two covers: current economic developments--recent economic developments and emerging policy issues in the ECAFE region, 1972/73; and current economic developments and policies in 28 countries of the ECAFE region.

ECONOMIC SURVEY OF LATIN AMERICA. Series.

Series of preliminary annual reports analyzing recent economic trends in individual Latin American countries. Each report presents detailed economic indicators, including GDP by sector, agricultural and industrial production by commodity, foreign trade, public and private sector finances, and prices. Also includes selected data on employment and earnings.

THE ECONOMIST. THE WORLD IN FIGURES. Third edition. New York: Facts on File, Inc., 1980.

Compendium of figures on economic, demographic, and sociopolitical aspects of over 200 countries of the world. The first part is a world section with information on population, national income, production, energy, transportation, trade, tourism, and finance. The second part is organized by country (grouped by main region), containing statistics on location, land, climate, time, measurement systems, currency, people, resources, production, finance and external trade, and politics and the economy. The data, from many sources, cover through 1976. Country name and "special focus" indices.

FACTS OF THE WORLD BANK. Monthly (current issues).

A compilation of figures on World Bank lending, giving cumulative amounts and amounts for the current fiscal year of commitments by number of projects and by sector, as well as for each country by region. Also gives figures on sales of parts of Bank loans and IDA credits and on World Bank borrowings by currency of issue, original and outstanding amounts, and number of issues.

IMF SURVEY. Biweekly.

Biweekly report on international financial and economic conditions; IMF activities; selected topics relating to exchange rates, international reserves, and foreign trade; and economic performance of individual countries and world areas.

MAIN ECONOMIC INDICATORS: HISTORICAL STATISTICS, 1960-1979. 1980, OECD, Sales No. 2750 UU-31 80 20 3.

Bilingual: E-F. Replaces previous editions. Base year for all indicators is 1970. Arranged in chapters by country, the tables cover the period 1960 to 1979, and are followed by short notes describing some major characteristics of the series, and, where applicable, indicating breaks in continuity. Note: Supplements the monthly bulletin Main Economic Indicators.

MONTHLY BULLETIN OF STATISTICS. Monthly.

Monthly report presenting detailed economic data including production, prices, and trade; and summary population data; by country, with selected aggregates for world areas and economic groupings, or total world. Covers population size and vital statistics; employment; industrial production, including energy and major commodities; construction activity; internal and external trade; passenger and freight traffic; manufacturing wages; commodity and consumer prices; and money and banking. Each issue includes special tables, usually on topics covered on a regular basis but presenting data at different levels of aggregation and for different time periods. Special tables are described and indexed in IIS as they appear.

POPULATION AND VITAL STATISTICS REPORT. Quarterly.

Quarterly report on world population, births, total and infant deaths, and birth and death rates, by country and territorial possession, as of cover date. Also shows UN population estimates for total world and each world region.

QUARTERLY BULLETIN OF STATISTICS FOR ASIA AND THE PACIFIC. Quarterly.

Quarterly report presenting detailed monthly and quarterly data on social and economic indicators for 38 ESCAP member countries. Includes data on population; births and deaths; employment; agricultural and industrial production; construction; transportation; foreign trade quantity, value, and direction; prices; wages; and domestic and international financial activity.

1978 REPORT ON THE WORLD SOCIAL SITUATION. 1979, UN, Sales No. E.79.IV.1.

Deals with the global issues of population trends and employment; growth and distribution of income and private consumption; the production and distribution of social services; and changing social concerns. A supplement reviews the patterns of recent governmental expenditures for social services in developing countries, developed market economies, and centrally planned economies.

STATISTICAL INDICATORS FOR ASIA AND THE PACIFIC. Quarterly.

Quarterly report presenting selected economic and demographic indicators for 26 Asian and Pacific countries. Covers, for most countries, population size, birth and death rates, family planning methods, industrial and agricultural production, construction, transport, retail trade, foreign trade, prices, money supply, currency exchange rate, and GDP.

STATISTICAL YEARBOOK, 1979/80. 1981, UN, Sales No. E/F.81.XVII.1.

A comprehensive compendium of the most important internationally comparable data needed for the analysis of socioeconomic development at the world, regional and national levels. Includes tables (200) grouped in two sections: (a) world summary by regions (17 tables); and (b) remaining tables of country-by-country data, arranged in chapters: population; manpower; agriculture; forestry; fishing; industrial production; mining and quarrying; manufacturing; construction; development assistance; wholesale and retail trade; external trade; international tourism; transport; communications; national accounts; wages and prices; consumption; finance; energy; health; housing; science and technology; and culture. For this first time, this issue contains three new tables on industrial property: patents, industrial designs, and trademarks and service marks. Note: This issue is a special biennial edition, covering data through mid-1980, and in some cases for 1980 complete.

STATISTICAL YEARBOOK FOR ASIA AND THE PACIFIC, 1978. 1979, UN, Sales No. E-F.79.II.F.4.

Eleventh issue. Contains statistical indicators for the ESCAP region and statistics for period up to 1978 available at the end of 1978 for 34 countries and territories members of ESCAP, arranged by country, covering, where available: population; manpower; national accounts; agriculture, forestry, and fishing; industry; consumption; transport and communication; internal and external trade; wages, prices, and household expenditures; finance; and social statistics.

STATISTICAL YEARBOOK FOR LATIN AMERICA, 1979. 1981, UN, Sales No. E/S.80.II.G.4.

In two parts. Part 1 presents indicators of economic and social development in Latin America for 1960, 1965, 1970 and 1975-1978, including: population; demographic characteristics; employment and occupational structure; income distribution; living levels; consumption and nutrition; health; education; housing; global economic growth; agricultural activities; mining and energy resources; manufacturing; productivity; investment; saving; public financial resources; public expenditure; structure of exports and imports; intra-regional trade; transport services; tourist services; and external financing. Part 2 contains historical series in absolute figures for the years 1960, 1965 and 1970-1978 on population; national accounts; domestic prices; balance of payments; external indebtedness; external trade; natural resources and production of goods; infrastructure services; employment; and social conditions.

SURVEY OF ECONOMIC AND SOCIAL CONDITIONS IN AFRICA, 1980-81 AND OUTLOOK FOR 1981-82: SUMMARY

Examines growth in GDP, agricultural and industrial production, trade and balance of payments, resource flows, energy production/consumption, and selected other economic indicators, 1979-80, with outlook for 1981-82 and trends from 1960's.

TECHNICAL DATA SHEETS

Provides up-to-date information about projects as they are approved for World Bank and IDA financing. In addition to a description of the project, its total cost, and the amount of Bank financing, each technical data sheet describes the goods and services that must be provided for the project's implementation and gives the address of the project's implementing organization. On the average, 250 such sheets will be issued annually. Requests for sample copies are to be addressed to: Publications Distribution Unit, World Bank, 1818 H St., N.W., Washington, D.C. 20433, U.S.A.

UNESCO STATISTICAL YEARBOOK, 1978-79, 1266 p. 1980, UNESCO.

Composite: E/F/S (introductory texts). Presents statistical and other information for 206 countries on education; science and technology; libraries; museums and related institutions; theater and other dramatic arts; book production; newspapers and other periodicals; film and cinema; radio broadcasting; and television. In this edition, the summary tables relating to culture and communications, previously given in the introduction to each of the corresponding chapters, have been grouped together in a separate chapter.

World Bank. ANNUAL REPORT, 1982. 1982, WBG.

Presents summary and background of the activities of the World Bank Group during the fiscal year ended 30 June 1982, covering: the International Bank for Reconstruction and Development (IBRD); the International Development Association (IDA); and the International Finance Corporation (IFC). Chapters cover: brief review of Bank operations in fiscal 1982; a global perspective of the economic situation; Bank policies, activities and finances for fiscal 1982; 1982 regional perspectives; and Executive Directors. Lists projects approved for IBRD and IDA assistance in fiscal 1982 by sector, region and purpose. Also reviews trends in lending by sector for 1980-82 and includes statistical annex.

WORLD BANK COUNTRY STUDIES. Series.

Series of studies, prepared by World Bank staff, on development issues and policies, and economic conditions in individual developing countries. Studies may focus on specific economic sectors or issues, or on general economic performance of the country as a whole.

The World Bank.
WORLD DEVELOPMENT REPORT, 1978. August 1978.

First volume in a series of annual reports designed to provide a comprehensive, continuing assessment of global development issues. After an overview of development in the past 25 years, the report discusses current policy issues and projected developments in areas of the international economy that influence the prospects of developing countries. Analyzes the problems confronting policy makers in developing countries, which differ in degree and in kind, affecting the choice of appropriate policy instruments, and recognizes that development strategies need to give equal prominence to two goals: accelerating economic growth and reducing poverty. Reviews development priorities for low-income Asia, sub-Saharan Africa, and middle-income developing countries.

The World Bank.
WORLD DEVELOPMENT REPORT, 1979. Washington, D.C.

Second in a series of annual reports designed to assess global development issues. Focuses on development in the middle income countries, with particular emphasis on policy choices for industrialization and urbanization. Part one assesses recent trends and prospects to 1990 and discusses capital flows, and energy. Part two focuses upon structural change and development policy relevant to employment, the balance between agriculture and industry, and urban growth. Part three reviews development experiences and issues in three groups of middle income countries: semi-industrialized nations; mineral primary-producing countries; and predominantly agricultural primary-producing countries. Maintains that progress toward expanding employment and reducing poverty in developing countries lies not only in internal policy choices but also in a liberal environment for international trade and capital flows.

The World Bank.
WORLD DEVELOPMENT REPORT, 1980. New York: Oxford University Press for the World Bank, 1980.

Third in a series of annual reports. Parts one examines economic policy choices facing both developing and developed countries and their implications for national and regional growth. Projects, to the year 2000 but particularly to the mid to late 1980's, growth estimates for oil-importing and oil-exporting developing countries; and analyzes the fundamental issues of energy, trade, and capital flows. Part two focuses on the links between poverty, growth, and human development. It examines the impact of education, health, nutrition, and fertility on poverty; reviews some practical lessons in implementing human development programs; and discusses the trade-offs between growth and poverty and the allocation of resources between human development and other activities. Stresses the views that growth does not obviate the need for human development and that direct measures to reduce poverty do not obviate the need for economic expansion. Concludes that world growth prospects have deteriorated in the past year, but higher oil prices have impoved the outlook [for the first half of the 1980's] for the fifth of the developing world's population that lives in oil-exporting countries; however, the four-fifths that live in oil-importing countries will experience slower growth for the first half of the decade. Includes a statistical appendix to part one; a bibliographical note; and a very lengthy annex of World Development Indicators.

The World Bank.
WORLD DEVELOPMENT REPORT, 1981. New York: Oxford University Press for the World Bank, 1981.

With the major focus on the international context of development, examines past trends and future prospects for international trade, energy, and capital flows and the effects of these on developing countries. Presents two scenarios for the 1980's, one predicting higher growth rates than in the 1970's and one lower. Analyzes national adjustments to the international economy, presenting in-depth case studies. Concludes that countries pursuing outward-oriented policies adjusted more easily to external shocks. Contends that whichever scenario prevails, income differentials will increase between the industrial and developing countries. Low income countries have fewer options and less flexibility of adjustment, therefore requiring continued aid from the more affluent countries. Advocates policies to channel increased resources to alleviate poverty.

The World Bank.
WORLD DEVELOPMENT REPORT, 1982.

The Report this year focuses on agriculture and food security. As in previous years there is also a section on global prospects and international issues, as well as the statistical annex of World Development indicators.

The World Bank.
WORLD TABLES 1980: FROM THE DATA FILES OF THE WORLD BANK. Second edition. Baltimore and London: Johns Hopkins University Press for the World Bank, 1980.

A broad range of internationally comparable statistical information drawn from the World Bank data files. Includes historical time series for individual countries in absolute numbers for most of the basic economic indicators for selected years (1950-77 when available); also presents derived economic indicators for selected periods of years and demographic and social data for selected years. Although the number of social indicators is fewer than those in the 1976 edition the quality of the data has been improved through the use of more uniform definitions and concepts, greater attention to population statistics, and better statistics on balance of payments and central government finance. Includes an index of country coverage.

WORLD ECONOMIC OUTLOOK: A SURVEY BY THE STAFF OF THE INTERNATIONAL MONETARY FUND. 1980, IMF.

An in-depth forecast of the world economy in 1980 and a preliminary summary for 1981. Chapters discuss: a profile of current situation and short-term prospects; global perspectives for adjustment and financing; industrial countries; developing countries--oil-exporting and non-oil groups; and key policy issues. Appendixes include country and regional surveys; technical notes on the world oil situation, estimated impact of fiscal balances in selected industrial countries, and monetary policy and inflation; and statistical tables.

WORLD ECONOMIC OUTLOOK: A SURVEY BY THE STAFF OF THE INTERNATIONAL MONETARY FUND. [1982 ed.] 1982, IMF.

A comprehensive analysis of economic developments, policies, and prospects through June 1981 for industrial, oil exporting and non-oil developing countries. It highlights persistent imbalances in the world economy, high inflation, rising unemployment, excessive rates of real interest, and unstable exchange rates. Appendix A includes supplementary notes providing information on selected topics in greater depth or detail than in the main body of the report: country and regional surveys; medium-term scenarios; fiscal development; monetary and exchange rate development; world oil situation; growth and inflation in non-oil developing countries; developments in trade policy; and commodity price developments and prospects. Appendix B presents statistical tables on: domestic economic activity and prices; international trade; balance of payments; external debt; medium-term projections; and country tables.

The World Bank.
WORLD BANK ATLAS. Fourteenth edition. 1979. Annual.

Presents estimates of gross national product (GNP) per capita (1977), GNP per capita growth rates (1970-77), and population (mid-1977), with population growth rates (1970-77) for countries with populations of one million or more in three global maps; a computer-generated map shows GNP per capita (1977) by major regions. Six regional maps give the same data for 184 countries and territories, as well as preliminary data for 1978. The base years 1976-78 have been used for the conversion of GNP for both 1977 and 1978. A Technical Note explains in detail the methodology used.

The World Bank.
WORLD DEVELOPMENT INDICATORS. June 1979. 71 pages.

A volume of statistics prepared in conjunction with and constituting the Annex to the World Development Report, 1979 to provide information of general relevance about the main features of economic and social development, reporting data for a total of 125 countries whose population exceeds one million. Countries are grouped in five categories and ranked by their 1977 per capita gross national product (GNP) levels. The volume contains 24 tables covering some 110 economic and social indicators. The choice of indicators has been based on data being available for a large number of countries, the availability of historical series to allow the measurement of growth and change, and on the relevance of data to the principal processes of development.

The World Bank.
WORLD ECONOMIC AND SOCIAL INDICATORS. Quarterly (current issues).

Presents most recent available data on trade, commodity prices, consumer prices, debt and capital flows, industrial production, as well as social indicators and select annual data (by countries where applicable). Each issue contains an article on topics of current importance. Strategies for improving the access to education of the disadvantaged rural poor by serving areas out of range of existing schools are discussed and programs in four projects financed by the World Bank are described.

WORLD ECONOMIC OUTLOOK: A SURVEY BY THE STAFF OF THE INTERNATIONAL MONETARY FUND. Annual. April 1982. (Occasional Paper No. 9)

Annual report on economic performance of major industrial and oil exporting and non-oil developing countries, 1970's-81 and forecast 1982-83, with some projections to 1986. Includes analysis of economic indicators for selected industrial countries, world economic groupings, and world areas, primarily for IMF member countries. Covers domestic economic activity, including prices, GNP, and employment; international trade; balance of payments; and foreign debt. Also includes financial indicators for selected industrial countries, including government budget surpluses and deficits, savings, money supply, and interest rates.

WORLD ECONOMIC SURVEY. 1978, UN, Sales No. E.78.II.C.1.

Provides an overview of salient developments in the world economy in 1977 and the outlook for 1978. Focuses on policy needs for improving the tempo of world production and trade. Examines in detail the course of production and trade and related variables in the developing economies, the developed market economies, and the centrally planned economies.

WORLD ECONOMIC SURVEY 1979-80. 1980, UN, Sales No. E.80.II.C.2.

A survey of current world economic conditions and trends, with chapters on salient features and policy implications; the growth of world output, 1979-1980; the accelerating pace of inflation; world trade and international payments; world economic outlook, 1980-1985; and adjustment policies in developing countries. Annexes cover external factors and growth in developing countries--the experience of the 1970's; supply and price of petroleum in 1979 and 1980; and prospective supply and demand for oil.

YEARBOOK OF NATIONAL ACCOUNTS STATISTICS, 1980. Annual. 1982.

Annual report presenting national income and product account balances for approximately 170 countries, and for world areas and economic groupings, selected years 1970-79, often with comparisons to 1960 and 1965. Data are compiled in accordance with the UN System of National Accounts (SNA) for market economies, and the System of Material Product Balances (MPS) for centrally planned economies. SNA data include GDP final consumption expenditures by type; production, income/outlay, and capital formation accounts, by institutional sector; and production by type of activity. MPS data include material and financial balances, manpower and resources, and national wealth and capital assets.

STATISTICAL NEWSLETTER. Quarterly. Approx. 15 p.

Quarterly newsletter on ESCAP statistical programs and activities, and major statistical developments in ESCAP countries. Includes brief descriptions of meetings, working groups, upcoming international statistical training programs, and regional advisory services; and an annotated bibliography of recent ESCAP and UN statistical publications.

INVENTORY OF DATA SOURCES IN SCIENCE AND TECHNOLOGY: A PRELIMINARY SURVEY. Monograph. 1982.

Directory of institutional data sources for research in agricultural sciences, renewable energy resources, and various related technical fields, by country; as of 1980-81. Also includes brief bibliographies of methodological and other publications in each subject area. Data sources: Inventory compiled by the Committee on Data for Science and Technology of the International Council of Scientific Unions.

II. STATISTICAL TABLES AND FIGURES

In this section, an attempt is made to provide the reader with an overview of global trends related to technology policy and development from a Third world perspective, based on an analysis of the country data, as it is sometimes difficult to form any such general impression when faced with a general body of highly detailed data. Statistical information in the following pages have been reproduced from the following sources:

WORLD TABLES 1980: FROM THE DATA FILES OF THE WORLD BANK, Baltimore: Johns Hopkins University Press for the World Bank, 1980. (Reprinted by permission of the World Bank and Johns Hopkins University Press.)

WORLD ECONOMIC AND SOCIAL INDICATORS, October 1978. Report No. 700/78/04. Washington, D.C: WORLD BANK.

PLANNING THE TECHNOLOGICAL TRANSFORMATION OF DEVELOPING COUNTRIES, Study by the UNCTAD Secretariat, New York: United Nations, 1981.

GLOBAL INDICATORS

TABLE 1. SOCIAL INDICATORS BY INCOME GROUP OF COUNTRIES

(ADJUSTED COUNTRY GROUP AVERAGES)

INDICATOR	DEVELOPING COUNTRIES EXCLUDING CAPITAL SURPLUS OIL EXPORTERS								
	LOW INCOME			LOWER MIDDLE INCOME			INTERMEDIATE MIDDLE INCOME		
	1960	1970	MOST RECENT ESTIMATE	1960	1970	MOST RECENT ESTIMATE	1960	1970	MOST RECENT ESTIMATE
GNP PER CAPITA (IN CURRENT US $)	67.4	107.4	162.0	136.1	239.6	398.6	225.6	410.9	817.9
POPULATION									
GROWTH RATE (%) - TOTAL	2.2	2.4	2.4	2.7	2.7	2.6	2.7	2.7	2.5
- URBAN	5.3	4.7	4.7	4.7	4.4	9.8	5.4	4.9	5.1
URBAN POPULATION (% OF TOTAL)	10.4	14.0	14.8	17.7	21.6	26.1	33.7	41.4	46.1
VITAL STATISTICS									
CRUDE BIRTH RATE (PER 1000)	47.5	46.9	45.2	47.1	45.5	42.6	44.6	41.2	38.2
CRUDE DEATH RATE (PER 1000)	26.1	21.7	18.2	21.4	16.1	12.7	18.6	13.5	11.1
GROSS REPRODUCTION RATE	2.9	3.1	3.1	3.4	3.2	3.3	3.0	2.8	2.6
EMPLOYMENT AND INCOME									
DEPENDENCY RATIO - AGE	0.8	0.9	0.9	0.9	0.9	0.9	0.9	0.9	0.9
- ECONOMIC	1.0	1.1	1.1	1.3	1.4	1.4	1.6	1.6	1.5
LABOR FORCE IN AGRICULTURE (% OF TOTAL)	65.4	62.0	59.3	70.4	65.9	62.5	62.8	54.5	47.0
UNEMPLOYED (% OF LABOR FORCE)	4.7	4.0	3.1	6.3	8.4	5.3	6.1	6.0	5.8
INCOME RECEIVED BY - HIGHEST 5%	24.5	23.3	20.3	25.1	23.1	25.5	31.5	27.0	19.3
- LOWEST 20%	4.6	5.1	6.5	4.8	4.9	4.8	4.4	3.9	5.7
HEALTH AND NUTRITION									
DEATH RATE (PER 1000) AGES 1-4 YEARS	43.6	33.0	33.0	9.3	6.8	7.5	8.5	3.6	2.7
INFANT MORTALITY RATE (PER 1000)	129.0	121.3	102.8	84.6	79.9	58.4	88.6	65.8	55.0
LIFE EXPECTANCY AT BIRTH (YRS)	39.2	43.8	46.0	45.0	50.8	53.2	51.1	57.0	59.1
POPULATION PER - PHYSICIAN	21790.7	15219.9	13235.9	16767.4	11977.3	10586.0	3299.7	2549.2	2412.7
- NURSING PERSON	8472.3	5215.0	4830.9	4078.2	1921.9	1683.8	3394.0	2205.1	1502.1
- HOSPITAL BED	1386.7	1267.8	1236.2	1037.6	815.3	793.1	721.8	629.2	507.1
PER CAPITA PER DAY SUPPLY OF:									
CALORIES (% OF REQUIREMENTS)	89.8	91.5	94.5	85.1	93.3	102.3	94.4	101.8	103.9
PROTEIN (GRMS) - TOTAL	50.5	51.6	53.9	47.4	53.0	56.9	54.4	58.7	60.6
- FROM ANIMALS & PULSES	14.9	14.4	16.4	17.7	18.1	18.8	21.8	22.1	23.0

(ADJUSTED COUNTRY GROUP AVERAGES)

INDICATOR	DEVELOPING COUNTRIES EXCLUDING CAPITAL SURPLUS OIL EXPORTERS								
	LOW INCOME			LOWER MIDDLE INCOME			INTERMEDIATE MIDDLE INCOME		
	1960	1970	MOST RECENT ESTIMATE	1960	1970	MOST RECENT ESTIMATE	1960	1970	MOST RECENT ESTIMATE
EDUCATION									
ADJ. ENROLLMENT RATIOS - PRIMARY	37.4	48.4	59.0	60.7	74.0	92.7	77.8	95.3	99.9
- SECONDARY	4.8	10.3	13.9	4.8	12.7	22.6	14.5	26.7	29.4
FEMALE ENROLLMENT RATIO (PRIMARY)	34.6	39.0	43.3	45.6	75.0	77.5	65.8	87.8	87.6
ADULT LITERACY RATE (%)	24.4	32.0	33.8	41.0	60.0	63.0	49.8	57.8	62.3
HOUSING									
PERSONS PER ROOM - URBAN	2.5	2.0	2.8	2.6	2.5	2.2	2.3	2.2	1.6
OCCUPIED DWELLINGS WITHOUT WATER	62.2	69.8	..	68.7	64.6	67.8	74.6	64.2	58.9
ACCESS TO ELECTRICITY (%) - ALL	..	..	..	17.3	23.3	40.4	28.4	49.6	71.9
- RURAL	..	..	..	..	..	..	5.6	26.7	34.1
CONSUMPTION									
RADIO RECEIVERS (PER 1000 POP.)	4.5	14.4	23.1	11.9	62.3	70.4	48.8	96.2	102.6
PASSENGER CARS (PER 1000 POP.)	1.3	2.5	3.0	3.0	6.5	8.6	4.2	7.5	11.1
ENERGY (KG COAL/YR PER CAPITA)	62.0	83.4	104.8	99.6	220.1	265.2	258.7	489.2	586.2
NEWSPRINT (KG/YR PER CAPITA)	0.2	0.4	0.3	0.6	0.8	0.8	1.1	1.8	2.4

(CONTINUED)

(ADJUSTED COUNTRY GROUP AVERAGES)

INDICATOR	DEV'G CTRIES. EXCL. CAP. SURP. OIL EXP.											
	UPPER MIDDLE INCOME			HIGH INCOME			CAP. SURP. OIL EXP.			INDUSTRIALIZED COUNTRIES		
	1960	1970	MOST RECENT EST.	1960	1970	MOST RECENT EST.	1960	1970	MOST RECENT EST.	1960	1970	MOST RECENT EST.
GNP PER CAPITA (IN CURRENT US $)	401.2	817.1	1648.7	689.4	1564.2	2911.1	1054.3	2858.9	5710.5	1417.4	3096.8	5297.7
POPULATION												
GROWTH RATE (%) - TOTAL	1.6	1.3	1.5	2.1	3.1	2.9	4.5	2.4	2.7	0.9	0.9	0.9
- URBAN	3.3	3.4	2.8	4.5	3.9	3.5	..	5.8	6.8	1.6	1.3	1.3
URBAN POPULATION (% OF TOTAL)	43.4	51.1	53.1	63.0	82.1	88.6	24.6	20.0	39.0	66.1	70.5	73.8
VITAL STATISTICS												
CRUDE BIRTH RATE (PER 1000)	26.2	28.5	20.8	41.7	37.4	33.6	48.3	49.4	45.0	21.3	20.0	18.7
CRUDE DEATH RATE (PER 1000)	10.4	9.1	8.9	9.6	8.3	8.0	21.2	22.8	14.7	9.7	9.0	8.8
GROSS REPRODUCTION RATE	1.7	1.8	1.8	2.3	2.5	1.8	..	3.5	3.3	1.3	1.3	1.2

TABLE 1. Social Indicators by Income Group of Countries (Continued)

(ADJUSTED COUNTRY GROUP AVERAGES)

INDICATOR	DEV'G CTRIES. EXCL. CAP. SURP. OIL EXP. UPPER MIDDLE INCOME			HIGH INCOME			CAP. SURP. OIL EXP.			INDUSTRIALIZED COUNTRIES		
	1960	1970	MOST RECENT EST.	1960	1970	MOST RECENT EST.	1960	1970	MOST RECENT EST.	1960	1970	MOST RECENT EST.
EMPLOYMENT AND INCOME												
DEPENDENCY RATIO - AGE	0.7	0.7	0.6	0.8	0.6	0.6	0.9	0.9	0.9	0.5	0.6	0.4
- ECONOMIC	1.3	1.7	1.6	1.2	1.2	1.2	1.8	1.7	1.7	0.9	0.9	0.8
LABOR FORCE IN AGRICULTURE (% OF TOTAL)	48.5	42.5	36.3	26.1	17.8	21.0	54.7	44.5	29.0	19.8	13.2	10.0
UNEMPLOYED (% OF LABOR FORCE)	7.4	3.3	4.0	9.0	5.4	5.1	7.4	2.0	3.0	2.1	1.5	1.9
INCOME RECEIVED BY - HIGHEST 5%	32.5	28.2	21.3	18.9	16.1	..	13.3	..	..	19.3	14.0	15.5
- LOWEST 20%	4.2	3.8	4.7	5.8	6.6	..	10.1	..	..	4.2	7.0	5.7
HEALTH AND NUTRITION												
DEATH RATE (PER 1000) AGES 1-4 YEARS	4.8	2.9	1.9	..	1.3	..	..	3.6	3.6	1.2	0.9	0.8
INFANT MORTALITY RATE (PER 1000)	74.4	51.3	37.9	44.9	27.8	23.2	..	134.3	80.3	27.9	17.0	15.0
LIFE EXPECTANCY AT BIRTH (YRS)	64.6	67.3	68.4	66.2	64.0	68.2	45.4	44.9	52.9	69.5	71.4	72.5
POPULATION PER - PHYSICIAN	1625.8	967.5	718.2	1117.5	888.9	756.9	9833.7	6323.4	1260.0	895.2	825.6	656.0
- NURSING PERSON	1690.7	1279.6	1028.5	1165.0	605.9	683.5	5140.0	2856.8	460.0	279.6	194.6	167.1
- HOSPITAL BED	209.9	180.4	185.8	170.0	162.5	170.0	1093.2	727.5	230.0	96.1	86.0	81.9
PER CAPITA PER DAY SUPPLY OF:												
CALORIES (% OF REQUIREMENTS)	104.5	114.4	111.5	106.3	107.2	113.6	83.9	90.3	104.9	118.7	118.7	119.5
PROTEIN (GRMS) - TOTAL	75.5	84.9	77.8	78.5	79.2	89.9	53.6	57.0	65.1	90.3	94.1	94.8
- FROM ANIMALS & PULSES	27.0	29.0	27.5	33.0	40.1	48.0	11.0	14.8	18.2	49.7	55.0	54.9
EDUCATION												
ADJ. ENROLLMENT RATIOS - PRIMARY	94.6	97.9	95.7	104.4	120.1	107.6	18.2	47.1	145.0	106.7	104.3	103.3
- SECONDARY	22.7	36.6	46.7	18.1	40.1	46.2	3.1	12.4	47.0	59.5	79.1	79.8
FEMALE ENROLLMENT RATIO (PRIMARY)	89.7	87.9	86.1	100.2	100.0	102.0	3.5	31.6	40.4	111.4	104.6	104.0
ADULT LITERACY RATE (%)	51.4	67.8	66.1	81.8	86.2	87.2	25.2	17.1	..	98.0	99.0	99.0
HOUSING												
PERSONS PER ROOM - URBAN	1.4	1.2	..	1.1	..	..	..	1.9	..	0.8	0.7	0.9
OCCUPIED DWELLINGS WITHOUT WATER	59.1	75.3	67.1	57.1	20.0	..	..	69.0	..	7.2	3.1	4.3
ACCESS TO ELECTRICITY (%) - ALL	50.6	47.4	59.8	79.3	91.0	..	..	24.0	..	97.3	98.9	99.1
- RURAL	26.9	..	..	57.0	58.0	..	..	..	..	91.4	95.2	94.8
CONSUMPTION												
RADIO RECEIVERS (PER 1000 POP.)	76.3	137.2	200.4	170.7	174.3	185.6	13.8	17.5	18.4	277.1	359.7	379.3
PASSENGER CARS (PER 1000 POP.)	11.2	29.2	42.3	14.1	41.3	54.4	7.6	16.6	113.7	90.7	233.3	266.5
ENERGY (KG COAL/YR PER CAPITA)	676.2	1426.1	1618.7	798.4	1755.1	2467.6	302.5	1003.1	1419.4	2624.7	4575.2	4997.3
NEWSPRINT (KG/YR PER CAPITA)	1.4	1.9	2.3	3.5	8.7	6.6	0.2	0.2	0.1	16.4	22.3	22.2

GLOBAL INDICATORS

	EUROPE			LATIN AMERICA & CARIBBEAN			N. AFRICA & MIDDLE EAST		
	1960	1970	MOST RECENT ESTIMATE	1960	1970	MOST RECENT ESTIMATE	1960	1970	MOST RECENT ESTIMATE
GNP PER CAPITA (IN CURRENT US $)	496.6	1018.0	2070.3	362.0	626.8	1015.6	307.7	579.0	1290.3
POPULATION									
GROWTH RATE (%) - TOTAL	1.0	0.8	0.9	2.5	2.7	2.6	2.7	2.9	3.0
- URBAN	3.8	2.8	2.1	4.2	4.1	4.2	6.4	4.5	5.1
URBAN POPULATION (% OF TOTAL)	32.1	40.7	38.7	48.3	54.3	58.5	33.8	39.6	44.3
VITAL STATISTICS									
CRUDE BIRTH RATE (PER 1000)	23.3	20.5	19.2	40.9	39.0	36.8	48.3	47.2	45.7
CRUDE DEATH RATE (PER 1000)	10.5	9.0	9.0	14.1	10.9	9.2	22.6	18.0	15.3
GROSS REPRODUCTION RATE	1.4	1.3	1.3	2.7	2.6	2.6	2.3	3.4	3.4
EMPLOYMENT AND INCOME									
DEPENDENCY RATIO - AGE	0.6	0.6	0.4	0.9	1.0	0.9	0.9	1.0	1.0
- ECONOMIC	1.0	1.1	1.0	1.6	1.5	1.7	1.6	2.0	1.9
LABOR FORCE IN AGRICULTURE (% OF TOTAL)	47.9	31.8	27.4	48.3	41.0	36.9	52.6	43.4	42.9
UNEMPLOYED (% OF LABOR FORCE)	3.0	4.0	6.0	7.6	6.2	8.8	6.3	3.4	4.1
INCOME RECEIVED BY - HIGHEST 5%	21.8	24.5	25.0	37.1	30.4	31.7	24.0	25.0	21.0
- LOWEST 20%	5.4	3.9	3.9	3.9	3.5	2.0	4.4	4.2	5.2
HEALTH AND NUTRITION									
DEATH RATE (PER 1000) AGES 1-4 YEARS	4.7	2.8	1.7	10.6	7.7	6.6	..	6.0	..
INFANT MORTALITY RATE (PER 1000)	60.4	39.7	34.5	77.4	67.3	56.2	127.8	111.6	97.8
LIFE EXPECTANCY AT BIRTH (YRS)	65.8	68.6	69.1	55.8	60.6	62.5	45.5	50.3	52.8
POPULATION PER - PHYSICIAN	1004.0	821.3	694.4	2058.1	1866.8	1796.9	5690.8	5760.2	4724.7
- NURSING PERSON	1343.2	653.9	339.2	4542.1	3389.5	2804.5	3286.6	2564.7	2383.1
- HOSPITAL BED	190.5	168.0	170.5	444.1	392.3	405.6	670.8	661.6	700.0
PER CAPITA PER DAY SUPPLY OF:									
CALORIES (% OF REQUIREMENTS)	109.3	118.0	118.0	97.6	103.2	105.5	80.9	91.0	96.0
PROTEIN (GRMS) - TOTAL	85.9	90.7	90.0	63.7	59.8	60.7	54.5	58.3	63.1
- FROM ANIMALS & PULSES	27.0	29.0	33.0	29.0	28.0	28.2	17.5	15.0	15.6
EDUCATION									
ADJ. ENROLLMENT RATIOS - PRIMARY	105.0	102.1	104.3	85.0	101.7	105.1	51.5	75.6	80.5
- SECONDARY	25.5	50.5	49.2	15.0	27.6	36.0	10.3	20.4	22.2
FEMALE ENROLLMENT RATIO (PRIMARY)	98.9	99.4	100.4	85.6	98.3	98.1	30.8	50.2	52.3
ADULT LITERACY RATE (%)	64.9	75.0	88.2	61.4	74.6	75.7	17.7	26.9	40.6
HOUSING									
PERSONS PER ROOM - URBAN	1.4	1.5	1.4	1.9	1.3	2.1	1.8	2.3	3.0
OCCUPIED DWELLINGS WITHOUT WATER	67.0	63.3	59.5	65.5	67.0	66.4	62.2	77.1	90.5
ACCESS TO ELECTRICITY (%) - ALL	51.4	46.3	57.9	44.4	54.2	53.1	40.1	31.0	39.1
- RURAL	18.1	20.9	33.8	9.3	12.5	12.6	..	..	..
CONSUMPTION									
RADIO RECEIVERS (PER 1000 POP.)	93.9	176.7	193.1	101.6	166.2	177.5	49.7	79.1	120.1
PASSENGER CARS (PER 1000 POP.)	6.8	49.9	90.0	11.7	22.1	29.9	7.5	11.5	15.4
ENERGY (KG COAL/YR PER CAPITA)	673.4	1334.0	1829.4	577.2	821.0	929.2	263.7	281.2	750.2
NEWSPRINT (KG/YR PER CAPITA)	2.3	3.2	3.3	2.2	3.1	3.0	0.3	0.3	0.3

TABLE 2. Social Indicators by Geographic Areas (Developing Countries), Continued

(ADJUSTED COUNTRY GROUP AVERAGES)

INDICATOR	AFRICA SOUTH OF SAHARA			SOUTH ASIA			EAST ASIA AND PACIFIC		
	1960	1970	MOST RECENT ESTIMATE	1960	1970	MOST RECENT ESTIMATE	1960	1970	MOST RECENT ESTIMATE
GNP PER CAPITA (IN CURRENT US $)	94.9	137.0	207.4	54.1	88.2	131.4	141.8	290.0	568.3
POPULATION									
GROWTH RATE (%) - TOTAL	2.2	2.4	2.6	2.2	2.6	2.1	3.0	2.8	2.3
- URBAN	5.6	6.0	6.0	5.2	4.1	4.3	5.4	5.0	5.2
URBAN POPULATION (% OF TOTAL)	9.1	12.5	13.5	7.8	9.8	12.4	28.1	27.1	38.1
VITAL STATISTICS									
CRUDE BIRTH RATE (PER 1000)	48.8	48.1	47.1	47.4	45.8	45.1	42.3	40.7	32.0
CRUDE DEATH RATE (PER 1000)	26.7	23.7	21.2	26.4	21.4	17.3	19.2	12.4	8.7
GROSS REPRODUCTION RATE	2.9	3.1	3.0	3.2	3.0	2.9	3.0	2.5	2.3
EMPLOYMENT AND INCOME									
DEPENDENCY RATIO - AGE	0.9	0.9	0.9	0.8	0.8	0.8	0.9	0.9	0.7
- ECONOMIC	1.1	1.1	1.1	1.5	1.4	1.2	1.4	1.4	1.3
LABOR FORCE IN AGRICULTURE (% OF TOTAL)	79.8	75.0	73.1	61.8	60.8	63.0	67.9	59.5	48.4
UNEMPLOYED (% OF LABOR FORCE)	5.1	4.6	5.1	..	..	11.0	5.1	5.1	4.1
INCOME RECEIVED BY - HIGHEST 5%	28.2	26.4	25.7	24.6	23.2	18.6	22.7	20.5	19.8
- LOWEST 20%	5.2	3.9	5.7	4.6	5.2	7.8	5.5	5.8	6.6
HEALTH AND NUTRITION									
DEATH RATE (PER 1000) AGES 1-4 YEARS	..	..	..	..	..	..	..	3.4	2.0
INFANT MORTALITY RATE (PER 1000)	153.9	129.6	127.5	136.2	124.3	104.0	61.2	31.1	27.4
LIFE EXPECTANCY AT BIRTH (YRS)	36.9	41.5	43.4	40.6	45.2	48.1	52.5	59.1	61.6
POPULATION PER - PHYSICIAN	31866.1	24906.5	21616.5	9920.9	8519.2	7412.6	3429.3	2268.9	2208.9
- NURSING PERSON	4558.4	3088.7	2496.5	14566.1	9168.6	8339.3	3096.6	1935.5	1465.5
- HOSPITAL BED	1234.7	819.9	799.9	2885.8	1998.5	1908.0	1270.8	921.7	662.1
PER CAPITA PER DAY SUPPLY OF:									
CALORIES (% OF REQUIREMENTS)	89.6	90.7	91.9	89.1	97.6	96.0	90.0	99.4	106.5
PROTEIN (GRMS) - TOTAL	56.6	59.0	60.6	47.8	53.2	50.8	48.1	53.4	55.6
- FROM ANIMALS & PULSES	19.2	19.8	23.1	15.0	16.0	15.5	19.0	22.1	22.1
EDUCATION									
ADJ. ENROLLMENT RATIOS - PRIMARY	27.7	42.4	50.0	36.9	47.9	55.2	95.0	105.7	110.0
- SECONDARY	1.9	5.6	6.9	9.1	15.5	20.0	17.3	26.9	51.1
FEMALE ENROLLMENT RATIO (PRIMARY)	21.7	37.4	43.2	22.2	53.8	44.5	88.5	102.0	104.7
ADULT LITERACY RATE (%)	9.8	17.4	18.4	16.0	20.0	21.0	47.7	66.4	72.6
HOUSING									
PERSONS PER ROOM - URBAN	2.7	2.4	1.7	..	..	..	2.5	2.3	..
OCCUPIED DWELLINGS WITHOUT WATER	..	..	..	..	..	..	83.7	69.5	60.3
ACCESS TO ELECTRICITY (%) - ALL	..	..	..	..	..	..	22.6	40.7	50.5
- RURAL	..	..	..	..	..	..	12.0	20.1	23.4
CONSUMPTION									
RADIO RECEIVERS (PER 1000 POP.)	5.2	21.2	25.4	6.0	14.3	17.4	34.6	97.5	90.3
PASSENGER CARS (PER 1000 POP.)	2.1	3.4	3.9	0.9	2.2	2.1	2.7	7.0	9.3
ENERGY (KG COAL/YR PER CAPITA)	37.4	73.1	75.0	63.7	89.9	133.3	140.9	382.7	432.9

TABLE 3. COMPARATIVE SOCIAL INDICATORS FOR DEVELOPING COUNTRIES (BY GEOGRAPHIC AREA AND COUNTRY)

(MOST RECENT ESTIMATE)

AREA AND COUNTRY	POPULATION & VITAL STATISTICS				EMPLOYMENT AND INCOME			HEALTH & NUTRITION			EDUCATION		
	POP. GROWTH RATE % (65-75)	URBAN POP. % OF TOTAL	CRUDE BIRTH RATE (/000)	CRUDE DEATH RATE (/000)	LABOR IN AGR. % OF TOTAL	INCOME RECD BY HIGHEST 5% HH	INCOME RECD BY LOWEST 20% HH	LIFE EXPECT. YRS AT BIRTH	CALORIE SUPPLY %/CAP REQD.	PROTEIN SUPPLY GR/DAY /CAP	PRIMARY SCHOOL ENROLL RATIO %	FEMALE ENROLL. RATIO PRIMARY	ADULT LITERACY RATE % OF TOTAL
EUROPE													
CYPRUS	0.6	42.2	22.2	6.8	34.0	12.1	7.9	71.4	113.0	86.0	71.0	72.0	85.0
GREECE	0.6	64.8	15.4	9.4	34.0	18.7	6.3	71.8	132.0	102.0	106.0	104.0	82.0
MALTA	0.2	94.3	17.5	9.0	6.0	..	..	69.6	114.0	89.0	109.0	109.0	87.0
PORTUGAL	0.3	28.8	19.2	10.5	32.5	56.3	7.3	68.7	118.0	85.0	116.0	94.0	70.0
ROMANIA	1.2	43.0	19.7	9.3	36.0	..	..	69.1	118.0	90.0	109.0	109.0	98.0
SPAIN	1.0	59.1	19.5	8.3	23.0	18.5	6.0	72.1	135.0	94.1	115.0	115.0	94.0
TURKEY	2.6	42.6	39.4	12.5	52.5	28.0	3.5	56.9	113.0	75.7	104.0	94.0	55.0
YUGOSLAVIA	0.9	38.7	18.2	9.2	39.0	25.1	6.6	68.0	137.0	97.5	97.0	93.0	85.0
ALL COUNTRIES - MEDIAN	0.8	42.8	19.4	9.3	34.0	18.6	6.5	69.4	118.0	89.5	107.5	99.0	85.0
LATIN AMERICA & CARIBBEAN													
ARGENTINA	1.4	80.0	21.8	8.8	15.0	21.4	5.6	68.3	129.0	107.1	108.0	109.0	93.0
BAHAMAS	3.6	57.9	22.4	5.7	7.0	20.7	3.4	66.7	100.0	87.0	135.0	..	93.0
BARBADOS	0.4	3.7	21.6	8.9	18.0	19.8	6.8	69.1	133.0	82.5	117.0	116.0	97.0
BOLIVIA	2.7	34.0	44.0	19.1	65.0	36.0	4.0	46.8	77.0	48.5	74.0	65.0	40.0
BRAZIL	2.9	59.1	37.1	8.8	37.8	35.0	3.0	61.4	105.0	62.1	90.0	90.0	64.0
CHILE	1.9	83.0	27.9	9.2	19.0	31.0	4.8	62.6	116.0	78.3	119.0	118.0	90.0
COLOMBIA	2.8	70.0	40.6	8.8	39.0	27.2	5.2	60.9	94.0	47.0	105.0	108.0	81.0
COSTA RICA	2.8	40.6	31.0	5.8	36.4	22.8	5.4	69.1	113.0	60.8	109.0	109.0	89.0
DOMINICAN REPUBLIC	2.9	45.9	45.8	11.0	53.8	26.3	4.3	57.8	98.0	45.4	104.0	105.0	51.0
ECUADOR	3.4	41.6	41.8	9.5	43.5	..	..	59.6	93.0	47.4	102.0	100.0	69.0
EL SALVADOR	3.4	39.4	42.2	11.1	55.0	38.0	2.0	65.0	84.0	50.3	75.2	69.0	63.0
GRENADA	1.0	..	27.4	6.8	30.8	..	..	..	89.0	57.0	99.0	..	85.0
GUATEMALA	3.2	37.3	42.8	13.7	56.0	35.0	5.0	54.1	91.0	52.8	62.0	56.0	47.0
GUYANA	2.1	40.0	32.4	5.9	30.9	18.8	4.3	67.9	104.0	58.0	114.0	114.0	85.0
HAITI	1.6	23.1	35.8	16.3	77.0	..	..	50.0	90.0	39.0	70.0	37.0	20.0
HONDURAS	2.7	31.4	49.3	14.6	60.3	28.0	2.5	53.5	90.0	56.0	90.0	88.0	53.0
JAMAICA	1.7	37.1	32.2	7.1	26.9	30.2	2.2	69.5	118.0	68.9	111.0	112.0	86.0
MEXICO	3.5	63.3	42.0	8.6	41.0	27.9	3.4	64.7	117.0	66.9	112.0	109.0	76.0
NICARAGUA	3.6	48.0	48.3	13.9	48.0	42.4	3.1	52.9	105.0	68.4	85.0	87.0	57.0
PANAMA	3.2	49.6	36.2	7.1	30.0	22.2	4.6	66.5	105.0	61.0	124.0	120.0	82.2
PARAGUAY	2.6	37.4	39.8	8.9	49.0	30.0	4.0	61.9	118.0	74.5	106.0	102.0	81.0
PERU	2.9	55.3	41.0	11.9	40.0	28.8	3.1	55.7	100.0	61.7	111.0	106.0	72.0
TRINIDAD & TOBAGO	1.0	25.1	27.3	5.9	13.5	..	..	69.5	114.0	66.0	111.0	111.0	90.0
URUGUAY	0.4	80.6	20.4	9.3	13.2	19.0	4.4	69.8	116.0	98.1	95.0	94.0	94.0
VENEZUELA	3.3	75.7	36.1	7.0	21.0	21.8	3.6	66.4	98.0	63.1	96.0	96.0	82.0
ALL COUNTRIES - MEDIAN	2.7	43.7	36.2	8.9	37.8	27.9	4.0	63.0	104.0	61.7	105.0	105.0	81.0

(CONTINUED)

TABLE 3. Comparative Social Indicators for Developing Countries (By Geographic Area and Country) **Continued**

(MOST RECENT ESTIMATE)

AREA AND COUNTRY	POPULATION & VITAL STATISTICS				EMPLOYMENT AND INCOME			HEALTH & NUTRITION			EDUCATION		
	POP. GROWTH RATE % (65-75)	URBAN POP. % OF TOTAL	CRUDE BIRTH RATE (/000)	CRUDE DEATH RATE (/000)	LABOR IN AGR. % OF TOTAL	INCOME RECD BY HIGHEST 5% HH	INCOME RECD BY LOWEST 20% HH	LIFE EXPECT. YRS AT BIRTH	CALORIE SUPPLY %/CAP REQD.	PROTEIN SUPPLY GR/DAY /CAP	PRIMARY SCHOOL ENROLL RATIO %	FEMALE ENROLL. RATIO PRIMARY	ADULT LITERACY RATE % OF TOTAL
NORTH AFRICA & MIDDLE EAST													
ALGERIA	3.3	39.9	48.7	15.4	42.8	..	..	53.3	88.0	57.2	77.0	72.0	35.0
BAHRAIN	3.3	78.1	49.6	18.7	..	..	..	44.5	..	..	..	..	..
EGYPT	2.4	44.6	37.8	14.0	43.9	21.0	5.2	52.4	113.0	70.7	72.0	55.0	40.0
IRAN	2.9	43.0	45.3	15.6	41.0	29.7	4.0	51.0	98.0	56.0	90.0	67.0	50.0
IRAQ	3.3	62.0	48.1	14.6	51.0	35.1	2.1	52.7	101.0	60.4	93.0	63.0	26.0
JORDAN	3.4	42.0	47.6	14.7	19.0	..	..	53.2	90.0	65.0	83.0	77.0	62.0
KUWAIT	7.7	88.0	45.4	..	2.0	..	..	64.0	..	..	90.0	86.0	55.0
LEBANON	2.8	60.1	39.8	9.9	17.8	26.0	4.0	63.3	101.0	67.9	132.0	125.0	68.0
LIBYA	4.2	30.5	45.0	14.7	19.5	13.3	10.1	52.9	117.0	62.0	145.0	135.0	27.0
MOROCCO	2.4	40.1	46.2	15.7	50.0	20.0	4.0	53.0	108.0	70.5	61.0	44.0	28.0
OMAN	3.1	5.0	49.6	18.7	48.0	..	..	47.0	..	..	44.0	..	20.0
QATAR	10.5	85.0	..	..	..	..	..	..	..	..	112.0	..	21.0
SAUDI ARABIA	1.9	17.9	50.2	24.4	61.0	..	..	42.0	86.0	56.0	34.0	27.0	15.0
SYRIAN ARAB REP.	3.1	46.2	45.4	15.4	49.9	17.0	5.0	56.0	104.0	66.7	102.0	81.0	40.0
TUNISIA	2.3	47.0	40.0	13.8	37.4	..	..	54.1	102.0	67.4	95.0	75.0	55.0
UNITED ARAB EMIRATES	13.1	80.0	..	..	..	..	..	..	..	..	75.0	..	21.0
YEMEN ARAB REP.	1.7	7.0	49.6	20.6	73.0	..	..	37.0	83.0	58.3	25.0	6.0	10.0
YEMEN PEOP. DEM. REP.	3.1	35.3	49.6	20.6	42.9	..	..	44.8	84.0	57.0	78.0	48.0	27.1
ALL COUNTRIES-MEDIAN	3.1	43.8	46.9	15.6	42.9	21.0	4.0	52.8	101.0	62.0	83.0	70.5	28.0
AFRICA SOUTH OF SAHARA													
BENIN PEOP. REP.	2.7	13.5	49.9	23.0	47.5	31.4	5.5	41.8	87.0	56.0	44.0	28.0	20.0
BOTSWANA	2.1	10.7	45.6	23.0	83.0	28.1	1.6	43.5	85.0	65.0	85.0	93.0	25.0
BURUNDI	2.0	3.7	48.0	24.7	86.0	..	..	39.0	99.0	62.0	23.0	17.0	10.0
CAMEROON	1.9	28.5	40.4	22.0	82.0	..	..	41.0	102.0	59.0	111.0	97.0	6.0
CENTRAL AFRICAN EMPIRE	2.2	35.9	43.4	22.5	91.0	..	..	41.0	102.0	49.0	79.0	53.0	15.0
CHAD	2.0	13.9	44.0	24.0	90.0	21.5	7.7	38.5	75.0	60.2	37.0	20.0	15.0
CONGO PEOP. REP.	2.3	38.0	45.1	20.8	56.0	..	..	43.5	98.0	44.0	153.0	140.0	50.0
EQUATORIAL GUINEA	1.3	..	..	..	..	..	..	..	..	..	..	..	..
ETHIOPIA	2.5	11.2	49.4	25.8	85.0	..	..	41.0	82.0	58.9	23.0	14.0	7.0
GABON	1.5	32.0	32.2	22.2	58.0	45.3	3.2	41.0	98.0	49.3	199.0	197.0	12.0
GAMBIA	2.3	14.0	43.3	24.1	79.6	..	..	40.0	98.0	64.0	32.0	21.0	10.0
GHANA	2.6	32.4	48.8	21.9	52.0	..	..	43.5	101.0	53.4	60.0	53.0	25.0
GUINEA	2.8	19.5	44.6	22.9	84.1	..	..	41.0	84.0	42.7	28.0	..	7.0
IVORY COAST	4.1	34.3	45.6	20.6	80.0	30.9	9.0	43.5	113.0	64.5	86.0	64.0	20.0
KENYA	3.4	13.0	48.7	16.0	84.0	20.2	3.9	50.0	91.0	59.6	109.0	101.0	40.0
LESOTHO	2.2	3.1	39.0	19.7	90.0	..	..	46.0	109.0	67.6	121.0	144.0	40.0
LIBERIA	3.3	27.6	43.6	20.7	72.0	61.7	5.3	43.5	87.0	39.0	62.0	44.0	15.0
MADAGASCAR	2.9	14.5	50.2	21.1	83.0	41.0	5.2	43.5	105.0	57.0	85.0	80.0	40.0
MALAWI	2.5	6.4	47.7	23.7	86.0	29.5	5.7	41.0	103.0	68.4	61.0	48.0	25.0
MALI	2.2	13.4	50.1	25.9	88.7	..	..	38.1	75.0	64.0	22.0	16.0	10.0

(MOST RECENT ESTIMATE)

AREA AND COUNTRY	POPULATION & VITAL STATISTICS				EMPLOYMENT AND INCOME			HEALTH & NUTRITION			EDUCATION		
	POP. GROWTH RATE % (65-75)	URBAN POP. % OF TOTAL	CRUDE BIRTH RATE (/000)	CRUDE DEATH RATE (/000)	LABOR IN AGR. % OF TOTAL	INCOME RECD BY HIGHEST 5% HH	INCOME RECD BY LOWEST 20% HH	LIFE EXPECT. YRS AT BIRTH	CALORIE SUPPLY %/CAP REQD.	PROTEIN SUPPLY GR/DAY /CAP	PRIMARY SCHOOL ENROLL RATIO %	FEMALE ENROLL. RATIO PRIMARY	ADULT LITERACY RATE % OF TOTAL
MAURITANIA	2.6	23.1	44.8	24.9	85.0	..	..	38.5	81.0	63.2	17.0	9.0	10.0
MAURITIUS	1.4	48.3	25.1	7.8	30.3	31.0	4.5	65.5	108.0	55.8	80.0	78.0	80.0
MOZAMBIQUE	2.2	55.0	43.3	21.4	73.0	..	..	41.0	94.0	41.0	46.0	..	..
NIGER	2.7	9.4	52.2	25.5	91.0	23.0	6.0	38.5	78.0	62.1	17.0	12.0	5.0
NIGERIA	2.5	26.0	49.3	22.7	62.0	..	..	41.0	88.0	46.3	49.0	39.0	25.0
RWANDA	2.8	3.8	50.0	23.6	93.0	..	..	41.0	90.0	51.3	58.0	54.0	23.0
SENEGAL	2.7	38.8	47.6	23.9	73.0	36.8	3.2	40.0	97.0	67.1	43.0	33.0	10.0
SIERRA LEONE	2.3	15.0	44.7	20.7	73.0	36.2	1.1	43.5	97.0	50.9	35.0	28.0	15.0
SOMALIA	2.4	28.3	47.2	21.7	77.0	..	..	41.0	79.0	55.1	58.0	41.0	50.0
SUDAN	2.2	13.2	47.8	17.5	66.5	20.9	5.1	48.6	88.0	60.4	40.0	27.0	15.0
SWAZILAND	3.2	14.3	49.0	21.8	83.0	..	..	43.5	89.0	..	103.0	102.0	50.0
TANZANIA	2.8	7.3	47.0	20.1	83.1	33.5	2.3	44.5	86.0	47.1	57.0	46.0	49.0
TOGO	2.7	15.0	50.6	23.3	75.0	..	..	41.0	96.0	52.1	98.0	68.0	12.0
UGANDA	3.1	8.4	45.2	15.9	86.0	20.0	6.2	50.0	90.0	54.0	44.0	43.0	25.0
UPPER VOLTA	2.2	12.1	48.5	25.8	89.0	..	..	38.0	78.0	59.2	14.0	11.0	7.0
ZAIRE	2.7	26.4	45.2	20.5	77.0	..	..	43.5	85.0	32.0	88.0	87.0	15.0
ZAMBIA	2.9	36.3	51.5	20.3	52.0	23.0	3.8	44.5	89.0	58.8	88.0	86.0	43.0
ALL COUNTRIES-MEDIAN	2.5	14.8	47.4	22.1	82.5	30.9	4.5	41.0	90.0	57.0	58.0	47.0	15.0
SOUTH ASIA													
AFGHANISTAN	2.2	14.3	51.4	30.7	52.9	..	..	40.3	83.0	61.5	23.0	7.0	14.0
BANGLADESH	2.3	8.8	49.5	28.1	78.0	16.7	7.9	45.0	93.0	58.5	73.0	51.0	23.0
BURMA	2.2	22.3	39.5	15.8	67.8	14.6	8.0	50.1	103.0	58.0	85.0	81.0	67.0
INDIA	2.2	20.6	37.0	17.0	69.0	25.0	4.7	49.5	89.0	48.0	65.0	52.0	36.0
NEPAL	2.1	4.8	42.9	20.3	94.4	..	..	43.6	95.0	50.0	59.0	10.0	19.2
PAKISTAN	2.9	26.0	47.4	16.5	54.8	17.3	8.4	49.8	93.0	54.0	51.0	31.0	21.0
SRI LANKA	2.0	24.3	28.2	7.9	55.0	18.6	7.3	67.8	97.0	48.0	77.0	77.0	78.1
ALL COUNTRIES-MEDIAN	2.2	20.6	42.9	17.0	67.8	17.3	7.9	49.5	93.0	54.0	65.0	51.0	23.0
EAST ASIA & PACIFIC													
CHINA REP.	2.6	51.1	23.0	4.7	35.0	13.3	8.8	68.6	111.0	68.0	104.0	..	82.0
FIJI	2.2	38.5	25.0	4.3	43.3	19.0	5.1	70.0	..	..	111.0	110.0	75.0
INDONESIA	2.3	18.2	42.9	16.9	69.0	33.7	6.8	48.1	98.0	43.8	81.0	75.0	62.0
KOREA	2.1	48.5	28.8	8.9	44.6	18.1	7.2	68.0	115.0	75.7	109.0	109.0	92.0
LAO P.D.R.	2.7	15.0	44.6	22.8	85.0	..	..	40.4	94.0	58.0	57.0	47.0	20.0
MALAYSIA	2.7	30.2	31.7	6.7	45.2	28.3	3.5	59.4	115.0	56.5	93.0	91.0	60.0
PAPUA NEW GUINEA	2.5	12.9	40.6	17.1	83.0	..	..	47.7	98.0	48.2	59.0	44.0	31.0
PHILIPPINES	2.9	29.8	43.8	10.5	52.6	28.8	5.5	58.5	87.0	50.0	105.0	103.0	87.0
SINGAPORE	1.7	90.2	21.2	5.2	2.8	..	..	89.5	122.0	74.7	111.0	108.0	75.0
THAILAND	3.1	16.5	37.6	9.1	76.0	22.0	5.6	58.0	107.0	50.0	78.0	75.0	82.0
VIET NAM	2.6	..	..	..	..	..	..	..	..	..	..	..	..
WESTERN SAMOA	1.9	21.5	36.9	6.7	67.0	..	..	58.0	100.0	52.7	91.0	..	97.8
ALL COUNTRIES-MEDIAN	2.6	29.8	36.9	8.9	52.6	22.0	5.6	58.5	103.5	53.6	93.0	91.0	75.0

TABLE 4. Selected Economic Development Indicators: Population and Production
(average annual real growth rates)

Income group/ region/country	Population				Gross domestic product				GDP per capita			
	1950-60	1960-65	1965-70	1970-77	1950-60	1960-65	1965-70	1970-77	1950-60	1960-65	1965-70	1970-77
Developing countries	2.2	2.4	2.5	2.4	4.9	5.6	6.4	5.7	2.7	3.1	3.8	3.2
Capital-surplus oil-exporting countries	2.3	3.2	3.7	4.1	. .	. .	11.0	6.7	. .	. .	7.2	3.0
Industrialized countries	1.2	1.2	0.9	0.8	3.8++	5.3	4.9	3.2	2.5++	4.0	4.0	2.4
Centrally planned economies	1.7	1.8	1.6	1.4	. .	6.2+	7.7+	6.4+	. .	4.8+	6.7+	5.6+
A. Developing countries by income group												
Low income	2.0	2.4	2.4	2.2	3.8	3.8	5.3	4.0	1.8	1.4	2.8	1.7
Middle income	2.4	2.5	2.5	2.5	5.3	6.1	6.6	6.0	2.8	3.5	4.0	3.4
B. Developing countries by region												
Africa south of Sahara	2.3	2.5	2.5	2.7	3.6	5.0	4.9	3.7	1.3	2.4	2.3	0.9
Middle East and North Africa	2.4	2.6	2.7	2.7	5.1	6.4	9.4	7.1	2.6	3.7	6.5	4.3
East Asia and Pacific	2.4	2.6	2.5	2.2	5.2	5.5	8.0	8.0	2.8	2.8	5.4	5.7
South Asia	1.9	2.4	2.4	2.2	3.8	4.3	4.9	3.2	1.8	1.9	2.4	1.0
Latin America and the Caribbean	2.8	2.8	2.7	2.7	5.3	5.2	6.1	6.2	2.4	2.3	3.3	3.4
Southern Europe	1.5	1.4	1.4	1.5	6.1	7.5	6.5	5.3	4.5	6.0	5.0	3.8
C. Developing countries by region and country												
Africa south of Sahara												
Angola	1.6	1.5	1.6	2.3	. .	5.9	3.2	-9.4	. .	4.3	1.6	-11.5
Benin	2.2	2.5	2.7	2.9	. .	3.1	2.7	2.7	. .	0.6	0.0	-0.1
Botswana	1.7	1.9	1.9	1.9	2.9	4.2	9.8	15.7	1.2	2.2	7.8	13.5
Burundi	2.0	2.3	2.4	1.9	-1.3	2.8	5.8	2.3	-3.2	0.5	3.3	0.4
Cameroon	1.4	1.7	1.9	2.2	1.7	2.9	7.3	3.4	0.3	1.2	5.3	1.2
Cape Verde	3.1	3.1	2.7	2.1	. .	. .	. .	. .	. .	. .	. .	. .
Central African Republic	1.4	2.2	2.2	2.2	2.6	0.4	3.5	3.1	1.1	-1.7	1.3	0.9
Chad	1.4	1.8	1.9	2.2	. .	0.5	1.6	1.2	. .	-1.3	-0.3	-0.9
Comoros	3.0	3.2	3.3	3.8	. .	9.5	3.2	-1.5	. .	6.1	-0.1	-5.2
Congo, People's Republic of the	1.6	2.1	2.2	2.5	1.1	2.7	3.4	3.9	-0.5	0.6	1.2	1.4
Equatorial Guinea	1.5	1.7	1.9	2.2	. .	13.8	2.0	-3.0	. .	11.8	0.1	-5.0
Ethiopia	2.1	2.3	2.4	2.5	3.9	5.1	3.7	2.6	1.7	2.7	1.2	0.1
Gabon	0.2	0.5	0.7	0.9	11.5	3.9	5.3	9.1	11.3	3.3	4.6	8.1

Income group/ region/country	Population				Gross domestic product				GDP per capita			
	1950-60	1960-65	1965-70	1970-77	1950-60	1960-65	1965-70	1970-77	1950-60	1960-65	1965-70	1970-77
Gambia, The	2.0	3.3	3.2	3.1	1.3	5.2	4.3	8.2	-0.7	1.9	1.1	4.9
Ghana	4.5	2.7	2.1	3.0	4.1	3.3	2.5	0.4	-0.4	0.6	0.4	-2.5
Guinea	2.2	2.7	3.0	3.0	. .	3.9	3.0	5.4	. .	1.2	0.0	2.4
Guinea-Bissau	0.2	-1.1	-0.2	1.6	. .	. .	. .	. .	. .	. .	. .	. .
Ivory Coast	2.1	3.7	3.8	6.0	3.6	10.1	7.4	6.5	1.5	6.1	3.5	0.5
Kenya	3.3	3.4	3.5	3.8	4.0	3.6	8.6	4.8	0.7	0.2	4.9	1.0
Lesotho	1.5	1.8	2.2	2.4	4.4	8.7	2.1	8.0	2.8	6.7	-0.1	5.5
Liberia	2.8	3.1	3.2	3.4	10.5	3.1	6.4	2.7	7.4	0.0	3.1	-0.6
Madagascar	1.8	2.1	2.3	2.5	2.3	1.4	4.9	-0.7	0.5	-0.7	2.5	-3.2
Malawi	2.4	2.7	2.9	3.1	. .	3.3	4.5	6.3	. .	0.6	1.5	3.1
Mali	2.1	2.5	2.4	2.5	3.2	3.2	2.9	4.7	1.0	0.7	0.5	2.1
Mauritania	2.2	2.5	2.6	2.7	. .	9.9	4.6	2.2	. .	7.2	2.0	-0.5
Mauritius	3.3	2.7	1.9	1.3	0.1	5.4	-0.3	8.2	-3.0	2.7	-2.2	6.8
Mozambique	1.4	2.1	2.3	2.5	3.1	2.3	8.3	-3.7	1.7	0.2	5.9	-6.1
Namibia	. .	2.4	2.6	2.8	. .	. .	. .	. .	. .	. .	. .	. .
Niger	2.3	4.0	2.7	2.8	. .	6.6	-0.3	1.2	. .	2.5	-2.9	-1.5
Nigeria	2.4	2.5	2.5	2.6	4.1	5.3	4.5	6.0	1.6	2.7	1.9	3.3
Réunion	. .	3.3	2.5	1.8	. .	. .	. .	. .	. .	. .	. .	. .
Rhodesia	4.1	4.2	3.7	3.3	. .	3.2	6.1	3.2	. .	-0.9	2.4	-0.1
Rwanda	2.3	2.5	2.7	2.9	1.0	-2.9	8.5	5.2	-1.2	-5.3	5.6	2.2
Senegal	2.2	2.4	2.5	2.6	. .	3.6	1.3	2.8	. .	1.1	-1.2	0.2
Sierra Leone	1.8	2.1	2.3	2.5	3.6	4.3	3.9	1.5	1.8	2.1	1.6	-1.0
Somalia	2.0	2.4	2.5	2.3	12.8	-0.5	3.4	1.2	10.6	-2.8	0.9	-1.1
South Africa	3.0	2.5	2.7	2.7	2.9	6.6	5.9	4.0	-0.1	4.0	3.1	1.2
Sudan	1.9	2.2	2.4	2.6	5.5	1.5	1.3	5.4	3.5	-0.7	-1.0	2.7
Swaziland	2.0	2.3	2.1	2.5	8.4	13.6	6.3	6.2	6.3	11.0	4.1	3.7
Tanzania, United Republic of	2.2	2.6	2.8	3.0	6.0	5.2	5.9	5.2	3.7	2.6	3.1	2.1
Togo	2.2	2.7	2.7	2.6	1.3	8.4	6.7	4.0	-0.9	5.6	3.9	1.4
Uganda	2.8	3.8	3.7	3.0	3.3	5.7	5.9	0.5	0.5	1.8	2.2	-2.4
Upper Volta	1.9	1.6	1.6	1.6	1.6	2.7	3.3	0.5	-0.3	1.0	1.6	-1.0
Zaire	2.3	1.9	2.1	2.7	3.4	3.7	4.3	1.0	1.1	1.7	2.2	-1.7
Zambia	2.4	2.8	2.9	3.0	5.6	5.4	2.8	2.8	3.1	2.6	-0.1	-0.2

See footnotes at end of table.

TABLE 4. COMPARATIVE ECONOMIC DATA (Continued)

Income group/ region/country	Population				Gross domestic product				GDP per capita			
	1950-60	1960-65	1965-70	1970-77	1950-60	1960-65	1965-70	1970-77	1950-60	1960-65	1965-70	1970-77
C. Developing countries by region and country (cont.)												
Middle East and North Africa												
Algeria	2.1	2.0	3.7	3.2	6.5	0.8	8.1	5.4	4.4	-1.2	4.2	2.2
Bahrain	3.5	4.2	2.9	7.1	. .	. .	. .	10.7[d]	. .	. .	. .	3.2[d]
Egypt, Arab Republic of	2.4	2.5	2.1	2.1	3.3	7.6	3.2	6.4	0.9	4.9	1.1	4.2
Iran	2.5	2.7	2.8	3.0	5.9	9.2	12.6	7.4	3.3	6.3	9.5	4.3
Iraq	2.8	3.1	3.2	3.4	9.9	7.7	4.1	8.1[e]	6.9	4.5	0.9	4.7[e]
Jordan	3.2	3.0	3.2	3.3	. .	. .	. .	7.0[f]	. .	. .	. .	3.6[f]
Lebanon	2.6	3.0	2.8	2.5	. .	. .	. .	. .	. .	. .	. .	. .
Morocco	2.7	2.5	2.9	2.7	2.0	4.2	5.7	6.4	-0.7	1.7	2.8	3.6
Syrian Arab Republic	2.7	3.1	3.3	3.3	. .	8.8	5.6	9.6	. .	5.4	2.3	6.2
Tunisia	1.8	1.9	2.1	2.0	. .	5.2[c]	4.9	8.5	. .	3.3[c]	2.8	6.4
Yemen, Arab Republic of	2.0	2.1	1.5	1.9	. .	. .	. .	8.4	. .	. .	. .	6.4
Yemen, People's Democratic Republic of	1.9	1.9	1.9	1.9	. .	. .	. .	6.8[d]	. .	. .	. .	4.8[d]
East Asia and Pacific												
Fiji	3.1	3.3	2.3	1.8	2.8	3.7	7.4	5.0	-0.3	0.3	5.0	3.1
Hong Kong	4.5	3.7	1.3	2.0	9.2	11.7	8.0	8.0	4.5	7.7	6.6	5.9
Indonesia	2.1	2.2	2.2	1.8	4.0	1.6	7.5	8.0	1.9	-0.6	5.2	6.1
Korea, Republic of	2.0	2.6	2.2	2.0	5.1	6.7	10.3	9.9	3.1	4.0	7.9	7.8
Malaysia	2.5	2.9	2.9	2.7	3.6	6.8	5.9	7.8	1.0	3.7	2.9	4.9
Papua New Guinea	1.8	2.3	2.4	2.4	4.8	6.4	5.7	5.0	2.9	4.0	3.3	2.5
Philippines	2.7	3.0	3.1	2.7	6.5	5.2	5.2	6.3	3.6	2.2	2.1	3.5
Singapore	4.8	2.8	2.0	1.6	. .	5.5	13.0	8.6	. .	2.6	10.7	6.9
Solomon Islands	2.6	2.6	2.6	3.5	10.7	3.7	2.5	5.4	7.9	1.1	-0.1	1.8
Taiwan	3.5	3.0	2.4	2.0	7.6	8.9	9.2	7.7	4.0	5.8	6.7	5.6
Thailand	2.8	3.0	3.1	2.9	5.7	7.4	8.4	7.1	2.8	4.2	5.1	4.0
South Asia												
Afghanistan	1.5	2.2	2.2	2.2	. .	1.7	2.3	4.5	. .	-0.4	0.2	2.2
Bangladesh	2.4	2.8	3.0	2.5	. .	4.6	3.4	2.3	. .	1.8	0.4	-0.1

Income group/ region/country	Population				Cross domestic product				GDP per capita			
	1950-60	1960-65	1965-70	1970-77	1950-60	1960-65	1965-70	1970-77	1950-60	1960-65	1965-70	1970-77
Bhutan	..	1.9	2.1	2.3	..	..	..	..	..	..	..	..
Burma	1.9	2.1	2.2	2.2	6.3	4.4	2.3	3.6	4.3	2.2	0.1	1.4
India	1.9	2.3	2.4	2.1	3.8	4.0	5.0	3.1	1.9	1.7	2.6	1.0
Nepal	1.2	1.9	2.2	2.2	2.4	2.7	2.2	2.6	1.2	0.8	0.0	0.4
Pakistan	2.3	2.7	2.9	3.1	2.4	7.2	6.9	3.8	0.1	4.4	3.8	0.7
Sri Lanka	2.6	2.5	2.3	1.7	3.9	4.0	5.8	2.9	1.3	1.5	3.4	1.1
Latin America and the Carribean												
Argentina	1.9	1.5	1.4	1.3	2.8	3.6	4.5	2.8	0.9	2.0	3.1	1.5
Bahamas	3.6	4.8	4.4	2.7	..	..	..	..	..	..	..	..
Barbados	0.9	0.3	0.3	0.5	5.9	4.5	7.9	2.0	4.9	4.1	7.6	1.4
Belize	3.1	2.8	2.8	0.9	..	..	..	6.0	..	..	..	5.0
Bolivia	2.1	2.5	2.6	2.7	..	5.0	4.8	5.9	..	2.5	2.1	3.1
Brazil	3.1	2.9	2.9	2.9	6.9	4.0	8.0	9.9	3.7	1.1	5.0	6.8
Chile	2.2	2.3	1.9	1.7	4.0	4.9	3.6	0.1	1.8	2.6	1.7	-1.6
Colombia	3.1	3.3	2.8	2.1	4.6	4.7	5.9	5.7	1.5	1.4	2.9	3.6
Costa Rica	3.7	3.7	3.2	2.5	..	5.3	6.9	6.0	..	1.5	3.6	3.3
Dominican Republic	2.9	2.9	2.9	3.0	5.8	4.6	6.6	8.0	2.8	1.6	3.6	4.9
Ecuador	2.9	3.0	3.0	3.0	..	..	5.8	9.2	..	..	2.7	6.0
El Salvador	2.8	3.4	3.5	3.1	4.4	6.7	4.3	5.3	1.5	3.2	0.7	2.1
Guatemala	2.8	2.8	2.9	2.9	3.8	5.5	5.9	6.0	1.0	2.7	2.9	3.0
Guyana	2.8	2.2	2.4	2.0	4.3	1.5	4.0	1.3	1.4	-0.8	1.5	-0.7
Haiti	1.9	1.5	1.6	1.7	..	0.6	0.1	3.8	..	-0.9	-1.5	2.1
Honduras	3.3	3.5	2.8	3.3	3.1	4.9	4.2	3.2	-0.2	1.4	1.4	-0.1
Jamaica	1.5	1.6	1.2	1.7	8.1	3.7	5.1	0.0	6.5	2.1	3.9	--1.7
Mexico	3.2	3.3	3.3	3.3	5.6	7.4	6.8	4.9	2.4	3.9	3.5	1.5
Netherlands Antilles	1.7	1.7	1.4	1.2	..	..	..	..	..	..	..	..
Nicaragua	2.9	3.0	3.0	3.3	5.2	10.4	4.1	5.8	2.2	7.2	1.1	2.4
Panama	2.9	3.1	3.1	3.1	4.9	7.9	7.8	3.5	1.9	4.7	4.6	0.4
Paraguay	2.6	2.6	2.7	2.9	2.7	4.5	4.3	7.0	0.1	1.9	1.6	4.0
Peru	2.6	2.9	2.9	2.8	4.9	7.1	4.3	4.5	2.2	4.1	1.4	1.7
Puerto Rico	0.6	1.6	1.3	2.8	5.3	8.1	7.0	3.6	4.7	6.5	5.6	0.7
Trinidad and Tobago	2.8	3.0	1.1	1.2	..	4.7	3.3	2.2	..	1.6	2.2	1.0
Uruguay	1.4	1.2	1.0	0.2	1.7	0.6	1.9	1.6	0.3	-0.6	1.0	1.3
Venezuela	4.0	3.6	3.3	3.4	8.0	7.4	4.9	5.6	3.8	3.7	1.6	2.2

TABLE 4. COMPARATIVE ECONOMIC DATA (Continued)

Income group/ region/country	Population				Gross domestic product				GDP per capita			
	1950-60	1960-65	1965-70	1970-77	1950-60	1960-65	1965-70	1970-77	1950-60	1960-65	1965-70	1970-77
C. Developing countries by region and country (cont.)												
Southern Europe												
Cyprus	1.5	0.6	0.8	0.7	4.0	3.7	8.1	1.0	2.5	3.0	7.2	0.3
Greece	1.0	0.5	0.6	0.7	6.0	7.7	7.2	4.6	5.0	7.2	6.6	3.9
Israel	5.3	3.9	3.0	2.8	11.3	9.8	8.7	5.0	5.6	5.6	5.5	2.2
Malta	0.5	-0.6	0.5	0.3	3.3	0.3	9.0	11.4	2.7	0.9	8.5	11.1
Portugal	0.7	0.2	-0.2	0.8	4.1	6.4	6.4	4.6	3.4	6.1	6.6	3.7
Spain	0.8	1.0	1.1	1.0	6.2	8.6	6.3	4.7	5.3	7.5	5.1	3.6
Turkey	2.8	2.5	2.5	2.5	6.3	5.3	6.3	7.3	3.4	2.8	3.7	4.6
Yugoslavia	1.2	1.1	0.9	0.9	5.6	6.6	6.2	6.2	4.4	5.4	5.2	5.2
D. Capital-surplus oil-exporting countries												
Kuwait	6.2	11.1	9.6	6.2	. .	4.7[h]	5.7	-0.1	. .	-5.8	-3.5	-6.0
Libyan Arab Republic	2.7	3.8	4.2	4.1	. .	. .	. .	. .	. .	. .	. .	. .
Oman	2.0	2.5	2.7	3.2	. .	5.7	39.7	6.8	. .	3.2	36.0	3.5
Qatar	2.4	7.1	6.9	10.3	. .	. .	. .	. .	. .	. .	. .	. .
Saudi Arabia	2.0	2.5	2.8	3.0	. .	. .	9.1	12.7	. .	. .	6.1	9.4
United Arab Emirates	2.4	9.3	13.7	16.7	. .	. .	. .	12.5[f]	. .	. .	. .	-2.1[f]
E. Industrialized countries												
Australia	2.3	2.0	1.9	1.7	4.7[i]	5.3	6.2	3.3	2.3[i]	3.2	4.2	1.6
Austria	0.2	0.6	0.5	0.2	5.6	4.3	5.1	4.0	5.4	3.7	4.6	3.8
Belgium	0.6	0.7	0.4	0.3	3.1[j]	5.2	4.8	3.7	2.5[j]	4.5	4.4	3.4
Canada	2.7	1.9	1.7	1.2	4.0[j]	5.8	4.8	4.7	1.3[j]	3.8	3.0	3.4
Denmark	0.7	0.8	0.7	0.4	3.6[j]	5.1	4.5	2.8	2.9[j]	4.3	3.7	2.4
Finland	1.0	0.6	0.2	0.4	4.9	5.0	5.1	3.4	3.9	4.4	4.9	3.0
France	0.9	1.3	0.8	0.7	4.8	5.9	5.3	3.8	3.8	4.5	4.5	3.1
Germany, Federal Republic of	1.0	1.3	0.6	0.2	7.3[j]	4.8	4.5	2.4	6.3[j]	3.5	3.9	2.2
Iceland	2.1	1.8	1.2	1.3	. .	7.1	1.1	4.6	. .	5.2	-0.2	3.2
Ireland	-0.5	0.3	0.5	1.2	. .	4.0	5.3	3.4	. .	3.7	4.7	2.3
Italy	0.7	0.7	0.7	0.7	5.5[g]	5.0	6.1	2.9	4.8[g]	4.3	5.4	2.2

Income group/ region/country	Population				Cross domestic product				GDP per capita			
	1950-60	1960-65	1965-70	1970-77	1950-60	1960-65	1965-70	1970-77	1950-60	1960-65	1965-70	1970-77
Japan	1.3	1.0	1.1	1.2	8.0[i]	10.1	12.4	5.0	6.6[i]	9.0	11.2	3.7
Luxembourg	0.6	1.1	0.4	0.8	..	3.3	3.6	2.3	..	2.2	3.2	1.5
Netherlands, The	1.3	1.4	1.2	0.9	4.7	4.9	5.7	3.2	3.3	3.5	4.4	2.3
New Zealand	2.2	2.1	1.4	1.7	..	5.0	2.6	2.9	..	2.9	1.2	1.2
Norway	0.9	0.8	0.8	0.6	3.4[i]	5.1	4.8	4.8	2.5[i]	4.3	3.9	4.1
Sweden	0.6	0.7	0.8	0.4	3.4	5.3	3.9	1.7	2.7	4.6	3.1	1.3
Switzerland	1.3	2.1	1.1	0.2	4.6	5.2	4.2	0.2	3.2	3.0	3.1	0.0
United Kingdom	0.4	0.7	0.3	0.1	2.8	3.2	2.5	1.9	2.3	2.4	2.2	1.7
United States	1.7	1.5	1.1	0.8	3.3	4.7	3.2	2.8	1.5	3.2	2.1	1.9
F. Centrally planned economies[k]												
Albania	2.8	3.0	2.8	2.5	..	..	..	..	..	..	..	..
Bulgaria	0.8	0.8	0.7	0.6	..	7.0	8.6	7.5	..	6.2	7.8	6.9
China	1.9	2.0	1.8	1.6	..	..	..	..	..	..	..	..
Cuba	2.6	2.1	1.9	1.6	..	..	..	..	..	..	..	..
Czechoslovakia	1.0	0.7	0.3	0.7	..	..	6.6[a]	5.2	..	..	6.3[a]	4.5
German Democratic Republic	-0.6	-0.3	0.0	-0.2	..	2.7[l]	5.5[l]	5.1[l]	..	5.9	5.5	5.3
Hungary	0.7	0.3	0.4	0.4	..	4.5	6.8	6.2	..	4.2	6.4	5.8
Korea, Democratic People's Republic of	0.8	2.8	2.8	2.6	..	..	..	..	..	..	..	..
Mongolia	2.2	2.8	3.1	3.0	..	..	..	..	..	..	..	..
Poland	1.8	1.3	0.6	1.0	..	6.0	6.0	8.7	..	4.6	5.4	7.6
Romania	1.2	0.7	1.4	0.9	..	8.8	7.7	10.7	..	8.0	6.2	10.1[m]
Union of Soviet Socialist Republics	1.8	1.5	1.0	0.9	..	6.6	8.2	6.1	..	5.0	7.1	5.5[m]

+ Weighted average of the country growth rates; GDP in US dollars were used as weights; these are not strictly comparable to other group averages. ++ 1955–60. a. 1966–70. b. 1967–70. differences in national accounting system. l. Based on NMP index (1960=100) constructed from 1975 constant price series. m. 1970–76.

TABLE 4. COMPARATIVE ECONOMIC DATA (Continued)

	Gross production							
	Agriculture				Manufacturing			
	1950-60	1960-65	1965-70	1970-77	1950-60	1960-65	1965-70	1970-77
Developing countries	3.9	2.6	3.4	2.7	4.9	7.6	7.5	7.4
Capital-surplus oil-exporting countries	..	..	2.2	4.4	..	..	..	..
Industrialized countries	2.3+	2.0	2.2	2.1	6.1+	5.9	5.8	2.8
Centrally planned economies	..	..	3.2+	2.4+	..	8.0+	8.3+	7.4+
A. Developing countries by income group								
Low income	..	1.6	4.1	2.2	..	8.4	3.3	5.2
Middle income	4.5	3.1	3.0	3.0	4.7	7.5	7.9	7.2
B. Developing countries by region								
Africa south of Sahara	4.8	2.6	2.4	1.3	..	8.3	6.6	5.6
Middle East and North Africa	..	1.3	3.6	2.8	..	10.0	6.9	12.1
East Asia and Pacific	4.8	4.6	3.4	4.1	..	4.8	11.9	11.6
South Asia	3.2	1.1	4.7	2.1	6.4	8.8	3.5	4.3
Latin America and the Caribbean	..	3.5	2.8	3.3	4.0	5.6	6.7	5.8
Southern Europe	4.4	3.2	3.2	3.1	8.4	11.4	9.3	6.1
C. Developing countries by region and country								
Africa south of Sahara								
Angola	..	3.2	2.0	-3.4	..	..	..	..
Benin	3.3	1.9	5.2	1.5	..	..	..	..
Botswana	..	2.4	0.4	4.5	..	..	..	..
Burundi	-1.5	1.7	1.7	2.0	..	..	..	..
Cameroon	3.7	7.2	4.4	1.8	..	..	10.7[a]	6.2
Cape Verde	..	..	..	..	..	..	..	..
Central African Republic	..	-1.2	3.3	2.1	..	4.1	8.1	6.0
Chad	..	0.9	-0.3	0.2	..	..	2.0[b]	5.7
Comoros	..	..	..	..	..	..	..	..
Congo, People's Republic of the	..	0.0	2.9	2.8	..	3.4	7.7	2.3
Equatorial Guinea	..	4.8[c]	-3.1	-4.9	..	..	..	..
Ethiopia	4.1	1.8	1.9	-0.1	..	..	..	..
Gabon	..	3.2[c]	1.9	1.3	..	..	..	..

TABLE 4. COMPARATIVE ECONOMIC DATA (Continued)

	Gross production							
	Agriculture				Manufacturing			
	1950-60	1960-65	1965-70	1970-77	1950-60	1960-65	1965-70	1970-77
Gambia, The	0.4	7.1	-0.3	1.7	..	3.1	8.9	4.3
Ghana	7.0	3.5	3.8	-0.4	4.6	10.8	12.5	-1.9
Guinea	2.7	1.9	3.5	0.0	..	..	..	..
Guinea-Bissau	..	..	..	..	..	..	..	..
Ivory Coast	9.3	11.3	3.7	4.5	..	14.2	8.0	7.5
Kenya	4.7	2.9	1.1	3.0	..	..	..	..
Lesotho	..	0.1	1.5	2.0	..	..	..	..
Liberia	2.6	3.0	3.0	2.3	..	..	14.1	5.5
Madagascar	3.9	4.1	1.6	2.7	..	..	..	1.2
Malawi	..	3.8	3.4	3.8	..	..	..	10.6
Mali	1.8	2.3	3.2	2.7	..	..	4.0	9.2
Mauritania	..	3.4	2.2	-3.3	..	..	35.1	2.9
Mauritius	..	3.1	-1.4	2.6	..	..	..	..
Mozambique	..	0.8	4.4	-1.6	..	..	..	..
Namibia	..	..	..	..	..	..	..	..
Niger	5.5	5.2	0.6	0.4	..	14.6	8.8	8.6
Nigeria	5.1	2.1	1.7	1.7	..	11.1	10.4	8.1
Réunion	..	..	..	..	..	..	..	..
Rhodesia	..	2.9	0.0	3.9	5.0	6.2	7.5	3.8
Rwanda	..	-1.6	8.4	3.6	..	..	..	..
Senegal	4.8	4.1	-4.5	5.0	..	0.4	1.2	8.2
Sierra Leone	0.3	4.7	2.2	2.0	..	..	2.7	3.6
Somalia	..	2.7	1.9	0.0	..	..	..	..
South Africa	4.1	1.5	3.9	2.2	5.0	8.4	6.7	3.2
Sudan	-0.4	5.4	6.0	2.3	..	..	..	..
Swaziland	..	4.6	7.0	3.6	..	17.1[c]	21.9	7.3
Tanzania, United Republic of	6.9	2.9	3.0	1.2	..	..	..	..
Togo	9.3	31.3	1.4	-5.5	..	..	6.7	11.3
Uganda	6.9	1.4	4.2	0.8	..	..	..	..
Upper Volta	-0.2	7.2	2.2	1.6	..	..	3.8[b]	7.1
Zaire	1.0	-2.2	2.3	1.9	..	..	0.3	1.1
Zambia	..	1.2	1.9	4.7	..	12.1	8.6	1.7

TABLE 4. COMPARATIVE ECONOMIC DATA (Continued)

	Gross production							
	Agriculture				Manufacturing			
	1950-60	1960-65	1965-70	1970-77	1950-60	1960-65	1965-70	1970-77
C. Developing countries by region and country (cont.)								
Middle East and North Africa								
Algeria	-0.5	-2.2	4.1	0.4	..	..	..	..
Bahrain	..	..	1.6	4.0	..	..	..	..
Egypt, Arab Republic of	3.4	3.6	3.5	0.7	8.8	20.0	3.8	6.3
Iran	2.1	2.7	4.1	4.2	..	9.5	13.4	16.0
Iraq	1.5	4.0	4.0	-2.3	..	..	..	..
Jordan	..	20.7	-15.3	-1.6	..	..	..	..
Lebanon	5.0	9.1	0.7	-0.5	..	..	..	..
Morocco	3.0	5.1	4.1	-2.0	5.2	3.3	5.8	6.5
Syrian Arab Republic	2.1	10.8	-1.7	9.8	6.3	9.2	4.8	8.3
Tunisia	4.5	3.6	-0.5	5.6	..	2.9	4.8	5.6
Yemen, Arab Republic of	..	1.2	-4.3	3.1	..	..	..	..
Yemen, People's Democratic Republic of	..	2.6	1.2	3.8	..	..	..	..
East Asia and Pacific								
Fiji	..	11.2[c]	3.3	-0.4	..	..	5.4	3.0
Hong Kong	..	0.9	-0.3	-10.3	..	12.4	18.6	4.0
Indonesia	2.6	3.9	3.3	3.1	..	1.0	7.9	12.5
Korea, Republic of	5.5	6.3	3.1	4.9	16.4	13.9	25.6	23.5
Malaysia	0.9	5.2	6.4	4.3	..	..	11.5	12.6
Papua New Guinea	..	4.2	2.8	3.1	9.7[g]	..	..	..
Philippines	3.3	2.8	3.2	5.8	..	6.2	4.4	5.2
Singapore	..	1.4	15.8	1.0	..	7.7	15.8	11.2
Solomon Islands	..	1.4	0.9	1.9	..	..	..	..
Taiwan	4.8	4.8	2.9	3.1	15.4	13.6	21.0	13.3
Thailand	3.8	6.6	3.1	4.7	6.4	10.8	10.2	11.8
South Asia								
Afghanistan	..	1.6	1.0	4.0	..	..	8.0	9.3
Bangladesh	..	3.1	3.1	1.8	..	5.8	7.4	5.7

TABLE 4. COMPARATIVE ECONOMIC [illegible]

	Gross production							
	Agriculture				Manufacturing			
	1950-60	1960-65	1965-70	1970-77	1950-60	1960-65	1965-70	1970-77
Bhutan	..	..	..	..	..	..	..	..
Burma	2.5	3.1	3.1	1.6	..	5.6	1.7	4.4
India	3.2	0.5	4.9	2.1	6.6	9.0	2.8	4.6
Nepal	..	0.9	1.7	1.2	..	..	..	..
Pakistan	..	4.6	5.9	2.6	..	12.1	9.4	2.6
Sri Lanka	2.3	3.6	2.1	2.4	-0.9	6.1	5.6	2.7
Latin America and the Carribean								
Argentina	..	3.0	3.0	3.3	0.4	3.9	6.1	3.0
Bahamas	..	..	..	..	..	..	..	..
Barbados	..	3.6	-2.5	-2.7	..	..	1.6[a]	6.5
Belize	..	13.7	5.4	2.6	..	..	..	..
Bolivia	..	4.1	4.2	4.4	1.3	7.0	3.4	7.1
Brazil	4.5	3.2	2.7	4.4	9.1	3.7	10.4	9.6
Chile	2.9	2.2	2.5	1.7	5.7	6.7	1.5	-4.4
Colombia	..	2.8	3.6	3.9	6.5	5.7	6.2	6.5
Costa Rica	..	3.4	5.6	3.5	..	9.2	8.0	8.1
Dominican Republic	..	-2.7	5.9	2.0	5.0	1.5	11.4	8.8
Ecuador	..	5.8	3.0	3.5	..	..	7.2	10.6
El Salvador	..	2.9	0.4	3.3	5.7	17.0	5.0	8.3
Guatemala	..	7.8	3.9	3.8	3.9	5.4	2.5	6.1
Guyana	..	-0.7	1.3	1.7	..	1.4	1.7	7.5
Haiti	..	3.5	0.8	0.6	0.3	0.9	0.8	7.0
Honduras	..	4.5	5.3	1.3	7.0	3.0	4.9	5.6
Jamaica	2.9	3.2	-1.3	1.1	..	7.6	3.0	0.6
Mexico	5.4	5.9	2.0	2.1	7.0	9.6	8.4	5.9
Netherlands Antilles	..	1.2	3.1	9.7	..	..	..	..
Nicaragua	1.1	13.0	1.2	5.0	..	..	..	..
Panama	..	4.3	6.5	3.5	10.7	12.6	9.6	1.2
Paraguay	..	4.9	2.4	4.6	..	3.5	5.9	3.1
Peru	3.6	2.4	2.3	0.7	7.3	9.7	6.7	5.8
Puerto Rico	..	..	..	..	..	..	..	..
Trinidad and Tobago	..	2.8	1.4	-0.5	..	..	6.1	-1.1
Uruguay	0.3	2.3	3.1	0.4	3.2	1.0	2.4	2.7
Venezuela	..	5.6	5.7	2.2	13.0	9.5	3.6	5.6

TABLE 4. COMPARATIVE ECONOMIC DATA (Continued)

	Gross production							
	Agriculture				*Manufacturing*			
	1950-60	*1960-65*	*1965-70*	*1970-77*	*1950-60*	*1960-65*	*1965-70*	*1970-77*
C. Developing countries by region and country (cont.)								
Southern Europe								
Cyprus	3.8	7.7	5.4	0.8	. .	2.5	10.1	1.4
Greece	4.7	6.5	2.4	3.8	7.9	7.9	8.7	7.7
Israel	12.0	8.1	5.8	4.9	11.6	13.6	11.6	6.1
Malta	. .	2.7	6.7	0.4	. .	. .	. .	. .
Portugal	2.1	3.7	1.1	-1.1	6.7	8.7	8.9	1.2
Spain	2.9	2.6	3.3	3.4	8.8	12.6	10.1	7.9
Turkey	4.7	2.9	4.0	3.7	8.6	12.7	11.6	10.1
Yugoslavia	6.2	2.5	3.1	4.6	10.4	11.7	6.1	8.2
D. Capital-surplus oil-exporting countries								
Kuwait	. .	. .	3.0	1.5	. .	. .	. .	. .
Libyan Arab Republic	3.9	7.2	-2.0	11.4	. .	. .	. .	. .
Oman	. .	. .	2.8	2.1	. .	. .	. .	. .
Qatar	. .	. .	. .	. .	. .	. .	. .	. .
Saudi Arabia	. .	2.8	2.9	2.8	. .	. .	11.9	4.0
United Arab Emirates	. .	. .	. .	. .	. .	. .	. .	. .
E. Industrialized countries								
Australia	3.8	0.5	3.6	1.9	6.3	6.1	4.9	1.3
Austria	4.7	0.7	3.6	1.9	7.5	4.5	6.4	3.3
Belgium	1.3	0.7	4.3	0.9	4.1	6.3	6.0	2.6
Canada	2.5	5.0	-1.2	2.8	3.5	6.3	4.9	3.7
Denmark	1.5	0.8	-1.1	1.2	3.5	6.9	4.6	1.8
Finland	3.4	2.4	1.4	1.4	6.4	6.2	7.5	3.3
France	3.2	2.7	1.6	1.3	6.8	5.5	9.6	2.8
Germany, Federal Republic of	3.3	0.8	3.7	0.4	9.8	5.7	6.5	1.6
Iceland	. .	. .	. .	3.2	. .	. .	. .	. .
Ireland	2.2	0.3	3.2	3.9	3.0	6.6	7.1	4.1
Italy	2.2	3.0	2.4	1.0	9.3	6.3	7.3	3.0

TABLE 4. COMPARATIVE ECONOMIC DATA (Continued)

	Gross production							
	Agriculture				Manufacturing			
	1950-60	1960-65	1965-70	1970-77	1950-60	1960-65	1965-70	1970-77
Japan	2.4	3.2	3.3	2.1	18.3	11.5	16.3	2.9
Luxembourg	1.3	0.7	4.5	0.9	4.4	1.8	4.5	0.0
Netherlands, The	2.9	0.1	6.6	3.0	5.9	6.2	7.4	2.1
New Zealand	3.0	3.0	2.2	1.0	..	..	..	..
Norway	0.6	-0.8	0.8	1.6	4.8	5.8	4.2	2.2
Sweden	-0.2	0.3	0.7	2.5	3.0	7.4	4.9	2.0
Switzerland	1.1	-0.9	2.5	1.9	6.0	5.1	5.8	-1.0
United Kingdom	2.7	3.1	1.1	0.8	3.5	3.4	3.0	0.5
United States	1.8	1.7	1.5	3.1	3.4	6.3	3.8	3.3
F. Centrally planned economies								
Albania	..	..	4.2	3.4	..	..	..	..
Bulgaria	..	..	1.5	1.6	..	11.0	11.0	8.4
China	..	..	2.5	3.0	..	..	..	..
Cuba	..	..	4.0	0.0	..	..	..	..
Czechoslovakia	..	..	3.8	2.9	..	4.5	6.8	6.7
German Democratic Republic	..	..	1.2	3.6	..	5.7	6.4	6.3
Hungary	..	..	2.7	3.7	..	7.7	5.9	6.2
Korea, Democratic People's Republic of	..	..	2.1	6.7	..	..	..	..
Mongolia	..	..	-2.0	2.3	..	..	..	..
Poland	..	..	1.6	1.3	..	8.7	8.9	10.7
Romania	..	..	-0.4	6.9	..	..	..	..
Union of Soviet Socialist Republics	..	..	4.1	1.8	..	8.4	8.6	7.2

c. 1961–65. d. 1973–77. e. 1970–75. f. 1972–77. g. 1951–60. h. 1962–65. i. 1952–60. j. 1953–60. k. GDP data are not strictly comparable to those of other countries because of

TABLE 5. **Selected indicators of technological diffusion, 1960-1976**

	Fertilizer consumption (kg/ha)[a]	*Per capita energy consumption*[b] *(kg of coal equivalent)*	*Per capita electricity consumption (kWh)*	*Number of commercial vehicles in use per 10,000 population*	*Railway freight traffic (Net ton-kilometres per capita)*[c]
Developing countries					
1960	6	211	97	26	124
1976	23	426	305	46	209
(1976 as percentage of 1960)	(383)	(202)	(314)	(177)	(169)
Socialist countries of Asia					
1960	13	552[d]	102	—	—
1976	51	719	189	—	—
(1976 as percentage of 1960)	(392)	(130)	(185)	(—)	(—)
Developed market economy countries					
1960	67	3 995	2 596	333	1 916
1976	114	6 388	6 077	1 189	2 270
(1976 as percentage of 1960)	(170)	(160)	(234)	(357)	(118)
Socialist countries of Eastern Europe					
1960	26	2 893	1 305	41[e]	5 428
1976	98	5 251	4 032	140[e]	10 034
(1976 as percentage of 1960)	(377)	(182)	(309)	(341)	(185)

Sources: For fertilizer consumption, FAO, *Annual Fertilizer Review, 1977* (Rome 1978); for energy consumption and electricity consumption: United Nations, *World Energy Supplies 1950-1974* (United Nations publication, Sales No. E.76.XVII.5, and *World Energy Supplies 1972-1976* (United Nations publication, Sales No. E.78.XVII.7); for commercial vehicles in use and net ton-kilometres of railway freight traffic: United Nations, *Statistical Yearbook 1966* (United Nations publication, Sales No. E/F.67.XVII.1) and *Statistical Yearbook 1977* (United Nations publication, Sales No. E/F.78.

TABLE 6. **Scientists and engineers engaged in research and development and expenditures on research and development, 1969-1975[a]**

	Scientists and engineers engaged in research and development			*Expenditures on research and development*				
	Period	*Total (Thousands of FTE)[b]*	*Per 10,000 of total population*	*Fiscal year*	*Total ($ million)*	*Per capita ($)*	*Per-centage of GNP*	*Annual percentage increase (At current prices)*
A. *Rough totals[c]*								
Developed market economy countries . .	Around 1973-1975	1 390	43[d]	Around 1973-1975	80 900	111	2.0	..
Socialist countries of Eastern Europe	Around 1973-1975	1 440	82[d]	Around 1973-1975	39 400	114	4.0	..
Developing countries[e] .	Around 1973-1975	210	1[d]	Around 1973-1975	1 900	1	0.3	..
B. *Selected developing countries*								
India	1973	97.0[f]	..	1972	256	0.5	0.4	18.1 (1969-1972)
Indonesia	1975	12.2	0.9	1975	47	0.3	0.2	80.5 (1972-1975)
Argentina	1974	8.1	3.2	1974	184	7.4	0.5	26.9 (1968-1974)
Brazil	1974	7.7	0.8	1974	346	3.4	0.4	10.6 (1973-1974)
Egypt	1973	6.9	1.9	1976	85	2.2	0.7	4.1 (1973-1976)
Republic of Korea . . .	1974	6.3	1.9	1976	128	3.6	0.7	20.7 (1969-1976)
Thailand	1974/75	6.1	1.5	1968	14	0.4	0.2	..
Mexico	1974	5.9	1.0	1973	102	1.8	0.2	18.8 (1970-1973)
Philippines	1965	5.6	1.8	1973	32	0.8	0.3	15.1 (1965-1973)
Iran	1972	4.9	1.6	1972	47	1.5	0.2	..
Pakistan	1973/74	4.2	0.6	1973	15	0.2	0.2	..
Ghana	1975	3.9	3.9	1971	21	2.4	0.7	..
Venezuela	1973	2.7	2.6	1973	67	6.0	0.3	43.6 (1970-1973)
Sudan	1974	2.7	1.6	1973	9	0.5	0.3	..
Nigeria	1970/71	2.1	0.4	1970	33	0.6	0.3	11.5 (1966-1970)
Peru	1975	2.1	1.3	1970	25	1.9	0.4	..
Cuba	1969	1.9	2.2	1969	92	11.0	2.2[g]	..
Bangladesh	1973/74	1.6	0.2	1974	24	0.3	0.3	..
Iraq	1974	1.5	1.4	1974	25	2.3	0.2	67.0 (1971-1974)
Colombia	1971	1.1	0.5	1971	10	0.5	0.1	..

Source: UNESCO, "Development in human and financial resources for science and technology: basic statistical tables showing the earliest and latest years for which data are available" (CSR-S-5), April 1978.

TABLE 7. **Indicative dynamics of technological transformation in developing countries**

Sector	*Ratio of per capita availability of goods and in services in developed market economy countries to availability in developing countries, 1975*[a]	*Annual percentage rates of per capita growth required to reach 1975 levels of developed market economy countries*		
		In 50 years (by 2025)	*In 35 years (by 2010)*	*In 25 years (by 2000)*
Agriculture	2.0	1.4	2.0	2.8
Mining	3.0	2.2	3.3	4.5
Manufacturing	16.0	5.7	8.2	11.7
Consumer goods	11.0	4.8	7.0	10.1
Intermediate goods	13.0	5.3	7.5	10.8
Capital goods	28.0	7.0	10.0	14.3
Services	14.0	5.4	7.8	11.1
Weighted average	11.0	4.8	7.0	10.1

Sources: UNCTAD, *Handbook of international trade and development statistics, Supplement 1977* (United Nations publication, Sales No. E/F.78.II.D.1), and United Nations, *Statistical Yearbook 1977* (United Nations publication, Sales No. E/F.78.XVII.1)

[a] Rounded figures. Developing countries exclude China and other socialist countries of Asia.

FIGURE 1. MAIN INFORMATION FLOWS BETWEEN GEOGRAPHIC AREAS, 1976–1978

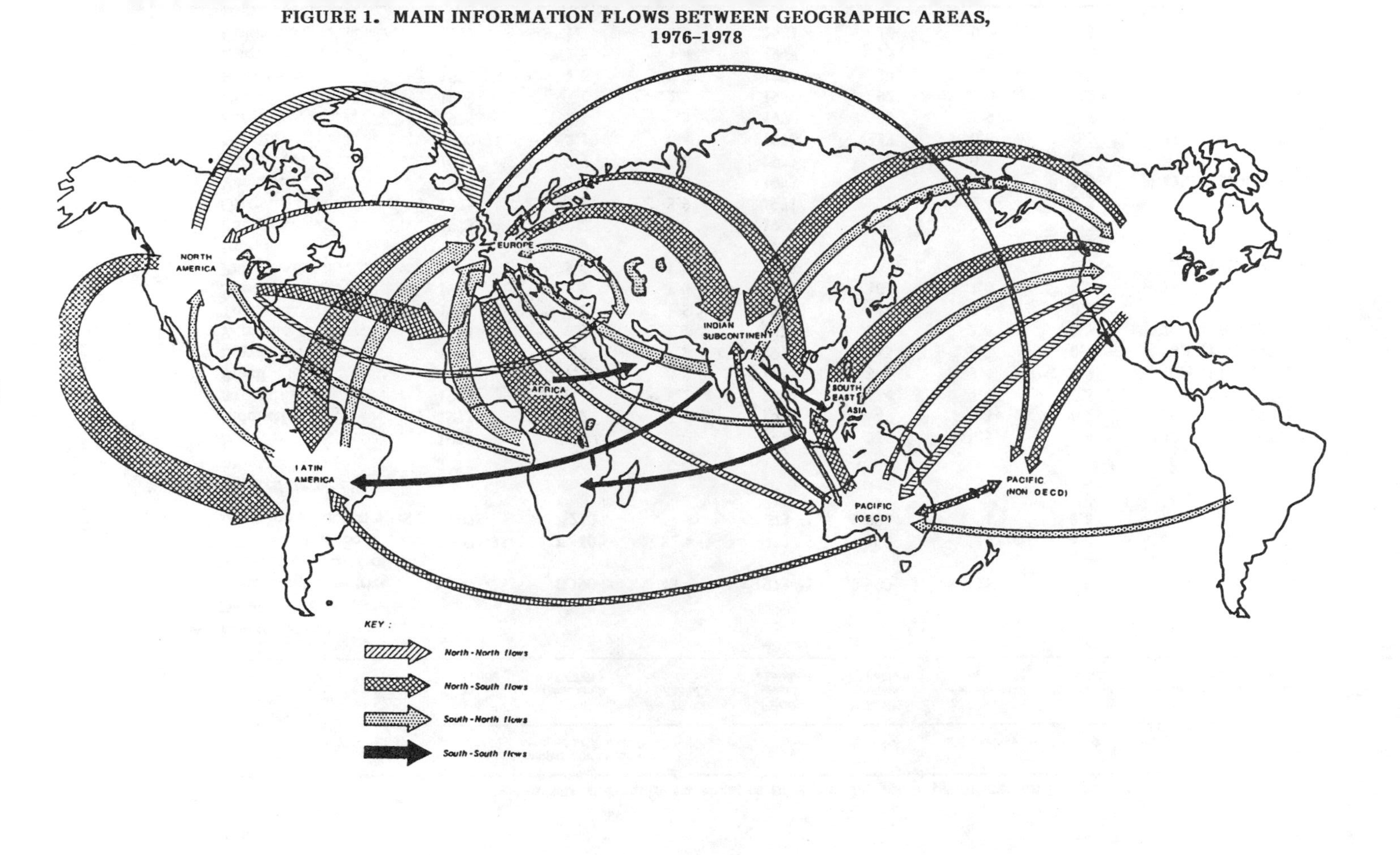

TABLE 8 Scientists and engineers engaged in R and D, and expenditures on R and D [a]

	Scientists and engineers engaged in R and D			Expenditure on R and D				
	Period	Total (thousands of FTE) [b]	Per 10,000 of total population	Fiscal year beginning	Total (millions of dollars)	Per capita (dollars)	Percentage of GNP	Annual percentage increase (at current prices)
A. *Rough totals* [c]								
Developed market-economy countries....	[c] 1973-75	1 390	43 [d]	[c] 1973-75	80 900	111	2.0	..
Socialist countries of Eastern Europe.......	[c] 1973-75	1 440	82 [d]	[c] 1973-75	39 400	114	4.0	..
Developing countries [e]...	[c] 1973-75	210	1 [d]	[c] 1973-75	1 900	1	0.3	..
B. *Selected developing countries*								
India..................	1973	97.0 [f]	—	1972	256	0.5	0.4	18.1 (1969-1972)
Indonesia..............	1975	12.2	0.9	1975	47	0.3	0.2	80.5 (1972-1975)
Argentina..............	1974	8.1	3.2	1974	184	7.4	0.5	26.9 (1968-1974)
Brazil.................	1974	7.7	0.8	1974	346	3.4	0.4	10.6 (1973-1974)
Egypt..................	1973	6.9	1.9	1976	85	2.2	0.7	4.1 (1973-1976)
Republic of Korea......	1974	6.3	1.9	1976	128	3.6	0.7	20.7 (1969-1976)
Thailand...............	1974/75	6.1	1.5	1968	14	0.4	0.2	..
Mexico.................	1974	5.9	1.0	1973	102	1.8	0.2	18.8 (1970-1973)
Philippines............	1965	5.6	1.8	1973	32	0.8	0.3	15.1 (1965-1973)
Iran...................	1972	4.9	1.6	1972	47	1.5	0.2	..
Pakistan...............	1973/74	4.2	0.6	1973	15	0.2	0.2	..
Ghana..................	1975	3.9	3.9	1971	21	2.4	0.7	..
Venezuela..............	1973	2.7	2.6	1973	67	6.0	0.3	43.6 (1970-1973)
Sudan..................	1974	2.7	1.6	1973	9	0.5	0.3	..
Nigeria................	1970/71	2.1	0.4	1970	33	0.6	0.3	11.5 (1966-1970)
Peru...................	1975	2.1	1.3	1970	25	1.9	0.4	..
Cuba...................	1969	1.9	2.2	1969	92	11.0	2.2 [g]	..
Bangladesh.............	1973/74	1.6	0.2	1974	24	0.3	0.3	..
Iraq...................	1974	1.5	1.4	1974	25	2.3	0.2	67.0 (1971-1974)
Colombia...............	1971	1.1	0.5	1971	10	0.5	0.1	..

Source: UNESCO, "Development in human and financial resources for science and technology: basic statistical tables showing the earliest and latest years for which data are available" (CSR-S-5), April 1978.

TABLE 9. GDP by sector (total and per capita) in developed market-economy countries and developing countries, 1975 [a]

Sector	GDP by sector: Total			Per head of total population [b]			Per head of economically active population [b]		
	Developed market-economy countries (billions of dollars) (1)	Developing countries [c] (billions of dollars) (2)	Ratio (1)/(2) (3)	Developed market-economy countries (dollars) (4)	Developing countries [c] (dollars) (5)	Ratio (4)/(5) (6)	Developed market-economy countries (dollars) (7)	Developing countries [c] (dollars) (8)	Ratio (7)/(8) (9)
Agriculture	179	185	0.96	230	100	2.3	5 190	430	12.2
Industry	1 844	337	5.5	2 320	180	13.0	14 230	2 370	6.0
of which:									
Mining	57	47	1.2	70	30	2.9	17 810	9 220	1.9
Manufacturing	1 154	168	6.9	1 450	90	16.3	13 640	1 870	7.3
Consumer goods [d]	374	81	4.6	470	40	11.0	10 600	1 290	8.2
Intermediate goods [e]	219	40	5.5	280	20	13.1	18 720	3 250	5.8
Capital goods [f]	561	47	11.9	710	30	28.2	14 960	3 070	4.9
Electricity, gas and water	102	16	6.4	130	10	16.0	36 430	7 270	5.0
Construction	241	51	4.7	300	30	11.2			
Transport and communications	290	54	5.4	370	30	12.6	12 610	2 650	4.8
Services	2 056	340	6.0	2 590	180	14.4			
Total	4 079	862	4.7	5 130	460	11.2	12 290	1 230	10.0

Sources: UNCTAD, *Handbook of International Trade and Development Statistics, Supplement 1977* (United Nations publication, Sales No. E/F.78.II.D.1), and United Nations, *Statistical Yearbook 1977* (United Nations publication, Sales No. E/F.78.XVII.1).

[a] Estimates of GDP by sector were derived by applying to the regional GDP figures (available in the UNCTAD source) the estimated percentage shares of sectors in each region. The latter were estimated on the basis of the sectoral weights used for the construction of GDP and world industrial production indices (1970 = 100), brought up to 1975 using the sectoral and subsectoral indices; the relevant figures are found in the United Nations source mentioned above. Total economically active population and the proportion in agriculture were taken from FAO, *Production Yearbook*, 1970 and 1977, and the proportion in industry was estimated on the basis of sectoral weights shown in industrial employment index tables in the United Nations *Statistical Yearbook* and the estimates made by ILO on manufacturing employment for certain years, in *Labour Force Estimates and Projections, 1950-2000*, vol. V, *World Summary* (Geneva, 1977). Economically active population in other sectors was derived as a residual. Because of these crude methods of estimating, and also because the country groupings vary slightly for the different series, the table provides only rough orders of magnitude.

[b] Rounded to the nearest $10.

[c] Not including China and other socialist countries of Asia.

[d] ISIC divisions 31, 32, 33, 34 and 39.

[e] ISIC divisions 35 and 36.

TABLE 10.

Pattern of limitations on access to technology by developing countries

Type of limitation	Replies as to whether the country faced the specified limitation	
	Yes	*No*
1. Tied purchases of imported inputs, equipment and spare parts	Argentina, Chile, Cyprus, Ecuador, Greece, Iran, Malta, Mexico, Nigeria, Pakistan, Peru, Sri Lanka, Turkey	Republic of Korea
2. Restriction of exports (total prohibition, partial limitation, geographical constraint)	Argentina, Chile, Cyprus, Ecuador, Greece, Iran, Malta, Mexico, Nigeria, Pakistan, Peru, Sri Lanka, Turkey	Singapore
3. Requirement of guarantees against changes in taxes, tariffs and exchange rates affecting profits, royalties and remittances	Cyprus, Nigeria, Turkey	Greece, Iran, Malta, Mexico, Singapore
4. Limitation of competing supplies by:		
(a) restriction of competing imports	Cyprus, Greece, Mexico, Nigeria, Peru	Iran, Malta, Pakistan, Republic of Korea, Singapore, Turkey
(b) preventing competition for local resources	Greece, Malta, Mexico	Iran, Nigeria, Pakistan, Republic of Korea, Singapore
(c) obtaining local patents to eliminate competitors	Ecuador, Malta, Nigeria	Greece, Iran, Singapore
5. Constraints limiting the dynamic effects of the transfer		
(a) excessive use of expatriate personnel	Argentina, Malta, Mexico Nigeria, Peru, Turkey	Singapore
(b) discouragement of the development of local technical and research and development capabilities	Argentina, Ecuador, Greece, Malta, Mexico, Nigeria, Turkey	

TABLE 11.

Direct costs of transfer of technology in comparison with other relevant foreign exchange flows of developing countries, 1968[a]

Flows	*Value (millions of dollars)*	*Proportion of direct payments for transfer of technology (per cent)*
Outflows		
1. Direct payments for transfer of technology (patents, licenses, know-how, trademarks, and management and other technical services)	1,500	100
2. Technology-related payments:		
(a) Imports (c.i.f.) of machinery and equipment (excluding passenger vehicles) and of chemicals	18,420	8
(b) Profit on direct foreign investment (excluding oil-producing countries)[b]	1,721	87
3. Service payments on external public debt	4,022	37
Inflows		
4. Non-petroleum exports (f.o.b.)	29,350	5
5. Total official flows	6,710	22
6. Direct foreign investment (including reinvested earnings)	2,700	56

Sources:

Line 1: UNCTAD secretariat estimates (see text).

Line 2 *(a)*: United Nations, *Monthly Bulletin of Statistics,* vol. XXVI, No. 7 (July 1972).

Line 2 *(b)*: "The outflow of financial resources from developing countries: note by the UNCTAD secretariat" (TD/118/Supp.5), *loc. cit.*

Line 3: IBRD/IDA, *Annual Report, 1972.*

Lines 4, 5 and 6: UNCTAD, *Handbook of International Trade and Development Statistics, 1972* (United Nations publication, Sales No. E/F.72.II.D.3).

[a] Data do not include Southern European countries.

[b] Including oil-producing countries: $4,934 million.

TABLE 12.

Payments[a] by developing countries for the transfer of technology and their relationship to GDP and exports

Country and region	Most recent year available	Payments for transfer of technology for: Patents, licences, know-how and trademarks	Payments for transfer of technology for: Management and other technical services	Payments for transfer of technology for: Total	GDP	Exports	Payments for transfer of technology as proportion of: GDP	Payments for transfer of technology as proportion of: Exports
		(1)	(2)	(3)	(4)	(5)	(6)	(7)
		(Millions of dollars)			*(Billions of dollars)*		*(Per cent)*	
Latin America:								
Argentina	1970	70.5	45.3	115.8	23.4[b]	1.8	0.49	6.5
Brazil	1970	..	..	104.0	35.3	2.7	0.29	3.8
Chile	1969	8.2	..	(8.2)	6.1	1.1	0.13	0.8
Colombia	1966	..	..	26.7	5.4	0.5	0.49	5.3
Mexico	1968	..	..	200.0	27.1	1.3	0.74	15.9
Peru	1971	9.9	1.1	11.0	5.8[b]	0.9	0.19	1.2
Venezuela	1966	14.8	..	(14.8)	8.8	2.7	0.17	0.5
Sub-total		..	..	(480.5)	111.9	10.9	0.43	4.4
Africa:								
Nigeria	1965	19.0	14.8	33.8	4.7	0.8	0.72	4.5

TABLE 12. (Continued)

Asia:								
India	1969	6.4	43.2	49.6	49.1	1.8	0.1	2.7
Indonesia	1968	25.0	..	(25.0)	11.0	0.7	0.23	3.6
Iran	1970	1.7	1.6	3.3	11.2	2.4	0.03	0.1
Israel	1961-1965[c]	1.6	2.3	3.9	2.6	0.3	0.15	1.2
Republic of Korea	1970	2.1	..	(2.1)	8.1	0.8	0.03	0.3
Pakistan	1965-1970[c]	2.1	(100)	(102.1)	14.5	0.6	0.7	15.7
Sri Lanka	1970	0.1	9.2	9.3	2.2	0.3	0.42	2.7
Sub-total		39.0	(156.3)	(195.3)	98.7	7.0	0.2	2.8
Southern Europe:								
Greece	1966	..	..	2.6	6.4	0.4	0.04	0.6
Spain	1970	81.6	52.2	133.8	32.4	2.4	0.41	5.6
Turkey	1968	..	..	49.1	12.6	0.5	0.39	9.9
Yugoslavia	1970	5.4	..	(5.4)	12.3	1.7	0.04	0.3
Sub-total		..	..	(190.9)	63.7	5.0	0.3	3.8
TOTAL, excluding Southern Europe		..	..	709.6	215.3	18.7	0.33	3.8
TOTAL, including Southern Europe		..	..	900.5	279.0	23.7	0.32	3.8

Sources: Replies to the UNCTAD secretariat's questionnaire and other sources shown in the annex to document TD/106, *loc. cit.* (cf. foot-note 4 above). For Venezuela: Oficina Central de Coordinación y Planificiación (CORDIPLAN), Departamento Industrial, *II Encuesta Industrial: Documento Básico* (Caracas, November 1968).

NOTE. Parentheses indicate that the information available is incomplete.

[a] In most cases payments refer to the foreign exchange cost (in dollars, at current prices) of the transfer. For further details, see the annex to document TD/106, *loc. cit.* (cf. foot-note 4 above).

[b] UNCTAD secretariat estimate.

[c] Annual average.

TABLE 13.

Relationship between increase in payments for transfer of technology, manufacturing output and GDP for selected countries

Country	Period	Payments for transfer of technology		Annual average growth rate of			Relationship of growth rates of	
		Initial year	End year	payments for transfer of technology (PTT)	manufacturing output (MO)	real GDP	PTT / MO (i.e. column 3 divided by column 4)	PTT / GDP (i.e. column 3 divided by column 5)
		(1)	(2)	(3)	(4)	(5)	(6)	(7)
		Millions of dollars		Per cent per year			Ratios	
eloping countries								
Nigeria	1963-1965	13.8	33.8	55.5	9.3[a]	4.0	6.0	13.9
Korea (Republic of)	1967-1970	0.7	2.1	43.0	24.2	12.5[b]	1.8	3.4
Sri Lanka	1965-1970	2.0	9.2	36.0	8.6[a,c]	3.9	4.2	9.2
Argentina	1965-1970	35.1	115.8	26.9	5.0	3.9[c]	5.4	6.9
Brazil	1965-1969	42.5	91.0	20.9	9.7[d]	6.2	2.1	3.4
India	1959-1969	12.0	49.6	15.2	5.8	9.2[e]	2.6	1.7
Mexico	1953-1968	14.7[f]	120.0[f]	15.0	8.5	6.7[g]	1.8	2.2
Iran	1965-1970	1.1	1.7	10.1	11.8	10.4[c]	0.9	1.0

TABLE 13. (Continued)

Country	Period	Payments for transfer of technology: Initial year (1)	Payments for transfer of technology: End year (2)	Annual average growth rate of: payments for transfer of technology (PTT) (3)	Annual average growth rate of: manufacturing output (MO) (4)	Annual average growth rate of: real GDP (5)	Relationship of growth rates of: PTT/MO (i.e. column 3 divided by column 4) (6)	Relationship of growth rates of: PTT/GDP (i.e. column 3 divided by column 5) (7)
ier technology-receiving countries								
Turkey	1964-1968	6.2	49.1	65.5	10.5[a]	6.6	6.2	9.9
Yugoslavia	1965-1970	0.6	5.4	50.5	6.3	5.3[c]	8.0	9.5
Ireland	1963-1969	0.2	2.2	49.0	6.6	4.3	7.4	11.4
Greece	1959-1966	0.7	2.6	19.8	8.6[h]	9.4[h]	2.3	2.1
Spain	1965-1969	79.9	133.0	13.6	11.0	6.5	1.2	2.1
veloped market-economy countries		*Receipts from developing countries for the transfer of technology*						
France[i]	1967-1969	23.2	32.2	17.8				
Germany (Federal Republic of)[j]	1963-1969	50.3	105.4	13.1				
Belgium	1966-1970	5.6	8.8	11.6				
United States of America	1960-1969	175.6	442.3	10.8				
United Kingdom[k]	1965-1969	19.6	29.3	10.6				
Sweden	1965-1970	0.2	0.2	1.9				
Japan[l]	1968-1969	12.4	11.3	–				

Sources: As for table 10.

[a] 1953-1967.

TABLE 14.

Receipts from the sale of technology to developing countries, by eight developed market-economy countries, 1965-1969

(Millions of dollars)

Country	*1965*	*1966*	*1967*	*1968*	*1969*
United States of America					
from Latin America	185	195	206	241	251
from other developing countries	130	135	162	179	191
Total	315	330	368	420	442
France					
from Latin America	..	..	6.0	3.6	5.3
from other developing countries	..	..	53.6	12.7	26.9
Total			59.6	16.3	32.2
Federal Republic of Germany					
from Latin America	..	..	..	20.6	19.7
from other developing countries	..	..	..	18.7	14.8
Total				39.3	34.5
*United Kingdom**					
from Latin America	..	..	..	..	..
from other developing countries	..	..	..	..	..
Total	19.6	26.2	20.7	27.2	29.3

Receipts from the sale of technology to developing countries, by eight developed market-economy countries, 1965-1969

(Millions of dollars)

Country	*1965*	*1966*	*1967*	*1968*	*1969*
Japan					
from Latin America	..	..	..	..	..
from other developing countries	..	..	..	12.4	11.3
Total				12.4	11.3
Belgium					
from Latin America	..	0.7	0.9	0.5	0.7
from other developing countries	..	4.9	2.4	3.5	2.6
Total		5.6	3.3	4.0	3.3
Denmark					
from Latin America	..	0.9	..	0.1	0.9
from other developing countries	1.7	0.2	1.5	4.4	2.3
Total	1.7	1.1	1.5	4.5	3.2
Sweden					
from Latin America	0.1	0.1	0.1	0.1	0.1
from other developing countries	0.1	0.1	0.2	0.1	0.3
Total	0.2	0.2	0.2	0.2	0.4

Source: Replies to UNCTAD questionnaire.

NOTE. All data exclude receipts from Southern European countries, except for the United Kingdom.

* Excluding receipts from petroleum producers.

Figure 2. Possible network of linkages of a national centre for technological development

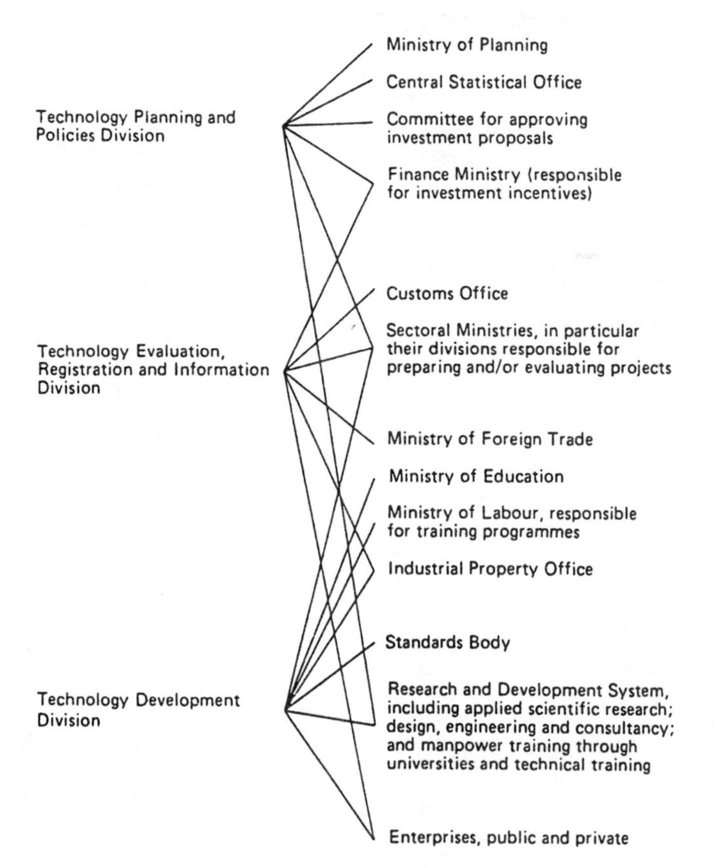

Source: Technological Transformation of Developing Countries, Discussion Paper No. 115 (Lund University, Research Policy Program, Sweden, 1978), p. 26.

PART III RESOURCE BIBLIOGRAPHY

This bibliography is entirely restricted to publications in English language and covers the literature since 1970. In a bibliography of this nature, it is essential that the material be as contemporary as possible, while at the same time it was thought desirable to provide a balanced weight of materials discussed over the last decade.

With respect to classification of the material, the form listed below seems to be appropriate. This classification is arbitrary, however much cross indexing is required.

A. Problems, Issues and Trends;
B. Analytical Methods;
C. Strategies and Policies; and
D. Country Studies.

Only the book section has been classified since the selected periodical articles and specialized publications dealt with specific categories from the above mentioned list of categories.

Many of the annotations in this section have been compiled from the Journal of Economic Literature, World Bank Publications, IMF-IBRD Joint Library Periodicals, Finance and Development, U.N. Documents and Publisher's Book Promotion Pamphlets.

I. BOOKS

DEVELOPMENT (GENERAL)

01 Abraham, M. Francis
A B PERSPECTIVES ON MODERNIZATION: TOWARD A GENERAL THEORY OF THIRD WORLD DEVELOPMENT
Washington, D.C.: University Press of America, 1980.

02 Adelman, Irma and Morris, Cynthia Taft
B ECONOMIC GROWTH AND SOCIAL EQUITY IN DEVELOPING COUNTRIES
Stanford, Calif.: Stanford University Press, 1973.

A quantitative investigation of the interactions among economic growth, political participation, and the distribution of income in noncommunist developing nations. The study is based on data (presented in the earlier study, Society, politics, and economic development) from 74 countries which is given in the form of 48 qualitative measures of the [countries] social, economic, and political characteristics, and it includes the use of discriminant analysis in an examination of the forces tending to increase political participation and the use of a stepwise analysis of variance technique in analyzing the distribution of income.

03 Albin, Peter S.
A B PROGRESS WITHOUT POVERTY; SOCIALLY RESPONSIBLE ECONOMIC GROWTH
New York: Basic Books, 1978.

Examines the relationship among important social tendencies, growth processes, and growth policies and argues for the return of the growth economy, with the caveat that social objectives and policy directions be reformulated to avert ecological disaster and to improve economic welfare. Using a dualistic imbalance framework, explores the style and impact of unbalanced growth in modern industrial capitalism, focusing on educational policy, income distribution, and the control of technology, poverty, and urban decay. Concludes with policy recommendations for a program of social and technical advance that is geared to the intelligent management of a growth economy and the renovation of its distributive mechanisms. An appendix presents a dualistic-imbalance model of modern industrial growth.

04
A B C
Alexander, Robert J.
A NEW DEVELOPMENT STRATEGY
Maryknoll, N.Y.: Orbis Books, 1976.

Focusing on the demand side of the development equation, this monograph concerns itself with an economic development strategy of import substitution where industries are established to manufacture products for which a home market has already been created by imports. Analyzing the effect on development of this assured demand, and exploring the limit to which this strategy can be used, the author, looks in detail at the prerequisites for the use of this method (substantial imports and protection for newly created industries) and discusses the priorities for private and public investment in this phase. Contends that this process provides a basis for developing countries to decide which projects should be undertaken first and which can be postponed until later.

05
A C
Alvarez, Francisco Casanova
NEW HORIZONS FOR THE THIRD WORLD
Washington, D.C.: Public Affairs Press, 1976.

Presents the factors leading to approval of the Charter of Economic Rights and Duties of States by the United Nations General Assembly on 12 December 1974. Shows that the charter, with the main objective of overcoming the injustice prevailing in economic relations between nations and [elimination of] the dependence of Third World countries on industrial nations, owes its origin and adoption to President Luis Echeverria of Mexico. Argues that the developing nations remain essentially colonized and dependent entities of the industrialized world. Concludes that the future world will be less unjust and less ridden with anxiety, more secure and better able to care for its own if we respect the principles of the charter.

06
A B C
Anell, Lars and Nygren, Birgitta
THE DEVELOPING COUNTRIES AND THE WORLD ECONOMIC ORDER
New York: St. Marin's Press, 1980.

Explores the possible form, functioning, and enforcement of a New International Economic Order (NIEO). Provides an account of the demands of developing countries for a better allocation of the world's resources and considers the early cooperation between developing and developed countries, particularly resolutions passed at various U.N. General Assembly sessions. Also analyzes and comments on the central NIEO demands. Among the possible actions the authors suggest developing countries could take are: (1) force industrialized countries to increase the flow and quality of aid by threatening trade discrimination; (2) establish a list of honest consultancy firms and a file of information on technology procurement; and (3) feel free to steal patents from big corporations and make use of copyrights without compensation.

07 Angelopoulos, Angelos T.
A C FOR A NEW POLICY OF INTERNATIONAL DEVELOPMENT
New York: Praeger, 1977.

08 Angelopoulos, Angelos T.
A C THE THIRD WORLD AND THE RICH COUNTRIES;
PROSPECTS FOR THE YEAR 2000
Translated by N. Constantinidis and C. R. Corner
New York: Praeger, 1972.

An examination and projection of the gap in incomes between the developed and underdeveloped countries of the world. The author brings data on and discusses the indicators of poverty, the population explosion in the developing world, the main causes of economic backwardness, the "myth" of development aid, the need for a new international development strategy, various strategies of development financing, precipitating factors in the emergence of the Third World, economic growth and forecasts of world income in the year 2000, and the possibilities of China becoming the spokesman for the Third World.

09 Arkhurst, Frederick S., ed.
B C D AFRICA IN THE SEVENTIES AND EIGHTIES;
ISSUES IN DEVELOPMENT
New York and London: Praeger in cooperation with the Adlai Stevenson Institute of International Affairs, 1970.

Eleven experts in various fields express their views in a symposium "Africa in the 1980's" which met in Chicago in early 1969 under the auspices of the Adlai Stevenson Institute of International Affairs. The purpose...was to attempt to draw a portrait of Africa in the 1980's on the basis of the experience of the past decade and, also, on the basis of current trends in the area of politics, economic development, population, agriculture, trade, education and law - all viewed as composite and interactive factors in the development process.

10 Arndt, H. W., et al.
A B C THE WORLD ECONOMIC CRISIS: A COMMONWEALTH PERSPECTIVE
London: Commonwealth Secretariat, 1980.

Report of a group of experts from Commonwealth countries on obstacles to structural change and sustained economic growth, with recommendations for specific measures by which developed and developing countries might act to reduce or eliminate such constraints. Focuses on the implications of the world economic crises - inflation, slowdown of economic growth, and staggering disequilibria in balance of payments - for the developing countries of the Third World. Stresses the need for collective action in view of the interdependence of the world economy.

11 Bairoch, Paul
A C D THE ECONOMIC DEVELOPMENT OF THE THIRD WORLD SINCE 1900
Translated from the fourth French edition by Cynthia Postan
Berkeley: University of California Press, 1975.

The author covers a wide range of factors important to development, namely population, agriculture, extractive industry, manufacturing industry, foreign trade, education, urbanization, the labor force and employment, and macroeconomic data. Particular attention is devoted to the development of agriculture. Comparison is drawn between the economic progress of Third World countries and developed countries at a similar stage of industrialization. Twenty-four countries were selected for the analysis, representing 80 percent of the population of the Third World. These include seven countries from each of Africa, Latin-America, and Asia respectively, and three countries from the Middle East.

12 Bairoch, Paul and Levy-Leboyer, Maurice, eds.
A B DISPARITIES IN ECONOMIC DEVELOPMENT SINCE THE INDUSTRIAL REVOLUTION
New York: St. Martin's Press, 1981.

Collection of thirty-five previously unpublished essays presented at the 7th International Economic History Congress in Edinburgh in August 1978. Main theme deals with disparities in economic development. Concerns differences in income at micro-regional and international levels. In four parts: (1) discussing economic disparities among nations (two papers on international disparities: ten on the Third World and five on the developed world); (2) covering regional economic disparities (eight essays on northern, western, and central Europe; three on France; two on Southern Europe and one on the Third World); (3) detailing relations between regional and national disparities (two papers); and (4) discussing the methodological aspects of measurement of economic disparities (two papers).

13 Baldwin, Robert E.
B C ECONOMIC DEVELOPMENT AND GROWTH
New York, London, Sydney and Toronto: John Wiley and Sons, Inc., 1972.

This short text seeks to provide "an analysis of economic development that in terms of breadth and sophistication lies between the usual elementary and advanced approaches to the development topic." It is organized around three themes, i.e., what the nature of growth problem is, what the main theories of growth and development are, and what the main policy issues facing less developed countries are. Therefore, the chapters deal with the characteristics of poverty, various classical development theories relatively more recent contributions to development theory, national and sectoral policies for growth, and issues in the financing of development.

14 Bauer, P.T.
B C DISSENT ON DEVELOPMENT. STUDIES AND DEBATES IN DEVELOPMENT ECONOMICS
Cambridge, Mass.: Harvard University Press, 1972.

A collection of previously published articles, essays, and book reviews, some of which have been rewritten and expanded, dealing with various theoretical and empirical issues in economic development. Part One ("Ideology and Experience") examines general problems of concept method, analysis, historical experience and policy in economic development, such as the vicious circle of poverty, the widening gap, central planning, foreign aid, Marxism, etc. Part Two ("Case Studies") features five of the author's studies on developing countries, particularly Nigeria and India. Part Three ("Review Articles") brings book reviews on such well known books as W. Arthur Lewis' The Theory of Economic Growth, Benjamin Higgins' Economic Development, Walt W. Rostow's The Stages of Economic Growth, Thomas Balogh's The Economics of Poverty, and other volumes by Gunnar Myrdal, John Pincus, Harry G. Johnson, E.A.G. Robinson, B.K. Madan and Jagdish Bhagwati.

15 Bauer, P.T.
A B EQUALITY, THE THIRD WORLD AND ECONOMIC DELUSION
Cambridge, Mass.: Harvard University Press, 1981.

Critique of methods and finding of contemporary economics, particularly development economics, arguing that there is a hiatus between accepted opinion and evident reality. All but four chapters are extended and/or revised versions of previously published articles. In the three parts: equality, the West and the Third World, and the state of economics. Criticizes economics and especially development economics for disregard of personal qualities and social and political arrangements as determinants of economic achievement and for ignoring the role of external contracts in extending markets. Notes that the benefits of mathematical economics have been bought at the cost of an uncritical attitude, which has led to inappropriate use and in some cases to an emphasis on form rather than substance.

16
A C

Berry, Leonard and Kates, Robert W., eds.
MAKING THE MOST OF THE LEAST
New York: Holmes and Meier Publishers, 1979.

The poverty faced by Third World countries today seriously challenges the stability of the world order. The contributors look torward the restructuring of the present economic order by establishing "harmonious linkages" between the industrialized and nonindustrialized worlds. A welcome addition to the literature on economic development.

17
A B C

Bhatt, V. V.
DEVELOPMENT PERSPECTIVES: PROBLEM, STRATEGY AND POLICIES
Oxford; New York: Sydney and Toronto: Pergamon Press, 1980.

Discusses the dynamics of the socioeconomic system in terms of the cumulative and cyclical changes in economic institutions, ideologies, and technology. Stresses the importance of: upgrading traditional technology and adapting modern technology to given situations; the financial system, since it affects savings and shapes the pattern of resource allocation; and upgrading of agricultural organization and technology. Sets forth as necessary for the development process: the stability of the international currencey and the international monetary system, which the author proposes be linked to prices of primary products; the shaping of the international monetary-financial-trade system to be consistent with LDC's development strategy; and viewing the process of socioeconomic development as an integral part of nation-building and of building the international community.

18
A C

Brown, Lester R.
THE GLOBAL ECONOMIC PROSPECT: NEW SOURCES OF ECONOMIC STRESS
Worldwatch Paper no. 20
Washington, D.C.: Worldwatch Institute; New York, 1978.

Considers the relationship between the expanding global economy and the earth's natural systems. Discusses the increase in fuel costs, suggesting that the world is running out of cheap energy; diminishing returns in grain production and to fertilizer use; overfishing; global inflation; capital shortages; unemployment; and the changing growth prospect. Concludes that future economic policies must shift from growth to sustainability; not advocating abandonment of growth as a goal, but with concern for carrying capacities of biological system. Fisheries, forests, grasslands, and croplands, require development of alternative energy sources and population policies consistent with resource availability.

19
A B C

Chenery, Hollis and Syrquin, Moises
PATTERNS OF DEVELOPMENT, 1950-1970
Assisted by Hazel Elkington
New York and London: Oxford University Press, 1975.

Examines principal changes in economic structure that normally accompany economic growth, focusing on resource mobilization and allocation, particularly those aspects needed to sustain further growth. These aspects are treated in a uniform econometric framework to provide a consistent description of a number of interrelated types of structural change and also to identify systematic differences in development patterns among countries that are following different development strategies. The major aim of the research is to separate the effects of universal factors affecting all countries from particular characteristics. The authors use data for 101 countries in the period 1950 to 1970. Countries are grouped into three categories: large country, balanced allocation; small country, industry specialization. Chapter 5 compares the results obtained from time-series data with those observed from cross-sectional data. Results are obtained from regression techniques, where income level and population are treated as exogenous variables. The demographic variables show how the movement of population from rural to urban areas and lowering of the birth rate and death rate have influenced demand and supply of labor. A technical appendix discusses the methods used, the problems encountered, and all the regression equations specified in this study.

20
A B C

Chenery, Hollis B., et al., eds.
STUDIES IN DEVELOPMENT PLANNING
Cambridge, Mass.: Harvard University Press, 1971.

Attempts to bring together the contributors' varied backgrounds in both field work and the use of quantitative techniques and show how modern methods can be used in operational development planning.

21
A B

Chodak, Szymon
SOCIETAL DEVELOPMENT: FIVE APPROACHES WITH CONCLUSIONS FROM COMPARATIVE ANALYSIS
New York: Oxford University Press, 1973.

A sociologist analyzes the development and change of societies using five different conceptual approaches, attempting to view the processes of development in society from a multidimensional synthesizing perspective. These five approaches are called: "Evolutionary Theories," "Development - The Growing Societal Systemness," "Development and Innovation in the Search for Security," "Economic and Political Development," and "Modernization." The author gives references to the societal development which has taken place in various parts of the world and under different political systems.

22
A B C

Colman, David and Nixson, Frederick
ECONOMICS OF CHANGE IN LESS DEVELOPED COUNTRIES
New York: Wiley, Halsted Press, 1978.

Analyzes the changes that are occurring in the less-developed countries (LDC's); considers the problems generated by change; and examines the agents of change. Emphasizes the internal (rather than the international) aspects of development and focuses on economic inequality within LDC's and the impact on the development process in agriculture and industry of different income distributions. Although recognizing the impact of transnational corporations on the nature and characteristics of development within the LDC's, the authors argue that it is the LDC government that is responsible for the economic policies pursued. Also outlines the concepts and measurement of development, and reviews the literature on economic theorizing about development. A final chapter discusses inflation and migration in LDC's. Authors note that too often policy recommendations ignore political acceptability and recommend that the economist should cooperate with the political scientist in the study of inflation and with the sociologist in the study of rural urban migration.

23
A B C

Corbet, Hugh and Jackson, Robert, eds.
IN SEARCH OF A NEW WORLD ECONOMIC ORDER
New York and Toronto: Wiley, Halsted Press, 1974.

Focuses on the reform of the international commercial systems for further liberalizations of world trade. Papers are grouped into four categories: (1) introduction, (2) general factors affecting negotiations, (3) outside issues of significance, (4) issues on the agenda.

24
A B C D

Fields, Gary S.
POVERTY, INEQUALITY, AND DEVELOPMENT
New York and London: Cambridge University Press, 1980.

Focuses on the distributional aspects of economic development and explores the impact of the rate and type of growth on poverty and inequality in poor countries. Findings show that in general growth reduces poverty, but a high aggregate growth rate is neither necessary nor sufficient for reducing absolute poverty or relative inequality. Uses case studies of distribution and development in Costa Rica, Sri Lanka, India, Brazil, the Phillippines, and Taiwan to examine which combinations of circumstances and policies led to differential performance. Concludes that a commitment to developing to help the poor does not guarantee progress, but it helps a great deal. In its absence, the flow of resources to the haves, with only some trickle down to the have-nots, will be perpetuated.

25 Finger, J. M.
A B D INDUSTRIAL COUNTRY POLICY AND ADJUSTMENT TO IMPORTS FROM DEVELOPING COUNTRIES
World Bank Staff Working Paper no. 470, July 1981.

A background study for World Development Report 1981. Reviews and interprets recent analyses of the policies established by industrial countries in response to increasing imports from developing countries.

26 Finger, Nachum
A C D THE IMPACT OF GOVERNMENT SUBSIDIES ON INDUSTRIAL MANAGEMENT: THE ISRAELI EXPERIENCE
New York: Praeger, 1971.

27 Fitzgerald, E. V.
A B PUBLIC SECTOR INVESTMENT PLANNING FOR DEVELOPING COUNTRIES
New York: Holmes and Meier, 1978.

28 Florence, P. Sargant
A B C ECONOMICS AND SOCIOLOGY OF INDUSTRY: A REALISTIC ANALYSIS OF DEVELOPMENT
Baltimore, Md.: Johns Hopkins University Press, 1969.

29 Frank, Andre Gunder
A B C CRISIS IN THE THIRD WORLD
New York: Holmes and Meier, 1981.

30 Frank, Andre Gunder
A B DEPENDENT ACCUMULATION AND UNDERDEVELOPMENT
New York and London: Monthly Review Press, 1979.

Explains underdevelopment by an analysis of the production and exchange relations of dependence. Distinguishes the three main stages or periods in this world embracing process of capital accumulation and capitalist development: mercantilist (1500-1770), industrial capitalist (1770-1870), and imperialist (1870-1930). Analyzes each period in terms of history, trade relations between the metropolis and the periphery, and transformation of the modes or relations of production, and the development of underdevelopment in the principal regions of Asia, Africa, and the Americas.

31
A B D

Frank, Charles R., Jr., and Webb, Richard C., eds.
INCOME DISTRIBUTION AND GROWTH IN THE LESS-DEVELOPED COUNTRIES
Washington, D.C.: Brookings Institution, 1977.

Fourteen previously unpublished essays representing part of the results of a project undertaken jointly by the Brookings Institution and the Woodrow Wilson School of Public and International Affairs at Princeton University, dealing with the relation between income distribution and economic growth in the developing countries. The first two articles present an overview of income distribution policy and discuss the causes of growth and income distribution in LDC's, respectively. The next nine examine the relation between income distribution and different economic policies and factors, including: industrialization, education, population, wage, fiscal, agricultural, public works, health and urban land policies.

32
A B

Gant, George F.
DEVELOPMENT ADMINISTRATION - CONCEPTS, GOALS, METHODS
Madison, Wisconsin: The University of Wisconsin Press, 1979.

Growth and modernization in the less developed countries (LDC's) during the past three decades has frequently depended upon the state's ability to plan and manage a range of developmental activities. Gant's study of development administration looks at some of the issues that could be of concern to managers in LDC's: in particular, coordination, budgeting, the selection of personnel, training, etc. He also delves into the administrative side of certain specific governmental concerns, such as family planning and education, drawing on examples from a number of Asian countries. This is not a book which goes into much technical detail. Nor does it tell one how to design an efficient administrative setup. Primarily for the general reader interested in an overview of these topics.

33
A B D

Garbacz, Christopher
INDUSTRIAL POLARIZATION UNDER ECONOMIC INTEGRATION IN LATIN AMERICA
Austin, Texas: Bureau of Business Research, Graduate School of Business, The University of Texas, 1971.

The author discusses the problem of increased disparities in the levels of regional economic development that tend to come about as a result of economic integration. The political and economic implications of industrial polarization are studies within the context and experience of the Central American Common Market and the Latin American Free Trade Association. Finally, the author considers the problem in the light of the planned Latin American Common Market, discussing the various measures that could be taken as well as the implications for the future.

34
A B C
Garzouzi, Eva
ECONOMIC GROWTH AND DEVELOPMENT: THE LESS DEVELOPED COUNTRIES
New York: Vantage Press, 1972.

Essays to consolidate into one readable text the whole of the economics of growth and development. Part I discusses the meaning and theories of economic development, outlines historical patterns of development, and summarizes the impact of capital, agriculture, industry, monetary and fiscal policies, international trade, and foreign aid on economic growth. Part II presents comparative analyses of developing regions, including Latin America, the Middle East and North Africa, Africa south of the Sahara, and Southeast Asia.

35
A B C D
Geithman, David T., ed.
FISCAL POLICY FOR INDUSTRIALIZATION AND DEVELOPMENT IN LATIN AMERICA
Gainesville: University Presses of Florida.

Collection of 10 previously unpublished papers (and related comments) presented at the Twenty-First Annual Latin American Conference held in February 1971. Central theme of the conference was the analysis and evaluation of the interaction among fiscal problems, fiscal tools, and fiscal systems in the industrializing economies of Latin America.

36
A B C
Ghai, D. P.
THE BASIC-NEEDS APPROACH TO DEVELOPMENT: SOME ISSUES REGARDING CONCEPTS AND METHODOLOGY
ILO, Geneva, 1977.

Contains five papers which discuss issues which arise in the formulation of criteria and approaches for the promotion of employment and the satisfaction of the basic needs of a country's population. Presents the first results of the research and conceptual work initiated by the ILO to help countries implement the basic needs-oriented strategy recommended by the World Employment Conference in 1976.

37
A B C
Gianaris, Nicholas V.
ECONOMIC DEVELOPMENT: THOUGHT AND PROBLEMS
North Quincy, Mass.: Christopher Publishing House, 1978.

Part one examines the process of development, the historical perspective, mathematical models, and modern theories of development; part two considers domestic problems of development, specifically land and other natural resources, human resources (particularly the role of education), capital formation and technological change, the allocation of resources, and the role of government and planning; part three discusses the international aspects of development (foreign trade, aid, investment, and multinationals) and current issues such as environmental problems, the status of women, income inequalities, and discrimination.

38
A B C D
Giersch, Herbert, ed.
INTERNATIONAL ECONOMIC DEVELOPMENT AND RESOURCE TRANSFER: WORKSHOP 1978
Tubingen, Germany: J. C. B. Mohr, 1979.

Twenty-four previously unpublished papers from a workshop held in June 1978 at the Institut fur Weltwirtschaft, Kiel University. Contributions organized under ten headings: Rural Industrialization, Employment and Economic Development; Choice of Techniques and Industries for Growth and Employment; Agricultural Patterns and Policies in Developing Countries; Hypotheses for the Commodity Composition of East-West Trade; The Relationship Between the Domestic and International Sectors in Economic Development; Patterns of Trade in Services and Knowledge; Changes in Industrial Interdependencies and Final Demand in Economic Development; Public Aid for Investment in Manufacturing Industries; Institutional and Economic Criteria for the Choice of Technology in Developing Countries; and Problems of Measuring the Production and Absorption of Technologies in Developing Countries.

39
A B C
Gierst, Friedrich and Matthews, Stuart R.
GUIDELINES FOR CONTRACTING FOR INDUSTRIAL PROJECTS IN DEVELOPING COUNTRIES
New York: United Nations Publications, 1975.

Designed to serve public and private organizations in developing countries as a guide in preparing contracts concerned with industrial investment projects. Examines various stages involved in the preparation of an industrial project and discusses the basic types of contacts involved (i.e. those with financial institutions, with consultants and with contractors).

40
A B C
Gill, Richard T.
ECONOMIC DEVELOPMENT: PAST AND PRESENT

Third Edition. Foundations of Modern Economics.
Englewood Cliffs, N.J.: Prentice-Hall, 1973.

Third edition of an introductory textbook with revisions of the discussions. The Green Revolution, two-gap analysis of foreign aid, Denison-Jorgenson-Griliches studies of factors affecting United States economic growth and Leibenstein's "X-efficiency" concept have been added. Statistical tables have been updated to include figures on Chinese economic growth. Six chapters cover: 1) General factors in economic development, 2) Theories of development, 3) Beginnings of development in advanced countries, 4) Growth of the American economy, 5) Problems of underdeveloped countries, and 6) Development in China and India.

41 Goulet, Denis
A B C THE CRUEL CHOICE: A NEW CONCEPT IN THE THEORY OF DEVELOPMENT
Cambridge, Mass.: Center for the Study of Development and Social Change, Atheneum, 1971.

This work is intended to probe moral dilemmas faced by economic and social development. Its central concern is that philosophical conceptions about the "good life" and the "good society" should be of more profound importance in assessing alternative paths to development than economic, political, or technological questions. The theoretical analysis is based on two concepts: "vulnerability" and "existence rationality." Vulnerability is defined as exposure to forces that can not be controlled, and is expressed in the failure of many low-income countries to attain their development goals, as well as in manifestations of mass alienation in certain societies where prosperity has already been achieved. Existence rationality denotes those strategies used by all societies to possess information and to make practical choices designed to assure survival and satisfy their needs for esteem and freedom. These strategies vary with a country's needs and are conditioned by numerous constraints.

42 Griffin, Keith
A B C INTERNATIONAL INEQUALITY AND NATIONAL POVERTY
New York: Holms & Meier, 1978.

Nine essays, seven previously published between 1970 and 1978. Challenges the classical assumption that unrestricted international intercourse will reduce inequality and poverty. Argues that forces creating inequality are automatic, and not due to malevolence of developed nations or corporations, but that the motor of change in the contemporary world economy is technical innovation. Since the advances tend to be concentrated in the developed countries where they are applicable to their technology, rich countries are able to extract supra-normal profits and rents from the poor countries through trade. The high level of factor earnings in rich countries attract the most valuable financial and human resources of the poor countries through induced international migration. Divided into two parts, part one deals with international inequality and discusses: the international transmission of inequality; multinational corporations; foreign capital, domestic savings, and economic development; emigration, and the New International Economic Order. The essays in part two focus on national poverty, discussing the facts of poverty in the Third World, analyzing models of development, and assessing the Chinese system of incentives.

43 Griffin, Keith B. and Enos, John L.
A B C PLANNING DEVELOPMENT
Reading, Mass.; Don Mills, Ontario; Sydney; London; and Manila: Addison-Wesley, 1971.

Part of a series intended to serve as guidebooks on development economics, this book deals with practical problems of planning and economic policy in underdeveloped countries. Consists of four parts: 1) the role of planning, 2) quantitative planning techniques, 3) sector policies, and 4) planning in practice with reference to Chile, Columbia, Ghana, India, Pakistan and Turkey.

44 Hagen, Everett E.
A B C THE ECONOMICS OF DEVELOPMENT
Revised Edition. The Irwin Series in Economics.
Homewood, Ill.: Irwin, 1975.

Revised edition with two new chapters added, one dealing with the earth's stock of minerals and economic growth, and the other on the relationships between economic growth and the distribution of income. Chapters on population and economic planning have been extensively revised, with the former focusing on the relationship of food supply to continued world growth. Additional changes include: reorganization of the discussion of growth theories; a considerably augmented discussion of entrepreneurhsip; and a reorganization of the chapters on import substitution versus export expansion and external finance.

45 Helleiner, G. K., ed.
B C A WORLD DIVIDED: THE LESS DEVELOPED COUNTRIES IN THE INTERNATIONAL ECONOMY
Perspectives on Development, no. 5
New York; London and Melbourne: Cambridge University Press, 1976.

Twelve papers discussing the new policies and instruments needed if the interests of poor nations are to be met. Within the realm of trade, consideration is given to the possibility of increased cooperation through: supply management schemes; bargaining capacity and power; closer ties with other less developed countries; and the development of alternative marketing channels and joint sales efforts. Relations between the less developed countries and transnational firms is then considered with special attention given to the factors affecting the bargaining position of the countries. Issues in international finance and monetary policy are: the borrowing of Eurodollars by less developed countries, internationally agreed upon principles for an honorable debt default, and interests of less developed countries in a new international monetary order. Another paper considers means by which a self-reliant but poor country can seek to conduct its economic affairs in the face of a most inhospitable and uncertain international environment. The concluding paper considers the implication of the new mood in the less developed countries for future international organisation.

46 Hermassi, Elbaki
A C D THE THIRD WORLD REASSESSED
Berkeley: University of California Press, 1980.

47 Horowitz, Irving Louis, ed.
A B C EQUITY, INCOME, AND POLICY: COMPARATIVE STUDIES IN THREE WORLDS OF DEVELOPMENT
New York and London: Praeger, 1977.

Ten previously unpublished papers by sociologists and economists on the multiple ideologies of development and the drive toward equity congruent with different social systems. Six essays address the problems of the "First World," i.e. those types of societies dominated by a free market and an open society, where the main problem would seem to be how to maintain growth and development while providing distributive justice. Two papers look at the "Second World" of socialism; these assume the central role of state power as imposing its will to produce equity. The remaining papers consider the Third World, examining in particular income distribution in Tanzania and economic equality and social class in general.

48 Jalan, Bimal
A B C ESSAYS IN DEVELOPMENT POLICY
Delhi: S. G. Wasani for Macmillan of India, 1975.

A common theme of the 11 essays (some previously published) is the explicit reference to political philosophies involved in the choices of means and objectives of development and social change. Essays include: discussion of self-reliance objectives; trade and industrialization policies; distribution of income; the project evaluation manual of Professors Little and Mirrlees; UNIDO guidelines for project evaluation; criteria for determination of appropriate terms of aid assistance; the definition and assessment of performance in developing countries; the history of the United Nations Capital Development Fund, the World Bank, and the International Development Association; and an analysis of the principal recommendations of the Pearson Commission Report (1969).

49
B C
Jumper, Sidney R.; Bell, Thomas L. and Ralston, Bruce A.
ECONOMIC GROWTH AND DISPARITIES: A WORLD VIEW
Englewood Cliffs, N.J.: Prentice-Hall, 1980.

The authors emphasize understanding of real world differences in levels of human development, rather than sophisticated analytical procedures. In seven parts: geographical concepts; the factors influencing variations in levels of development; world food supplies; minerals; factors affecting intensity of manufacturing development; the service industries; and a summary of the role of geographers in facing these development problems.

50
A B C
Kahn, Herman
WORLD ECONOMIC DEVELOPMENT: 1979 AND BEYOND
With the Hudson Institute.
Boulder: Westview Press, 1979.

Examines economic prospects focusing on the period 1978-2000, and particularly the earlier part of the period. In two parts, part one presents the general historical framework, concepts, and perspectives on economic growth and cultural change. Part two examines the major trends and problems of the real world, focusing on the elements of change and continuity in both the advanced and developing economies. Rejects attempts by some to stop the world and argues for and suggests strategies for rapid worldwide economic growth, for Third World industrialization, and for the use of advanced (or at least appropriate) technology.

51
A B C
Kasdan, Alan R.
THE THIRD WORLD: A NEW FOCUS FOR DEVELOPMENT
Cambridge, Mass.: Schenkman Publishing, 1973.

52
B
Kindleberger, Charles P. and Herrick, Bruce
ECONOMIC DEVELOPMENT
Third Edition. Economics Handbook Series.
New York; London; Paris and Tokyo: McGraw-Hill, 1977.

Textbook that survey[s] the present panorama of international poverty, the applications to it of economic analysis, and the policies for improvement that the analysis implies. This edition which has been completely rewritten and updated, includes new chapters on: population, urbanization, collective international action, employment, income distribution, and the theories of economic development.

53 Leipziger, Danny M., ed.
A B C BASIC NEEDS AND DEVELOPMENT
Foreword by Paul P. Streeten
Cambridge, Mass.: Oelgeschlager, Gunn & Hain, 1981.

Five previously unpublished essays discuss the potential contribution of the basic needs approach to developmental theory and practice. Michael J. Crosswell gives his views in two essays on a development planning approach and on growth, poverty alleviation, and foreign assistance. Maureen A. Lewis discusses sectional aspects of the linkages among population, nutrition, and health. Danny M. Leipziger writes about policy issues and the basic human needs approach. Martha de Melo presents a case study of Sri Lanka focusing on the effects of alternative approaches to basic human needs. The authors are all economists.

54 Leontief, Wassily, et al.
A B C THE FUTURE OF THE WORLD ECONOMY: A UNITED NATIONS STUDY
New York: Oxford University Press, 1977.

Investigates the interrelationships between future world economic growth and availability of natural resources, pollution, and the impact of environmental policies. Includes a set of alternative projections of the demographic, economic, and environmental states of the world in the years 1980, 1990, and 2000 with a comparison with the world economy of 1970. Constructs a multiregional input-output economic model of the world economy. Investigates some of the main problems of economic growth and development in the world as a whole, with special accent on problems encountered by the developing countries. The findings include: (1) target rates of growth of gross product in the developing regions...are not sufficient to start closing the income gap between the developing and the developed countries; (2) the principal limits to sustained economic growth and accelerated development are political, social and institutional in character rather than physical; (3) the necessary increased food production is technically feasible, but dependent on drastically favorable public policy measure; (4) pollution is not an unmanageable problem.

55 Lin, Ching-Yuan
A C D DEVELOPING COUNTRIES IN A TURBULENT WORLD: PATTERNS OF ADJUSTMENT SINCE THE OIL CRISIS
New York: Praeger, 1981.

Examines national authorities' policy reactions to changes in the world economy since 1973, to determine whether differences in national economic performances can be explained in terms of differences in their policy reactions. Investigates global patterns of absorption, production, and adjustment since the oil crisis; global expenditure flows before and after the crisis; and international bank transactions and world trade. Reviews the experiences of developing countries during the period, focusing on non-oil countries. Finds that collectively the non-oil developing countries experienced a much milder contraction of domestic demand and real ouput than the more developed countries after the disturbances in 1973-75, although individual experiences varied; however, inflation remains persistent. Argues that most developing countries did not pursue demand management policies early enough to counteract sharp changes in external demand.

56 Madhava, K. B., ed.
A B C D INTERNATIONAL DEVELOPMENT, 1969: CHALLENGES TO PREVALENT IDEAS ON DEVELOPMENT
Dobbs Ferry: Oceana for Society for International Development, 1970.

Contains the proceedings of the 11th World Conference of the Society for International Development held in 1969 in New Delhi. The theme "Challenges to Prevalent Ideas on Development" was carried out through roundtable discussions centering on: the redefinition of goals; foreign aid; manpower, education, and development; population communication; social communication; political and social-cultural requisites; and challenges to theorists and strategists.

57 May, Brian
A C D THE THIRD WORLD CALAMITY
London and Boston: Routledge & Kegan Paul, 1981.

Assessment of social conditions, politics, economics, and cultural barriers in the Third World, with particular reference to India, Iran, and Nigeria. Contends that the "chronic socio-economic stagnation" that characterizes these countries is not attributable to Western imperialism, maintaining that fundamental change in Third World countries was and is blocked by psychological and cultural facts. Compares relevant factors in Europe and in the three countries to show the constraints that block significant socioeconomic change.

58 McGreevey, William Paul, ed.
A B C THIRD-WORLD POVERTY: NEW STRATEGIES FOR MEASURING DEVELOPMENT PROGRESS
Lexington, Mass.: Heath, Lexington Books, 1980.

Five previously unpublished papers on the problems of measuring progress in alleviating poverty in the Third World, originally part of a series of seminars (1976-79) sponsored by the Agency for International Development. Editor McGreevey reviews the development progress from both a human capital and poverty alleviation standpoint; Gary S. Fields looks at absolute-poverty measures (i.e., those not depending on income distribution considerations); Harry J. Bruton considers the use of available employment and unemployment data in assessing government poverty policy, and G. Edward Schuh and Robert L. Thompson discuss measures of agricultural progress and government commitment to agricultural development. The fifth paper by Nancy Birdsall is a summary of discussion in two seminars on time-use surveys and networks of social support in LDC's. The authors find in part that: (1) existing data are inadequate to judge progress; (2) the best data gathering method is multipurpose household surveys; and (3) networks of social support are important (and unmeasured) means of income transfer between households.

59 McHale, John and McHale, Magda C.
A B C BASIC HUMAN NEEDS: A FRAMEWORK FOR ACTION
New Brunswick, N.J.: Rutgers University, Transaction Books, 1978.

60 Meadows, Dennis L., ed.
A B C ALTERNATIVES TO GROWTH--I: A SEARCH FOR SUBSTAINABLE FUTURES: PAPERS ADAPTED FROM ENTRIES TO THE 1975 GEORGE AND CYNTHIA MITCHELL PRIZE AND FROM PRESENTATIONS BEFORE THE 1975 ALTERNATIVES TO GROWTH CONFERENCE, HELD AT THE WOODLANDS, TEXAS
Cambridge, Mass.: Lippincott, Ballinger, 1977.

Seventeen previously unpublished interdisciplinary papers on the transition from growth to a steady-state society, i.e., a society with a constant stock of physical wealth and a constant stock of people. In four parts: the relation between population and food or energy; economic alternatives; the rationales, mechanisms, and implications of various long-term planning proposals; and analysis of the determinants, nature, and implications of current paradigms, norms, laws, and religion.

61 Melady, Thomas Patrick and Suhartono, R. B.
A B DEVELOPMENT -- LESSONS FOR THE FUTURE
Maryknoll, New York: Orbis Books, 1973.

Investigation of what determines, economically, which countries are developing, based on examination of characteristics of nations agreed to be undergoing this experience. The study examines such facets of development as the nonhomogeneity of the developing countries; factors affecting economic growth, the sectoral aspect of growth (industry and agriculture), measurements of the phenomenon, and the applicability of economic theory in this work; and the effects of economic development on man and his role in society.

62 Morawetz, David
A B D TWENTY-FIVE YEARS OF ECONOMIC DEVELOPMENT, 1950 TO 1975
Johns Hopkins University Press for IBRD, 1977.

Assesses development programs of developing countries and global development targets adopted by international organizations over the past 25 years. Chapters cover: a) changing objectives of development; b) growth in GNP per capita, population and the gap between rich and poor countries; c) reduction of poverty, including employment, income distribution, basic needs, nutrition, health, housing and education; d) self-reliance and economic independence; and e) conclusions, hypotheses, and questions.

63 Morgan, Theodore
B C ECONOMIC DEVELOPMENT: CONCEPT AND STRATEGY
New York and London: Harper & Row, 1975.

Textbook in economic development with emphasis on policy, its appropriate definition, its targets, and its improvement of application. Diverts focus from GNP and average income growth rates and into issues such as income distribution, nutrition, disease, climate, and population increases and their effects on development. Surveys existing theoretical literature. Discusses development planning and the importance of the statistical foundation of decision-making, and planning techniques such as cost-benefit analysis. Provides sporadic data for less-developed countries, mostly for the post-World War II period, on various national variables.

64 Ramati, Yohanan, ed.
A B C ECONOMIC GROWTH IN DEVELOPING COUNTRIES--MATERIAL AND HUMAN RESOURCES: PROCEEDINGS OF THE SEVENTH REHOVOT CONFERENCE
Praeger Special Studies in International Economics and Development
New York and London: Praeger in cooperation with the Continuation Committee of the Rehovot Conference, 1975.

Collection of 49 papers presented in September 1973. The papers are grouped into five sections following the structure of the conference. Part I includes papers setting the framework to analyze natural and human resources as factors in development and problems of planning and the quality of life. Part II includes papers on resources, technology, and income distribution. Part III deals with external constraints on development. Part IV examines planning and implementation. Part V contains the very brief closing addresses by Simon Kuznets and Abba Eban. Participants included 99 experts and policy makers for developing countries in Africa, Latin America, and Southeast Asia.

65
A B
Rubinson, Richard, ed.
DYNAMICS OF WORLD DEVELOPMENT
Political Economy of the World-System Annuals, vol. 4
Beverly Hills and London: Sage, 1981.

Twelve previously unpublished papers, almost all by sociologists, presented at the Fourth Annual Political Economy of the World-System conference at Johns Hopkins University, June 1980. Papers are based on the assumption that the world's history is the history of capitalist accumulation; and that capitalist development is the development of a single...modern world-system. Papers cover: development in peripheral areas; development in semiperipheral states; development and state organization; cycles and trends of world system development; theooretical issues; and dynamics of development of the world economy.

66
A C
Sachs, Ignacy
THE DISCOVERY OF THE THIRD WORLD
Cambridge, Mass., and London: M. I. T. Press, 1976.

Focusing on a redefinition of development theory, discusses the role of ethnocentrism and domination by European and Western ideas in such areas as science, technology, and economics. Argues that discussions regarding economic development strategies attempt to apply Western theories and ignore the fact that Third World growth, unlike capital-intensive European growth, must be based on the use of labor. Proposes a general development theory to bridge the gap between European theory and Third World practice and discusses problems such as economic surplus and economic aid. Recommends that the U.N. assess Western nations and funnel the money to Third World nations on a "no-strings" basis.

67
A B
Shafei, Mohamed Z.
THREE LECTURES ON ECONOMIC DEVELOPMENT
Beirut, Lebanon: Beirut Arab University, 1970.

The first lecture focuses on the characteristics of developing countries. The second traces the process of economic development in the U.A.R. (Egypt) since 1952. The third is on the foreign assistance needs of developing countries.

68
A C

Singer, H. W.
THE STRATEGY OF INTERNATIONAL DEVELOPMENT: ESSAYS IN THE ECONOMICS OF BACKWARDNESS
Edited by Sir Alec Cairncross and Mohinder Puri
White Plains, N.Y.: International Arts and Sciences Press, 1975.

A collection of 13 papers by the author, all published in past years, dealing with some of the central problems of economic development and development policy. Papers cover such issues as gains distribution among borrowing and investing countries, dualism, international aid, trade and development, employment problems, income distribution, science and technology transfers, etc. Introduction to the author's work and career by editor Sir Alec Cairncross.

69
A

Singer, Hans W. and Ansari, Javed A.
RICH AND POOR COUNTRIES
Baltimore and London: Johns Hopkins University Press, 1977.

Examines the changes that are required if the relationship between rich and poor countries is to make a more effective contribution to the development of the poor countries. Part one describes the structure of international economy and the nature of development process. Part two discusses the importance of the international trade sector to development in the poorer countries and reviews the trade policies of the rich and poor countries. Part three deals with the role of aid in the development process; and part four is concerned with international factor movement. Stresses the need for the formulation of an international development strategy...by the rich countries (both old and new), providing assistance in an increasing flow of resources through trade, aid capital and the transfer of skills and technology to the poor countries. Argues that such a strategy first must provide for some discrimination in international trade in favor of poor countries to provide more resources and secondly to enable the importation of more appropriate technologies.

70
A B

Spiegelglas, Stephen and Welsh, Charles J., eds.
ECONOMIC DEVELOPMENT; CHALLENGE AND PROMISE
Englewood Cliffs, N.J.: Prentice-Hall, 1970.

A collection of 33 reprinted readings, each representing an outstanding contribution, controversial issue, or synthesis of ideas in economic development. Major sections include: an introduction; nature and techniques of planning; strategy and policy; and trade or aid. The selection of topics in these sections reflects recent increased emphasis on practical development problems, particularly on human resources development and the need to create exportable manufactured goods. A matrix showing how each selection fits into the scheme and sequence of the seven widely used development textbooks is included.

71
A B C D
T. N. Srinivasan
DEVELOPMENT, POVERTY, AND BASIC HUMAN NEEDS: SOME ISSUES
World Bank Reprint Series, 76
IBRD, 1977.

Reprinted from Food Research Institute Studies, vol. XVI, no. 2 (1977), pp. 11-28. Deals with the raising of standard of living of the poorest sections of the population in developing countries. Discusses aid problems, distributional aspects of economic growth, employment goals, and the new perceptions of development.

72
A C D
Stein, Leslie
ECONOMIC REALITIES IN POOR COUNTRIES
Sydney, London and Singapore: Angus and Robertson, 1972.

This book surveys the problems of growth faced by the developing countries of the world. The first part of the book describes the economic and social characteristics of Third World countries and presents some theories of development, including Baran's Marxian view, W. W. Rostow's non-Marxist alternative, balanced growth theory, and Myrdal's view which considers non-economic as well as economic factors of growth. Succeeding chapters discuss population growth, problems of education, the role of agriculture and industrial development, obstacles to trade, and government plans which have been used in developing countries. Designed for use as a text or for the layman.

73
A B D
Streeten, Paul
DEVELOPMENT PERSPECTIVES
New York: St. Martin's Press, 1981.

A combination of 17 previously published articles and 7 new chapters, in five parts: concepts, values, and methods in development analysis; development strategies; transnational corporations; the change in emphasis from the growth approach to the basic needs approach; and two miscellaneous chapters on taxation and on Gunnar Myrdal. Newly written chapters cover: the results of development strategies for the poor, alternatives in development, the New International Economic Order, the basic needs approach, human rights and basic needs, the search for a basic-needs yardstick (with Norman Hicks), and transnational corporations and basic needs.

74
A C
Thomson, W. Scott, ed.
THE THIRD WORLD: PREMISES OF U.S. POLICY
San Francisco: Institute for Contemporary Studies, 1978.

75
A B
Tinbergen, Jan
THE DESIGN OF DEVELOPMENT
The Johns Hopkins University Press, 1958.

Formulates a coherent government policy to further development objectives and outlines methods to stimulate private investments.

76
A B C
Todaro, Michael P.
ECONOMIC DEVELOPMENT IN THE THIRD WORLD: AN INTRODUCTION TO PROBLEMS AND POLICIES IN A GLOBAL PERSPECTIVE
London and New York: Longman, 1977.

In four parts: Part one discusses the nature of underdevelopment and its various manifestations in the Third World, and parts two and three focus on major development problems and policies, both domestic (growth, income distribution, population, unemployment, education, and migration) and international (trade, balance of payments, and foreign investment). The last part reviews the possibilities and prospects for Third World development.

77
A B
Todaro, Michael P.
DEVELOPMENT PLANNING: MODELS AND METHODS
Series of undergraduate teaching works in economics, Volume V.
London, Nairobi, and New York: Oxford University
Press, 1971.

This is the last in a series of undergraduate teaching works in economics developed at Makere University, Uganda. This book is an introduction to development planning, with emphasis on plan formulation rather than implementation.

78
United Nations Department of Economic and Social Affairs
THE INTERNATIONAL DEVELOPMENT STRATEGY: FIRST OVER-ALL REVIEW AND APPRAISAL OF ISSUES AND POLICIES. REPORT OF THE SECRETARY-GENERAL
New York: United Nations, 1973.

Deals with the issues and policies in the field of economic and social development...of prime concern in the first two years of the Second United Nations Development Decade. Emphasis is upon changes in the following areas: priorities of objectives, techniques of production, trade and aid relationships, and the external environment in which economic and social development takes place.

79
A C

United Nations Department of Economic and Social Affairs
SHAPING ACCELERATED DEVELOPMENT AND INTERNATIONAL CHANGES
New York: United Nations Publications, 1980.

Contains views and recommendations of the UN Committee for Development Planning relating to the international development strategy for a third UN development decade. Chapters cover general premises and basic objectives; priority areas for action; means and implementation; and key goals and needed changes.

80
A C

United Nations Department of Economic and Social Affairs
DEVELOPMENT IN THE 1980'S: APPROACH TO A NEW STRATEGY; VIEWS AND RECOMMENDATIONS OF THE COMMITTEE FOR DEVELOPMENT PLANNING
New York: United Nations Publications, 1978.

Reviews development issues for the 1980's with a discussion of the current situation and preliminary comments relating to a development strategy for the 1980's. Discusses economic cooperation among developing countries, covering trade, economic integration and other arrangements for economic cooperation.

81
A B C

United Nations Industrial Development Organization
INDUSTRIALIZATION FOR NEW DEVELOPMENT NEEDS
New York: United Nations Publication, 1974.

Emphasizes the reshaping of industrial development in the light of new development needs that the pervasive problems of unemployment, maldistribution of income, and poverty in general have brought to the fore in the developing countries.

82
A B C

UNRSID
THE QUEST FOR A UNIFIED APPROACH TO DEVELOPMENT
UNRSID: 1980.

Provides background information on UNRSID's efforts to formulate a unified approach to development analysis and planning, an approach which would bring together all the different aspects of development into a set of feasible objectives and policy approaches. Chapters cover: styles of development--definitions and criteria; strategies; the findings of the Expert Group; an assessment by Marshall Wolfe, former Chief of the Social Development Division of UN ECLA; and an annex containing the final report on the project by the UN Commission for Social Development, covering questions of diagnosis, monitoring, indicators, and planning and capicitation.

83 Uri, Pierre
A B C DEVELOPMENT WITHOUT DEPENDENCE
New York: Praeger for the Atlantic Institute for International Affairs, 1976.

Monograph on foreign aid. Contends that the aid programs of the 1950's and 1960's were lopsided and failed to address the needs of the truly poor. According to Bundy, the author argues that although effective transfer of resources and skill remains a vital part of the need...such nation-to-nation aid...can only help to foster the very feelings of dependence...that are the deepest grievance of the developing world. Discusses control of population growth, the role and necessary scale of official foreign aid, stabilization of the raw materials market so as to assist consumers and producers alike, and the types of industries the developing countries should strive to build as a part of a rational world division of labor. Examines the control and regulation of multinational corporations and, focusing on Latin America, the extent to which regional cooperation can be developed. Recommends that development planning be based on future population growth and distribution.

84 Varma, Baidya Nath
A B C THE SOCIOLOGY AND POLITICS OF DEVELOPMENT: A THEORETICAL STUDY
International Library of Sociology Series
London and Boston: Routledge & Kegan Paul, 1980.

The author critically examines theories of development and presents his own theory. Considers general criteria used for evaluating the modernization process; describes a model for a general paradigm of modernization; surveys other models encompassing ideological, social scientific, anthropological and activistic theories; and discusses theoretical problems of planning and national reconstruction. Summarizes views of theorists in various social science disciplines and features of modernization in terms of guidance provided for economic, political, educational, and bureaucratic decision-making in a developing country. Concludes that both the socialist and capitalist systems of modernization are viable models for Third World countries.

85 Vogeler, Ingolf and De Souza, Anthony R., eds.
A C DIALECTICS OF THIRD WORLD DEVELOPMENT
Montclair: Allanheld, Osmun, 1980.

Collection of previously published (some revised) papers designed for use by students of economics, political science, and development. Representing a variety of ideas and arguments relevant to Third World underdevelopment, the readings discuss climate and resources, cultural traditions, European colonialism (i.e., plantation agriculture), population, tourism, and imperialism. An appendix provides "awareness" exercises.

86
A B
Wallman, Sandra, ed.
PERCEPTIONS OF DEVELOPMENT
New York: Cambridge University Press, 1977.

87
A C
Ward, Richard J.
DEVELOPMENT ISSUES FOR THE 1970'S
New York and London: Dunellen, 1973.

An assessment of key issues and problems which emerged from the Decade of Development and which will continue to absorb the attention of students of development in the present decade. The author, former Chief of Planning of the U.S. Agency for International Development, presents much data which has not been previously released and which is unavailable elsewhere. The book is divided into three parts: "Food and Human Welfare," "Development Problems for This Decade," and "Planning Programs and Strategies." The chapters specifically discuss such issues as labor absorption in agriculture, means of population control, the burden of debt service, the role of foreign aid, big-push development, etc.

88
A B
Waterston, Albert
DEVELOPMENT PLANNING; LESSONS OF EXPERIENCE
The Johns Hopkins University Press, 1979.

Analyzes the success of the development planning experience in over 100 countries in Asia, Africa, Europe, and the Americas. In two parts. Part 1 describes and analyzes the problems associated with the implementation of planning programs, the provision of basic data, the role of national budget, and administrative obstacles. Part 2 contains an extensive and comparative discussion of the experience of the countries under review in setting up organizations and administrative procedures for preparing and implementing development projects; the distribution of planning functions, types of central planning agencies, and subnational regional and local planning bodies.

89
A B
Watts, Nita, ed.
ECONOMIES OF THE WORLD
New York: Oxford University Press.

The purpose of this new series is to provide a brief review of economic development during the post-war period in each of a number of countries which are of obvious importance in the world economy, or interesting because of peculiarities of their economic structure or experience, or illustrative of widespread economic development problems. The series will be of interest to economists in universities, and in business and government.

90
A B

Wilber, Charles K., ed.
THE POLITICAL ECONOMY OF DEVELOPMENT AND UNDERDEVELOPMENT
New York: Random House, 1973.

Emphasis in approach and content is on political economy in the sense of attempting to incorporate such noneconomic influences as social structures, political systems, and cultural values as well as such factors as technological change and the distribution of income and wealth. Readings are radical in that they are willing to question and evaluate the most basic institutions and values of society. Divided into eight groups concerned with methodological problems, historical perspective, trade and imperialism, agricultural and industrial institutions and strategies, comparative models of development, the human cost of development, and indications for the future.

91
A C

Worsley, Peter
THE THIRD WORLD
Chicago: University of Chicago Press, 1972.

92
A C

Wriggins, W. Howards and Adler-Karlsson, Gunnar
REDUCING GLOBAL INEQUALITIES
New York: McGraw-Hill, 1978.

Two papers, plus an introduction on the role that developing countries themselves take to reduce the gap between rich nations and poor and to eliminate mass poverty within their own societies. W. Howard Wriggins, U.S. ambassador to Sri Lanka and formerly professor of political science at Columbia University, analyzes the various bargaining strategies open to developing countries such as developing commodity or regional coalitions, or a variety of threats to developed countries. The future is likely to see continued efforts at coalition building, but also periodic outbreaks of irregular violence against local opponents, neighbors, or Northern centers of power.

93
A B C

Zuvekas, Clarence, Jr.
ECONOMIC DEVELOPMENT: AN INTRODUCTION
New York: St. Martin's Press, 1979.

Text written from an interdisciplinary perspective stressing policy and empirical findings rather than an overall development theory. Aims at balance between theory and policy, including historical development and empirical evidence. After discussing the terminology of and the obstacles to development, the author examines population growth, trade and development, and the role of government. Also covers: the problems of agriculture and industry, income distribution, employment, mobilization of domestic and foreign savings, manipulation of trade to the advantage of the developing country, and with the limits to growth controversy. Presumes acquaintance with basic macro and micro theory.

TECHNOLOGY, TECHNOLOGY POLICY

94 Abdel-Malek, Anouar; Blue, Gregory and Pecujlic, Miroslav, eds.
A B C SCIENCE AND TECHNOLOGY IN THE TRANSFORMATION OF THE WORLD
Tokyo: United Nations University, 1982.

Contains papers presented at the First International Seminar on the Transformation of the World, sponsored by UNU and held in Belgrade, Yugoslavia, 22-26 October 1979. Papers are arranged by the topic of each session: (a) science and technology as formative factors of contemporary civilization--from domination to liberation; (b) technology generation and transfer--transformation alternatives; (c) biology, medicine and the future of mankind; (d) the control of space and power; and (e) from intellectual dependence to creativity. Individual papers address such topics as: the technology of repression and repressive technology; human aspects of medical sciences; legal aspects of the transfer of technology in modern society; and science and the making of contemporary civilization.

95 Ahmad, Yusuf J. and Muller, Frank G., eds.
A B C INTEGRATED PHYSICAL, SOCIO-ECONOMIC AND ENVIRONMENTAL PLANNING
Dublin: Tycooly International Publishing for UNEP, 1982.

This book is a series of case studies on the environmental impacts of socioeconomic activities in the context of an integrated physical planning policy. It examines scientific and technological development, and identifies and defines social objectives which take into account the impact of technological innovations in improving living standards and the need for populations to adapt to changes in the availability of resources and increased leisure time.

96 Asian Productivity Organization
A B INTERNATIONAL SUB-CONTRACTING: A TOOL OF TECHNOLOGY TRANSFER
Bangkok: APO, 1978.

Deals with the extent, practice and potential of international subcontracting and its impact on technological standards and productivity. In two parts: (a) survey report and (b) country report.

97 Asian Productivity Organization
A C D TECHNOLOGY DEVELOPMENT FOR SMALL INDUSTRY IN SELECTED ASIAN COUNTRIES
Bangkok: APO, 1982.

In four parts: (a) an overview of the role of public research and development departments in technology for small-scale industry; (b) review of the situation in eight Asian countries; (c) expert papers discussing the process of technology development and transfer, the role and strategies of public research institutes for small and medium enterprises in Japan, and problems and solutions from the U.S. experience in technology development for small industries; and (d) report and main findings of an APO Symposium held on 15 to 22 December 1980 in Bangalore, India.

98 Asian Productivity Organization
A C D TECHNOLOGY TRANSFER IN SOME ASIAN COUNTRIES: SOME DIMENSIONS ON INDIGENOUS DEVELOPMENT AND TRANSFER
Bangkok: APO, 1979.

Contains some of the papers prepared for the Symposium on Technology Transfer, India, February 1978. Topics cover: technology and its transfer and role of productivity organization; technology development and adaptation; intra-national technology development and transfer; technology development and transfer from laboratory to field and to small industry; and a model for setting technology transfer priorities.

99 Balasubramanyam, V. N.
A B D INTERNATIONAL TRANSFER OF TECHNOLOGY TO INDIA
New York: Praeger, 1973.

100 Baranson, Jack
A B C NORTH-SOUTH TECHNOLOGY TRANSFER: FINANCING AND INSTITUTION BUILDING
Mt. Airy, MD: Lomand Publications, 1981.

101 Baranson, Jack
B TECHNOLOGY FOR UNDERDEVELOPED AREAS: AN ANNOTATED BIBLIOGRAPHY
Oxford, New York: Pergamon Press, 1967.

102 Barrio, S.
A B C SCIENCE AND TECHNOLOGY FOR DEVELOPMENT
STPI Module 7: Policy Instruments to Define the Pattern of Demand for Technology
IDRC, 1980.

This report constitutes a part of the Main Comparative Report of the Science and Technology Policy Instruments (STPI) project that examines the design and implementation of science and technology policies in 10 developing countries. Topics covered in this report includes industrial programming and mechanisms for setting priorities for industry; industrial financing mechanisms; state purchasing power; fiscal measures; price controls; export promotion measures; and other mechanisms.

102 Barrio, S.
B C SCIENCE AND TECHNOLOGY FOR DEVELOPMENT
STPI Module 9: Policy Instruments for the Support of Industrial Science and Technology Activities
IDRC, 1980.

This module contains four groups of policy instruments that provide support to the science and technology activities in industries and industry-related institutions. These four groups are policy instruments relating to technical norms and standards; information centers; manpower development programs; and consulting and engineering activities.

103 Behari, Bepin
A C D ECONOMIC GROWTH AND TECHNOLOGICAL CHANGE IN INDIA
Delhi: Vikas Publishing and Portland, Ore.: International Scholarly Book, 1974.

104 Behrman, Daniel
A B C SCIENCE AND TECHNOLOGY IN DEVELOPMENT: A UNESCO APPROACH
UNESCO, 1979.

A useful introduction to development problems, this book gives a popular, readable account of Unesco's contributions to development through its programs in science and technology.

105 Behrman, Jack N.
A B C INDUSTRY TIES WITH SCIENCE AND TECHNOLOGY POLICIES IN DEVELOPING COUNTRIES
Cambridge, Mass.: Oelgeschlager, Gunn & Hain for the Fund for Multinational Management Education, 1980.

Focuses on S & T policies toward the industrial sector and the potential contributions that transnational corporations in the industrial sector can make to the development of indigenous science and technology capabilities. Data come from interviews conducted in 1978 and 1979 in Korea, Indonesia, Mexico, Brazil, Egypt, Malaysia, India, the Philippines, and Iran. Concludes that the creation of an appropriate S & T policy requires the education of the industrial and the consumer world as to the applicability of new technologies and their usefulness in meeting developmental goals, while at the same time building the infrastructure needed to supply the technologies desired.

106
A B C
Behrman, Jack N. and Fischer, William A.
SCIENCE AND TECHNOLOGY FOR DEVELOPMENT: CORPORATE AND GOVERNMENT POLICIES AND PRACTICES
Cambridge, Mass.: Oelgeschlager, Gunn & Hain for the Fund for Multinational Management Education, 1980.

A study package of investigations into research and development (R & D) activities overseas and science and technology (S & T) policy issues in LDC's.

107
B C
Bello, Joseph A. and Iyanda, Olukunle
APPROPRIATE TECHNOLOGY CHOICE AND EMPLOYMENT CREATION BY TWO MULTINATIONAL ENTERPRISES IN NIGERIA
Geneva: ILO, 1981.

Topics discussed: legal and socioeconomic environment of manufacturing businesses in Nigeria; comparative employment creation by two manufacturing multinational enterprises (MNE's) in Nigeria; determinants of the two MNE's employment effects; peripheral technologies and employment creation; and appropriate technology, training process and staff motivation.

108
A B C
Bernard, Harvey R. and Pelto, Pertti J.
TECHNOLOGY AND SOCIAL CHANGE
New York: MacMillan, 1972.

109
A B C
Bhagwati, Jagdish N., ed.
THE NEW INTERNATIONAL ECONOMIC ORDER: THE NORTH-SOUTH DEBATE
Cambridge, Mass.: The MIT Press, 1977.

Papers prepared at a 1975 workshop at the Massachusetts Institute of Technology that sought to examine specific proposals which could form the concrete content of a reformed world economy. The wide range of international economic issues, the distinguished panel of international and development economists, and the lively and informed disagreements among them make this book a useful survey of the dominant international economic issues today. The papers are grouped under: (i) resource transfers; (ii) international trade; (iii) world food problems; (iv) technology transfer and diffusion; and (v) a panel discussion of general North-South issues.

110
A B D
Bhalla, A. S.
TECHNOLOGY AND EMPLOYMENT IN INDUSTRY: A CASE STUDY APPROACH
Geneva: ILO, 1975.

Consists of a collection of case studies mainly concerned with identifying and analyzing alternative techniques of production, and examining their implications for specific policy decisions. In two parts: (a) deals with conceptual issues and questions of measurement; and (b) contains eight empirical case studies from developing countries in Africa, Asia, and Latin America. Concluding chapter by the editor provides a synthesis of the findings of studies and draws lessons for policy making.

111
Bhatrasali, B. N.
TRANSFER OF TECHNOLOGY AMONG THE DEVELOPING COUNTRIES
Tokyo: Asian Productivity Organization, 1972.

112
A B
Bliss, Charles and Reedy, John H.
APPROPRIATE TECHNOLOGIES IN CIVIL ENGINEERING WORKS IN DEVELOPING COUNTRIES: AN EXPLORATORY APPRAISAL OF THE ART
New York: United Nations, 1976.

Discusses the selection of the civil engineering technology most suitable to a developing country from the range of available choices (labor-intensive to capital-intensive). Topics covered: (a) the meaning and measurement of appropriateness; (b) recent construction practices; (c) efforts to advance the use of appropriate technologies; (d) obstacles to adoption of appropriate technologies.

113
A B C
Boon, Gerard K.
TECHNOLOGY AND SECTOR CHOICE IN ECONOMIC DEVELOPMENT
Alphen aan den Rijn, The Netherlands: Sijthoff & Noordhoff, 1978.

Presents studies in economic development, specifically in metal-working and grain-production. Discusses the methods for evaluating and selecting technology. Organized in two parts: microeconomic analyses based on engineering and intercountry statistical production data; and sectoral and macro analyses, utilizing Mexican statistical data. Assesses Mexico's present and future economic situation and forecasts the labor absorptive capacity of the Mexican economy for 1982 and 2000.

114
A B C
Boserup, Ester
POPULATION AND TECHNOLOGICAL CHANGE: A STUDY OF LONG-TERM TRENDS
Chicago: University of Chicago Press, 1981.

115
A C
Boucher, Wayne I., ed.
THE STUDY OF THE FUTURE: AN AGENDA FOR RESEARCH
Washington, D.C.: U.S.G.P.O., 1977.

discusses the strengths and limitations associated with futures research type of forecasting. The four sections consider: a new perspective on forecasting methodology, the validity of the systems, and the nature of unforeseen development; attitudes towards forecasting in the various social sciences; the functions, forms, and critical issues in the futures field; and an agenda for futures research. Results from a survey of current forecasting efforts appended.

116
A B C
Buzzati-Traverso, Adriano
THE SCIENTIFIC ENTERPRISE, TODAY AND TOMORROW
UNESCO, 1977.

Provides a comprehensive overview of science, its disciplines and institutions in the postwar decades. It takes a hard look at the development of current science and technology, including their organization and their effects on culture. It assesses the changing role of the scientist, and discusses the sciences in relation to changing moral values. The author asks many provocative questions which require the reader to think about science in its broadest context, and to consider options open to future research.

117
A C
Bykov, A. N.; Letenko, A. V. and Strepetova, M. P.
SOVIET EXPERIENCE IN TRANSFER OF TECHNOLOGY TO INDUSTRIALLY LESS DEVELOPED COUNTRIES
UNITAR, 1973.

Survey of the experience of the U.S.S.R. in the transfer of technology to developing countries carried out for UNITAR by a team of experts at the institute of Economics of World socialist Systems, Moscow. Considers technology transfer to socialist countries and to other developing countries, with emphasis on the examples of India, Egypt, and Iran.

118
B C
Chudson, Walter A. and Wells, Louis T., Jr.
THE ACQUISITION OF TECHNOLOGY FROM MULTINATIONAL CORPORATIONS BY DEVELOPING COUNTRIES
United Nations, 1974.

Constitutes part of the documentation supplied to the Group of Eminent Persons to Study the Impact of Multinational Corporations on Development and on International Relations, appointed by the UN Secretary-General in 1972. Also supplements the report Multinational Corporations in World Development. The study examines the multinational corporation as a vehicle for supplying technology to developing countries, comparing it with certain alternatives and considering the issues that arise and possible solutions.

119
A C
Contreras, Carlos
TECHNOLOGICAL TRANSFORMATION OF DEVELOPING COUNTRIES: SOME ISSUES FOR DISCUSSION AND PRELIMINARY IDEAS FOR ACTION AT THE NATIONAL AND INTERNATIONAL LEVELS IN THE 1980'S
Research Policy Program, University of Luno, 1978.

120
A B C
Cooper, Charles, ed.
SCIENCE, TECHNOLOGY AND DEVELOPMENT: THE POLITICAL ECONOMY OF TECHNICAL ADVANCE IN UNDERDEVELOPED COUNTRIES
London: Frank Cass, 1973.

121
Council on Environmental Quality and the Department of State
A B C
THE GLOBAL 2000 REPORT TO THE PRESIDENT. VOLUME ONE: ENTERING THE TWENTY-FIRST CENTURY. VOLUME TWO: THE TECHNICAL REPORT.
Harmondsworth, Middlesex: Allen Lane and Penguin Books, 1982.

This study prepared by the U.S. Government presents a projection of probable changes in the world's population, natural resources, and environment through the end of the century to serve as the foundation of our longer-term planning. An impressive amount of material is presented (more is available in the original three volumes) on demographic, climate, mineral, food and energy trends, the economic framework being provided by an adaptation of the World Bank's SIMLINK model. The projections generate a gloomy outlook in terms of both human poverty and biological overload. This piece of futurology was commissioned by Carter in 1977, but by its completion in 1980 its approach was already out of date in both economic and political terms. Concern is now for unemployment, low growth rates and deteriorating terms of trade on the one hand, and inflationary wages and insufficient investment on the other. The Keynesian politics of global cooperation, even under the benevolent guidance of the USA, have also passed in favour of the mercantilism of Reagan. But who knows what the rest of the century will bring?

122
B C
Damjanovic, Zvonimir
SCIENCE AND TECHNOLOGY AS AN ORGANIC PART OF CONTEMPORARY CULTURE
Tokyo: UNU, 1980.

Presents six theses on the effects of science and technology on modern culture and comments on them.

123
A B C
Dasgupta, Ajit K.
ECONOMIC FREEDOM, TECHNOLOGY AND PLANNING FOR GROWTH
New Delhi: Associated Publishing House, 1973.

Development study contending "it is...impossible to increase the rate of economic growth by increasing rate of domestic saving" and "that only by resorting to factors which are exogenous to the [nation's] economic system can an acceleration of the economic growth process be obtained." Technological development is identified as an exogenous factor sufficient to initiate the development process in a nearly closed economy. Arguments, based on a work flow analysis of the economic system, described in the appendix, proceeds ad seriatim with discussions of economic dynamics and planning, employment and growth, the diffusion of technological improvement, the establishment of the proper growth conditions in an economy, industrial planning, and the commodity sector as an indicator of growth in the service sector.

124 De Gregori, Thomas R.
A B C D TECHNOLOGY AND THE ECONOMIC DEVELOPMENT OF THE TROPICAL AFRICAN FRONTIER
Cleveland: The Press of Case Western Reserve University, 1969.

Outlines sub-Saharan Africa's technological history and potential; discusses the inistitutional barriers to technological diffusion. Discusses traditional and Marxian theories of capital movements in the context of frontier economics, generally, and of the African economy, in particular. Finds that capital transfers have occurred, in many cases, from the less developed to the more developed economies in contradiction to traditional theory. Contains data on African foreign trade by country and major project.

125 Diwan, Romesh K. and Livingston, Dennis
A B ALTERNATIVE DEVELOPMENT STRATEGIES AND APPROPRATE TECHNOLOGY: SCIENCE POLICY FOR AN EQUITABLE WORLD ORDER
New York: Pergamon Press, 1979.

Advocates the establishment of a new international economic order, which would reduce and eventually eliminate inequalities, armaments, price biases, and technological inappropriateness, arguing that is in the interest of both developed and developing countries. Stresses the role of sophisticated science and technology in the existing inequality among nations and maintains that soft technologies appropriate for use by small groups or individuals would reduce inequalities. In four parts: part one describes the exisiting international order composed of developed countries, high-income developing countries, high technology developing countries, and other developing countries; part two assesses the Conventional Development Strategy (CDS) followed in both developed, and developing countries during the 1950's and 1960's and based upon the maximization of the rate of growth in the GNP and leading to the desired state of development defined by the West; part three advances Alternative Development Strategies (ADS) with emphasis on basic needs and appropriate technology (AT); part four discusses policies for both developed and developing countries that are consistent with ADS and AT, which are designed to promote cooperation and contribute to an equitable world order.

126 Dunn, Peter D
A B D APPROPRIATE TECHNOLOGY: TECHNOLOGY WITH A HUMAN FACE
New York: Schocken Books, 1979.

Examines in non-technical language development programs that stress the application of technology appropriate to combatting poverty, especially rural poverty, in developing countries. Aims at job creation with small-scale, low cost methods. Covers not only the techniques but also the social aspects of development and technology. Devoted mostly to the actual practice of appropriate technology in the areas of food, water and health, energy, medical services, building services, small industry, and education and research. Appendix contains related statistical and technical material.

127 Eckhaus, Richard S.
A B C APPROPRIATE TECHNOLOGIES FOR DEVELOPING COUNTRIES
Washington: National Academy of Sciences, 1977.

Analyzes the interrelationships between technological choices and economic, social, and political aspects of the development process in developing countries. Examines several issues including: alternative criteria of appropriateness of technological decision; technological opportunities and transfer of technical information; technological choices in agriculture, services, and small-scale enterprise; and policies for promoting choices of appropriate technologies. Concludes that no panaceas are to be found in the choice of particular types of intermediate technologies. Recommends: (1) the formation of resource and product price policies that will encourage the use of efficient labor-intensive methods; (2) the identification of particular products and processes for which technological and economic research is successful; (3) the investigation of the conditions that stimulate the adoption of efficient labor-intensive technologies.

128 Edwards, Alfred L.; Oyeka, Ikewelugo C. and Wagner, Thomas eds.
A B C D NEW DIMENSIONS OF APPROPRIATE TECHNOLOGY: SELECTED PROCEEDINGS OF THE 1979 SYMPOSIUM SPONSORED BY THE INTERNATIONAL ASSOCIATION FOR THE ADVANCEMENT OF APPROPRIATE TECHNOLOGY FOR DEVELOPING COUNTRIES
Ann Arbor: University of Michigan, Graduate School of Business, 1980.

129 el-Kholy, Asama A.
A B TOWARD A CLEARER DEFINITION OF THE ROLE OF SCIENCE AND TECHNOLOGY IN TRANSFORMATION
Tokyo: UNU, 1980.

Discusses the role of science and technology in bringing about significant changes in society. In three parts: (a) a view of the problem from within; (b) the view from without; and (c) toward a clearer definition of the role of science and technology in transformation.

130 FAO
A B APPROPRIATE TECHNOLOGY IN FORESTRY. REPORT OF THE CONSULTATION ON INTERMEDIATE TECHNOLOGY IN FORESTRY, HELD IN NEW DELHI AND DEHRA DUN, 18 OCTOBER - 7 NOVEMBER 1981.
FAO, 1982.

A compilation of technical papers and country reports presented to the Consultation, covering: trends in forestry management; forestry and employment; education and training; ergonomics and safety; energy from the forest; basic logging tools; charcoal production; project forestry; and timber harvesting.

131 Galtung, Johnan
A B DEVELOPMENT, ENVIRONMENT AND TECHNOLOGY: TOWARDS A TECHNOLOGY FOR SELF-RELIANCE
New York: U.N., 1979.

In four chapters: (a) conceptual framework; (b) technological transfer as a process--a theory and some cases; (c) technology and ecodevelopment; and (d) the technology of self-reliance.

132 Ghatak, Subrata
A B TECHNOLOGY TRANSFER TO DEVELOPING COUNTRIES: THE CASE OF THE FERTILIZER INDUSTRY
Greenwich, Conn.: JAI Press, 1981.

133 Giarini, Orio and Louberge, Henri
B C THE DIMINISHING RETURNS OF TECHNOLOGY: AN ESSAY ON THE CRISIS IN ECONOMIC GROWTH
Systems Science and World Order Library: Explorations of World Order
Oxford; New York; Toronto and Paris: Pergamon Press, 1978.

Account of the "marriage" of technology and science in the nineteenth century, which allowed for increases in both growth and welfare, and their recent "divorce" caused by diminishing returns of technology. Looks at the underlying causes of economic growth and explores the possibility that technological advance, as the principal component of economic growth, is subject to diminishing returns; if so, the study of economic growth and eventual decline is transposable to technical progress in the aggregate. Concludes that due to diminishing returns of technology emphasis is no longer on scale economics and concentration but on a new "federalism," i.e., a pluralist organization of autonomous memeber units.

134 Golebiowski, Janusz W.
A B SOCIAL VALUES AND THE DEVELOPMENT OF TECHNOLOGY
Tokyo: UNU, 1982.

Coverage includes: culture, social order and technology; why modern technology is not socially and culturally neutral; assessment of technology--resolving the problem of humanization of technology; and dignity, partnership and emancipation linked with new prospects for technology.

135 Gordon, J. King, ed.
B C CANADA'S ROLE IN SCIENCE AND TECHNOLOGY FOR DEVELOPMENT: PROCEEDINGS OF A SYMPOSIUM...TORONTO, CANADA, 10-13 MAY 1979
Ottawa: IDRC, 1979.

Consists of the speeches of the participants of the Symposium and record of the discussions. Covers policy, politics, technology transfer and other aspects of Canada's contribution to science and technology for development.

136 Hawthorne, Edward P.
B C THE TRANSFER OF TECHNOLOGY
Paris: Organisation for Economic Co-operation and Development, 1971.

137 Hetzler, Stanley, A.
B C APPLIED MEASURES FOR PROMOTING TECHNOLOGICAL GROWTH
International Library of Science
London and Boston: Routledge and Kegan Paul, 1973.

Addresses the developmental problems facing technologically retarded societies. The treatment is based on the thesis that social change is the consequence of a forerunning technological change. Solutions to the problems are offered, and the author deals, in detail, with the areas of industrial development, agricultural development, the raising of finance, and the methods for organizing and staffing the national planning facilities required for development programming. The study is divided into four main parts: orientation to the problem, general mechanization and industrialization, domestic maintenance administration, and supporting structures.

138 Hill, Patricia Madden

B TECHNOLOGICAL CHANGE IN LESS DEVELOPED COUNTRIES: A GUIDE TO SELECTED LITERATURE
Ames: Department of Electrical Engineering, Iowa State University, 1981.

139 Holland, Susan S., ed.
B C CODES OF CONDUCT FOR THE TRANSFER OF TECHNOLOGY: A CRITIQUE
New York: Council of the Americas and Fund for Multinational Management Education, 1976.

Five papers that set out some of the key issues and questions being discussed concerning codes of conduct for technology transfer between developed and developing countries. Following an exploration of the background of these codes and an examination of the changing nature of the technology transfer process, two papers look at the concerns of developing countries and the feasibility of many of the code regulations they wish to enact. A final paper presents the partial findings from a research project on the lack of understanding of the different parties of each other's objectives and views.

140 IDRC
C D ANDEAN PACT TECHNOLOGY POLICIES
Ottawa: IDRC, 1976

This publication describes basic policy decisions made to achieve the development of an integrated technology policy for the six Andean Pact countries which would fulfill specific economic and social needs and provide technological autonomy for the region. Topics covered: foreign direct investment; technology; trademarks; patents; licensing agreements; royalties; and industrial property.

141 IDRC
A C D PRIORITIES FOR SCIENCE AND TECHNOLOGY POLICY RESEARCH IN AFRICA. REPORT OF A SEMINAR HELD AT THE UNIVERSITY OF IFE, IFE-IFE, NIGERIA, 3-6 DECEMBER 1979.
Ottawa: IDRC, 1981.

Reviews the present situation and discusses ways in which technology policy research in Africa might be developed.

142 IDRC
A B D SCIENCE AND TECHNOLOGY FOR DEVELOPMENT
STPI Module 6: Policy Instruments for the Regulation of Technology Imports.
Ottawa: IDRC, 1980.

This module is concerned with the effects of the policy instruments for the regulation of technology imports in Brazil, Mexico, Venezuela, the Republic of Korea, Peru, and Colombia. It deals with tariffs, foreign investments, foreign exchange control, patents, licensing, legal aspects, and joint ventures.

143 IDRC
A B D SCIENCE AND TECHNOLOGY POLICY IMPLEMENTATION IN LESS-DEVELOPED COUNTRIES: METHODOLOGICAL GUIDELINES FOR THE STPI PROJECT
Ottawa: IDRC, 1976.

Discusses methods of relating national science and technology investment priorities to national industrial development objectives, within the framework of the Science and Technology Polciy Instruments (STPI) Project. The report focuses on countries in Latin America, the Middle East, southern Europe, and Asia.

144 IDRC
A B D TECHNOLOGY POLICY AND ECONOMIC DEVELOPMENT: A SUMMARY REPORT ON STUDIES UNDERTAKEN BY THE BOARD OF THE CARTAGENA AGREEMENT FOR THE ANDEAN PACT INTEGRATION PROCESS
Ottawa: IDRC, 1976.

This report summarizes results of studies by the Board of the Cartagena Agreement on the relationships between technological development policy and economic and social progress. The report covers studies made in Europe, South American and Asia on: technology policy in various industries; the role of national institutions; finance and taxation; and, in the Andean Pact countries, technology commercialization, consultancy needs and services, and the integration of technology and economic development.

145 ILO
B C TECHNOLOGY, EMPLOYMENT AND BASIC NEEDS
Geneva: ILO, 1978.

Constitutes the preliminary ILO position statement regarding the items on the agenda of the United Nations Conference of 1979. In three parts: (a) technology for development-the choice of technology for development, and policies for appropriate technology; (b) institutions for appropriate technology--institutions at the national level; institutions at the international level; and (c) the role of the ILO.

146 International Labour Office
A B C AUTOMATION IN DEVELOPING COUNTRIES
Round-Table Discussion on the Manpower Problems Associated with the Introduction of Automation and Advanced Technology in Developing Countries
Geneva: Author, 1972.

Contains seven previously unpublished papers (by professional economists, academic economists, and others) examining the increased use of advanced technology and automation in industry in developing countries and their effects on manpower. The papers, presented at a round-table discussion in Geneva (July 1-3, 1970), look at advanced technology as a strategy of economic development; compare the reactions of developed and underdeveloped countries to advanced technology, now and in the past; review the experiences of specific countries with automation and advanced technology; examine the organizational requirements and technological skills needed for the introduction of advanced technology and conditions required for its effective use; examine and compare the development of trade unionism in difference countries. Case studies are used in many of the papers.

147 Ishikawa, Shigeru
A B C D ESSAYS ON TECHNOLOGY, EMPLOYMENT AND INSTITUTIONS IN ECONOMIC DEVELOPMENT: COMPARATIVE ASIAN EXPERIENCE
Tokyo: Kinokuniya, 1981.

The studies deal with: labor absorption in developing Asian countries; technological change in agricultural production and its impact on agrarian structure, presenting a Japanese case study in prewar agricultural development; and institutional arrangements for resource allocation in the developing economies. Discusses how pre-war Japan solved the issue of identifying and developing technologies in spite of competition from Western industrialized countries, including a post-script on appropriate technologies in tractor and power-tiller industries in Southeast Asia.

148 Jequier, Nicholas
B D APPROPRIATE TECHNOLOGY DIRECTORY
Paris: OECD, 1979.

A directory of the institutions involved in the appropriate technology field in 78 countries.

149 Jequier, Nicholas
A B APPROPRIATE TECHNOLOGY: PROBLEMS AND PROMISES
Paris: OECD, 1976.

Examines the structural problems facing the appropriate technology movement from the standpoint of a national innovation policy to further the development process. In two parts, Part 1 reviews the major policy issues in six chapters: (a) the origins and meanings of appropriate technology; (b) the innovation system in appropriate technology; (c) the information networks; (d) the role of the universities; (e) building up new industries; and (f) policies for appropriate technology. Part 2 contains 19 papers presenting the points of view of practitioners of low-cost technology from both developed and developing countries.

150 Jones, Graham
A B C THE ROLE OF SCIENCE AND TECHNOLOGY IN DEVELOPING COUNTRIES
New York: Oxford University Press, 1971.

A chemist explores the different ways modern science and technology are applied to promote economic and social growth in poor but developing countries. Intended to be of practical value to scientists and technologists working in these countries. Other subjects covered are agricultural development, industrialization and research, education, and manpower.

151 Junta del Acuerdo de Cartagena
A B C D TECHNOLOGY POLICY AND ECONOMIC DEVELOPMENT: A SUMMARY REPORT ON STUDIES UNDERTAKEN BY THE BOARD OF THE CARTAGENA AGREEMENT FOR THE ANDEAN PACT INTEGRATION PROCESS.
Ottawa: International Development Research Centre, 1976.

Summarizes the results of studies undertaken by political and industrial economists, engineers, and sociologists in the early 1970's: on national and sectoral technology policy in Italy, Yugoslavia, Japan, and Czechoslovakia; on the process of commercialization of foreign technology in the Andean Pact countries; on the importation of the technology package; on engineering and consulting needs and services in the Andean subregion; and on the international search for technology in the iron and steel industry in Mexico, Japan, India, Spain, Germany, and Sweden. Concludes with the first part of a position paper on the necessary policies for technological development in the Andean Pact countries, relating the issues of technological dependence to the general problem of economic development. Major conclusions are that (1) internal technological development has been too heavily directed toward the natural and medical sciences and requires reorientation toward economic and social problems; (2) importing of foreign technology is indiscriminate and may not be absorbed by the internal technological infrastructure; and (3) the technological infrastructure of the Subregion is isolated from the government and other economic sectors, thus restraining the development process.

152 Kalbermatten, John M.; Julius, DeAnne S. and Gunnerson, Charles G.
B C APPROPRIATE TECHNOLOGY FOR WATER SUPPLY AND SANITATION; VOLUME I: TECHNICAL AND ECONOMIC OPTIONS
1980.

Reports technical, economic, health, and social findings of the research project on appropriate technology and discusses the program planning necessary to implement technologies available to provide socially and environmentally acceptable low-cost water supply and waste disposal.

153 Kalbermatten, John M.; Julius, DeAnne S.; Mara, D. Duncan and Kunnerson, Charles G.
B C APPROPRIATE TECHNOLOGY FOR WATER SUPPLY AND SANITATION; VOLUME 2: A PLANNER'S GUIDE
1980

Provides information and instructions on how to design and implement appropriate technology projects based on the findings reported in Volume I: Technical and Economic Options.

154 Kalbermatten, John M.; Julius, DeAnne S. and Gunnerson, Charles G.
B C NUMBER 1: APPROPRIATE SANITATION ALTERNATIVES: A TECHNICAL AND ECONOMIC APPRAISAL
The Johns Hopkins University Press, 1982.

This volume summarizes the technical, economic, environmental, health, and sociocultural findings of the World Bank's research program on appropriate sanitation alternatives and discusses the aspects of program planning that are necessary to implement these findings. It is directed primarily toward planning officials and sector policy advisers for developing countries.

155 Kalbermatten, John M.; Julius, DeAnne S.; Gunnerson, Charles G. and Mara, D. Duncan
B C NUMBER 2: APPROPRIATE SANITATION ALTERNATIVES: A PLANNING AND DESIGN MANUAL
The Johns Hopkins University Press, 1982.

This manual presents the latest field results of the research, summarizes selected portions of other publications on sanitation program planning, and describes the engineering details of alternative sanitation technologies and how they can be upgraded.

156 Kintner, William R.
A C TECHNOLOGY AND INTERNATIONAL POLITICS: THE CRISIS OF WISHING
Lexington, Mass.: Lexington Books, 1975.

157 Knight, Peter T.
B D BRAZILIAN AGRICULTURAL TECHNOLOGY AND TRADE. A STUDY OF FIVE COMMODITIES.
Praeger Special Studies in Internatonal Economics and Development
New York and London: Praeger, 1971.

A study of technological change, public policy, and economic behavior in Brazil's agricultural sector, with emphasis on their implications for international trade. Five agricultural commodities (rice, wheat, corn, beef, and soybeans) important in international trade are studied to provide answers to broad questions of Brazilian development planners in formulating policy regarding export performance, import substitution, productivity increases, and marketing systems for principal agricultural commodities. Supplemented by fifty tables and nine figures.

158 Kowalski, Gregory S.
B TECHNOLOGY AND SOCIAL CHANGE BIBLIOGRAPHY: 1976
Monticello, Ill.: Council of Planning Librarians, 1977.

159 Lim, Linda and Fong, Pan Eng
B C D TECHNOLOGY CHOICE AND EMPLOYMENT CREATION: A CASE STUDY OF THREE MULTINATIONAL ENTERPRISES IN SINGAPORE
Geneva: ILO, 1981.

Analyzes the experience of three multinational enterprises (MNE's) in the electronics industry in Singapore (1 European, 1 American, 1 Japanese) relevant to technology, employment and linkages, and technology choice and employment creation.

160 Long, Franklin A. and Oleson, Alexandra, eds.
B C APPROPRIATE TECHNOLOGY AND SOCIAL VALUES - A CRITICAL APPRAISAL
Cambridge, Mass.: Harper & Row, Ballinger, in association with the American Academy of Arts and Sciences, 1980.

Ten papers originally presented at an international symposium held in June 1978 in Racine, Wisconsin, under the auspices of the International Pugwash Council and the U.S. Pugwash Committee. The four papers in part one explore ambiguities in the concept of appropriate technology and discuss the role and impact of social values on the selection of technologies. The essays in part two examine technology choice in developing nations, including a general paper, case studies of Korea, Ghana, India, and China, and a final paper by Kenneth Boulding, which consider the philosophical question--"On Being Rich and Being Poor: Technology and Productivity."

161 Lucas, Barbara A.
B C TECHNOLOGICAL CHOICE AND CHANGE IN DEVELOPING COUNTRIES: INTERNAL AND EXTERNAL CONSTRAINTS
New York: Unipub, 1982.

162 Majumdar, Atreyi
A B C D STRUCTURAL TRANSFORMATION AND ECONOMC DEVELOPMENT
Atlantic Highlands, N.J.: Humanities Press, 1980.

Monograph on problems and prospects of the economic growth of India, comparing India's experience with that of other developed and developing countries. Emphasizes the need for fundamental changes in technology, institutions, and attitudes in the agrarian economy of India to effect transformation to an industrial economy. Focuses on the role of the manufacturing sector in absorbing surplus labor released from agriculture and attaining higher productivity, examining the pattern in a variety of developed and developing countries. Finds that structural stagnation in India has been the outcome of inadequate growth in output and employment.

163 Manser, William A. P.
B D TECHNOLOGY TRANSFER TO DEVELOPING COUNTRIES
London: Royal Institute of International Affairs, 1979.

164 Martino, Joseph P.
B C TECHNOLOGICAL FORECASTING FOR DECISION MAKING
New York: Elsevier Publishing Company, 1972.

165 McInerney, John P.
B C THE TECHNOLOGY OF RURAL DEVELOPMENT
World Bank, 1978.

Examines the nature of technical innovation in rural development, the conditions required for successful innovation in agriculture, and the implications for development activities. Attempts to develop a conceptual framework for the design, preparation, and implementation of projects in the rural sector.

166 McMains, Harvey and Wilcox, Lyle, eds.
A B C ALTERNATIVES FOR GROWTH: THE ENGINEERING AND ECONOMICS OF NATURAL RESOURCES DEVELOPMENT
Cambridge, Mass.: Ballinger Publishing, 1978.

167 Mehra, Shakuntla
B D INSTABILITY IN INDIAN AGRICULTURE IN THE CONTEXT OF THE NEW TECHNOLOGY
Washington D.C.: International Food Policy Research Institute, Research Report 25, 1981.

This monograph explores relationships between the growth of crop production, particularly using new seeds, and the variability of production. It is shown that in India in the period 1967/68 to 1977/78 the standard deviation and the coefficient of variation of production for virtually all crops examined was larger than in the period 1949/50 to 1964/65. Most of the variability in production was accounted for by variations in yield. Standard deviation of yield increased particularly for foodgrains. The data available make it difficult to compare directly the variability of yields of high-yielding varieties with that of traditional varieties. However, the data are consistent with the contention that yield variability has increased with the use of high-yielding varieties, especially in areas where irrigation is not generally assured. Where irrigation is good, variability is not greatly affected by the use of the new varieties. But, if states like the Punjab which are well supplied with irrigation are excluded, a high correlation is found between the increase in the standard deviation of yield and the area of a crop sown with high-yielding varieties.

168 Mensch, Gerhard
B C STALEMATE IN TECHNOLOGY: INNOVATIONS OVERCOME THE DEPRESSION
Cambridge, Mass.: Harper & Row, Ballinger, 1979.

Maintains that socioeconomic stagnation is the result of a lack of basic innovations, leading to discrepancies between the need structure of the populace's changing life style and the supply structure, and that innovations tend to occur in clusters (e.g., 1825, 1886, and 1935). Part one examines the fundamentals of economic change and stagnation, and part two investigates the fluctuations between innovative abundance and scarcity, the origins of the surges, and particularly the vehemence of these innovative spurts during technological stalemates of the past. The last part considers the future, where two scenarios are possible: industrial evolution with technological innovation in large-scale operations ("hard" transiton) or social innovation in small-scale, participative operations ("soft" transition). Concludes that the 1980's should see another basic spurt in the hard direction, i.e., towards hyperindustrialization.

169 Meyer, K. Rudy
B C THE TRANSFER OF TECHNOLOGY TO DEVELOPING COUNTRIES--THE PULP AND PAPER INDUSTRY
UNITAR, 1974.

Describes types of technology transferred in the pulp and paper industry and the methods of transfer. Discusses motives and problems of suppliers of technology, of the developing countries in receiving it, and the options open to the developing countries. This is the ninth and final UNITAR study in the field of transfer of technology and skills to developing countries.

170 Michelana, Jose A. Silva
A B C SCIENCE, TECHNOLOGY, AND POLITICS IN A CHANGING WORLD
Tokyo: UNU, 1980.

Highlights the main points of relatioships between science, technology, and politics in the changing world.

171 Mishan, E. J.
A B C TECHNOLOGY AND GROWTH; THE PRICE WE PAY
New York: Praeger, 1970.

172 Montgomery, John D.
A B C TECHNOLOGY AND CIVIC LIFE: MAKING AND IMPLEMENTING DEVELOPMENT DECISIONS
M.I.T. Studies in Comparative Politics
Cambridge, Mass. and London: MIT Press, 1974.

The Third World development experience is looked upon from the point of view of the effects of technological improvements and modernization on the people, whose welfare was the original, yet forgotten, goal. In an introduction and six chapters (aspirants to modernization, technology as modernization, modernizing behavior, agents of change, systems of change, decision), the author reviews the Third World experience during the past 25 years, analyzing technology, organization, and administrative systems and their interaction with civic life in what constitutes a new approach in development studies.

173 Moravcsik, Michael J.
A B C SCIENCE DEVELOPMENT: TOWARD THE BUILDING OF SCIENCE IN LESS DEVELOPED COUNTRIES
Bloomington: Indiana University, International Development Research Center, 1975.

174 Morehouse, Ward, ed.
B C SCIENCE, TECHNOLOGY AND THE SOCIAL ORDER
New Brunswick, N.J.: Transaction Books, 1979.

175 Morgan, Robert P.
A B C SCIENCE AND TECHNOLOGY FOR DEVELOPMENT: THE ROLE OF U.S. UNIVERSITIES
New York: Pergamon Press in cooperation with the Center for Development Technology, Washington University, St. Louis, 1979.

Reports on a National Science Foundation program in preparation for the 1979 United Nations Conference on Science and Technology for Development; investigates the past, present, and future roles of U.S. universities in helping to build indigenous science and technology base in developing countries. Analyzes key policy issues and options, as well as mechanisms for future university involvement in the fields of engineering, agriculture, and science. Concludes, in part, that U.S. university personnel have contributed to building an indigenous science and technology base in LDC's and that some shift of emphasis from large-scale, institution-building projects toward cooperative research and development can be expected.

176 Morgan, Robert P. and Icerman, Larry J.
A B C RENEWABLE RESOURCE UTILIZATION FOR DEVELOPMENT
New York: Pergamon Press in cooperation with the Center for Development Technology, Wshington University, St. Louis, 1981.

Investigative study by nine authors on possible efforts that the United States might undertake in light-capital technology for aiding developing countries in the utilization of renewable resources. These cover steps, policies, and programs for undertaking, adoption, or support by the United States. Center around five principal project study areas: wind energy for agriculture, water lifting, rotary power, and electricity; processes for direct biomass utilization for cooking; direct solar energy for grain crop and timber drying; useful food and feed products from agriculture by-products and wastes; and useful non-food and non-feed products.

177 Mori, Yuji
B C RESTRUCTURING A FRAMEWORK FOR ASSESSMENT OF SCIENCE AND TECHNOLOGY AS A DRIVING POWER FOR SOCIAL DEVELOPMENT
Tokyo: UNU, 1980.

This study explains and analyzes the basic ideas of biosociology and investigates science and technology as the driving forces for social development. Topics include the Darwinian and Neo-Darwinian systems; how to view humans and their society; three levels of production and consumption; needs; science and technology as cultural phenomena; and the turning point of social development: space and time.

178 Moyibi, Amoda and Tyson, Cyril D.
A D TECHNOLOGICAL DEVELOPMENT IN NIGERIA
Lagos and New York: Third Press International, 1979.

179 Claire, Nader and Zahlan, A. B., eds.
A B C D SCIENCE AND TECHNOLOGY IN DEVELOPING COUNTRIES
New York: Cambridge University Press, 1969.

The proceedings of an international conference held in Beirut at the end of 1967 which discussed the role of science and technology in the agricultural and economic advancement of the developing world. Eleven papers surveyed the national organizations for science and technology in Turkey and the Arab world, while other papers discussed more general matters such as the "brain drain", attitudes to science and health needs. An edited version of the discussion following most papers in included.

180 Nair, K. N. S.
A B C TECHNOLOGICAL CHANGES IN AGRICULTURE: IMPACT ON PRODUCTIVITY AND EMPLOYMENT
New Delhi: Vision Books for Birla Institute of Scientific Research, Economic Research Division, 1980.

An attempt to investigate the pattern of technological progress, land and labour productivity, and employment in the agricultural sector of twenty-one major countries of the world. The countries were selected based on their relative importance in world agriculture and data availability for the period 1960-77. These countries account for about 70 percent of world population, 74 percent of agricultural land, and 81 percent of total food grain production in 1977. Presents a comparative analysis of how progress in technology in the sixties and seventies has affected agricultural yield and of the economic structure of population in these countries in different development stages. Findings indicate an important growth rate in food production in most countries, but to a considerable extent, the faster rate of population growth in the developing countries has wiped out the per capita gain; also finds that mechanization initially does not dispace agricultural labor and does so only after attainment by a country of an advanced stage of development; and that agricultural technological progress has mainly taken place through the use of irrigation, fertilizers, pesticides, tractors, and high-yielding seed varieties.

181 National Academy of Sciences
B C RESOURCE SENSING FROM SPACE: PROSPECTS FOR DEVELOPING COUNTRIES
Washington, D.C.: National Academy of Sciences, 1977.

182 Norabhoompipat, Thanet
B C STRUCTURE OF TECHNOLOGY: A MODEL FOR RURAL DEVELOPMENT
Tokyo: UNU, 1982.

Focusing on rural technology, this study models technology as an information system. The model provides a comprehensive, operational framework for identifying the various technological attributes. Chapters cover: introduction; background; action information; initial-state and end-state information; appraisal; and summary and conclusions.

183 Norman, Colin
A B C THE GOD THAT LIMPS: SCIENCE AND TECHNOLOGY IN THE EIGHTIES
New York: W. W. Norton, 1981.

184 OECD
B C SCIENCE AND TECHNOLOGY IN THE MANAGEMENT OF COMPLEX PROBLEMS
Paris: OECD, 1976.

Contains the principal reports prepared for the Meeting of the OECD Committee for Scientific and Technological Policy at Ministerial Level (Paris, 24th and 25th June 1975). In five parts: (a) introduction, which consists of the Report of the Committee submitted for discussion by Ministers; (b) background report covering the social dimension of science and technology; (c) background paper on science policy and the management of natural resources;

(d) analysis of trends of research and development expenditures since 1971, the date of the preceding Meeting of Ministers of Science of OECD countries; and (e) the Communique containing the conclusions arrived at by the Ministers.

185 OECD
B C TECHNICAL CHANGE AND ECONOMY POLICY
Paris: OECD, 1980.

Describes and evaluates the links between research, technology and the economy. Examines how scientific research, innovation and technical change influence economic performance today as compared to the 1960's and how, in reverse, economic slack and inflation affect research and innovation. In four parts: (a) the new economic and social context; (b) trends in R & D and innovation; (c) technical change and the economy; and (d) conclusions and recommendations.

186 OECD
B C TECHNOLOGY ON TRIAL
Paris: OECD, 1979.

Examines OECD Member countries' experience regarding public participation in government decision-making related to science and technology. Presents a preliminary assessment of the nature of different national experiences; provides a clearer understanding of the participatory phenomena; and identifies the various problems and areas of present concern as a guide for future action.

187 O'Kelly, Elizabeth
B C D SIMPLE TECHNOLOGIES FOR RURAL WOMEN IN BANGLADESH
Paris: UNESCO, 1978.

Presents documentation about different technologies and simple hand-operated machines that can easily be made or are available for sale in developing countries, with particular emphasis on their use by women in Bangladesh. This handbook was produced by and is available from the Women's Development Programme, UNICEF Country Office, CPO Box 58, Dacca-5, Bangladesh.

188 Okolie, Charles Chukwuma
B C LEGAL ASPECTS OF THE INTERNATIONAL TRANSFER OF TECHNOLOGY TO DEVELOPING COUNTRIES
New York and London: Praeger, 1975.

Examines the obstacles to the transfer of technology and resources to developing countries and suggests ways of improving the transfer. The study, an outgrowth of work done by the United Nations Institute for Training and Research, discusses the legal and moral problems of international technology transfer, relevant aspects of international law, United States attitude to the transfer, the role of Soviet corporations and the legal norms for Soviet transfer of technology, and the role of the multinational company in the transfer of technology.

189
B C

Onoh, J. K.
STRATEGIC APPROACHES TO CRUCIAL POLICIES IN ECONOMIC DEVELOPMENT. A MACRO LINK STUDY IN CAPITAL FORMATION, TECHNOLOGY AND MONEY
Rotterdam: Rotterdam University Press; Portland, Oregon: International Scholarly Book Services, 1972.

The book surveys current economic development concepts and strategies, concentrating attention on the problems related to capital allocation, to acquisition of technological capability, and the money supply in a developing economy. Using data for several African nations, the author evaluates recent trends in fixed capital formation, acquisition of technology, money supply and capital formation, and monetary expansion in general.

190
B C

Pecujlic, Miroslav and Vidakovic, Zoran
ON THE EDGE OF A RAZOR BLADE: THE NEW HISTORICAL BLOCS AND SOCIO-CULTURAL ALTERNATIVES IN EUROPE
Tokyo: UNU, 1980.

Describes and analyzes the effects of the modern science and technology on the developed and developing parts of the world. Chapters discuss: the new Janus--two faces of science and technology; the pathology of power and science; the new protagonist--social movements and organic intelligentsia; dramatic birth of alternatives; and self-reliance and solidarity (autonomy and new universality).

191
B C

Peng, Khor Kok
TRADITIONAL TECHNOLOGY, A NEGLECTED COMPONENT OF APPROPRIATE TECHNOLOGY: CHARACTERISTICS AND PROBLEMS--A MALAYSIAN CASE STUDY
Tokyo: UNU, 1980.

This paper analyzes the working of traditional technology in the dynamics of social structures and social change in Malaysia. It makes a critical appraisal of the concept of traditional technology and its usefulness in fulfilling basic needs.

192
B C

Perlmutter, Howard V. and Sagafi-Nejad, Tagi
INTERNATIONAL TECHNOLOGY TRANSFER: GUIDELINES, CODES AND A MUFFLED QUADRILOGUE
New York; Oxford; Sydney and Paris: Pergamon Press, 1981.

First in a three-volume series on technology transfer, discusses this process between advanced countries and less industrialized countries. Proposes guidelines or codes of conduct for international technology transfer, emphasizing the need to establish some mandatory code authority, which would accelerate the reduction in the gap between the advanced market economies and the LDC's and the attainment by the latter of basic needs and minimal standards of living. Analyzes the views, perceptions, attitudes, and feelings of persons representing supplier countries, supplier firms, recipient countries, and recipient firms in developing an internationanl control structure that would benefit all four groups. Records some major themes that underlie the mistrust among the four groups.

193 Poats, Rutherford M.
A B C TECHNOLOGY FOR DEVELOPING NATIONS: NEW DIRECTIONS FOR U.S. TECHNICAL ASSISTANCE
Washington, D.C.: Brookings Institution, 1972.

194 Radhakrishna, S.; Brenner, P. and Scott, D.
A B C TECHNOLOGICAL EDUCATION AND NATIONAL DEVELOPMENT
Bangalore: Indian Institute of Science, 1978.

195 Ramesh, Jairam and Weiss, Charles, eds.
A B C MOBILIZING TECHNOLOGY FOR WORLD DEVELOPMENT
Foreword by Barbara Ward
New York: Praeger, for the International Institute for Environment and Development and the Overseas Development Council.

An exploration into the common interest and mutual needs of rich and poor countries in mobilizing science and technology for development. Presents the Report of the Jamaica Symposium organized by the International Institute for Environment and Development for 25 scientists, technologists, business executives, economists, bankers, and political leaders in Jamaica in January 1979 and 18 papers (some published elsewhere in somewhat different versions). The papers illustrate the symposium message of "strategic pragmatism," or practical actions that countries can take to solve national and global development problems through technology and science; stresses the need for development strategy in the agricultural area, adapted to the socioeconomic needs of the area.

196 Rao, C. H. Hanumantha
B C D TECHNOLOGICAL CHANGE AND DISTRIBUTION OF GAINS IN INDIAN AGRICULTURE
Institute of Economic Growth, Studies in Economic Growth, No. 17
Delhi: Macmillan, 1975.

Analyzes the economics of technological change in Indian agriculture from 1949 to 1971. The first part assesses the magnitude of the technological changes and the factors accounting for such changes, as well as estimating costs and returns associated with the use of modern inputs. The second part of the book shows how the gains from technological changes have been distributed between different regions as well as between different classes of farm people within regions experiencing change. Finally, the author discusses socio-political factors and agricultural policies and trends in technological change, agricultural growth, and distribution of income.

197 Richardson, Jacques, ed.
B C INTEGRATED TECHNOLOGY TRANSFER
Mt. Airy, MD: Lomond Publications, Inc., 1979.

Collections of articles on the transfer of technology from developed to developing countries. The first addresses only the problems arising in the formulation of the code of conduct for the transfer of technology. The authors--a group of French civil servants and scholars--draw essentially from the work of UNCTAD V (1979, Manilla), the UN Conference on Science and Technology (1979, Vienna), and other UN conferences on the code.

198 Robinson, Austin, ed.
A B C D APPROPRIATE TECHNOLOGIES FOR THIRD WORLD DEVELOPMENT: PROCEEDINGS OF A CONFERENCE HELD BY THE INTERNATIONAL ECONOMIC ASSOCIATION AT TEHERAN, IRAN
New York: St. Martin's Press, 1979.

Fifteen conference papers, with discussions of nine, and a summing up of the proceedings. The focus is on the failure of developing countries to use appropriate technologies and the reasons why they do not. Views of appropriateness differ on whether they are "efficient" or fit the world economic conditions; the general development strategy or political implications also underlie the discussion. Topics include the use of technologies of Mainland China and of Japan; contrasts in the experiences of Taiwan and the Philippines; the role of donor institutions in the employment and development of appropriate technologies; two case studies of the choice of technology by multinational corporations; some aspects of technologies used in Turkey and Iran; and a re-examination by the State Department of the technologies program of United States Agency for International Development (USAID), including the reprinting of two background studies prepared for USAID. Authors are economists and other social scientists, government officials, engineers, and representative of international agencies.

199
A B C

Rosenblatt, Samuel M., ed.
TECHNOLOGY AND ECONOMIC DEVELOPMENT: A REALISTIC PERSPECTIVE
Boulder, CO: Westview Press; in cooperation with the International Economic Studies Institute, 1979.

Five previously unpublished papers focused on technology's role in the development process and the problem of technological transfer. The editor cites a concensus of the authors on the recognition that: (1) the solutions found in technology for LDC's are limited; (2) the latest technology is not invariably inappropriate nor the most basic invariably appropriate; (3) that adaptations of the technology and adaptation of their economic and social policies to the technology is a responsibility of the LDC's; (4) transnational enterprise is neither hero nor villain but a necessary if unfortunate conveyor of technology; (5) "goodwill, incentives,...[or] political decisions can [not] substitute for the basic disciplines of competitive economic forces in the developing countries themselves" in the transfer of technology.

200
B C

Rosenberg, Nathan, ed.
THE ECONOMICS OF TECHNOLOGICAL CHANGE
New York: Penguin Books, 1971.

201
A B C

Rostow, W. W.
THE STAGES OF ECONOMIC GROWTH: A NON-COMMUNIST MANIFESTO
New York and London: Cambridge University Press, 1971.

The second edition of the well-known work. Changes are in the preface and in an appendix in which the author aduces evidence, in large part derived from what he terms Kuznets' earlier approach, to answer some of his critics. Rostow concludes, after lengthy consideration and in spite of the attacks of many critics, that the stages of growth approach remains a useful tool in viewing the past as well as the present developments.

202
A B C

Rothwell, Roy and Zegveld, Walter
TECHNICAL CHANGE AND EMPLOYMENT
New York: St. Martin's Press, 1979.

Report of a project (the Six Countries Programme on Aspects of Government Policies toward Technological Innovation in Industry). Studies past and future trends of employment and unemployment in mature industrialized societies, focusing on technical change and structural adjustments. Observes that the rate of increase in manufacturing employment in those countries had slowed well before 1973, which suggests that this change cannot simply be attributed to the OPEC crises and the resultant shocks to the world economy. Argues that failure of an industrial country to sustain technical progress would be far more likely to cause, under the present liberal trade conditions, serious structural unemployment problems, than a high rate of technical change. Notes direction of technical change is also important and that there is a need for product and service innovation.

203 Rybczynski, Witold; Polprasert, Chongrak and McGarry, Michael
B LOW-COST TECHNOLOGY OPTIONS FOR SANITATION--A STATE-OF-THE ART REVIEW AND ANNOTATED BIBLIOGRAPHY
World Bank/International Development Research Centre Publication, 1978.

A comprehensive bibliography that describes alternative approaches to the collection, treatment, reuse, and disposal of wastes.

204 Sagafi-Nejad, Tagi; Moxon, Richard W. and Perlmutter, Howard
B C CONTROLLING INTERNATIONAL TECHNOLOGY TRANSFER: ISSUES, PERSPECTIVES AND POLICY IMPLICATIONS
New York: Pergamon Press, 1981.

205 Sagasti, Francisco
B C SCIENCE AND TECHNOLOGY FOR DEVELOPMENT: MAIN COMPARATIVE REPORT OF THE SCIENCE AND TECHNOLOGY POLICY INSTRUMENTS PROJECT
Ottawa: IDRC, 1978.

Summary of the main results of the 3½ year STPI research project undertaken in 10 less-developed countries: Argentina, Brazil, Colombia, Egypt, India, Mexico, Peru, South Korea, Venezuela, and Yugoslavia. The project investigated the implementation of science policy in developing countries; studied the role of science and technology in economic development, particularly in the industrialization process; and examined the mechanisms of policy formulation, decision making, and policy implementation, the factors affecting technological change, and industrial administration.

206 Sagasti, Francisco
B C SCIENCE AND TECHNOLOGY FOR DEVELOPMENT; STPI MODULE 1: A REVIEW OF SCHOOLS OF THOUGHT ON SCIENCE, TECHNOLOGY, DEVELOPMENT, AND TECHNICAL CHANGE
Ottawa: IDRC, 1980.

Examines different approaches to the study of the interrelations between science, technology, development, industrialization, and technical change, including neoclassical theory; Rostow's stages-of-growth theory; the structuralist view; concepts of dependence and technological dependence; oligopoly and the role of technical progress; technology and capital accumulation at the international level; views of technological change at the firm level; and policy implications of different schools of thought.

207 Sagasti, F.
B C SCIENCE AND TECHNOLOGY FOR DEVELOPMENT; STPI MODULE 8: POLICY INSTRUMENTS TO PROMOTE THE PERFORMANCE OF S AND T ACTIVITIES IN INDUSTRIAL ENTERPRISES
Ottawa: IDRC, 1980.

This report constitutes a part of the Main Comparative Report of the Science and Technology Policy Instruments (STPI) project that examines the design and implementation of science and technology policies in 10 developing countries. Topics covered in this report include special credit lines, tax incentives, and administrative measures.

208 Sen, Amartya
A B EMPLOYMENT, TECHNOLOGY AND DEVELOPMENT: A STUDY PREPARED FOR THE INTERNATIONAL LABOUR OFFICE WITHIN THE FRAMEWORK OF THE WORLD EMPLOYMENT PROGRAMME
New York: Oxford University Press, 1975.

The author discusses the choice of technology as it relates to employment policy. The book is primarily analytical, with particular attention given to India. Stresses the use of existing technologies through an adequate institutional and incentives structure and pricing policies. Four appendices deal with the empirical problems of unemployment measurement and employment policy in India. Appropriate unemployment policies must be based on an assessment of the extent to which these structural relations can be deliberately influenced, including technological possibilities, institutional features, political feasibilities, and behavioral characteristics.

209 Sercovich, F.
B C D SCIENCE AND TECHNOLOGY FOR DEVELOPMENT: STPI MODULE 10: TECHNICAL CHANGES IN INDUSTRIAL BRANCHES
Ottawa: IDRC, 1980.

The material in this module examines the patterns of technical change at the industrial branch level, focusing on the impact of certain key policy instruments and related factors. The five case studies included cover the machine tools industry (Argentina and Brazil), the powder metallurgy industry (Korea), the electronics industry (India), and the capital good industries in general (Venezuela).

210 Sercovich, F.
B C D SCIENCE AND TECHNOLOGY FOR DEVELOPMENT: STPI MODULE 11: TECHNOLOGY BEHAVIOUR OF INDUSTRIAL ENTERPRISES
Ottawa: IDRC, 1980.

The 8 studies presented in this module focus on the technological behaviour of enterprises in selected industrial branches, analyzing the characteristics of change and assessing the relative importance of the various factors that affect them. The two Colombian studies refer to agricultural implements and fertilizers; the two Argentinian studies deal with state enterprises in the gas and electricity generation fields; the two Brazilian studies also deal with state enterprises, but in the electric power and flat steel products industries; the Venezuelan report covers the technological behaviour of mixed enterprises in the petrochemical industry; and the last study in the module refers to the contribution of transnational corporations to the technological development of Mexican industry.

211
B C D

Sercovich, F.
SCIENCE AND TECHNOLOGY FOR DEVELOPMENT: STPI MODULE 12: STUDIES ON TECHNICAL CHANGE
Ottawa: IDRC, 1980.

This report contains three studies on technological innovation. The first of the reports examines three case studies of diffusion of innovations in the Brazilian industry: shuttleless looms in the textile industry, special presses in the paper industry, and the dry method process in cement manufacturing. The Mexican report examines the orientation of technical change in three industrial branches: capital goods, petrochemicals, and food processing. The third report, the Venezuelan study, deals with some atypical cases of technological innovation.

213
A B C

Singer, Hans
TECHNOLOGIES FOR BASIC NEEDS
Geneva: ILO, 1977.

This study suggests new criteria for establishing socially oriented technology policies in the developing economies and demonstrates how technology can be related to satisfying basic human needs. Chapters cover: (a) appropriate technology and basic needs; (b) determinants of a country's technology mix; (c) selection, transfer and dissemination of existing technology; (d) adaptation; (e) new technologies; (f) some problems of education and training for appropriate technology; (g) institutional, administrative and planning requirements for appropriate technology; and (h) conclusions. Appendixes contain: text of the Declaration of Principles and Programme of Action adopted by the Tripartite World Conference on Employment, Income Distribution and Social Progress, and the International Divison of Labour, Geneva, 4-17 June 1976; and list of institutions dealing with appropriate technology.

214 Soedjarwo, Anton
B C TRADITIONAL TECHNOLOGY--OBSTACLE OR RESOURCE? BAMBOO-CEMENT RAIN-WATER COLLECTORS AND COOKING STOVES
Tokyo: UNU, 1981.

Chapters cover: introduction; environmental conditions; traditional technology concerning water problems in Gunung Kidul, Indonesia; traditional technology concerning energy; and general conclusions.

215 Solo, Robert A.
A B C ORGANIZING SCIENCE FOR TECHNOLOGY TRANSFER IN ECONOMIC DEVELOPMENT
East Lansing: Michigan State University Press, 1975.

Reconsiders some suppressed earlier (1963) work done by the author on how government agencies and research establishments in France, the United Kingdom, W. Germany, and the Netherlands organized and allocated their scientific resources through their government programs, to promote economic development to various less-developed country groups. Deals specifically with technical aid. The study is based on information gathered through visits to the relevant government agencies and research development centers, and through interviews with some of their officials. Some per capita income and aid contributions and receipts for donor and recipient countries and some charts.

216 Solo, Robert A. and Rogers, Everett M., eds.
B C INDUCING TECHNOLOGICAL CHANGE FOR ECONOMIC GROWTH AND DEVELOPMENT
East Lansing, MI: Michigan State University Press, 1972.

A collection of ten original papers written mostly by economists but also by sociologists and anthropologists and dealing with the processes and agencies of technological change particularly in relation to economic development.

217 Stanley, Manfred
A B C THE TECHNOLOGICAL CONSCIENCE: SURVIVAL AND DIGNITY IN AN AGE OF EXPERTISE
New York: Free Press, 1978.

218 Stark, Oded
B C D TECHNOLOGICAL CHANGE AND RURAL-TO-URBAN MIGRAITON OF LABOUR: A MICRO-ECONOMIC CAUSAL RELATIONSHIP IN THE CONTEXT OF LESS DEVELOPED ECONOMIES
IUSSP Papers, no. 11
Liege, Belgium: International Union for the Scientific Study of Population, 1979.

Stressing that no one general comprehensive microeconomic model of rural-to-urban migration of labor (RUMOL) will do, the author examines the decision-making unit (in this case the family unit producing food on its own small holding), the change in its welfare over time, and the impetus to change technology. The internal and external factors affecting technology, as well as the risk incurred, are explored. Also considers the relationship between RUMOL and the alternative social welfare criteria likely to prevail. Some empirical evidence with regards to RUMOL are presented. Concludes that since RUMOL serves to facilitate technological change, alternative polcies might be introduced to facilitate the same technological change, which are preferable in their externalities.

219 Starr, Chauncey and Ritterbush, Philip C., eds.
A B C SCIENCE, TECHNOLOGY AND THE HUMAN PROSPECT: PROCEEDINGS OF THE EDISON CENTENNIAL SYMPOSIUM
Pergamon Policy Studies on Science and Technology
Oxford; Toronto and Sydney: Pergamon Press, 1980.

Ten commissioned papers and eight summaries of "workshop" evaluations from a 1979 symposium. Includes two papers by economists: "Science, Technology, and Economic Growth," by Edwin Mansfield and "Science and Technology in Global Development," by Sumitro Djojohadikusumo. The editors are Vice-Chairman of the Electric Power Research Institute and Director of a center for higher education programs, respectively.

220 Stewart, Frances
A B C TECHNOLOGY AND UNDERDEVELOPMENT
Boulder, CO: Westview Press, 1977.

221 Stewart, Frances
B C INTERNATIONAL TECHNOLOGY TRANSFER: ISSUES AND POLICY OPTIONS
Washington, D.C.: World Bank, 1979.

A background study for World Development Report, 1979. Surveys four issues raised by technology transfer to developing countries--the costs of the transfer, the appropriateness of products and techniques that are transferred, the effects of transfer on learning and technological development in the developing countries, and the effects of independence. Considers the consequence of the transfer, as well as the range of policies that might be adopted, in relation to each of these issues, and argues that appropriate policies will vary according to the developmental stage, technological capacity, and objectives of each country.

222 Strassmann, W. Paul
B C D HOUSING AND BUILDING TECHNOLOGY IN DEVELOPING COUNTRIES
MSU International Business and Economic Studies
East Lansing: Michigan State University, Graduate School of Business Administration, Divison of Research, 1978.

Examines the role that building technology plays in economic development. Analyzes the relationship between conventional technology, construction wages, and employment; surveys innovations and analyzes their cost effects; explores the concept of industrialized systems building for developing countries; assesses the impact on technology of changes in demand; and considers the effect of technology on employment.

223 Street, James H. and James, Dilmus D., eds.
A C D TECHNOLOGICAL PROGRESS IN LATIN AMERICA: THE PROSPECTS FOR OVERCOMING DEPENDENCY
Westview Special Series on Latin America and the Caribbean
Boulder, Colo.: Westview Press, 1979.

Twelve papers and case studies, four previously published, appraising the prospects for internal technological development in Latin America; selected from papers presented at recent meetings. Four papers review the nature of technological dependency and examine the possibilities of stimulating indigenous research and development. The remaining papers constitute case studies of successful internal technological diffusion in various sectors and countries. Contributors, most of them American economists, employ empirical techniques to establish their conclusions.

224 Streeton, Paul
A B C THE FRONTIERS OF DEVELOPMENT STUDIES
New York: John Wiley & Sons, 1972.

A collection of mostly previously published (but all to some degree revised) essays on issues in the theory and practice of economic development. The papers included deal with development theories, the use of economic models, problems in the planning of growth, the role of international trade, investment, and aid, problems in the world transfer of technology, the relationship between the lesser and more developed countries, etc.

225 Subrahmanian, K. K.
C D IMPORT OF CAPITAL AND TECHNOLOGY. A STUDY OF FOREIGN COLLABORATIONS IN INDIAN INDUSTRY.
New Delhi: People's Publishing House, 1972.

An assessment of some of the contributions of foreign investors to the economic growth of the Indian economy. It begins by reviewing the theoretical aspects of foreign investment as it affects a developing economy and by providing a profile of the growth of private foreign investment in India. The bulk of the book is an empirical investigation, working with both aggregate data and individual case-studies, of the technical collaboration provided by foreign firms and of the working of collaboration arrangements. Finally, there is an overall evaluation of the contribution of private foreign investment to and policy implications for the economy of India.

226
A B C
Szyliowicz, Joseph S.
TECHNOLOGY AND INTERNATIONAL AFFAIRS
New York: Praeger, 1981.

227
B C D
Thebaud, Schiller
STATISTICS ON SCIENCE AND TECHNOLOGY IN LATIN AMERICA: EXPERIENCE WITH UNESCO PILOT PROJECTS, 1972-1974
Paris: UNESCO, 1976.

Reports on four UNESCO pilot projects in Latin America on statistics for science and technology. These projects studied the methods adopted for national statistical surveys of science and technology with a view to facilitating international comparison of their results. Also reproduces a document on the preparation of science and technology statistics used as a basis for discussion at a symposium held in Caracas in September 1974.

228
B C D
Thomas, D. Babatunde
IMPORTING TECHNOLOGY INTO AFRICA: FOREIGN INVESTMENT AND THE SUPPLY OF TECHNOLOGICAL INNOVATIONS
New York: Praeger, 1976.

229
B C D
Thomas, D. Babatunde and Wionczek, Miguel S., eds.
INTEGRATION OF SCIENCE AND TECHNOLOGY WITH DEVELOPMENT: CARIBBEAN AND LATIN AMERICAN PROBLEMS IN THE CONTEXT OF THE UNITED NATIONS CONFERENCE ON SCIENCE AND TECHNOLOGY FOR DEVELOPMENT
Pergamon Policy Studies, no. 22
New York; Oxford; Toronto and Sydney: Pergamon Press, 1979.

Nineteen previously unpublished papers by nongovernmental participants plus a summary statement on the science and technology (S & T) problems of LDC's in the West, most originally presented at a conference sponsored by the Institute of Social and Economic Research, University of the West Indies, and the Institute of Development Studies, University of Guyana, held at Florida International University (Miami) in April 1978. Considers the problems in building S & T capability, infrastructure and technology transfer, and technological problems in the Caribbean. Also examines S & T policies in Latin

America and preparatory work for the then forthcoming U.S. conference held in August 1979. Concludes in part that: (1) building the needed technology is a long-term proposition; (2) we need to investigate the impact of multinationals on technology transfer; (3) priority should be given to policies to harness nonproprietary technology for development; and (4) new methods are needed for the integration of scientific and technological development goals.

230
A C
Timmer, C. Peter, et al.
THE CHOICE OF TECHNOLOGY IN DEVELOPING COUNTRIES: SOME CAUTIONARY TALES
Harvard University, 1975.

Argues that the use of less capital-intensive methods of production than those used in advanced countries could be one solution to the problems of unemployment in developing countries. C. P. Timmer discusses the sources of labor displacement in a developing economy, showing methods of evaluating government projects and the role of the market economy in allocating labor. using Indonesian data, he discusses the choice of technique in rice milling. John Woodward Thomas examines the choice of technology for irrigation tube wells in East Pakistan, describing pilot projects and evaluating tube well alternatives. He discusses the factors affecting the choice of technology. Louis T. Wells discusses the behavior of businessmen in Indonesia, particularly the choice of technology at the firms level. Finally, David Morawetz discusses import substitution, employment, and foreign exchange in Colombia, comparing investment in the petrochemical industry with potential alternative uses of the same funds in labor-intensive export-oriented industries. The studies are mainly empirical.

231
A B C
Tomovic, Rajko
TECHNOLOGY AND SOCIETY; HUMAN AND SOCIAL DEVELOPMENT PROGRAMME: PROJECT ON SOCIO-CULTURAL DEVELOPMENT ALTERNATIVES IN A CHANGING WORLD
Tokyo: UNU, 1980.

A brief discussion of past, present and future interrelationships between technology and society.

232
A B C
United Nations
SCIENCE, TECHNOLOGY AND GLOBAL PROBLEMS: THE UNITED NATIONS ADVISORY COMMITTEE ON THE APPLICATION OF SCIENCE AND TECHNOLOGY FOR DEVELOPMENT
Elmsford, NY: Pergamon, 1979.

233
B C
United Nations
AN INTERNATIONAL CODE OF CONDUCT ON TRANSFER OF TECHNOLOGY
New York: UN, 1975.

Part One contains an analytical description of the main issues to be considered in the preparation of any draft outline of an international code of conduct in transfer of technology: objectives and principles; scope of application of the code; ownership and control; relations among suppliers; restrictive practices related to the acquisition of technology for production; practices relating to distribution; pricing and costs of technology; development of national technologies and scientific capabilities; and special preferences for developing countries. Part Two discusses various considerations relevant to the formulation of the code: legal nature and possible form; approaches to formulation; machinery for implementation; and applicable law and settlement disputes.

234 United Nations
B C CASE STUDIES IN THE ACQUISITION OF TECHNOLOGY
New York: UN, 1981.

Provides a fairly detailed description of actual cases of successful and unsuccessful transfer of technology both from industrialized to developing countries and among industrialized countries. Contains case histories of licensing agreements; acquisition and assimilation of technology; and transfer and absorption of technology.

235 United Nations
B C DEVELOPMENT AND TRANSFER OF INDUSTRIAL TECHNOLOGY
New York: UN, 1978.

Brief outline of UNIDO activities and its cooperation with UN agencies and international organizations to promote implementation of an integrated program of development and transfer of industrial technology and to strengthen national technological capabilities in developing countries.

236 United Nations
B C DEVELOPMENT, ENVIRONMENT AND TECHNOLOGY: TOWARDS A TECHNOLOGY FOR SELF-RELIANCE
New York: UN, 1979.

This study was prepared by Johan Galtung at the request of UNCTAD and the United Nations Environment Programme as a contribution to their joint study project on transfer of technology and interrelated environmental problems. It discusses aspects of the social and environmental impact of Western technology on developing countries.

237 United Nations
B C DEVELOPMENT OF APPROPRIATE TECHNOLOGY FOR SMALL-SCALE PRODUCTION OF PORTLAND CEMENT IN LEAST DEVELOPED COUNTRIES AND REGIONS
New York: UN, 1976.

Coverage includes: (a) capital investment for new cement plants; (b) rotary kiln with preheater and calciner; and (c) shaft kiln cement mini-plants. Includes expert findings and recommendations.

238 United Nations
B C DRAFT INTERNATIONAL CODE OF CONDUCT ON THE TRANSFER OF TECHNOLOGY
New York: UN, 1981.

Contains text of a Draft International Code of Conduct on the Transfer of Technology. Prepared by UNCTAD, it deals with national regulation of transfer of technology transactions; guarantees, responsibilities, obligations; special treatment for developing countries; international collaboration; international institutional machinery; and applicable law and settlement of disputes.

239 United Nations
B C D SCIENCE AND TECHNOLOGY FOR DEVELOPMENT: AFRICAN GOALS AND ASPIRATIONS IN THE UNITED NATIONS CONFERENCE
New York: UN, 1978.

Report of a Symposium held in Arusha, Tanzania, January 30-February 4, 1978, which brought together prominent African thinkers to discuss: (a) the main issues underlying the objectives of the UN Conference on Science and Technology for Development (UNCSTD: Vienna, August 1979); (b) the relation of the Conference to the estimation of a new international economic order; (c) what African Member States and Africa as a continent should aim to achieve at the Conference; and (d) some of the crucial issues underlying the application of science and technology to African development.

240 United Nations
B C D NATIONAL APPROACHES TO THE ACQUISTION OF TECHNOLOGY
New York: UN, 1977.

Discusses many of the principles, business practices, and laws that govern licensing, and more generally, transfer of technology in the world at large. It includes an examination of broad principles and of specific problems in specific parts of the world. Chapters cover: (a) transfer of technology through licensing; (b) the technology for licensing; (c) legal considerations as affected by governments, including antitrust regulation in Western Europe; (d) terms of license agreements; (e) considerations when licensing in a developing country; (f) licensing in specific industries (mechanical, electrical, chemical and pharmaceutical); and (g) conclusions.

241
B C

United Nations
THE APPLICATION OF MODERN TRANSPORT TECHNOLOGY TO MINERAL DEVELOPMENT IN DEVELOPING COUNTRIES
New York: UN, 1976.

Evaluates the technical and economic changes in the transport and handling of minerals that took place in the 1960's and the effect these may have had on mineral production in developing countries. Briefly describes the significant technical characteristics of each major mode of transport: unitized cargo, containers and pallets; belt conveyors, aerial ropeways and related systems; the Marconaflo system; pipeline transportation of solids; and ocean, coastal, inland waterway, railway, road and air transport. Gives examples of the application of each mode and discusses criteria governing the application of each mode as well as its selection from among alternatives. Emphasizes the interrelationship among modes, stresses the importance of integration and considers in detail applications to problems of the developing nations.

242
A B C

United Nations
THE APPLICATION OF COMPUTER TECHNOLOGY TO DEVELOPMENT
New York: UN, 1973.

Contains recommendations concerning national policy in developing countries, education, and institutional arrangements within the UN system, and a brief review of the current situation with respect to computers and computer technology since the Secretary-General's first report. Includes some comparative data on developed and developing countries. The Secretary-General was assisted by an ad hoc Panel of Experts and information supplied by international agencies and governments. This report updates and amplifies, but does not replace, the first report, The Application of Computer Technology for Development.

243
B C

United Nations
SCIENCE AND TECHNOLOGY FOR DEVELOPMENT: TERMINOLOGY BULLETIN
New York: UN, 1979.

A list of 2827 relevant terms in English alphabetical order.

244
B C

United Nations
SCIENCE AND TECHNOLOGY IN AN ERA OF INTERDEPENDENCE: A REPORT OF A NATIONAL POLICY PANEL
New York: UN, 1975.

A report on the proceedings of a National Policy Panel that met with experts on such topics as food and nutrition, energy, ocean affairs, and meteorology, as well as with specialists in international organization affairs. The panel focused on the international economic and political implications of scientific and technological developments. The report present an appraisal of the limitations and potentialities of international organizations and a number of recommendations that can provide these institutions with the tools, as well as the responsibilities appropriate to the changing world situation.

245 United Nations
B C TECHNOLOGY ASSESSMENT FOR DEVELOPMENT: BANGALORE, INDIA, 30 OCTOBER - 10 NOVEMBER 1978
New York: UN, 1979.

In three parts: (a) report of the Seminar; (b) Rapporteur's notes; and papers presented to the Seminar on various aspects of technology assessment.

246 United Nations
B C TECHNOLOGY FOR SOLAR ENERGY UTILIZATION
New York: UN, 1978.

Presents a review of the technology of exploitation of solar energy for the benefit of the developing countries, based primarily on contributions made to the Expert Group Meeting on the Existing Solar Technology and the Possibilities of Manufacturing Solar Equipment in Developing Countries, organized by UNIDO in cooperation with the Austrian Solar and Space Agency (ASSA) and held at Vienna, 14-18 February 1977. In three parts: (a) utilization of solar energy in developing countries; (b) summaries of country and institutional programs on solar energy; and (c) technical papers, covering such topics as conversion of solar into mechanical or electrical energy, solar flat-plate collectors, solar timber kilns, solar refrigeration, solar distillation, solar agricultural driers, solar cooking, and assessment of solar applications for technology transfer. Appendixes include lists of recommendations of the Expert Group Meeting, current International Energy Agency (IEA) projects in solar energy, solar energy information systems, and institutions involved in solar energy development.

247 United Nations
B C D TECHNOLOGIES FROM DEVELOPING COUNTRIES
New York: UN, 1980.

Contains information on technologies from developing countries appropriate for use or adaptation by other developing countries. In two sections: Part 1 describes the technologies, and Part 2 the research and development institutes where the products and processes were developed. The technologies covered include chemicals and metalworking; drugs and pharmaceuticals; textiles; cement and building materials; food storage and processing;

agricultural machinery and implements; light engineering and rural workshops; oils and fats; paper products and small pulp mills; energy for rural requirements; and low-cost transport for rural areas.

248 United Nations
B C TECHNOLOGICAL SELF-RELIANCE OF THE DEVELOPING COUNTRIES: TOWARDS OPERATIONAL STRATEGIES
New York: UN, 1981.

Reviews some of the problems associated with the promotion of national technological self-reliance and discusses the design and implementation of strategies promoting indigenous technological capabilities. It analyzes the extent and nature of the constraints imposed on the development of autonomous capacities by the present international technology system and identifies the principal components of technology development programs.

249 United Nations
B C D TECHNICAL CO-OPERATION UNDER THE COLOMBO PLAN: REPORT OF THE COLOMBO PLAN COUNCIL FOR TECHNICAL CO-OPERATION IN SOUTH AND SOUTHEAST ASIA FOR THE YEAR 1 JULY 1976 TO 30 JUNE 1977.
New York: UN, 1977.

Contains a brief history of the Colombo Plan; brief surveys of the work of the Council, Bureau and Consultative Committee; figures and trends for 1976; role of Colombo Plan assistance in the promotion of rural development; examples of technical cooperation in economic and social development; information activities; and drug advisory program. An appendix covers statistical data on technical cooperation for the calendar year 1976.

250 United Nations
A B C THE REVERSE TRANSFER OF TECHNOLOGY: A SURVEY OF ITS MAIN FEATURES, CAUSES, AND POLICY IMPLICATIONS
New York: UN, 1979.

This study discusses some recent trends and future prospects of the flow of highly skilled manpower from developing to developed countries. It examines the causes of this migration and attempts to estimate its imputed capitalized value in the international resource flow accounting system. It also surveys current national and international policy issues and suggests future cooperative solutions.

251 United Nations
A B C D THE TRANSFER AND ADAPTATION OF TECHNOLOGY
New York: UN, 1978.

Discusses problems relating to the transfer and adaptation of technology to and among Member countries of the Colombo Plan region with special reference to technical cooperation among developing countries. In three parts: (a) extract from the 23rd report of the Consultative Committee; (b) 22 country papers; and (c) consultant's working paper. An annex contains the report by the Ministry of Overseas Development Working Party on Appropriate Technology.

252
B C D
United Nations
HANDBOOK ON THE ACQUISITION OF TECHNOLOGY BY DEVELOPING COUNTRIES
New York: UN, 1978.

This handbook discusses: (a) the relationship of development and the acquisition of technology; (b) international technology transactions; (c) objectives in negotiations; (d,e) acquiring technology through foreign investment and public sector enterprises; (f) development of domestic technological capabilities; (g) integrating national policies and institutional arrangements; national legislative framework; (h) cooperation among developing countries; and (i) the role of UNCTAD.

253
B C
United Nations
MAJOR ISSUES ARISING FROM THE TRANSFER OF TECHNOLOGY TO DEVELOPING COUNTRIES
New York: UN, 1975.

In two parts: part 1 (the experience of technology-receiving countries) covers: contractual arrangements by industry and by organizational form; limitations on access to technology; foreign exchange cost of the transfer of technology; and institutions and policies for the transfer of technology. Part 2 surveys the experience of the technology-supplying countries of the developed market economies and the socialist countries of Eastern Europe.

254
B C
United Nations
TRANSFER OF TECHNOLOGY: ITS IMPLICATIONS FOR DEVELOPMENT AND ENVIRONMENT
New York: UN, 1978.

This study seeks to identify the major economic, social, and environmental issues arising from the transfer of technology to the Third World. The report examines the world economic and social environment, including the unequal distribution of income and resources consumption; technology, man, and environment; the economic costs of recent industrialization policies; the appropriateness of technology; and environmental and energy implications for future development.

255 UNCRD
B C D NEW TECHNOLOGY AND AGRICULTURAL TRANSFORMATION: A COMPARATIVE STUDY OF PUNJAB, INDIA, AND PUNJAB, PAKISTAN
UNCRD, 1979.

Studies the comparative and contrasting features of the two Punjabs in regard to the rural structural changes that have taken place as a result of the green revolution. In two parts: Part 1 describes the green revolution in Punjab, India and the resulting rural structural changes in nonmetropolitan regions. Part 2 comprises a case study of Pakistan Punjab.

256 UNESCO
B C A STUDY ON THE APPLICATION OF SCIENTIFIC AND TECHNOLOGICAL INFORMATION IN DEVELOPMENT: FIELD STUDY IN TANZANIA
Paris: UNESCO, 1978.

Reviews techniques currently used to transfer information to all sectors involved in the development process as well as to identify weaknesses in the existing procedures. Provides guidelines for the design of appropriate information systems particularly in developing countries. Covers: user needs; communication techniques; application of information in production; examination of information transfer chain.

257 UNESCO
A B C AN INTRODUCTION TO POLICY ANALYSIS IN SCIENCE AND TECHNOLOGY Paris: UNESCO, 1979.
Paris: UNESCO, 1979.

A synthesis of considerations on science and technology policy that have emerged over the last decade from various UNESCO sources. In four parts: (a) Part 1 deals with some fundamental concepts and principles as well as organizational practices in the field of science and technology policies; (b) Part 2 concerns the major resources which characterize a nation's scientific and technological potential--manpower, finances, information, and facilities; (c) Part 3 presents the processes governing the transfer of operative technologies in the productive sectors of the economy; and (d) Part 4 addresses the problems encountered by policy-makers in regard to international scientific and technological cooperation.

258 UNESCO
B C D APPROPRIATE TECHNOLOGIES IN THE CONSERVATION OF CULTURAL PROPERTY
Paris: UNESCO, 1981.

A collection of five papers discussing the application of appropriate technology to the conservation of cultural property: (a) traditional crafts and modern conservation methods in Nepal; (b) appropriate technologies and restoration of historic monuments; (c) appropriate Indian technology for the conservation of museum collections; (d) the restoration of mud-brick structures in historical monuments of the Andean region in Peru; and (e) appropriate technology?--covering materials, transfer of technology, physical organization, style and identity.

259 UNESCO
B C D METHOD FOR PRIORITY DETERMINATION IN SCIENCE AND TECHNOLOGY
Paris: UNESCO, 1978.

Describes the method devised by UNESCO consultants and experts for assessing institutional needs and priorities in science and technology in developing countries. Covers such fields as research and training so that responsible governing bodies may be assisted in planning and budgeting for the future.

260 UNESCO
A B THE ROLE OF SCIENCE AND TECHNOLOGY IN ECONOMIC DEVELOPMENT
New York: Unipub, 1970.

This eighteenth issue of the UNESCO series "Science Policy Studies and Documents" details the economic problems associated with scientific and technological development. Problems discussed include: integrating scientific and economic plans into general or over-all planning; methods of financing scientific and technological research in various countries; and the economic impact of scientific progress.

261 UNESCO
B C D SCIENCE AND TECHNOLOGY IN AFRICAN DEVELOPMENT
Paris: UNESCO, 1974.

Presents the results of the 1974 Dakar Conference of Ministers of African States Responsible for the Application of Science and Technology to Development. Contains the Conference's final report, main working document, and adopted recommendations. Subject areas covered include: government structures for science and technology planning and policy making; the national research and development system; the transfer of new technologies and their implantation in Africa; objectives of science and technology policy; and technologically feasible futures for Africa.

262 UNESCO
B C D SCIENCE AND TECHNOLOGY IN THE DEVELOPMENT OF THE ARAB STATES
Paris: UNESCO, 1977.

This publication is based on the Conference of Ministers of Arab States Responsible for the Application of Science and Technology to Development (CASTARAB), organized by UNESCO with the cooperation of the Arab Educational, Cultural and Scientific Organization (ALESCO) in Rabat, Morocco, from 16 to 25 August 1976. In two parts: (a) Final Report of the Conference, giving highlights of the debates and the texts of the Rabat Declaration and the resolutions adopted; and (b) main working document of the Conference, covering the present situation and future prospects of science and technology policies in Arab States, projects of regional cooperation in scientific and technological research, and measures to be taken for following up the decisions of the CASTARAB Conference.

263 UNESCO
B C THE SOCIAL IMPLICATIONS OF THE SCIENTIFIC AND TECHNOLOGICAL REVOLUTION: A UNESCO SYMPOSIUM
Paris: UNESCO, 1981.

This publication reports on the proceedings of an international symposium on the scientific and technological revolution and the social sciences which was held in Prague in September 1976. This book, consisting of 23 essays written by leading scholars from the world over, assesses the present state of knowledge about the interaction between science, technology and society. Papers are arranged in five topical sections: the scientific and technological revolution as a social process; philosophical problems of technological advancement; social functions of science and science policy; science, technology and development; and social sciences and social change.

264 UNESCO
B C SOCIETAL UTILIZATION OF SCIENTIFIC AND TECHNOLOGICAL RESEARCH
Paris: UNESCO, 1981.

Reveiws major issues concerning the production, dissemination and utilization of knowledge, and proposes a framework for further investigation on societal utilization of scientific and technological research. Other topics covered: existing models and their implications; an alternative approach; global problems of societal relevance and their corresponsding research areas; and international problems and national solutions.

265 UNESCO
B C TECHNOLOGIES FOR RURAL DEVELOPMENT; BASED ON AN EXPERT MEETING OF NEW MODALITIES FOR THE ACTION OF UNESCO IN THE FIELD OF TECHNOLOGIES FOR RURAL DEVELOPMENT, BRUSSELS, MAY 1980.
Paris: UNESCO, 1981.

A collection of expert papers covering a wide range of experience in the application of technology in the rural areas of developing countries. Coverage includes: an overview of the current situation; the experience of a number of development institutions, including the Intermediate Technology Development Group (London), The International Association of Rural Development (Brussels), The Technology Consultancy Centre (Ghana), Dian Desa (Indonesia), and Third World Environment and Development, ENDA (Senegal); impact of rural technology on women; science education for rural innovation; and linkages for promoting application of technology in developing countries.

266 UNIDO
A C APPROPRIATE INDUSTRIAL TECHNOLOGY FOR AGRICULTURAL MACHINERY AND IMPLEMENTS
Vienna: UNIDO, 1979.

267 UNDIO
A C APPROPRIATE INDUSTRIAL TECHNOLOGY FOR BASIC INDUSTRIES
New York: UN Publications, 1981.

Reviews issues and considerations related to appropriate technology for basic industries, with selected papers on: basic materials industries--aspects of technology choice and industrial location; choice and adaptation of alternative technology for the iron and steel industry; appropriate technology for the capital goods industry (machine tools) and the chemical industry; the role of the engineering industry; technology for oil and gas based industries--the case of Kuwait; and the fertilizer industry in India. List of documents.

268 UNIDO
A C APPROPRIATE INDUSTRIAL TECHNOLOGY FOR CONSTRUCTION AND BUILDING MATERIALS
New York: UN Publications, 1980.

Contains 15 papers on appropriate building technologies and materials industry. Presents case studies on rural areas and such countries as India, Indonesia, Iran, Cameroon, and Nepal. Other topics cover building materials and components; recent trends in integrated stone development planning; non-cement-based hydraulic binders; strategies for development of and allied industries in developing countries; appropriate technologies for small-scale production of cement and cementitious materials; proposal and feasibility study for a 25 T/D mini cement plant; sisal-fiber concrete for roofing sheets and other purposes; and mechanization of construction and choice of appropriate technology in civil engineering.

269
A C
UNIDO
APPROPRIATE INDUSTRIAL TECHNOLOGY FOR DRUGS AND PHARMACEUTICALS
Vienna: UNIDO, 1980.

270
A C
UNIDO
APPROPRIATE INDUSTRIAL TECHNOLOGY FOR ENERGY FOR RURAL REQUIREMENTS
Vienna: UNIDO, 1979.

271
A C
UNIDO
APPROPRIATE INDUSTRIAL TECHNOLOGY FOR FOOD STORAGE AND PROCESSING
Vienna: UNIDO, 1980.

272
A C
UNIDO
APPROPRIATE INDUSTRIAL TECHNOLOGY FOR LIGHT INDUSTRIES AND RURAL WORKSHOPS
New York: UN Publications, 1980.

Topics covered include: (a) light-engineering workshops for rural areas; (b) light-industry technologies and rural development; (c) rural workshops in developing countries; (d) appropriate technology for rural industries; (e) the light-engineering industries of the Philippines; (f) Swedish experience in small-scale industry--the role of government policies and institutional mechanisms; (g) small-scale rural industries--light-engineering workshops; (h) light-engineering and rural workshops in Egypt; and (i) establishment of small-scale rural workshops (for light-engineering goods) in East Africa.

273
A C
UNIDO
APPROPRIATE INDUSTRIAL TECHNOLOGY FOR LOW COST TRANSPORT FOR RURAL AREAS
Vienna; UNIDO, 1979.

274
A C
UNIDO
APPROPRIATE INDUSTRIAL TECHNOLOGY FOR OILS AND FATS
Vienna: UNIDO, 1979.

275
A C
UNIDO
APPROPRIATE INDUSTRIAL TECHNOLOGY FOR PAPER PRODUCTS AND SMALL PULP MILLS
New York: UN Publications, 1979.

Examines pulp and paper industry, inter alia, in the Philippines, Brazil, Egypt, and India. Some of the other topics cover: the search for an appropriate technology for the paper and board industry of the United Kingdom; pulping technology and requirements and potentialities of developing countries; paper, paper products and pulp mills; and small pulp and paper mills in developing countries and their recovery systems.

276 UNIDO
A C APPROPRIATE INDUSTRIAL TECHNOLOGY FOR SUGAR
New York: UN Publications, 1980.

Discusses appropriate technology in cane-sugar production; technological choices in sugar processing; cane-sugar production techniques in developing countries; "mini" sugar technology in India; appropriate technology for production of sugar and other sweetening agents; technology planning factors in the cane-sugar industry; and by products of the sugar industry in Cuba.

277 UNIDO
A C APPROPRIATE INDUSTRIAL TECHNOLOGY FOR TEXTILES
Vienna: UNIDO, 1979.

278 UNIDO
A C APPROPRIATE INDUSTRIAL TECHNOLOGY FOR LOW-COST TRANSPORT FOR RURAL AREAS
New York: UN Publications, 1979.

Aims to provide a basis for a better understanding of the concept and use of appropriate industrial technology and thereby contribute to increased cooperation between developing and developed countries and among the developing countries themselves. Covers appropriate transport facilities for the rural sector in developing countries; rural transport in the Philippines; and modernizing the bullock-cart: a case of appropriate technology in India.

279 UNIDO
A B C APPROPRIATE TECHNOLOGY FOR DEVELOPING COUNTRIES: UNIDO'S CO-OPERATIVE PROGRAMME OF ACTION FOR NATIONAL AND REGIONAL PROGRESS
New York: UN Publications, 1980.

Briefly describes UNIDO's Co-operative Programme of Action for National and Regional Progress.

280 UNIDO
A C CONCEPTUAL AND POLICY FRAMEWORK FOR APPROPRIATE INDUSTRIAL TECHNOLOGY
New York: UN Publications, 1979.

Consists of the report of and selected papers presented at the Ministerial-Level Meeting of the International Forum on Appropriate Technology. Topics covered include activities to promote appropriate technology; dualism, sectoral planning and integration of the modern industrial and dispersed traditional sectors; an integrated approach to appropriate industrial technology; and institutional development of appropriate industrial technology in developing countries: R and D policies and programs.

281 UNIDO
A C TOWARDS A STRATEGY FOR INDUSTRIAL GROWTH AND APPROPRIATE TECHNOLOGY
New York: UN Publications, 1977.

Brief review of current alternative industrial and development of strategies, including appropriate technology and development of technological capability.

282 UNIDO
A C INDUSTRIAL AND TECHNICAL CO-OPERATION AMONG DEVELOPING COUNTRIES
New York: UN Publicaitons, 1977.

Discusses cooperation among developing countries in achieving the industrial goals set forth in the Lima Declaration and Plan of Action. Reviews the experience of India in planning and implementing industrial development programs and establishing the requisite infrastructural and technological capacity over the last three decades, with special emphasis on the engineering industries and small-scale and rural industrial developmentprograms.

283 UNIDO
A C INTERNATIONAL INDUSTRIAL CO-OPERATION
New York: UN Publications, 1974.

Report prepared by the UNIDO Secretariat on the basis of discussions and conclusions arrived at by the group of experts who participated in the meeting. Topics covered include: general perspectives on the role of industrializaiton in the development process; principles and objectives of a new international economic structure in industry; and ways and means towards a new world economic structure in industry.

284 UNIDO
A C UNIDO ABSTRACTS ON TECHNOLOGY TRANSFER: STUDIES AND REPORTS ON THE DEVELOPMENT AND TRANSFER OF TECHNOLOGY (1970-1976)
New York: UN, 1977.

Presents a selective compilation of abstracts of documents issued by UNIDO on the subject of transfer of technology. Includes only abstracts from 1970 onward using major descriptors in the computer program. In two parts: (a) subject index, by title, arranged alphabetically by descriptors; and (b) bibliographical abstracts.

285 UNU
B C SHARING OF TRADITONAL TECHNOLOGY: PROJECT MEETING REPORT, JAPAN, SEPTEMBER 1977.
Tokyo: UNU, 1979.

Covers the procedures of a meeting on the Project on the Sharing of Traditional Technology and discusses the concept of traditional technology as well as the objectives and implementation of the Project. Appendixed with 8 pilot studies and a list of participants.

286 UNU
B C TRANSFORMATION AND DEVELOPMENT OF TECHNOLOGY IN THE JAPANESE COTTON INDUSTRY
Tokyo: UNU, 1980.

Examines the technical transformation in the Japanese cotton-spinning industry. In four parts: (a) coping with technological transformation; (b) the recruitment of operatives and their training; (c) Japanese ways of rationalization and international competitiveness; and (d) the position of the cotton industry in the Japanese economy.

287 Vacca, Roberto
A B C D MODEST TECHNOLOGIES FOR A COMPLICATED WORLD
Pergamon International Library series
Oxford; New York; Tokyo and Frankfort: Pergamon Press, 1980.

First published in Italian in 1978, this discussion of "modest technology," which features low investment per employee, focuses on determining what conditions which must be satisfied and what precautions must be taken, so that the introduction of new modest technologies in the third world--as well as in the first or second--can be implemented on the basis of concrete and pragmatic evaluations. With concern for the survival of the planet and the attainment of equilibrium in the socio-political situation, argues that both advanced and modest--sometimes called intermediate or alternative--technologies are necessary; defines the contemporary problem as the finding of the right mix of the two. Some topics are: international aid and the New International Economic Order, lack of innovation as a cause of economic crisis, the neutrality of science and technology, and the costs of risks.

288 Vidakovic, Zoran
B C THE TECHNOLOGY OF REPRESSION AND REPRPESSIVE TECHNOLOGY: THE SOCIAL BEARERS AND CULTURAL CONSEQUENCES
Tokyo: UNU, 1980.

Deals with armament and its scientific research and technological potential. Discusses the vicious circle of repressive technology and the main social figures of such technology; militarization of the economy and science--the birthplace of the metropolitan technocracy; and the genesis of the technocratic elite in dependent societies.

289
B C D

Vyasulu, Virod
TECHNOLOGICAL CHOICE IN THE INDIAN ENVIRONMENT: PAPERS AND PROCEEDINGS OF A SEMINAR ORGANIZED BY THE INDIAN INSTITUTE OF MANAGEMENT, BANGALORE
New Delhi: Sterling, 1980.

290
A B C D

Wallender, Harvey W., III
TECHNOLOGY TRANSFER AND MANAGEMENT IN THE DEVELOPING COUNTRIES: COMPANY CASES AND POLICY ANALYSES IN BRAZIL, KENYA, KOREA, PERU, AND TANZANIA
Cambridge, Mass.: Harper & Row, Ballinger, 1979.

Focuses on the problems related to seeking, receiving, and using technology by user firms in developing countries. Reviews prevailing theories of the transfer of technology and examines the management behavior of developing country firms in using technology through the analysis of case studies and environmental analyses prepared by a study team's inquiry into 67 consulting projects in Brazil, Peru, Korea, Tanzania, and Kenya and a survey of 405 similar projects in 43 other developing countries. Concludes that the major obstacle in the technology transfer process is at the receiving end, primarily the weakness in local managerial capabilities to handle technology--to diagnose needs, to devise solutions, and to seek assistance, when necessary, in implementing solutions.

291
B C

Weiss, Charles, Jr.
MOBILIZING TECHNOLOGY FOR DEVELOPING COUNTRIES
Washington, D.C.: WBG, 1979.

Reprinted from Science, Vol. 203 (March 16, 1979), pp. 1083-89. Deals with a new problem in technology policy: the need for technology suited to creating productive jobs and providing minimum public services at a cost and level of sophistication within the reach of poor people in developing countries. Examines overall and sectoral development objectives; economic and manpower resources; and the local institutional and sociocultural context. Discusses the need for hardware innovation, such as low-cost alternatives to waterborne sewage, and social ("software") innovation, such as training large numbers of supervisors to implement improved technologies for labor-intensive civil works.

292
A B C

Weiss, Frank Dietmar
ELECTRICAL ENGINEERING IN WEST GERMANY: ADJUSTING TO IMPORTS FROM LESS DEVELOPED COUNTRIES
Tubingen: J. C. B. Mohr (Paul Siebeck), 1978.

Analyzes the causes for the declines in competitiveness of West Germany's electrical engineering trade vis-a-vis the LDC's; for purposes of comparison the experiences of the U.S. and Sweden with LDC competition are used.

Measures the competitiveness of the electrical engineering industry in these three developed countries as a whole vis-a-vis the LDC's and the total world since 1960. Using a disaggregated approach, examines the structure of competitiveness within the industry employing results of a survey conducted especially for this analysis of West German electrical engineering firms. Discusses demand considerations in the adjustment process of developed countries and projects the overall employment effects in West Germany, Sweden, and the United States resulting from increased electrical engineering imports from LDC's. Provides a qualitative prognosis of those subindustries likely to be hit hardest by LDC competition.

293 Wells, Louis T.
A B C TECHNOLOGY AND THIRD WORLD MULTINATIONALS
Geneva: International Labor Office, 1982.

294 Wignaraja, Ponna
B C A FRAMEWORK FOR RETHINKING THE CONCEPT OF APPROPRIATE TECHNOLOGY FOR DEVELOPMENT
New York: UN, 1978.

Presents the author's view of the main objectives and techniques of an alternative approach to development in developing countries: an integrated process of total mobilization involving raising people's consciousness, inculcating democratic values, transforming labor power into the means of production, fully utilizing local natural resources, and the systematic development of appropriate technology.

295 Wilczynski, J.
B C D TECHNOLOGY IN COMECON: ACCELERATION OF TECHNOLOGICAL PROGRESS THROUGH ECONOMIC PLANNING AND THE MARKET
New York: Praeger, 1975.

296 Wilson, George W.
B C TECHNOLOGICAL DEVELOPMENT AND ECONOMIC GROWTH
School of Business, Indiana University, 1971.

297 World Bank
B C MOBILIZING RENEWABLE ENERGY TECHNOLOGY IN DEVELOPING COUNTRIES: STRENGTHENING LOCAL CAPABILITIES AND RESEARCH
World Bank, 1981.

Focuses on the research required to develop renewable energy resources in the developing countries and on the need to strengthen the developing countries' own technological capabilities for using renewable energy. (One of three publications dealing with renewable energy resources and issues in developing countries. See Alcohol Production from Biomass in the Developing Countries and Renewable Energy Resources in the Developing Countries.)

298 Wynne-Roberts, C. R.
B C MANAGEMENT AND THE TRANSFER OF TECHNOLOGY
New York: UN, 1976.

Report of a study that attempted to determine (a) what steps have been taken in selected firms, public and private, in Afghanistan, India, and Indonesia to match management practices to the requirements of the technologies employed; (b) how far owners and managers are aware of the need to match management with technological practice and the steps taken to meet the need; (c) assistance necessary to enable them to understand and meet this problem; and (d) what steps UNIDO can take to provide technical cooperation.

299 Yap, K. H.
A C ON THE ESTABLISHMENT OF AN INDUSTRIAL TECHNOLOGY DEVELOPMENT POLICY
New York: UN Publications, 1978.

Paper presented at meeting of the Second Consultative Group on Appropriate Industrial Technology, Vienna, Austria, 26-30 June 1978. Discusses the creation of a long-term technology development framework in developing countries. Topics include range of technologies needed, heavy and light industry development, geographic dispersal of industrial activities, and functional aspects of technology development.

300 Yudelman, Montague; Butler, Gavan and Banerji, Ranadev
B C D TECHNOLOGICAL CHANGE IN AGRICULTURE AND EMPLOYMENT IN DEVELOPING COUNTRIES
Development Centre Studies, Employment Series: No. 4
Paris: Organisation for Economic Cooperation and Development, 1971.

"...the fundamental premise...[is] that there must be greater provision of remunerative employment...within the agricultural sector of almost every poor or developing country" because the labor force in each developing country is growing rapidly, industrialization is providing too few employment opportunities, poverty and malnutrition affect large portions of the rural population, the high rate of rural-urban migration and consequent urbanization which yields open unemployment that is extensive in urban areas, especially among the young.

301 Zaleski, Eugene and Wienert, Helgard
B C D TECHNOLOGY TRANSFER BETWEEN EAST AND WEST
Paris: OECD, 1980.

Reviews and assesses the literature concerned with East-West technology transfer, and provides an overall analysis of the problems posed by technology transfer in the context of East-West economic relations. Chapters cover: (a) East-West trade and technology transfer in historical perspective; (b) statistical evaluation of technology transfer; (c) the forms of technology transfer; (d) eastern and western policies toward technology transfer; (e) the influence of technology transfers of eastern economies; (f) effect of economic factors on East-West technology transfer on western economies.

II. SELECTED PERIODICAL ARTICLES

302 Abiodun, A. Ade
Technology development in Africa. AFRICA no. 113:56-57, January, 1981.

303 Ahmad, Aqueil
Science and technology in development; policy options for India and China. ECONOMIC AND POLITICAL WEEKLY (BOMBAY) 13:2079-90, Dec. 23/30, 1978.

304 Alam, G. and Langrish, J.
Non-multinational firms and transfer of technology to less developed countries. WORLD DEVELOPMENT 9:383-7, April 1981.

305 Allen, Philip M.
The technical assistance industry in Africa: A case for nationalization. INTERNATIONAL DEVELOPMENT REVIEW 12 (1970): 8.

306 Alleyne, D. H. N.
The State petroleum enterprise and the transfer of technology. NATURAL RESOURCES FORUM (DORDRECHT) 3:5-17, Oct. 1978.

This article examines the potential of State Petroleum Enterprises in the developing countries for taking a lead in transfer of technology.

307 Altschul, Aaron M.
Proposals for activating the interface between governments and technology. FOOD TECHNOLOGY 24 (March 1970): 224-25, 228, 230.

308 Alviar, Nelly G.
Technology transfer in rice production: the Philippine case. KAJIAN EKONOMI MALAYSIA (KUALA LUMPUR) 16:307-23, June/Dec. 1979.

309 Ba, Boubakar
The problem of transferring technology to the least industrialized countries. LABOUR AND SOCIETY (GENEVA) 1:121-26, July/Oct. 1976.

Suggests that the underdeveloped countries should aim at reorganising economies in order to have better control of all national riches and outlines the plan of action which might assist in attaining this objective.

310 Bagchi, Amiya Kumar
On the political economy of technological choice and development. CAMBRIDGE JOURNAL OF ECONOMICS (LONDON) 2:215-32, June 1978.

Examines some illustrative contrasts between technical choice in Third World countries and in the socialist framework of China.

311 Balasubramanyam, V. N.
Transfer of technology; the UNCTAD arguments in perspective. THE WORLD ECONOMY (AMSTERDAM) 1:69-80, Oct. 1977.

This article has focused on some of the issues pertaining to international transfer of technology and foreign private investment in the new international economic order.

312 Baranson, Jack
The international transfer of automotive technology to developing countries. FOREIGN AFFAIRS REVIEW no. 13723. Washington, D.C.: U.S. Department of State, Office of External Research. 95 pp.

Paper for United Nations Institute for Training and Research, March 1971.

313 Barbiroli, Giancarlo and Ballini, Vladimiro
The effect of technological development on renewable raw materials; a look at the international situation. MONTE DEI PASCHT DI SIENA, ECONOMIC NOTES (SIENA) 10, No. 1:104-21, 1981.

314 Barker, Randolph, et al.
Employment and technological change in Philippine agriculture. INTERNATIONAL LABOUR REVIEW 106 (Aug.-Sept. 1972): 111-39.

315 Barraclough, Solan and Schatan, Jacobo
Technological policy and agricultural development. LAND ECONOMICS 49 (May 1973): 175-96.

316 Basiuk, Victor
The impact of technology in the next decades. ORBIS (Spring 1970).

317 Beg, M. Arshad Ali
The role of technology transfer. PAKISTAN ECONOMIST (KARACHI) 20:33-38, Aug. 16, 1980.

Suggests that technological dependence in developing countries has given rise to inappropriate technologies and inappropriate products and gives examples from Pakistan.

318 Bell, Daniel
Technology, nature and society: The vicissitudes of three world views and the confusion of realms. THE AMERICAN SCHOLAR 42 (Summer): 385-404, 1973.

319 Belshaw, Deryke
Taking indigenous technology seriously; the case in inter-cropping techniques in East Africa. IDS BULLETIN, INSTITUTE OF DEVELOPMENT STUDIES AT THE UNIVERSITY OF SUSSEX (BRIGHTON) 10:24-27, Jan. 1979.

320 Berlinguet, Louis
Science and technology for development. SCIENCE (WASHINGTON) 213:1073-76, Sept. 1981.

Focuses on the need to convince governments in the developed countries that money invested in science and technology in the Third World is for the mutual benefit of all.

321 Berrios, Ruben
The commercialization of technology and problems with its inappropriateness for less developed countries. ECONOMIC PAPERS (WARSAW) No. 11:19-33, 1980.

322 Bhalla, A. S.
Technology and employment: Some conclusions. INTERNATIONAL LABOUR REVIEW 113 (March-April): 189-214, 1976.

323 Bhalla, G. S.
Transfer of technology and agricultural development in India. ECONOMIC AND POLITICAL WEEKLY (BOMBAY) 14:A130-42, Dec. 22, 1979.

324 Bhat, B. A. and Prendergast, C. C.
Some aspects of technology choice in the iron foundry industry. WORLD DEVELOPMENT 5 (Sept.-Oct.): 791-801, 1977.

325 Bhatt, V. V.
Financial institutions and technology policy [underdeveloped states]. WORLD DEVELOPMENT 8:813-22, Oct. 1980.

326 Bhatt, V. V.
Indigenous technology and investment licensing; the case of the Swaraj tractor. JOURNAL OF DEVELOPMENT STUDIES (LONDON) 15:320-30, July 1979.

It is the purpose of this case study of the Swaraj tractor to illustrate how the decision making structure of the investment licensing system (ILS) in India resulted in decisions that tended to frustrate the very objectives that the system was designed to serve.

327 Bhattacharya, Debesh
Development and technology in the third world. JOURNAL OF CONTEMPORARY ASIA 6 (no. 3): 314-22, 1976.

328 Biggs, Huntley H.
Alternative agricultural technologies for developing countries. TECHNOS 1 (July-Sept.): 27-32, 1972.

329 Biggs, Stephen D.
Planning rural technologies in the context of social structures and reward systems. JOURNAL OF AGRICULTURAL ECONOMICS (ASHFORD, KENT) 29:257-77, Sept. 1978.

330 Binswanger, Hans P.
Income distribution effects of technical change; some analytical issues. YALE UNIVERSITY. ECONOMIC GROWTH CENTER. CENTER DISCUSSION PAPER (NEW HAVEN) No. 281: 1-39, May 1978.

Reviews partial and general equilibrium approaches to analysis of distributional consequences of technical change.

331 Blumenthal, Tuvia and Teubal, Morris
The role of future oriented technology in Japan's economic development. HITOTSUBASHI JOURNAL OF ECONOMICS (TOKYO) 20:33-43, June 1979.

332 Bronger, Dick
The dilemma of developing country research. INTERECONOMICS 9/10:245-250, Sept.-Oct. 1977.

333 Brooks, Harvey
The technology of zero growth. DAEDALUS 102 (Fall 1973): 139-52.

334 Buechler, Rose M.
Technical aid to upper Peru: The Nordenflicht expedition. JOURNAL OF LATIN AMERICN STUDIES 5 (May 1973): 37-77.

335 Bueno, G. M.
A new order in financial and technological relations with the Third World? TIJDSCHRIFT VOOR ECONOMIE EN MANAGEMENT (LOUVAIN) 26, No. 1:95-116, 1981.

336 Burkner, Hans-Paul and Park, Sung-Jo
Optimal technology policy and autonomous development: Theoretical and strategic relections with special reference to the Chinese development planning and technology policy.

CULTURES ET DEVELOPPEMENT 8 (1976): 494-516.

337 Byres, T. J.
The new technology, class formation and class action in the Indian countryside. JOURNAL OF PEASANT STUDIES (LONDON) 8:405-54, July 1981.

338 Canadian Hunder Foundation and Bruce Research Institute
Appropriate technology primer. EKISTICS 43:360-7, June 1977.

339 Carew, Joy Gleason
A note on women and agricultural technology in the Third World. LABOUR AND SOCIETY (GENEVA) 6:279-85, July/Sept. 1981.

340 Chandavarkar, Anand G.
Technical cooperation with the Third World. FINANCE AND DEVELOPMENT 9 (Dec. 1972): 17-22.

341 Charney, J. I.
Technology and international negotiations. AMERICAN JOURNAL OF INTERNATIONAL LAW 76:78-118, Jan. 1982.

342 Chase-Dunn, Christopher
The effects of international economic dependence on development and inequality: A cross-national study. AMERICAN SOCIOLOGICAL REVIEW 40:720-738, Dec. 1975.

343 Chen, Edward K. Y.
The empirical relevance of the endogenous techncial progress function. KYKLOS (BASLE) 29, No. 2:256-71, 1976.

Tests some endogenous technical progress hypotheses in the light of the experience of the manufacturing sector of five fast-growing Asian economies, viz., Hong Kong, Japan, South Korea, Singapore, and Taiwan. The period under consideration is 1955-1970 for Japan and Taiwan, and 1960-70 for Hong Kong, Korea, and Singapore.

344 Chico, Leon V.
Sharing in technological development. INTERNATIONAL DEVELOPMENT REVIEW: FOCUS 19 (No. 4): 7-10, 1977.

345 Chidzero, B. T. G.
The management of multilateral technical assistance. JOURNAL OF WORLD TRADE LAW 4 (March-April 1970): 192-206.

346 Childers, E.
Technical co-operation among developing countries: History and prospects. JOURNAL OF INTERNATIONAL AFFAIRS 33:19-42, Spring 1979.

347 Chisti, Sumitra
Monitoring import of technology into India. ECONOMIC AND POLITICAL WEEKLY (BOMBAY) REVIEW OF MANAGEMENT, 10: M39-M52, May 31, 1975.

This paper examines the trends in technology imports since Independence, and traces ways in which these have influenced the direction and pattern of Indian economic development.

348 Cho, Youngnae
Technology, labour market segmentation, and earnings distribution in developing countries. LABOUR AND SOCIETY (GENEVA) 3:184-98, April 1978.

Endeavours to explore the functional relationship between factors affecting income distribution with particular emphasis on the role of the technology and the structure of the labour market. The labour market structure in Korea is also examined.

349 Coates, Joseph F.
The role of formal models in technology assessment. TECHNOLOGICAL FORECASTING AND SOCIAL CHANGE 9 (No. 1-2): 139-190, 1976.

350 Cohen, R. S.
Science and technology in global perspective. INTERNATIONAL SOCIAL SCIENCE JOURNAL 34 (No. 1): 61-70, 1982.

351 Collier, William L. et al.
Agricultural technology and institutional change in Java. FOOD RESEARCH INSTITUTE STUDIES 13 (1974): 169-94.

352 Colson, R. F.
Brazil; technological transfer and economic growth. BANK OF LONDON & SOUTH AMERICAN REVIEW (LONDON) 10: 118-26, March 1976.

Discusses the role of field research and the need to tie research firmly in with local knowledge and conditions and with consumer response, and assesses the growth of a strong local technology that will ultimately render imports of technical knowledge unnecessary except in extremely specialized fields.

353 Constantino, Renato
Global enterprises and the transfer of technology. JOURNAL OF CONTEMPORARY ASIA 7 (No. 1): 44-55, 1977.

354 Cooper, Charles
Science, technology and production in the underdeveloped countries: An introduction. JOURNAL OF DEVELOPMENT STUDIES 9:1-18, October 1972.

355 Cooper, Charles
The transfer of industrial technology to the underdeveloped countries. BULLETIN OF THE INSTITUTE OF DEVELOPMENT STUDIES 3 (Oct. 1970): 3-7.

356 Cooper, Charles
Choice of techniques and technological change as problems in political economy. INTERNATIONAL SOCIAL SCIENCE JOURNAL 25 (No. 3): 293-304, 1973.

357 Cooper, Charles
Science policy and technological change in underdeveloped economies. WORLD DEVELOPMENT 2 (March): 55-64.

358 Coundrey, Gordon
The commonwealth fund for technical co-operation : A new concept in multilateral aid. ROUND TABLE 72 (Jan. 1972): 93-99.

359 Courtney, William H. and Leipziger, Danny M.
Multinational corporations in LDC's: The choice of technology. OXFORD BULLETIN OF ECONOMICS AND STATISTICS 37:297-304, November 1975.

360 Crane, Diana
Technological innovation in developing countries: A review of the literature. RESEARCH POLICY 6:374-395, October 1977.

361 Curnow, Ray
A note on some issues raised for science and technology policy by the increase in oil prices. INSTITUTE OF DEVELOPMENT STUDIES BULLETIN 6 (October 1974): 81-90.

362 Dahlberg, Kenneth A.
The technological ethic and the spirit of international relations. INTERNATIONAL STUDIES QUARTERLY 17 (No. 1): 55-88, 1973.

363 Dahlman, Carl J. and Westphal, Larry E.
The meaning of technological mastery in relation to transfer of technology. AMERICAN ACADEMY OF POLITICAL AND SOCIAL SCIENCE, ANNALS (PHILADELPHIA) 458:12-26, Nov. 1981.

364 Damodaran, G. R.
Technical education in India: Looking back and forward. EDUCATION QUARTERLY 24 (Jan. 1973): 5-9.

365 Davies, Howard
Technology transfer through commercial transactions. JOURNAL OF INDUSTRIAL ECONOMICS (OXFORD) 26:161-75, Dec. 1977.

Examines the hypothesis that the quality and the quantity of assistance provided may depend on the profitability of the project and upon the technology "donor's" ability to secure control over those profits. The author draws on a study of collaboration agreements between British and Indian firms.

366 Dean, Genevieve C.
Innovation in a choice of techniques context: The Chinese experience, 1958-1970. BULLETIN [Institute of Development Studies] 4 (June 1972): 39-48.

367 de Bandt, Jacques
Optimal use of existing technology versus new technology. JOURNAL OF INDUSTRIAL ECONOMICS (OXFORD) 26:69-80, Sept. 1977.

Seeks to evaluate the potential contributions to providing needed technical progress in various industries of transferring technologies already in use elsewhere as against undertaking new R & D programs.

368 De Gregori, Thomas R.
Technology and economic dependency; an institutional assessment. JOURNAL OF ECONOMIC ISSUES (UNIVERSITY PARK, PA) 12:467-76, June 1978.

Concludes that ultimately, technological dependence can only be lessened by restructuring international economic and political relations.

369 Dellimore, J. W.
Select technological issues in agro-industry (I). SOCIAL AND ECONOMIC STUDIES (MONA, JAMAICA) 28:54-96, March 1979.

Examines some technological issues relevant to the development of agro-based industries and presents information and some principles for decision-making and action on these issues, which could result in more appropriate technological choices and more effective use of available resources.

370 Development Center of the Organization for Economic Cooperation and Development
Choice and adaptation of technology in developing countries. Paris: OECD, 1974. 240 pp.

371 Dima, S. A. and Amann, V. F.
Small holder farm development through intermediate technology. EASTERN AFRICA JOURNAL OF RURAL DEVELOPMENT 8 (1975): 215-45.

372 Dinkelspiel, John R.
Technology and tradition: Regional and urban development in the Guayana. INTER-AMERICAN ECONOMIC AFFAIRS 23 (Spring

1970): 47-79.

373 Djojohadikusumo, Sumitro
Technology, economic growth and environment. INDONESIAN QUARTERLY (DJAKARTA) 5:17-33, Jan. 1977.

The following observations are an attempt to review to state of the debate on technology, economic growth and the environment and to identify their interconnecting links.

374 Dobrov, Gennady M.
Technology as a form of organization. INTERNATIONAL SOCIAL SCIENCE JOURNAL (PARIS) 31, No. 3:585-605, 1979.

375 Dommen, Arthur J.
The bamboo tube well: A note on an example of indigenous technology. ECONOMIC DEVELOPMENT AND CULTURAL CHANGE 23 (April 1975): 483-89.

376 Dreitzel, Hans Peter
Social science and the problem of rationality: Notes on the sociology of technocrats. POLITICS AND SOCIETY 2 (Winter, 1972): 175-182.

377 Drissell, J. P.
Why the world will shift to intermediate technology: An interview with G. McRobie. FUTURIST 11:83-6, April 1977.

378 D'Souza, Henry
Technical education in Kenya; some problems. AFRICAN STUDIES REVIEW (WALTHAM, MA) 19:33-41, Dec. 1976.

379 Dylander, B.
Technology assessment--as science and as a tool for policy. ACTA SOCIOLOGICA 23 (No. 4): 217-37, 1980.

380 Edquist, C. and Edquist, O.
Social carriers of techniques for development. JOURNAL OF PEACE RESEARCH 16 (No. 4): 313-31, 1979.

381 Egea, Alejandro Nadal
Multinational corporations in the operation and ideology of international transfer of technology. STUDIES IN COMPARATIVE INTERNATIONAL DEVELOPMENT 10 (Spring): 11-29, 1975.

382 Eilers, William L.
Doing more with less: The future of technology in the third world; dwindling resources and mounting debt force developing countries to seek modern technology. AGENDA 3:24-8, June 1980.

383 Eltis, W. A.
The rate of technical progress. ECONOMIC JOURNAL (Sept. 1971), pp. 502-25.

384 El-Zaim, Issam
Problems of technology transfer; a point of view from the Third World. VIENNA INSTITUTE FOR DEVELOPMENT. OCCASIONAL PAPER (VIENNA) No. 6:1-50, 1978.

385 Emmerson, Donald K.
Introducing technology: The need to consider local culture. INTERNATIONAL DEVELOPMENT REVIEW: FOCUS 19 (No. 1): 17-20, 1977.

386 Erber, Fabio Stefano
Science and technology policy in Brazil: A review of the literature. LATIN AMERICAN RESEARCH REVIEW (CHAPEL HILL) 16, No. 1:3-56, 1981.

387 Ernst, Dieter
Technology policy for self-reliance: Some major issues. INTERNATIONAL SOCIAL SCIENCE JOURNAL (PARIS) 23, No. 3:466-80, 1981.

388 Evenson, Robert E.
Benefits and obstacles to appropriate agricultural technology. AMERICAN ACADEMY OF POLITICAL AND SOCIAL SCIENCE, ANNALS (PHILADELPHIA) 458:54-67, Nov. 1981.

389 Evenson, Robert
International diffusion of agrarian technology. THE JOURNAL OF ECONOMIC HISTORY 34:51-90, March 1974.

390 Ewing, Arthur and Koch, Gloria-Veronica
Some recent literature on development. THE JOURNAL OF MODERN AFRICAN STUDIES 15 (No. 3): 456-480, 1977.

391 Eze, O.
Patents and the transfer of technology, with special reference to the East African community. EASTERN AFRICA LAW REVIEW 5 (1972): 127-40.

392 Farrell, Trevor M. A.
A tale of two issues: nationalization, the transfer of technology and the petroleum multinationals in Trinidad-Tobogo. SOCIAL AND ECONOMIC STUDIES (MONA, JAMAICA) 28:234-81, March 1979.

393 Findlay, Ronald
Some aspects of technology transfer and direct foreign investment. AMERICAN ECONOMIC REVIEW, PAPERS AND PROCEEDINGS (NASHVILLE) 68:275-79, May 1978.

Attempts to build a very simple model of technology transfer that explicitly incorporates interdependence between output, costs, and technology.

394 Fisher, J. C. and Pry, R. H.
A simple substitution model of technological change. TECHNOLOGICAL FORECASTING AND SOCIAL CHANGE 3 (No. 1): 75-88, 1971.

395 Flit, Issias
Struggling for self-reliance in science and technology: The Peruvian case; ITINTEC. DEVELOPMENT DIALOGUE (UPPSALA) No. 1:39-45, 1979.

Describes the policies pursued, the instruments used and the choices made by Peru in its efforts to develop an autonomous capacity in science and technology.

396 Flit S., Issias and Gustavo Flores G.
A structure for industrial technology policy in Peru; the development of ITINTEC. COMERCIO EXTERIOR (MEXICO) 23:266-74, July 1977.

This paper has presented a Peruvian case study on the formulation and implementation of industrial technology policies.

397 Forje, John W.
Science and technology as the social dynamics of development in the Third World. MARGA QUARTERLY JOURNAL (COLOMBO) 4, No. 3:110-17, 1977.

Recommends the application and adaptation of western science and technology to the cultural and environmental surroundings of the African continent to activate rural development.

398 Forsyth, David J. C.
Appropriate technology in sugar manufacturing. WORLD DEVELOPMENT 5:189-202, March 1977.

399 Fortner, Robert S.
Strategies for self-immolation; the Third World and the transfer of advanced technologies. INTERAMERICAN ECONOMIC AFFAIRS (WASHINGTON) 31:25-50, Summer 1977.

400 Freeman, C., ed.
Technical innovation and long waves in world economic development [symposium]. FUTURES 13:238-338, August; 347-88, October 1981.

401 Frohman, Alan L.
Technology as a competitive weapon. HARVARD BUSINESS REVIEW (BOSTON) 60:97-104, Jan./Feb. 1982.

The author concludes that although investments in R & D are crucial to developing technology, without other essential ingredients, they probably won't pay off.

402 Galtung, John
Technology and dependence. CERES 7:45-50, Sept.-Oct. 1974.

403 Garmany, J. W.
Technology and employment in developing countries. JOURNAL OF MODERN AFRICAN STUDIES 16:549-64, December 1978.

404 Gerster, Richard
Men and machines in the Third World. SWISS REVIEW OF WORLD AFFAIRS (ZURICH) 29:12-17, Nov. 1979.

Examines the problem of technologies for road building in developing countries, using examples from various countries including Nepal, Kenya and Cameroon.

405 Ghandour, Marwan and Muller, Jurgen
A new approach to technological dualism. ECONOMIC DEVELOPMENT AND CULTURAL CHANGE (CHICAGO) 25:629-37, July 1977.

The objective of this essay is to suggest a new interpretation of the technological dualism observed in many less-developed countries (LDC's). This interpretation will involve a view of the nature of the dualistic economy and its development.

406 Gifford, Adam, Jr.
Pollution, technology, and economic growth. SOUTHERN ECONOMIC JOURNAL (October 1973).

407 Gillette, Robert
Latin America: Is imported technology too expensive? SCIENCE 181:41-44, July 1973.

408 Gilpin, Anthony C.
New dimensions of technical assistance; the evolving role of UNDP. INTERNATIONAL DEVELOPMENT REVIEW (WASHINGTON) Focus: Technical Cooperation, 17, No. 2:13-16, 1975.

Three major changes in the approach to technical assistance should be considered if the United Nations Development Programme is to perform a valid and effective role.

409 Girling, R. K.
Technology and dependent development in Jamaica: A case study. SOCIAL AND ECONOMIC STUDIES 26:169-89, June 1977.

410 Girvan, Norman, ed.

Essays on science and technology policy in the Caribbean. SOCIAL AND ECONOMIC STUDIES (MONA, JAMAICA) 28:1-336, March 1979.

411 Girvan, Norman
The approach to technology policy studies. SOCIAL AND ECONOMIC STUDIES (MONA, JAMAICA) 28:1-53, March 1979.

412 Gottstein, Klaus
Science and technology for the third world; the United Nations conference on science and technology for development [August 1979]. ECONOMICS (TUBINGEN) 21:136-51, 1980.

413 Goulet, Denis
The paradox of technology transfer. THE BULLETIN OF THE ATOMIC SCIENTISTS 31:29-46, June 1975.

414 What price technology? AMERICAS (WASHINGTON) 31: 30-36, June/July 1979.

Attempts to describe a few working concepts that can help to decide which elements of technology assessment can contribute to Latin America's development efforts.

415 Goulet, Denis
Suppliers and purchasers of technology: A conflict of interests. INTERNATIONAL DEVELOPMENT REVIEW 18 (1976): 14-20.

416 Goulet, Denis
Can values shape third world technology policy? JOURNAL OF INTERNATIONAL AFFAIRS 33:89-109, Spring 1979.

417 Gouri, Gangadhar S.
Planning for technological development; the role of international cooperation. VIENNA INSTITUTE FOR DEVELOPMENT. OCCASIONAL PAPER (VIENNA) No. 78/2:1-9, 1978.

418 The grass-roots approach to technology for development. EUROPEAN COMMUNITIES. COMMISSION. COURIER (BRUSSELS) 55:73-78, May/June 1978.

From a paper presented by the Pan-African Institute for Development and the environment agency ENDA to the African regional conference on the UNCSTD, June 1978.

419 Greeley, Martin
Rural technology, rural institutions and the rural poorest: The case of rice processing in Bangladesh. THE BANGLADESH DEVELOPMENT STUDIES (DACCA) 8:143-59, Winter/Summer 1980.

420 Grynspan, Devora

Technology transfer patterns and industrialization in LDC's. INTERNATIONAL ORGANIZATION 36:795-806, Autumn 1982.

421 Guess, George M.
Technical and financial policy options for development forestry. NATURAL RESOURCES JOURNAL (ALBUQUERQUE) 21: 37-55, Jan. 1981.

422 Gulati, I. S. and Bansal, Swaraj K.
Export obligation, technology transfer and foreign collaboration in electronics. ECONOMIC AND POLITICAL WEEKLY (BOMBAY) 15:1246-56, Oct. 1980.

423 Haas, E. B.
Technological self-reliance for Latin America: The OAS contribution. INTERNATIONAL ORGANIZATION 34:541-70, Autumn 1980.

424 Hamdani, Khalil A. and Mohmood, M. A.
Analytics of the technology transfer cost issue. PAKISTAN DEVELOPMENT REVIEW (ISLAMABAD) 15:154-70, Summer 1976.

Focuses on transfer costs as these determine the relationship between technology transfer and technological dependence.

425 Havens, A. Eugene and Flinn, William
Green revolution technology and community development: The limits of action programs. ECONOMIC DEVELOPMENT AND CULTURAL CHANGE 23 (April 1975): 469-81.

426 Hendry, Peter
Who's afraid of advanced technology? CERES 3 (July-Aug. 1970): 45-48.

427 Herrera, Amilcar O.
The generation of technologies in rural areas. WORLD DEVELOPMENT (OXFORD) 9:21-35, Jan. 1981.

Proposes a research methodology that, while generating technologies appropriate for rural areas, contributes at the same time to the building of a new system of paradigms adequate for the needs and conditions of developing societies.

428 Heston, A. W. and Pack, H.
Technology transfer, new issues, new analysis. AMERICAN ACADEMY OF POLITICAL AND SOCIAL SCIENCE 458:7-186, November 1981.

429 Hochschild, Steven F.
Technical assistance and international development: A need for fundamental change. INTERNATIONAL DEVELOPMENT REVIEW: FOCUS 14 (No. 4): 15-19, 1972.

430 Hoelscher, Harold E.
Technology transfer; a complex problem imperfectly understood. ARAB ECONOMIST (BEIRUT) 10:15-17, Dec. 1978.

Shows how developing nations can adapt foreign technology to meet local needs.

431 Hoelscher, Harold E.
Some perspectives on industrialization and technology transfer. TECHNOS 3 (July-Sept. 1974): 53-60.

432 Hoogvelt, Ankie
Indigenization and technological dependency. DEVELOPMENT AND CHANGE (BEVERLY HILLS, CALIF., ETC.) 11:257-72, April 1980.

Offers a contribution to the debate on the effective control of private foreign investment in developing countries.

433 Howe, James W. and Knowland, William E.
Energy for rural Africa: The potential of small-scale renewable energy techniques. INTERNATIONAL DEVELOPMENT REVIEW: FOCUS 19 (No. 3):18-21.

434 Howes, Michael
The uses of indigenous technical knowledge in development. IDS BULLETIN, INSTITUTE OF DEVELOPMENT STUDIES AT THE UNIVERSITY OF SUSSEX (BRIGHTON) 10:12-23, Jan. 1979.

Attempts to draw together a sample of the literature dealing with the scope, nature and propensity for change of the technical knowledge embedded in various third world social systems and to explore ways in which such knowledge might be used in combination with established sciences and technology.

435 Huq, M. M. and Aragaw, H.
Technical choice in developing countries: The case of leather manufacturing. WORLD DEVELOPMENT 5:777-789, Sept.-Oct. 1977.

436 Ilunkamba, Ilunga
Copper, technology and dependence in Zaire: Towards the demystification of the new white magic. NATURAL RESOURCES FORUM (DORDRECHT) 4:147-56, April 1980.

Deals with three broad aspects of technological dependence: in the provision of technical personnel for the mining enterprises, in the processing of minerals and in marketing. In each case, it can be shown that the potential independence of Zaire is considerably greater than what has actually been achieved to date.

437 Importing technology into a Third World nation. FUTURIST 11:77, April 1977.

438 Ireson, W. Grant
Some problems in technical assistance to LDC universities. TECHNOS (FORT COLLINS, CO) 4:13-23, Jan./March 1975.

Outlines some of the more common problems that are encountered in rendering technical assistance.

439 Ishikawa, Shigeru
The Chinese method of technological development; the case of the agricultural machinery and implement industry. DEVELOPING ECONOMIES (TOKYO) 13:430-58, Dec. 1975.

Studies the economic processes leading to remarkable technological development in the People's Republic of China, especially since the middle 1960's, on a basis which makes possible comparison with the experiences of other contemporary developing nations in technological development.

440 Islam, Rizwanul
Some constraints on the choice of technology. THE BANGLADESH DEVELOPMENT STUDIES (DACCA) 5:255-84, July 1977.

The purpose of this paper is to see whether the choice of technology in LDC's like Bangladesh is constrained by the lack of alternatives or by some other factors.

441 Jacoby, Erich
World Bank policy and the peasants in the Third World. DEVELOPMENT AND CHANGE (BEVERLY HILLS, CA, ETC.) 10:489-94, July 1979.

Recommends that the Bank should design a new type of technology based on necessary institutional reforms, using the natural assets of the country and particularly its abundant labour force in the service of development.

442 James, Jeffrey
Growth, technology and the environment in less developed countries: A survey. WORLD DEVELOPMENT (OXFORD) 6:937-65, July/Aug. 1978.

Contains a discussion of the relationship of growth and technology to each other and to the environment in the context of the countries of the Third World. With a comment by Ignacy Sachs, p. 967-69.

443 Jedlicka, Allen D.
An experiment on the institutional development of appropriate technology in Colombia. TECHNOS 4:49-55, April-June 1975.

444 Joshi, N.
Technological choice and socio-economic imperative: A case study of textile technologies in India. RESEARCH POLICY 6:202-213, July, 1977.

445 Kahane, Reuven and Starr, Laura
The impact of rapid social change on technological education: An Israeli example. COMPARATIVE EDUCATION REVIEW 20 (June 1976): 265-78.

446 Kanbur, S. M. Ravi
Technology policy and development planning: A guide for planners in developing countries. PRINCETON UNIVERSITY. WOODROW WILSON SCHOOL OF PUBLIC AND INTERNATIONAL AFFAIRS. RESEARCH PROGRAM IN DEVELOPMENT STUDIES. DISCUSSION PAPER No. 99:1-74, Oct. 1981.

447 Kaplinsky, Raphael
Accumulation and the transfer of technology: Issues of conflict and mechanisms for the exercise of control. WORLD DEVELOPMENT 4:197-224, March 1976.

448 Kennedy, C. and Thirlwall, A. P.
Technical progress: A survey. ECONOMIC JOURNAL 82 (March 1972): 11-72.

449 Khindaria, Brij
Transfer of technology; working without a code. DEVELOPMENT FORUM BUSINESS EDITION (GENEVA) No. 70:3, Jan. 16, 1981.

450 Khindaria, Brij
Transfer of technology; rocky road to building a code. DEVELOPMENT FORUM BUSINESS EDITION (GENEVA) No. 78: p. 3, May 16, 1981.

451 Kiang, Wan-lin
Management of technological innovation; some strategic alternatives for Taiwan industries. INDUSTRY OF FREE CHINA (TAIPEI) 50:19-29, Aug. 25, 1978.

452 Kim, Linsu
Stages of development of industrial technology in a developing country: A model. RESEARCH POLICY (AMSTERDAM) 9:254-77, July 1980.

Provides a skeletal description of the development pattern of industrial technology as it appears to have taken place in productive units in the electronics industry in Korea.

453 Koppel, Bruce
Technology assessing; a view from Asia. TECHNOS (FT. COLLINS, CO) 5:62-70, Oct./Dec. 1976.

Reports technology assessment's prospects in six nations of Asia (Iran, Nepal, India, Philippines, Taiwan, Japan).

454 Kuitenbrouwer, J. B. W.
Some reflections on the uses of science and technology in Indonesia. INSTITUTE OF SOCIAL STUDIES. OCCASIONAL PAPERS (THE HAGUE) No. 72:1-27, Aug. 1979.

455 Kuo, Shirley W. Y.
Technology assessment and planning. INTERNATIONAL COMMERCIAL BANK OF CHINA, ECONOMIC REVIEW (TAIPEI) No. 179:4-10, Sept./Oct. 1977.

Examines the presently undeveloped methodologies which are urgently needed for technology assessment and discusses utilization of resources, policies on R & D, evaluation of alternative technologies and the mechanism of planning in mixed economies.

456 Kuuya, P. Masette
Transfer of technology; an overview of the Tanzania case. AFRICA DEVELOPMENT (DAKAR) 2:47-72, April/June 1977.

Examines the problem attendant to the transfer of technology to LDC's with Tanzania's experience being used to illustrate some of the author's contentions.

457 Laing, N. F.
Technological uncertainty and the pure theory of optimum growth. AUSTRALIAN ECONOMIC PAPERS (ADELAIDE) 18:131-37, June 1979.

458 Langrish, John
The changing relationship between science and technology. NATURE 250:614-16, August 1974.

459 Lall, S.
Developing countries and the emerging international technological order. JOURNAL OF INTERNATIONAL AFFAIRS 33:77-88, Spring 1979.

460 Lall, Sanjaya
Developing countries as exporters of industrial technology. RESEARCH POLICY (AMSTERDAM) 9, No. 1:24-52, 1980.

This paper presents a preliminary attempt to describe and assess the emergence of domestic enterprises in the Third World as exporters of capital and technology.

461 Lall, Sanjaya
International technology market and developing countries. ECONOMIC AND POLITICAL WEEKLY (BOMBAY) 15:311-32, Feb. 1980.

462 Lall, Sanjaya
Indian technology exports and technological development. AMERICAN ACADEMY OF POLITICAL AND SOCIAL SCIENCE, ANNALS (PHILADELPHIA) 458:151-62, Nov. 1981.

463 Lansing, Richard M.
On technical progress and the speed of adjustment. ECONOMICA (LONDON) 42:394-400, Nov. 1975.

This paper contends that greater embodiment of technical progress need not always lead to faster adjustment given an initial disequilibrium within a neoclassical growth model.

464 Lasserre, Philippe
Training: Key to technological transfer. LONG RANGE PLANNING 15:51-60, June 1982.

465 Lavania, G. S. and Bhalerao, M. M.
Modern technology and human labour utilization in developing countries; a sample study in India. ECONOMIC AFFAIRS (CALCUTTA) 22:352-60, Sept./Oct. 1977.

Aims at finding out the impact of modern technology on: (1) the crop-wise employment of labour on the farm, (2) composition of human labour employment on the farm and (3) seasonal utilization of human labour.

466 Layton, Edwin T., Jr.
Technology as knowledge. TECHNOLOGY AND CULTURE 15:31-41, Jan. 1974.

467 Lecraw, Donald J.
Technological activities of less-developed-country-based multinationals. AMERICAN ACADEMY OF POLITICAL AND SOCIAL SCIENCE, ANNALS (PHILADELPHIA) 458:163-74, Nov. 1981.

468 Lecraw, Donald J.
Choice of technology in low-wage countries; a non-neoclassical approach. QUARTERLY JOURNAL OF ECONOMICS (CAMBRIDGE, MA) 93:631-54, Nov. 1979.

Attempts to answer two questions: (1) What is the extent of this noncost-minimizing behavior? (2) What rewards did the owners and managers of these firms receive in exchange for their forgone profits? The replies are based on the choice of technology of 400 firms that invested in the light manufacturing sector in Thailand over the period 1962-1974.

469 Leff, Nathaniel H.
Technology transfer and U.S. foreign policy: The developing countries. ORBIS 23:145-65, Spring 1979.

470 Lele, U. and Mellor, J. W.
Technological change, distributive bias and labor transfer in a two sector economy. OXFORD ECON PA 33:426-41, November 1981.

471 Lele, Uma
Technology and rural development in Africa. TECHNOS (FT. COLLINS, CO) 4:4-12, July/Sept. 1975.

Points out the nature of some of the problems encountered in developing agricultural technology suited to small farm conditions and the steps necessary to ameliorate this major constraint.

472 Li, K. T.
Two-way technology transfer between the United States and the Republic of China. INDUSTRY OF FREE CHINA (TAIPEI) 50:2-8, July 25, 1978.

473 Livingston, Dennis
International technology assessment and the United Nations system. AMERICAN JOURNAL OF INTERNATIONAL LAW 64 (Sept. 1970): 163-71.

474 Long, F.
Role of social scientific inquiry in technology transfer. AMERICAN JOURNAL OF ECONOMICS AND SOCIOLOGY 38:261-74, July 1979.

475 Loveday, Anthony J.
Aspects of external and technical assistance to developing libraries. OVERSEAS UNIVERSITIES (Sept. 1974), pp. 11-13.

476 Macioti, M.
Technology and development: The historical experience. IMPACT OF SCIENCE ON SOCIETY 28:313-19, Oct. 1978.

477 MacPherson, George and Jackson, Dudley
Village technology for rural development: Agricultural innovation in Tanzania. INTERNATIONAL LABOUR REVIEW 111 (Feb. 1975): 97-118.

478 Magat, Wesley A.
Technological advance with depletion of innovation possibilities; implications for the dynamics of factor shares. ECONOMIC JOURNAL (LONDON) 89:614-23, Sept. 1979.

479 Magdoff, H.
Capital, technology and development. MONTHLY REVIEW 27:1-11, January 1976.

480 Magee, Stephen P.
The appropriability theory of the multinational corporation. AMERICAN ACADEMY OF POLITICAL AND SOCIAL SCIENCE, ANNALS (PHILADELPHIA) 458:123-35, Nov. 1981.

481 Maitra, Priyatosh
Technology transfer, population growth and the development gap since the Industrial Revolution. UNIVERSITY OF OTAGO. ECONOMICS DISCUSSION PAPERS (DUNEDIN) No. 7707:1-28, Aug. 1978.

482 Malaysia takes the plunge: Heavy industry plus technology. ASIAN FINANCE (HONGKONG) 7:59-87, Oct. 15, 1981.

483 Malecki, Edward J.
Science, technology, and regional economic development: Review and prospects. RESEARCH POLICY (AMSTERDAM) 10: 312-34, Oct. 1981.

Reviews the literature on the locational incidence and regional effects of science and technology with an emphasis on R & D activities in the USA.

484 Mason, R. Hal
The multinational firm and the cost of technology to developing countries. CALIFORNIA MANAGEMENT REVIEW 15 (Summer): 5-13, 1973.

485 Mason, R. Hal
Some observations on the choice of technology by multinational firms in developing countries. REVIEW OF ECONOMICS AND STATISTICS 55 (August): 349-55, 1973.

486 Mathur, H. M.
Social aspects of administering technical aid programmes. INDIAN JOURNAL OF PUBLIC ADMINISTRATION 18 (June 1972): 245-53.

487 McCulloch, R. and Yellen, J.
Can capital movements eliminate the need for technology transfer? JOURNAL OF INTERNATIONAL ECONOMICS 12:95-109, F 1982.

488 McCulloch, Rachel
Technology transfer to developing countries: Implications of international regulation. AMERICAN ACADEMY OF POLITICAL AND SOCIAL SCIENCE, ANNALS (PHILADELPHIA) 458:110-22, Nov. 1981.

489 McRobie, George
Rural industry and low-cost technology. EKISTICS 28 (December 1969): 425-27.

490 McRobie, George
Intermediate technology; small is successful. THIRD WORLD QUARTERLY (LONDON) 1:71-86, April 1979.

491 Menck, Karl W.
Economic and technical co-operation among the developing countries. ECONOMICS (TUBINGEN) 24:87-108, 1981.

492 Menck, Karl-Wolfgang
The concept of appropriate technology. INTER-ECONOMICS (January 1973), pp. 8-10.

493 Menck, Karl W.
Fundamental concepts of the transfer of technology to the developing countries. ECONOMICS (TUBINGEN) 15:30-9, 1977.

494 Mendis, D. L. O.
Science, technology and the liberation of the Third World. MARGA QUARTERLY JOURNAL (COLOMBO) 3, No. 3:75-88, 1976.

495 Menon, Usha
World Bank and transfer of technology; case of Indian fertiliser industry. ECONOMIC AND POLITICAL WEEKLY (BOMBAY) 15:1437-43, Aug. 23, 1980.

496 Miller, Debra Lynn
An economic appraisal of the proposed code of conduct for technology transfer. AUSSENWIRTSCHAFT (ST. GALLEN) 36:121-42, June 1981.

497 Mitra, Tapan and Zilcha, Itzhak
On optimal economic growth with changing technology and tastes: Characterization and stability results. INTERNATIONAL ECONOMIC REVIEW (OSAKA) 22:221-38, Feb. 1981.

498 Modern technology: Problem or opportunity? DAEDALUS: JOURNAL OF THE AMERICAN ACADEMY OF ARTS AND SCIENCES (BOSTON) p. 1-190, Winter 1980.

499 Morehouse, Ward
Is technology helping the developing countries? STANDARD CHARTERED REVIEW p. 2-7, May 1977.

The spread of the scientific revolution poses new problems and raises the question whether technological development has advanced or harmed the quality of life in the third world.

500 Morgan, Robert P.
Technology and international development: New directions needed. CHEMICAL AND ENGINEERING NEWS 55:31-39, Nov. 14, 1976.

501 Moritani, M.
International co-operation in technological development--can it revitalize the world economy? JAPAN QUARTERLY 29:432-36, Oct./Dec. 1982.

502 Morley, S. A. and Smith, G. W.
Choice of technology: Multinational firms in Brazil. ECONOMIC DEVELOPMENT AND CULTURAL CHANGE 25:239-64, Jan. 1977.

503 Moshi, H. P. B.
Industrialisation and technological policy in Tanzania: An overview. AFRICA DEVELOPMENT (DAKAR) 6:80-96, Jan./April 1981.

504 Muqtada, M.
The seed-fertilizer technology and surplus labour in Bangladesh agriculture. BANGLADESH DEVELOPMENT STUDIES 3 (Oct. 1975): 403-28.

505 Mytelka, Lynn Krieger
Licensing and technology dependence in the Andean group. WORLD DEVELOPMENT (OXFORD) 6:447-59, April 1978.

Explores the extent to which licensing promotes technology transfer and future technological self-reliance in 47 metal working and 43 chemical firms located in Peru, Ecuador and Colombia.

506 Mytelka, Lynn K.
Technological dependence in the Andean group. INTERNATIONAL ORGANIZATION (MADISON) 32:101-39, Winter 1978.

It is suggested here that ownership structure, product sector, and licensing interact with choice of machinery imports and research and development activites in such a way as to produce a technological dependence syndrome; illustrated by data from Colombia, Ecuador and Peru.

507 Naur, M.
Transfer of technology--a structural analysis. JOURNAL OF PEACE RESEARCH 17 (No. 3): 247-59, 1980.

508 Nayudama, Y.
Endogenous development: Science and technology. VIENNA INSTITUTE FOR DEVELOPMENT. OCCASIONAL PAPER (VIENNA) No. 78/3:1-24, 1978.

509 Ndongko, W. A. and Anyang, S. O.
Concept of appropriate technology: An appraisal from the Third World. MONTHLY REVIEW 32:35-43, F 1981.

510 Nelson, Richard R.
Less developed countries--Technology transfer and adaptation: The role of the indigenous science community. ECONOMIC DEVELOPMENT AND CULTURAL CHANGE 23 (Oct. 1974): 61-78.

511 Nelson, Richard R. and Winter, Sidney G.
Technical change in an evolutionary model. QUARTERLY JOURNAL OF ECONOMICS (CAMBRIDGE, MA) 90:90-118, Feb. 1976.

The heart of this paper is an evolutionary model of the processes of technological advance and economic growth, a rough calibration of this model with data on U.S. economic growth, and a comparison with neoclassical models of the sort initiated by Solow in his classic 1957 paper.

512 Nilsen, Svein Erik
The use of computer technology in some developing countries. INTERNATIONAL SOCIAL SCIENCE JOURNAL (PARIS) 31, No. 1:513-28, 1979.

513 Njoku, Athanasius O.
The impact of technological change in a rural economy. CIVILISATIONS (BRUSSELS) 25, No. 1/2:52-61, 1975.

Presents a situation in traditional African agriculture where an increase in capital (in forms of new technology) has increased the labor and land components of agricultural production. The introduction raised agricultural output and led to rural immigration.

514 Norman, Colin
Soft technologies, hard choice. WORLDWATCH INSTITUTE (Paper no. 21), June 1978, 48 pp.

Problems in choosing development technologies appropriate to the social, economic and ecological needs of each country.

515 Nwosu, Emmanuel J.
Some problems of appropriate technology and technological transfer. THE DEVELOPING ECONOMIES 13:82-93, March 1975.

516 O'Brien, Peter
Third world industrial enterprises; export of technology and investment. ECONOMIC AND POLITICAL WEEKLY (BOMBAY) 15:1831-43, Oct. 1980.

Examines some of the causes and consequences of the growth in international operations of LDC enterprises.

517 Olaloye, A. O.
Technology transfer and employment in Nigerian manufacturing industries. QUARTERLY JOURNAL OF ADMINISTRATION 12:167-76, Sept. 1978.

Shows that the transfer of technology has been found to lead to unemployment in Nigerian manufacturing industries.

518 Ono, Akira
Borrowed technology in the iron and steel industry: A comparison between Brazil, India and Japan. HITOTSUBASHI JOURNAL OF ECONOMICS (TOKYO) 21:1-18, Feb. 1981.

519 Otsuka, Katsuo
Technological choice in the Japanese silk industry; implication for development in LDC's. INTERNATIONAL DEVELOPMENT CENTER OF JAPAN. IDCJ WORKING PAPER (TOKYO) No. A-05:1-31, March 1977.

520 Pack, Howard
Appropriate industrial technology: Benefits and obstacles. AMERICAN ACADEMY OF POLITICAL AND SOCIAL SCIENCE, ANNALS (PHILADELPHIA) 458:27-40, Nov. 1981.

Suggests that the systematic adoption of appropriate rather than advanced industrial technology in the modern manufacturing sector of less-developed countries could increase ouput and employment by substantial amounts.

521 Page, John M., Jr.
Technical efficiency and economic performance: Some evidence from Ghana. OXFORD ECONOMIC PAPERS (LONDON) 21:319-29, June 1980.

Seeks to clarify the relationship between technical (or managerial) efficiency, the choice of technique, and economic performance, using DRC estimates of economic efficiency for firms in three Ghanaian industries.

522 Pal, S. K. and Bowonder, B.
High technology and development in India. FUTURES 10: 337-41, August 1981.

523 Parker, John B.
New technology in India's agriculture. Washington, D.C.: U.S. Department of Agriculture, Economic Research Service, June 1972. 48 pp.

524 Parthasarathi, Ashok
Technological bridgeheads for self-reliant development. DEVELOPMENT DIALOGUE (UPPSALA) No. 1:33-38, 1979.

525 Patel, Surendra J.
The technological dependence of developing countries. JOURNAL OF MODERN AFRICAN STUDIES 12 (March 1974): 1-18.

526 Patel, Surendra J.
The cost of technological dependence. CERES 6 (March-April 1973): 16-19.

527 Pavitt, Keith
Technology, international competition, and economic growth: Some lessons and perspectives. WORLD POLITICS 25 (22 January 1973): 183-205.

528 Pavitt, Keith and Ward, Salomon
A factor for progress: Technological innovation. OECD Observer (Oct. 1971): pp. 10-13.

529 Phillips, David A.
Choice of technology and industrial transformation: The case of the United Republic of Tanzania. INDUSTRY AND DEVELOPMENT (NEW YORK) No. 5:85-106, 1980.

530 Pickett, James
The choice of industrial technology. DEVELOPMENT RESEARCH DIGEST (BRIGHTON) No. 3:18-20, Spring 1980.

Presents a summary account of a continuing series of investigations into the choice of industrial technology in developing countries undertaken by the David Livingstone Institute of Overseas Development Studies.

531 Pickett, James; Forsyth, D. J. C. and McBain, N. S.
The choice of technology, economic efficiency and employment in developing countries. WORLD DEVELOPMENT 2:47-54, March 1974.

532 Pitt, Mark M. and Lee, Lung-Fei
The measurement and sources of technical inefficiency in the Indonesian weaving industry. JOURNAL OF DEVELOPMENT ECONOMICS (AMSTERDAM) 9:43-64, Aug. 1981.

533 Powelson, John P.
Planning from below with an intermediate technology model. CULTURES ET DEVELOPPEMENT (LOUVAIN) 11, No. 3: 465-77, 1979.

Suggests a model that will be understandable, not only by econometricians, but widely among middle- and high-level civil servants, and shows how a variant of it was used in the macro-economic projections for the Kenya five-year plan (1974-78).

534 Pray, Carl E.
The Green revolution as a case study in transfer of technology. AMERICAN ACADEMY OF POLITICAL AND SOCIAL SCIENCE, ANNALS (PHILADELPHIA) 458:68-80, Nov. 1981.

535 Pugel, Thomas A.
Endogenous technological change and international technology transfer in a Ricardian trade model. U.S. BOARD OF GOVERNORS OF THE FEDERAL RESERVE SYSTEM. INTERNATIONAL FINANCE DIVISION. INTERNATIONAL FINANCE DISCUSSION PAPERS No. 167:1-37, Sept. 1980.

536 Radhu, G. M.
Transfer of technical knowhow through multinational corporations in Pakistan. PAKISTAN DEVELOPMENT REVIEW 12:361-374, Winter 1973.

537 Rajan, J. V. and Seth, N. D.
Transfer of indigenous technology--some Indian cases. RESEARCH POLICY 10:173-94, April 1981.

538 Ramaswamy, G. S.
The transfer of technology among developing countries. FOCUS (1976), pp. 7-10.

539 Ranis, Gustav
Technology choice and employment in developing countries; a synthesis of economic growth center research. YALE UNIVERSITY. ECONOMIC GROWTH CENTER. CENTER DISCUSSION PAPER No. 276:1-37, Feb. 1978.

Summarizes the major findings of research financed under contract AID/OTR C-1326, including the conclusions for behavior and policy which emanate from that research.

540 Ranis, Gustav
Science, technology and development; a retrospective view. YALE UNIVERSITY. ECONOMIC GROWTH CENTER. CENTER DISCUSSION PAPER No. 268:1-49, Oct. 1977.

Attempts to illuminate the relationship among science, technology and development, using the experience of seven countries: Great Britain, Germany, the United States, Japan, Hungary, Brazil and Ghana.

541 Rao, K. N.
University-based science and technology for development: New patterns of international aid. IMPACT SOC SCI 28: 117-25, April 1978.

542 Reddy, Amulya Kumar N.
Alternative technology: A viewpoint from India. SOCIAL STUDIES OF SCIENCE 5:331-342, Aug. 1975.

543 Retz, Joachim
Problems of international development taxes; illustrated by the financing system for science and technology for development. INTERECONOMICS, REVIEW OF INTERNATIONAL TRADE AND DEVELOPMENT (HAMBURG) No. 1:36-42, Jan. 1982.

544 Rifkin, Susan B.
The Chinese model for science and technology; its relevance for other developing countries. DEVELOPMENT AND CHANGE (THE HAGUE) 6:23-40, Jan. 1975.

Discusses the Chinese model in the development of science and technology and its relevance to other developing countries.

545 Riskin, Carl
Intermediate technology in China's rural industries. WORLD DEVELOPMENT (OXFORD) 6:1297-1311, Nov./Dec. 1978.

546 Rittberger, Volker
The role of science and technology in the New international order. INTERECONOMICS, Nos. 11/12:279-86, Nov./Dec. 1978.

Describes the preparations for the United Nations Conference on Science and Technology for Development (UNCSTD) to be held in Vienna, 20-31 August 1979.

547 Robertson, Andrew
Introduction: Technological innovations and their social impact. INTERNATIONAL SOCIAL SCIENCE JOURNAL (PARIS) 23, No. 3:431-46, 1981.

548 Rondinelli, Dennis A. and Ruddle, Kenneth
Appropriate institutioins for rural development: Organizing services and technology in developing countries. PHILIPPINE JOURNAL OF PUBLIC ADMINISTRATION 21:35-52, Jan. 1977.

549 Rosenberg, Nathan
The direction of technological change: Inducement mechanisms and focusing devices. ECONOMIC DEVELOPMENT AND CULTURAL CHANGE (Oct. 1969).

550 Rosenberg, Nathan
Economic development and the transfer of technology: Some historical perspectives. TECHNOLOGY AND CULTURE 11:550-575,Oct. 1970.

551 Ruttan, Vernon W. and Hayami, Yujiro
Technology transfer and agricultural development. TECHNOLOGY AND CULTURE 14:119-151, April, 1973.

552 Sabato, Jorge A.
Technological development in Latin America and the Caribbean. CEPAL REVIEW No. 10:81-94, April 1980.

553 Sabourin, Louis
The contribution of science and technology to development. VIENNA INSTITUTE FOR DEVELOPMENT. OCCASIONAL PAPER (VIENNA) No. 78/4:1-13, 1978.

An address delivered at the Vienna Institute for Development 24 April 1978.

554 Sachs, Ignacy
Controlling technology for development. DEVELOPMENT DIALOGUE (UPPSALA) No. 1:24-32, 1979.

555 Sagasti, Francisco R.
Science and technology policies for development: A review of problems and issues. HABITAT INTERNATIONAL (OXFORD) 5, No. 3/4:573-86, 1980.

556 Sagasti, Francisco R.
Technology, planning, and self-reliant development: A Latin American view. New York: PRAEGER, 1979.

557 Sagasti, Francisco R.
Towards endogenous science and technology for another development. DEVELOPMENT DIALOGUE (UPPSALA) No. 1:13-23, 1979.

Argues for the strengthening of the autonomous S and T capacity of the Third World and gives special attention to the need for a recovery and development of traditional technologies.

558 Sato, Masumi
Technological development in China viewed through the electronics industry: An engineer's view. DEVELOPING ECONOMIES 9 (Sept. 1971): 315-31.

559 Schiesser, Walter
Soft technology for the Third World. SWISS REVIEW OF WORLD AFFAIRS (ZURICH) 27:20-23, July 1977.

Describes the Village Technology Unit near Nairobi, jointly established by Kenya's Ministry of Housing and Social Welfare and UNICEF.

560 Schiel, T. W.
Appropriate technology: Some concepts, some ideas, and some recent experiences in Africa. EASTERN AFRICAN JOURNAL OF RURAL DEVELOPMENT 7 (No. 1-2):77-121, 1974.

561 Schmidt, Jean-Louis
The transfer of technology from industrialized to developing countries. EUROPEAN COMMUNITIES. COMMISSION. COURIER (BRUSSELS) No. 39:39-39, Sept./Oct. 1976.

562 Schroeder, Dennis
In search of an appropriate technology. INTERNATIONAL DEVELOPMENT REVIEW: FOCUS 18 (No. 3): 3-7, 1976.

563 Schumacher, E. F.
The work of the Intermediate Technology Development Group in Africa. INTERNATIONAL LABOUR REVIEW 106 (July 1972): 75-92.

564 Science and technology for development. EUROPEAN COMMUNITIES. COMMISSION. COURIER (BRUSSELS) 55:62-98, May/June 1978.
Looks at some of the main topics to be discussed at the UN Conference on Science and Technology for Development (UNCSTD), to be held in Vienna from 20-31 August.

565 Science in the underdeveloped countries: World plan of action for the application of science and technology to development. MINERVA 9 (Jan. 1971): 101-21.

566 Seaborg, Glenn T.
Science, technology, and development. SCIENCE 181 (6 July 1973): 13-19.

567 Seifert, Hubertus
Direct investments and technology transfer in developing countries. ECONOMICS (TUBINGEN) 20:80-95, 1980.

568 Selwyn, Percy
Conflict in technical assistance. BULLETIN [Institute of Development Studies] 3 (Aug. 1971): 44-47.

569 Sen, Amartya
Employment, institutions and technology: Some policy issues. INTERNATIONAL LABOUR REVIEW 112 (July 1975): 45-73.

570 Sepulveda, Sergio
The effects of modern technologies on income distribution: A case of integrated rural development in Colombia. DESARROLLO RURAL EN LAS AMERICAS (SAN JOSE) 12:105-23, May/Aug. 1980.

571 Sewell, John W.
Can the rich prosper without the progress of the poor? INTERNATIONAL DEVELOPMENT REVIEW (ROME) FOCUS: TECHNICAL COOPERATION 20, No. 2:17-21, 1978.

Examines the proposition that continued growth and achievement in the rich countries depend to a much larger degree than realized heretofore on the growth and prosperity of the developing countries.

572 Shapiro, Kenneth H. and Muller, Jurgen
Sources of technical efficiency; the roles of modernization and information. ECONOMIC DEVELOPMENT AND CULTURAL CHANGE (CHICAGO) 25:293-310, Jan. 1977.

This paper addresses the issue of improved specification by analyzing the roles of information and modernization in the production process on cotton farms in Tanzania.

573 Sheldon, Ronald A.
Capital-energy substitution: Implications for technology transfer. TECHNOS 6 (Jan.-March 1977): 46-51.

574 Shen, T. Y.
Macro development planning in tropical Africa; technocratic and non-technocratic causes of failure. JOURNAL OF DEVELOPMENT STUDIES (LONDON) 13:413-27, July 1977.

The analysis of the causes of implementation failure shows that most of the causes lie outside of the competence of the planners. Alternative methods to calculate plan targets are not likely to meet with superior fulfilment results.

575 Sigurdson, Jon
Labour allocation and technology choice in China. TECHNOS 5:4-21, July/Sept. 1976.

Attempts to explain a couple of the major issues facing the labour force and indicates the role of technology and agro-industrial communities in attempted and future solutions.

576 Singer, Hans W.
Appropriate technology for a basic human needs strategy. INTERNATIONAL DEVELOPMENT REVIEW 19 (No. 2): 8-11, 1977.

577 Singer, Hans W.
The foreign company as an exporter of technology. BULLETIN OF THE INSTITUTE OF DEVELOPMENT STUDIES 3 (Oct. 1970): 8-15.

578 Singer, H. W. and Reynolds, L.
Technological backwardness and productivity growth. ECONOMIC JOURNAL (Dec. 1975).

579 Solo, R. A.
Dilemmas of technology: A review article. JOURNAL OF ECONOMIC ISSUES 13:733-42, S 1979.

580 Some issues of technology; from a conference held in Cambridge, Massachusetts, April 30, 1979. DAEDALUS: JOURNAL OF THE AMERICAN ACADEMY OF ARTS AND SCINECES p. 1-24, Winter 1980.

Discussion by Harvey Brooks, Edward E. David, Jr., Stephen R. Graubard, and others.

581 Spencer, Dunstan S. C. and Byerlee, Derek
Technical change, labor use, and small farmer development; evidence from Sierra Leone. AMERICAN JOURNAL OF AGRICULTURAL ECONOMICS 58:874-80, Dec. 1976.

Highlights some of the interrelationships between improved technology and labor use.

582 Srivastava, Uma K. and Heady, Earl O.
Technological change and relative factor shares in Indian agriculture: An empirical analysis. AMERICAN JOURNAL OF AGRICULTURAL ECONOMICS 55 (Aug. 1973): 509-14.

583 Stevenson, Rodney
Measuring technological bias. AMERICAN ECONOMIC REVIEW 70:162-73, March 1980.

584 Stewart, Frances
Choice of technique in developing countries. JOURNAL OF DEVELOPMENT STUDIES 9:99-121, Oct. 1972.

585 Stewart, Frances
Technology and employemnt in LDC's. WORLD DEVELOPMENT 2:17-46, March 1974.

586 Stewart, Frances
Inequality, technology and payments systems. WORLD DEVELOPMENT 6:275-93, March 1978.

The paper considers the way in which the interaction between technology, population growth and the payments system is responsible for growing inequality in many poor countries.

587 Stewart, Frances
Transfer of technology; is the price too high? DEVELOPMENT FORUM BUSINESS EDITION 45:p. 5, Dec. 31, 1979.

Adapted from a World Bank study entitled "International technology transfer: issues and policy options," issued as World Bank Staff Working Paper, No. 344.

588 Stewart, Frances
Arguments for the generation of technology by less developed countries. AMERICAN ACADEMY OF POLITICAL SCIENCE, ANNALS (PHILADELPHIA) 458:97-109, Nov. 1981.

589 Stout, David K.
The impact of technology on economic growth in the 1980's. DAEDALUS: JOURNAL OF THE AMERICAN ACADEMY OF ARTS AND SCIENCES p. 159-67, Winter 1980.

590 Strassmann, W. Paul
Can technology save the cities of developing countries? JOURNAL OF ECONOMIC ISSUES (UNIVERSITY PK, PA) 12:457-65, June 1978.

Attempts to see which view of technology makes the best tool for coping with major problems, using the phenomenal growth of cities in developing countries as the illustrative problem. With comments by John Adams, p. 497-500.

591 Street, James H.
The technological frontier in Latin America; creativity and productivity. JOURNAL OF ECONOMIC ISSUES (UNIVERISTY PK, PA) 10:538-58, Sept. 1976.

Content: Latin America as a great frontier; technological transfers and domestic creativity; the role of education on the technological frontier; conclusion.

592 Street, James H.
The internal frontier and technological progress in Latin America. LATIN AMERICAN RESEARCH REVIEW (CHAPEL HILL) 12, No. 3:25-56, 1977.

Describes the impact of the OPEC oil crisis and the recession in Latin America, in order to draw attention to the severity of their effects.

593 Streeten, Paul
Technology gaps between rich and poor countries. SCOTTISH JOURNAL OF POLITICAL ECONOMY 19:213-230, Nov. 1972.

594 Streeten, Paul P.
The multinational enterprise and the theory of development policy. WORLD DEVELOPMENT 1:1-14, Oct. 1973.

595 Streeten, Paul P.
Social science research on development: Some problems in the use and transfer of an intellectual technology. JOURNAL OF ECONOMIC LITERATURE 12 (Dec 1974): 1290-1300.

596 Sunkel, O.
Underdevelopment, the transfer of science and technology, and the Latin American University. HUMAN RELATIONS 24 (1971): 1-18.

597 Sutter, Rolf
Technology transfer into LDC's. INTERECONOMICS (Dec. 1974), pp. 380-84.

598 Svejnar, Z.
Technology and factor proportions in investment decisons; African aspects. EASTERN AFRICA ECONOMIC REVIEW (NAIROBI) 7:1-24, June 1975.

By examining a sample of plants in tropical Africa, the paper aims to throw some additional light on the widely discussed issue of factor proportions in the less developed countries.

599 Technology and development. HABITAT INTERNATIONAL (OXFORD) 5, No. 34:265-586, 1980.

Content: Section 1: Construction and development; Section 2: Infrastructure and services; Section 3: Energy considerations; Section 4: Science and technology in development.

600 Technology and development: Problems and responses. DEVELOPMENT RESEARCH DIGEST (BRIGHTON) No. 3:4-12, Spring 1980.

601 Technology and development in Africa. AFRICA DEVELOPMENT (DAKAR) 2:1-151, April/June 1977.

602 Technology and the new international order [symposium]. JOURNAL OF INTERNATIONAL AFFAIRS 33:1-148, Spring 1979.

603 Technology and the future [symposium]. PUBLIC MANAGEMENT 60:2-18, July 1978.

604 Teece, D. J.
Technology transfer by multinational firms: The resource cost of transferring technological know-how. THE ECONOMIC JOURNAL 87:242-261, June 1977.

605 Teitel, Simon
Towards an understanding of technical change in semi-industrialized countries. RESEARCH POLICY (AMSTERDAM) 10:127-41, April 1981.

606 Thompson, Dennis
The UNCTAD code on transfer of technology. JOURNAL OF WORLD TRADE LAW 16:311-37, July/Aug. 1982.

607 Thompson, Kenneth W.
Higher education for national development: One model for technical assistance. Occasional Paper no. 5. New York: INTERNATIONAL COUNCIL FOR EDUCATIONAL DEVELOPMENT, 1972. 24 pp.

608 Tyler, William G.
Technical efficiency in production in a developing country: An empirical examination of the Brazilian plastics and steel industries. OXFORD ECONOMIC PAPERS (LONDON) 31: 477-95, Nov. 1979.

609 Uchendu, V. C.
The role of intermediate technology in East African agricultural development. EASTERN AFRICA JOURNAL OF RURAL DEVELOPMENT 8 (1975): 182-90.

610 United Nations Conference of Trade and Development
Technological dependence: Its nature, consequences and policy implications; report by the UNCTAD Secretariat. AFRICA DEVELOPMENT 2:27-45, April/June 1977.

611 United Nations
National approaches to the acquisition of technology. UNITED NATIONS INDUSTRIAL DEVELOPMENT ORGANIZATION, 1977.

612 Urquidi, Victor L.
Technology, planning and Latin American development. INTERNATIONAL DEVELOPMENT REVIEW 13 (1971): 8-12.

613 Vaitsos, Constantine V.
Bargaining and the distribution of returns in the purchase of technology by developing countries. BULLETIN OF THE INSTITUTE OF DEVELOPMENT STUDIES 3 (Oct. 1970): 16-23.

614 Venkataswami, Thurai S.
Production functions and technical change in Indian manufacturing industries. INDIAN JOURNAL OF ECONOMICS (ALLAHABAD) 55:435-64, April 1975.

The purpose of this study is the statistical estimation of production functions and technical change in the Indian manufacturing sector as a whole and in twenty-eight individual industries.

615 Vessuri, Hebe M. C.
Technological change and the social organization of agricultural production. CURRENT ANTHROPOLOGY (CHICAGO) 21:315-27, June 1980.

Attempts to develop some ideas on technological change in agriculture, focusing on contemporary Latin American societies.

616 Villegas, B. M.
For an appropriate technology that uses the resources of the Third World by substituting labour for capital. CERES 7:44-47, May-June 1974.

617 Volkov, Mo
The developing countries: The choice of technology. PROBLEMS OF ECONOMICS 18:3-25, August 1975.

618 Wallender, Harvey W. III
Developing country orientations toward foreign technology in the eighties: Implications for new negotiation approaches. COLUMBIA JOURNAL OF WORLD BUSINESS (NEW YORK) 15:20-27, Summer 1980.

619 Weinberg, Alvin M.
Technology and ecology--is there a need for confrontation? BIOSCIENCE 23:41-45, January 1973.

620 White, L. J.
Appropriate technology, x-inefficiency, and a competitive environment: Some evidence from Pakistan. QUARTERLY JOURNAL OF ECONOMICS 90:575-589, November 1976.

621 Whitehead, Judy A.
Select technological issues in agro-industry. SOCIAL AND ECONOMIC STUDIES (MONA, JAMAICA) 28:139-88, March 1979.

622 Wilbanks, Thomas J.
Accessibility and technological change in northern India. ANNALS OF THE ASSOCIATION OF AMERICAN GEOGRAPHERS 62:427-36, September 1972.

623 Winiecki, Jan
Japan's imports of technology. INTERECONOMICS, MONTHLY REVIEW OF INTERNATIONAL TRADE AND DEVELOPMENT (HAMBURG) Nos. 3/4:77-81, March/April 1978.

624 Winner, Langdon
On criticizing technology. PUBLIC POLICY 20:35-60, Winter 1972.

625 Wionczek, Miguel
Technology transfer: What do the developing countries want? INTERECONOMICS (March 1976), pp. 76-79.

626 Wionczek, Miguel
Notes on technology transfer through multinational enterprises in Latin America. DEVELOPMENT AND CHANGE (THE HAGUE) 7:135-55, April 1976.

Reviews the state of knowledge about the transfer of technology through multinational firms to Latin America.

627 Wofsey, Marvin M. and Dickie, Paul M.
Computers in less developed economies. FINANCE AND DEVELOPMENT 8 (March 1971): 2-6.

628 Wood, Harold A.
Technology and productivity: Examples from Mexico and Central America. PROFESSIONAL GEOGRAPHER 22 (May 1970): 147-51.

629 Young, Earl C.
Designing programs to educate engineers for development in LDC's. TECHNOS 5:15-28, April-June 1976.

630 Young, J. T.
Technicians of the world. EDUCATION AND TRAINING 14 (August-September 1972): 263-65.

631 Youngs, A. J.
Developing countries need food technology extension. FOOD TECHNOLOGY 28 (April 1974): 97-104.

632 Zevin, L. Z.
Integrated approach to technology transfer: Soviet co-operation with developing countries. IMPACT SOC SCI 28:183-92, April 1978.

III. SPECIALIZED PUBLICATIONS (REPORTS, DOCUMENTS AND DIRECTORIES)

633 Adoption of Modern Agricultural Technology by Small Farm Operators, by J. Paul Leagans, and Trends and Fluctuations

in the Agricultural Exports of Less-Developed Countries, 1950-76, by David Blandford, are the latest Cornell International Agriculture Mimeographs. Single copies are available from the Department of Agricultural Economics, New York State College of Agriculture and Life Sciences, Cornell University, Ithaca, New York 14853, USA.

634 Building National Institutions for Science and Technology in Developing Countries contains the Proceedings of a workshop held in April, 1979, conducted by the American Association for the Advancement of Science for the U.S. Department of State, as a contribution to U.S. preparations for the United Nations Conference on Science and Technology for Development. A limited number of copies of the document are available from the Office of International Science, AAAS, 1776 Massachusetts Avenue, N.W., Washington, D.C. 20036, USA.

635 Computer Applications and Facilities for Science and Technology in the Asian and Pacific Region is an addition to the directory series compiled by the Registry of Scientific and Technical Services for the Asian and Pacific Region in Canberra, Australia. The directory contains two parts. Part 1 lists agencies that maintain computers in Australia, Republic of China, Japan, Republic of Korea, Malaysia, New Zealand, Philippines, and Thailand. Part 2 lists groups in the same countries that use these computer facilities. The directory is available from the Registry of Scientific and Technical Services for the Asian and Pacific Region, P.O. Box 62, Jamison Centre, Canberra, A.C.T. 2614, Australia.

636 Economics and the Design of Small-Farmer Technology, edited by Alberto Valdes, contains eleven papers on the design of new technologies for small farmers in developing countries. Copies are available from the Iowa State University Press, South State Avenue, Ames, Iowa 50010, USA.

637 The Effects of Modern Technology on Labor Needed for Producing Crops on Small Farms in Two Integrated Rural Development Districts in Colombia, by Sergio Sepulveda Silva, and Howard E. Conklin, is a mimeograph prepared under auspices of Cornell University's Department of Agricultural Economics. Copies are available from the Program in International Agriculture, 252 Roberts Hall, Cornell University, Ithaca, New York 14853, USA.

638 An Inquiry into the Uses of Instructional Technology, published by the Ford Foundation, reports on how new technologies of teaching and learning are being used in several countries. These technologies include

television, videotape, film, radio, programmed instruction, computers, and new kinds of books. The report explores the current and potential uses for technology in schools and universities, and describes how the U.S. Agency for International Development, the World Bank, Unesco, the United Nations Development Programme, and the Ford Foundation are assisting projects throughout the world. The report is available from the Ford Foundation, P.O. Box 1919, New York, New York 10001, USA.

639 Planning for Technological Development--The Role of International Cooperation, by Gangadhar S. Gouri, is a publication of the Vienna Institute for Development. Other recent Institute papers are: Indigenous Development--Science and Technology, by T. Nayudamma; and The Contribution of Science and Technology to Development, by Louis Sabourin. The studies are available from the Vienna Institute for Development, Karntnerstrasse 25, A-1010 Vienna, Austria.

640 Rural Women: Their Integration in Development Programmes and How Simple Intermediate Technologies can Help Them, by Elizabeth O'Kelly, discusses women in Asian and African societies and the particular technologies that are suitable for them. Copies of the booklet are available from the author, 3 Cumberland Gardens, Lloyd Square, London WC1X 9AF, England.

641 Science, Technology, and the Global Equity Crisis: New Directions for United States Policy, by Ward Marehouse, is a publication of The Stanley Foundation. Copies may be obtained from The Stanley Foundation, Stanley Building, Muscatine, Iowa 52761, USA.

642 Science, Technology and the Social Order, edited by Ward Morehouse, contains seven papers on the impact of modern technology on social and political institutions. The authors are well known scholars in the field. Copies of the book are available from The Institute for World Order, 7777 U.N. Plaza, New York, New York 10017, USA.

643 Technical Co-operation under The Colombo Plan (July 1, 1976 to June 30, 1977) is the latest report of the Colombo Plan Council for Technical Co-operation in South and South-East Asia. Copies may be obtained from the Colombo Plan Bureau, Colombo, Sri Lanka.

644 Technologies from Developing Countries, prepared by the United Nations Industrial Development Organization, contains information on 138 technologies from developing countries. Each technology is described along with its distinctive features. Copies are available from the Industrial Information Section, UNIDO, P.O. Box 707, A-1011, Vienna, Austria.

IV. BIBLIOGRAPHIC SUBJECT INDEX

PART IV

DIRECTORY OF INFORMATION SOURCES

I. UNITED NATIONS INFORMATION SOURCES

AUDIO MATERIALS LIBRARY
United Nations, Department of Public Information, Radio and Visual Services Division, United Nations Plaza, New York, NY 10017.

COMPREHENSIVE INFORMATION SYSTEM ON TRANSNATIONAL CORPORATIONS
United Nations, Information Analysis Division, Centre on Transnational Corporations, 605 Third Avenue, New York, NY 10017.

BIBLIOGRAPHY ON TRANSNATIONAL CORPORATIONS
United Nations, Centre on Transnational Corporations, Information Analysis Division, 605 Third Avenue, New York, NY 10017.

SURVEY OF RESEARCH IN TRANSNATIONAL CORPORATIONS
United Nations, Centre on Transnational Corporations, Information Analysis Division, 605 Third Avenue, New York, NY 10017.

DAG HAMMARSKJOLD LIBRARY
United Nations, Department of Conference Services, United Nations Plaza, New York, NY 10017.

UNITED NATIONS BIBLIOGRAPHIC INFORMATION SYSTEM
United Nations, Department of Conference Services, Dag Hammarskjold Library, United Nations Plaza, New York, NY 10017.

UNBIS DATA BASE
United Nations, Department of Conference Services, Dag Hammarskjold Library, United Nations Plaza, New York, NY 10017.

DEVELOPMENT INFORMATION SYSTEM
United Nations, Department of International Economic and Social Affairs, Information Systems Unit, Room DC 594, New York, NY 10017.

DEVELOPMENT INFORMATION SYSTEM DATA BASE
United Nations, Department of International Economic and Social Affairs, Information Systems, Unit, Room DC 594, NY, NY 10017.

REFERENCE UNIT OF THE OFFICE FOR DEVELOPMENT RESEARCH AND POLICY ANALYSIS
United Nations, Department of International Economic and Social Affairs, Office for Development Research and Policy Analysis, New York, NY 10017.

REFERENCE UNIT OF THE OFFICE FOR SCIENCE AND TECHNOLOGY
United Nations, Department of International Economic and Social Affairs, Office for Science and Technology, New York, NY 10017.

UNITED NATIONS PHOTO LIBRARY
United Nations, Department of Public Information, Radio and Visual Services Division, United Nations Plaza, New York, NY 10017.

UNITED NATIONS VISUAL MATERIALS LIBRARY
United Nations, Department of Public Information, Radio and Visual Services Division, United Nations Plaza, New York, NY 10017.

UNCRD LIBRARY AND DOCUMENTATION SERVICES
United Nations Centre for Regional Development, Marunouchi 2-4-7, Naka-ku, Nagoya 460 Japan.

UNITED NATIONS LIBRARY AT GENEVA
United Nations Office at Geneva, 8-14, avenue de la Paix, Palais des Nations, 1211 Geneva 10, Switzerland.

UNRISD LIBRARY AND DOCUMENTATION UNIT
United Nations Research Institute for Social Development, Palais des Nations, CH-1211 Geneva 10, 16 Avenue Jean Trembley, 1209 Geneva, Switzerland.

CAFRAD LIBRARY AND DOCUMENTATION CENTRE
United Nations, African Training and Research Centre in Administration for Development (CAFRAD), BP 310, Tangier, Morocco.

ESCAP LIBRARY
Economic and Social Commission for Asia and the Pacific, United Nations Building, Rajadamnern Avenue, Bangkok 2, Thailand.

ESCAP DOCUMENTATION INFORMATION SYSTEM
Economic and Social Commission for Asia and the Pacific, Library, United Nations Building, Rajadamnern Avenue, Bangkok 2, Thailand.

ESCAP LIBRARY SERIALS INFORMATION SYSTEM
Economic and Social Commission for Asia and the Pacific, Library, United Nations Building, Rajadamnern Avenue, Bangkok 2, Thailand.

ESCAP LIBRARY SERIALS DATA BASE
Economic and Social Commission for Asia and the Pacific, ESCAP Library, United Nations Building, Rajadamnern Avenue, Bangkok 2, Thailand.

NETWORK FOR TECHNOLOGICAL INFORMATION ON AGRO-INDUSTRIES
Economic and Social Commission for Asia and the Pacific,

ESCAP/UNIDO Division of Industry, Housing and Technology, United Nations Building, Rajadamnern Avenue, Bangkok 2, Thailand.

CARIBBEAN DOCUMENTATION CENTRE
Economic Commission for Latin America, CEPAL Office for the Caribbean, P.O. Box 1113, Room 300, Salvatori Building, Port of Spain, Trinidad and Tobago.

LATIN AMERICAN CENTRE FOR ECONOMIC AND SOCIAL DOCUMENTATION
Economic Commission for Latin America, Casilla 179-D, Edificio Naciones Unidas, Avenida Dag Hammarskjold, Santiago, Chile.

CLADBIB DATA BASE
Economic Commission for Latin America, Latin American Centre for Economic and Social Documentation, Casilla 179-D, Edificio Naciones Unidas, Avenida Dag Hammarskjold, Santiago, Chile.

CLADIR DATA BASE
Economic Commission for Latin America, Latin American Centre for Economic and Social Documentation, Casilla 179-D, Edificio Naciones Unidas, Avenida Dag Hammarskjold, Santiago, Chile.

CLAPLAN DATA BASE
Economic Commission for Latin America, Latin American Centre for Economic and Social Documentation, Casilla 179-D, Edificio Naciones Unidas, Avenida Dag Hammarskjold, Santiago, Chile.

CENTRAL AMERICAN CENTRE FOR ECONOMIC AND SOCIAL DOCUMENTATION
Economic Commission for Latin America/Central American Higher University Council, Apartado 37, Universidad de Costa Rica, San Jose, Costa Rica.

INFORMATION SYSTEM OF THE HUMAN SETTLEMENTS TECHNOLOGY PROGRAMME
Economic Commission for Latin America, Human Settlements Section, Apartado 6-718, President Masarik 29, Mexico 5, D.F. Mexico.

JOINT CEPAL/ILPES LIBRARY BIBLIOGRAPHICAL INFORMATION SYSTEM
Economic Commission for Latin America, Latin American Institute for Economic and Social Planning, Casilla 179-D, Edificio Naciones Unidas, Avenida Dag Hammarskjold, Santiago, Chile.

ECA LIBRARY
Economic Commission for Africa, P.O. Box 3001, Africa Hall, Addis Ababa, Ethiopia.

ECWA LIBRARY
Economic Commission for Western Asia Administration, P.O. Box 4656, United Nations Building, Bir Hassan, Beirut, Lebanon.

INDUSTRIAL INFORMATION SYSTEM
United Nations Industrial Development Organization, Industrial Information Section, P.O. Box 300, Vienna International Centre, Wagramer Strasse 5, A-1400 Vienna, Austria

INDUSTRIAL INFORMATION SYSTEM DATA BASE--INDUSTRIAL DEVELOPMENT ABSTRACTS
United Nations Industrial Development Organization, Industrial Information Section, P.O. Box 300, Vienna International Centre, Wagramer Strasse 5, A-1400 Vienna, Austria.

INDUSTRIAL AND TECHNOLOGICAL INFORMATION BANK
United Nations Industrial Development Organization, Industrial Information Section, P.O. Box 300, Vienna International Centre, Wagramer Strasse 5, A-1400 Vienna, Austria.

TECHNOLOGICAL INFORMATION EXCHANGE SYSTEM
United Nations Industrial Development Organization, Technology Group, P.O. Box 300, Vienna International Centre, Wagramer Strasse 5, A-1400 Vienna, Austria.

INFORMATION SERVICE OF THE INDUSTRY AND ENVIRONMENT GROUP
United Nations Environment Programme, Industry and Environment Office, 17, rue Margueritte, 75017 Paris, France.

INDUSTRY AND ENVIRONMENT DATA BASE
United Nations Environment Programme, Industry and Environment Office, 17, rue Margueritte, 75017 Paris, France.

UNEP LIBRARY AND DOCUMENTATION CENTRE
United Nations Environment Programme, P.O. Box 30552, Nairobi, Kenya.

PROJECT INSTITUTIONAL MEMORY
United Nations Development Programme, Bureau for Programme Policy and Evaluation, One United Nations Plaza, New York, NY 10017.

PROJECT INSTITUTIONAL MEMORY DATA BASE
United Nations Development Programme, Bureau for Programme Policy and Evaluation, One United Nations Plaza, New York, NY 10017.

UNITAR LIBRARY
United Nations Institute for Training and Research, 801 United Nations Plaza, New York, NY 10017.

ABSTRACTS OF SELECTED SOLAR ENERGY TECHNOLOGY
United Nations University, Natural Resources Programme, 29th Floor, Toho Seimei Building, 15-1 Shibuya 2-chome, Shibuya-ku, Tokyo 150, Japan.

CINTEFOR DOCUMENTATION AND INFORMATION SERVICE
International Labour Office, Inter-American Centre for Research and Documentation on Vocational Training, Casilla de Correo 1761, San Jose 1092, Montevideo, Uruguay.

DEVELOPMENT EDUCATION EXCHANGE SERVICE
Food and Agriculture Organization of the United Nations, Development Department, Via delle Terme di Caracalla, 00100 Rome, Italy.

UNESCO DATA BASE
United Nations Educational, Scientific and Cultural Organization, Division of Library, Archives and Documentation Services, 7, Place de Fontenoy, 75700 Paris, France.

DARE DATA BASE
United Nations Educational, Scientific and Cultural Organization, Social Science Documentation Centre (SSDC), 7 Place de Fontenoy, 75700 Paris, France.

INTERNATIONAL DATA BASE ON RESEARCH PROJECTS, STUDIES AND COURSES IN SCIENCE AND TECHNOLOGY POLICY
United Nations Educational, Scientific and Cultural Organization, Division of Science and Technology Policies, 7, Place de Fontenoy, 75700 Paris, France.

UNESCO STATISTICAL DATA BANK SYSTEM
United Nations Educational, Scientific and Cultural Organization, Office of Statistics, 7, Place de Fontenoy, 75700 Paris, France.

LIBRARY OF THE UNESCO REGIONAL OFFICE FOR SCIENCE AND TECHNOLOGY FOR AFRICA
United Nations Educational, Scientific and Cultural Organization, Regional Office for Science and Technology for Africa, P.O. Box 30592, Bruce House, Standard Street, Nairobi, Kenya.

REGIONAL DATABANK ON TECHNOLOGIES IN AFRICA
United Nations Educational, Scientific and Cultural Organization/ National Science and Technology Development Agency, Publication and Information Division, PMB 12695, 8 Strachan Street, Lagos, Nigeria.

APPROPRIATE TECHNOLOGY FOR HEALTH INFORMATION SYSTEM
World Health Organization, Division of Strengthening of Health Services, ATH Unit, 20 avenue Appia, 1211 Geneva 27, Switzerland.

HEALTH LITERATURE SERVICES OF THE WHO REGIONAL OFFICE FOR AFRICA
World Health Organization, Regional Office for Africa, P.O. Box 6, Brazzaville, Congo.

DOCUMENTATION REFERRAL SERVICE
World Bank, Records Management Division, Document Acquisition and Control, 1818 H Street, N.W., Washington, D.C. 20433.

INTERACTIVE BIBLIOGRAPHIC INDEXING SYSTEM
Records Management Division, Document Acquisition and Control, 1818 H Street, N.W., Washington, D.C. 20433.

JOINT BANK-FUND LIBRARY
International Monetary Fund, 700 19th Street, N.W., Washington, D.C. 20431.

II. BIBLIOGRAPHY OF BIBLIOGRAPHIES

ANNOTATED BIBLIOGRAPHY OF COUNTRY SERIALS is a listing of periodicals, annuals and other serials containing information of economic, business or trade interest. The listing is organized on a regional and country basis. Copies are available from the Documentation Service, International Trade Centre UNCTAD/GATT, 1211 Geneva 10, Switzerland.

BASIC-NEEDS APPROACH: A SURVEY OF ITS LITERATURE, edited by M. Rutjes, contains a brief analysis of the concept of basic needs, its targets, its strategy and implications, followed by a concise bibliography related to the topic. Copies may be obtained from the Centre for the Study of Education in Developing Countries, Badhuisweg 251, The Hague, The Netherlands.

DEVELOPMENT PLANS AND PLANNING - BIBLIOGRAPHIC AND COMPUTER AIDS TO RESEARCH, by August Schumacher, is arranged in three parts. The first contains more than 100 selected bibliographies on development plans and planning, the second is concerned with a new source of empirical materials for the development planner - the automated documentation centre, and the third analyzes recent work on computer aids for the research library. The publication is available from Seminar Press Ltd., 24-28 Oval Road, London NW1, England.

BIBLIOGRAPHY ON DEVELOPMENT EDUCATION lists books, manuals, resource materials, magazines, and articles in the field of development education. The listing was prepared by the Dutch Central Bureau of Catholic Education. Copies are available from the Central Bureau of Catholic Education, G. Verstijnen, Secretary Foreign Department, Bezuidenhoutseweg 275, The Hague, Netherlands.

BIBLIOGRAPHY OF GERMAN RESEARCH ON DEVELOPING COUNTRIES, prepared by the German Foundation for International Development, is divided into two sections: Part A contains an index of research institutes, author index, subject-matter index, and a geographical index. Part B contains specific information on each of the studies listed. The text is in German with explanatory notes in German, English, French and Spanish. Copies may be obtained from the Deutsche Stiftung fur Internationale Entwicklung (DSE), Endenicher Strasse 41, 53 Bonn, Federal Republic of Germany.

BIBLIOGRAPHY OF SELECTED LATIN AMERICAN PUBLICATIONS ON DEVELOPMENT is a listing of over 200 titles in Latin American development literature, including subject and author indexes. The document was prepared by the Institute of Development Studies Library. Copies are available from the Librarian, Institute of Development Studies, University of Sussex, Brighton BN1 9RE, England.

CANADIAN DEVELOPMENT ASSISTANCE: A SELECTED BIBLIOGRAPHY 1950-70, compiled by Shirley B. Seward and Helen Janssen, covers Canada's foreign aid programs and policies from 1950 to 1970. Copies are available from the Distribution Unit, International Development Research Centre, P.O. Box 8500, Ottawa, Canada KIG 3H9.

DEVINDEX CANADA is a bibliography of literature on social and economic development in Third World countries, which originated in Canada in 1975. Copies may be obtained from the International Development Research Centre, Box 8500, Ottawa, Canada KIG 3H9.

The UNESCO Division of Scientific Research and Higher Education has compiled **A DIRECTORY AND BIBLIOGRAPHY ON THE THEME "RESEARCH AND HUMAN NEEDS"**, listing organizations, journals, newsletters, reports and papers, information services and data banks. The bibliographical section includes headings such as food and nutrition, health, housing and sanitation, environment, energy, technology. For copies contact "Research and Human Needs", Division of Scientific Research and Higher Education, UNESCO, Place de Fontenoy, 75007 Paris, France.

GUIDE TO CURRENT DEVELOPMENT LITERATURE ON ASIA AND THE PACIFIC is published every two months by the Library and Documentation Centre of the Asia Pacific Development Information Service. For more information write to the Centre, United Nations Asian and Pacific Development Institute, P.O. Box 2-136, Sri Aydudhya Road, Bangkok, Thailand.

Hald, Marjorie W.
A SELECTED BIBLIOGRAPHY ON ECONOMIC DEVELOPMENT AND FOREIGN AID, rev. ed., Santa Monica, CA: The Rand Corporation, 1958.

Hazelwood, Arthur
THE ECONOMICS OF "UNDERDEVELOPED" AREAS: AN ANNOTATED READING LIST OF BOOKS, ARTICLES, AND OFFICIAL PUBLICATIONS. London: Oxford University Press for the Institute of Colonial Studies, 1954. 623 titles.

THE ECONOMICS OF DEVELOPMENT: AN ANNOTATED LIST OF BOOKS AND ARTICLES PUBLISHED 1958-1962. London: Oxford University Press, for the Institute of Commonwealth Studies, 1964.

INTERNATIONAL BIBLIOGRAPHY, INFORMATION DOCUMENTATION (IBID) provides bibliographic details and annotations necessary to identify the full range of publications prepared by the United Nations and its related agencies, plus those of ten organizations outside the UN system. IBID is published quarterly by Unipub. Available from Unipub, Box 433, Murray Hill Station, New York, New York 10016, USA.

THE 1978/79 PUBLICATIONS LIST OF THIRD WORLD PUBLICATIONS contains over 300 titles of pamphlets, books and teaching materials about the Third World. The listing is available from Third World Publications, Ltd., 151 Stratford Road, Birmingham B11 1RD, England.

A list of 200 books on **NORTH-SOUTH WORLD RELATIONS** has been compiled by the Developing Country Courier. The listing is organized by subject and region. For copies write to the Courier, P.O. Box 239, McLean, Virginia 22101, USA.

United States Agency for International Development
A PRACTICAL BIBLIOGRAPHY FOR DEVELOPING AREAS. Washington, D.C., 1966. 2 vols. (Vol. 1 - A selective, annotated and graded list of United States publications in the social sciences. 202 pp.) (Vol. 2 - A selective, annotated and graded list of United States publications in the physical and applied sciences. 332 pp.)

PUBLIC ADMINISTRATION--A SELECT BIBLIOGRAPHY, prepared by the British Ministry of Overseas Development Library is the second supplement to the 1973 revised edition. Copies may be obtained from Eland House, Stag Place, London SW1E 5DH, England.

PUBLIC ADMINISTRATION--A SELECT BIBLIOGRAPHY, prepared by the Library of the British Ministry of Overseas Development, is a supplement to the revised edition which appeared in 1973. It includes material published in the period 1972-1975 with 1,600 references. Copies may be obtained from the Library, British Ministry of Overseas Development, Eland House, Stag Place, London SW1E 5DH, England.

The OECD Development Centre has gathered together in the catalog **PUBLICATION AND DOCUMENT, 1962-1979** all the books and documents it has published since its establishment in 1962 up to August 1979. Copies available from OECD Development Centre, 94 rue Chardon Lagache, 75016 Paris, France.

REGISTER OF RESEARCH PROJECTS IN PROGRESS IN DEVELOPMENT STUDIES IN SELECTED EUROPEAN COUNTRIES was prepared by the Centre for Development Studies of the University of Antwerp at the request of the European Association of Development Research and Training Institutes. Copies are available from the Centre, St. Ignatius

Faculties, University of Antwerp, 13 Prinsstraat, 2000 Antwerp, Belgium.

Re Qua, Eloise and Statham, Jane
THE DEVELOPING NATIONS: A GUIDE TO INFORMATION SOURCES CONCERNING THEIR ECONOMIC, POLITICAL, TECHNICAL AND SOCIAL PROBLEMS. Detroit: Gale Research Company, 1965.

The East African Academy has published two new bibliographies. **SCIENCE AND TECHNOLOGY IN EAST AFRICA** contains more than 5,000 titles about research in the agriculture, medical technological, and related fields in East Africa, with short summaries on the problems and progress of research in these areas. **TANZANIA EDUCATION SINCE UHURU: A BIBLIOGRAPHY--1961-1971** was compiled by Dr. George A. Auger of the University of Dar es Salaam. Both publications are available from the East African Academy, RIPS, P.O. Box 47288, Nairobi, Kenya.

SELECTIVE ANNOTATED BIBLIOGRAPHY ON BRAZILIAN DEVELOPMENT has been prepared by the SID Sao Paulo Chapter. This first issue contains only references that have appeared in 1975. Copies are available from the Society for International Development, Sao Paulo Chapter, Caixa Postal 20.270-Vila Clementino, 04023-Sao Paulo-S.P. Brazil.

A SELECTED ANNOTATED BIBLIOGRAPHY: INDIGENOUS TECHNICAL KNOWLEDGE IN DEVELOPMENT, compiled by Liz O'Keefe and Michael Howes, is contained in the January 1979 IDS BULLETIN. This issue of the BULLETIN is devoted to the importance of indigenous technical knowledge in rural areas. Single copies of the BULLETIN are from the Communications Office, Institute of Development Studies, University of Sussex, Brighton N1 9RE, United Kingdom.

SELECTED BIBLIOGRAPHY OF RECENT ECONOMIC DEVELOPMENT PUBLICATIONS covers a period of one year, from July 1977 to June 1978 and contains two main sections, one for general and theoretical works, the other for literature related to regions and countries. For copies write to the Graduate Program in Economic Development, Vanderbilt University, Nashville, Tennessee 37235, USA.

International Bank for Reconstruction and Development; Economic Development Institute
SELECTED READINGS AND SOURCE MATERIALS ON ECONOMIC DEVELOPMENT. A list of books, articles, and reports included in a small library assembled by the Economic Development Institute, Washington, D.C., 1961.

SOCIAL AND ECONOMIC DEVELOPMENT PLANS - MICROFICHE PROJECT is a cumulative catalogue listing the holdings of Inter Documentation Company AG on social and economic development plans around the world. About 1400 plans from over 180 countries are included. Copies of the catalogue and other catalogues of IDC's microfiche projects are free on request from Inter Documentation Company AG, Poststrasse 14, 6300 Zug-Switzerland.

Powelson, John
A SELECT BIBLIOGRAPHY ON ECONOMIC DEVELOPMENT. Boulder, Colorado: Westview Press, 1979.

THIRD WORLD BIBLIOGRAPHY AND RESOURCE GUIDE features a wide range of material on Third World issues. It is designed for students and general readers. Copies may be obtained from the Development Education Library Project, c/o OSFAM/Ontario, 175 Carlton Street, Toronto, Canada.

The United Nations Asian and Pacific Development Institute has prepared a **SPECIAL BIBLIOGRAPHY ON ALTERNATIVE STRATEGIES FOR DEVELOPMENT WITH FOCUS ON LOCAL LEVEL PLANNING AND DEVELOPMENT** in connection with a UNAPDI meeting, held in Bangkok, October 31 - November 4, 1978. Copies are available from the APDI Library and Documentation Centre. UNAPDI, P.O. Box 2-136, Sri Ayudhya Road, Bangkok, Thailand.

Vente, Role and Dieter Seul
MACRO-ECONOMIC PLANNING: A BIBLIOGRAPHY. Nomos Verlagsgesellshaft, Baden-Baden, 1970.

Volunteers in Technical Assistance (VITA) has published its 1979 **CATALOGUE OF BOOKS, BULLETINS AND MANUALS.** The listing contains VITA documents related to appropriate technology, as well as materials published by other development organizations around the world. Copies are available from VITA, 2706 Rhode Island Avenue, Mt. Ranier, Maryland 20822, USA.

Agajanian, A. H.
MOSFET TECHNOLOGY: A COMPREHENSIVE BIBLIOGRAPHY. Plenum Publishers, 1980.

Alsmeyer, D. and Atkins, A. G.
GUIDE TO SCIENCE AND TECHNOLOGY IN THE ASIA PACIFIC AREA. Longman.

Baranson, Jack
TECHNOLOGY FOR UNDERDEVELOPED AREAS: AN ANNOTATED BIBLIOGRAPHY. New York: Pergamon Press, 1967.

BIBLIOGRAPHIC GUIDE TO TECHNOLOGY: 1976. Bibliographic Guides; G. K. Hall, 1977.

BIBLIOGRAPHIC GUIDE TO TECHNOLOGY: 1979. Bibliographic Guides; G. K. Hall, 1980.

BIBLIOGRAPHIC GUIDE TO TECHNOLOGY: 1980. Bibliographic Guides; G. K. Hall, 1981.

BIBLIOGRAPHIC GUIDE TO TECHNOLOGY: 1981. Bibliographic Guides; G. K. Hall, 1982.

FIVE-YEAR INDEX TO ASTM TECHNICAL PAPERS: 1966-70. ASTM, 1971.

GUIDE TO INFORMATION SOURCES ON ENVIRONMENTAL SCIENCE AND TECHNOLOGY contains an annotated listing of more than 470 books, 280 periodicals, 130 reference works, over 100 individual bibliographies and nearly 100 current secondary literature sources, all dealing with environmental science and technology. The Guide was compiled by the Scientific Staff of the Center for Industrial Information. Copies may be obtained from the Flemish Economic Association, Tavernierkaai 4/4, B-2000 Antwerp, Belgium.

Houghton, Bernard
TECHNICAL INFORMATION SOURCES: A GUIDE TO PATENT SPECIFICATIONS, STANDARDS AND TECHNICAL REPORTS LITERATURE. Shoe String, 1972.

Malinowsky, H. Robert and Richardson, Jeanne M.
SCIENCE AND ENGINEERING LITERATURE: A GUIDE TO REFERENCE SOURCES. Libraries Unlimited, 1980.

McGraw-Hill Editors
BASIC BIBLIOGRAPHY OF SCIENCE AND TECHNOLOGY. McGraw-Hill, 1966.

New York Public Library Research Library
BIBLIOGRAPHIC GUIDE TO TECHNOLOGY. Bibliographic Guides, 1979.

Ocran, Emanuel B.
SCIENTIFIC AND TECHNICAL SERIES: A SELECT BIBLIOGRAPHY. Scarecrow, 1973.

Research Libraries of the New York Public Library and Library of Congress
BIBLIOGRAPHIC GUIDE TO TECHNOLOGY: 1979. Bibliographic Guides, 1980.

Research Libraries of the New York Public Library and Library of Congress
BIBLIOGRAPHIC GUIDE TO TECHNOLOGY: 1980. Bibliographic Guides, 1981.

Swanson, Gerald
TECHNOLOGY BOOK GUIDE, 1974. Bibliographic Guides, 1974.

Taube, Utz Friedebert
BIBLIOGRAPHY OF PERIODICALS LITERATURE ON THE STATE OF THE ART FOR THE AREAS OF TECHNOLOGY. K. G. Saur, 1978.

AN ANNOTATED BIBLIOGRAPHY ON THE RELATIONSHIP BETWEEN THE TECHNOLOGICAL CHANGE AND EDUCATIONAL DEVELOPMENT. By Tom Whiston et al. IIEP, 1980. This annotated bibliography is concerned with the reciprocal relationships and interactions between technology and education. It includes publications which address themselves both to education and technological change or technological factors in the same paper. Contents cover effects of technology on education; society, technology and education; technological change, training and retraining; technological change and specific industries and professions; planning and development; geographical, regional and development issues; educational technology; and publications of international organizations.

APPROPRIATE TECHNOLOGY: A BIBLIOGRAPHY. Washington, D.C.: Department of Energy, 1981.

Carr, Marilyn
ECONOMICALLY APPROPRIATE TECHNOLOGIES FOR DEVELOPING COUNTRIES: AN ANNOTATED BIBLIOGRAPHY. London: Intermediate Technology Publications, 1976. Intended use is by individuals and groups concerned with intermediate technology and the choice of appropriate technologies. The references are listed by subject categories under such general headings as: agriculture, low-cost housing and building materials, and manufacturing and are weighted toward application rather than theory. Entries are indexed by author, country and subject.

French, David
APPROPRIATE TECHNOLOGY IN SOCIAL CONTEXT: AN ANNOTATED BIBLIOGRAPHY. Mt. Ranier, MD: VITA, 1977. The author maintains that appropriate technology can be successful if attention is paid to locally based development institutions and to specific innovation and its effect on people in a given environment. Also considered relevant is the need for local participation in decision making.
Morehouse, Cynthia
SCIENCE AND TECHNOLOGY FOR DEVELOPMENT: INTERNATIONAL CONFLICT AND COOPERATION: A BIBLIOGRAPHY. Lund, Sweden: Research Policy Program, University of Lund, 1978.

DEVELOPMENT--A BIBLIOGRAPHY, was compiled by Vaptistis-Titos Patrikios (Rome: FAO, 1974) and updates the first edition, published in 1970, to cover the 1970/73 period. Contains eight sections relating to development: theories and problems; perspectives of the Third World countries; population and food production; aid, trade and international cooperation; agriculture; manpower and employment; education; and environment. A ninth section lists bibliographies.

III. DIRECTORY OF PERIODICALS

ACTUEL DEVELOPPEMENT, English Digest Edition, Paris.

AFRICA, London, Africa Journal, Ltd.

AFRICA INSTITUTE, Pretoria, Africa Institute.

AFRICA QUARTERLY, New Delhi, India Council for Africa.

AFRICA RESEARCH BULLETIN, Exeter, Eng. Africa Research, Ltd.

AFRICA, SOUTH OF THE SAHARA, London, Europa Publications.

AFRICA TODAY, New York, American Committee on Africa.

AFRICAN AFFAIRS, London, Journal of the Royal African Society.

AFRICAN DEVELOPMENT, London.

AFRICAN DEVELOPMENT BANK, Annual Report, Ibadan.

AFRICAN ENVIRONMENT, Dakar, United Nations Environmental Program.

AFRICAN STATISTICAL YEARBOOK, Addis Ababa, Economic Commission for Africa.

AFRICAN STUDIES REVIEW, Stanford, Boston, East Lansing, African Studies Association.

AFRICAN URBAN STUDIES, East Lansing, Mich., African Studies Center.

AGENDA, Washington, D.C, U.S. Agency for International Development.

APPROPRIATE TECHNOLOGY, London, Intermediate Technology Publications, Ltd.

APPROTECH, Ann Arbor, Mich., International Association for the Advancement of Appropriate Technology for Developing Countries.

ARTHA VIJNANA, Poona, Gokhale Institute of Politics and Economics.

ASIA AND THE WORLD MONOGRAPHS, Taipei, Asia and the World Forum.

ASIA YEARBOOK, Hong Kong, Far Eastern Economic Review.

ASIAN AFFAIRS, London, Royal Central Asian Society.

ASIAN DEVELOPMENT BANK, Annual Report, Manila.

ASIAN REGIONAL CONFERENCE OF THE INTERNATIONAL LABOR ORGANIZATION, Proceedings, Geneva, ILO.

ASIAN SURVEY, Berkeley, Institute of International Studies.

BANGLADESH DEVELOPMENT STUDIES, Dhaka, Bangladesh Institute of Development Studies.

BANGLADESH ECONOMIC REVIEW, Dhaka, Bangladesh Institute of Development Economics.

BULLETIN OF INDONESIAN ECONOMIC STUDIES, Canberra, Dept. of Economics, Australian National University.

CEPAL REVIEW, Santiago, Chile.

CANADIAN JOURNAL OF AFRICAN STUDIES, Montreal, Loyola College.

COMMUNITY DEVELOPMENT JOURNAL, Manchester, U.K., Oxford University Press.

DEVELOPING ECONOMIES, Tokyo, The Institute of Asian Economic Affairs.

DEVELOPMENT, Rome, Society for International Development.

DEVELOPMENT CENTER STUDIES, OECD, Paris.

DEVELOPMENT AND CHANGE, Beverly Hills, Calif.: Sage Publications.

DEVELOPMENT CO-OPERATION, Paris, OECD.

DEVELOPMENT DIGEST, Washington, D.C., U.S. Agency for International Development.

DEVELOPMENT DIALOGUE, Uppsala, Sweden, Dag Hammarskjold Foundation.

EASTERN AFRICA ECONOMIC REVIEW, Nairobi, Oxford University Press.

ECONOMIC DEVELOPMENT AND CULTURAL CHANGE, Chicago, University of Chicago Press.

ETHIOPIAN JOURNAL OF DEVELOPMENT RESEARCH, Addis Ababa, Institute of Development Research.

FAR EASTERN ECONOMIC REVIEW, Hong Kong.

FINANCE AND DEVELOPMENT, Washington, D.C.

IDS BULLETIN, Institute of Development Studies, University of Sussex, U.K.

IMPACT OF SCIENCE ON SOCIETY, Paris, UNESCO.

INDIAN JOURNAL OF INDUSTRIAL RELATIONS, New Delhi, India.

INDUSTRY AND DEVELOPMENT, Vienna, UNIDO.

INTERNATIONAL DEVELOPMENT REVIEW, Rome, Society for International Development.

INTERNATIONAL LABOR REVIEW, Geneva, ILO.

INTERNATIONAL STUDIES QUARTERLY, San Francisco.

JOURNAL OF AFRICAN STUDIES, Los Angeles, UCLA African Studies Center.

JOURNAL OF DEVELOPING AREAS, Macomb, IL, Western Illinois Univ.

JOURNAL OF DEVELOPMENT ECONOMICS, Amsterdam, North Holland Publishing Co.

JOURNAL OF DEVELOPMENT STUDIES, London, U.K.

JOURNAL OF ECONOMIC DEVELOPMENT, JOURNAL OF INTERNATIONAL AFFAIRS, New York, Columbia University.

JOURNAL OF MODERN AFRICAN STUDIES, New York, Cambridge University Press.

LATIN AMERICAN RESEARCH REVIEW, Chapel Hill, North Carolina.

MODERN ASIAN STUDIES, New York, Cambridge University Press.

MONOGRAPH, DEVELOPMENT STUDIES CENTER, AUSTRALIAN NATIONAL UNIVERSITY.

MONOGRAPH, OVERSEAS DEVELOPMENT COUNCIL, Washington, D.C.

ODI REVIEW, Overseas Development Institute, London, U.K.

OXFORD ECONOMIC PAPERS, Oxford, U.K.

PAKISTAN DEVELOPMENT REVIEW, Karachi, Pakistan.

PUBLIC ADMINISTRATION AND DEVELOPMENT, Sussex, U.K., Royal Institute of Public Administration.

THIRD WORLD QUARTERLY, London, Third World Foundation for Social and Economic Studies.

WORLD BANK STAFF WORKING PAPER, IBRD, Washington, D.C.

WORLD DEVELOPMENT, Pergamon Press, N.Y.

NOTE:

For more information on relevant periodicals please consult:

1. **DIRECTORY OF UNITED NATIONS INFORMATION SYSTEMS**

2. **REGISTER OF UNITED NATIONS SERIAL PUBLICATIONS**

Published by **Inter-Organization Board for Information Systems,** IOB Secretariat, Palais des Nations, CH-1211 Geneva 10, Switzerland.

IV. RESEARCH INSTITUTIONS

INTERNATIONAL (GENERAL)

AFRICAN INSTITUTE FOR ECONOMIC DEVELOPMENT AND PLANNING
United Nations Economic Commission for Africa, Dakar, Senegal.

AFRO-ASIAN ORGANIZATION FOR ECONOMIC CO-OPERATION
Cairo Chamber of Commerce Building, Midan el-Falsky, Cairo, Egypt.

ASIAN ASSOCIATION OF DEVELOPMENT RESEARCH AND TRAINING INSTITUTES
P.O. Box 2-136, Sri Ayudhya Road, Bangkok, Thailand.

ASIAN DEVELOPMENT CENTER
11th Floor, Philippine Banking Corporation Building, Anda Circle, Port Area, Manila, Philippines.

ASIAN INSTITUTE FOR ECONOMIC DEVELOPMENT AND PLANNING
P.O. Box 2-136, Sri Ayudhya Road, Bangkok, Thailand.

ATLANTIC INSTITUTE FOR INTERNATIONAL AFFAIRS
120, rue de Longchamp, 75016 Paris, France.

CARIBBEAN STUDIES ASSOCIATION
Inter-American University of Puerto Rico, P.O. Box 1293, Hato Rey, Puerto Rico 00919.

CENTRE FOR STUDIES AND RESEARCH IN INTERNATIONAL LAW AND INTERNATIONAL RELATIONS
The Hague Academy of International Law, The Hague, Netherlands.

CENTRE FOR THE CO-ORDINATION OF SOCIAL SCIENCE RESEARCH AND DOCUMENTATION IN AFRICA SOUTH OF THE SAHARA
B.P. 836, Kinshasa XI, Zaire.

CLUB OF ROME
Via Giorgione 163, 00147 Roma, Italy.

COMMITTEE ON SOCIETY, DEVELOPMENT AND PEACE
Oecumenical Centre, 150, route de Ferney, 1211 Geneve 20, Suisse.

COUNCIL FOR ASIAN MANPOWER STUDIES
P.O. Box 127, Quezon City, Philippines.

COUNCIL FOR THE DEVELOPMENT OF ECONOMIC AND SOCIAL RESEARCH IN AFRICA
B.P. 3186, Dakar, Senegal.

EAST AFRICAN ACADEMY RESEARCH INFORMATION CENTRE
Regional Building of East African Community, Ngong Road, (rooms 359-60), Nairobi, Kenya.

EASTERN REGIONAL ORGANIZATION FOR PLANNING AND HOUSING
Central Office: 4a, Ring Road, Indraprastha Estate, New Delhi, India.

EASTERN REGIONAL ORGANIZATION FOR PUBLIC ADMINISTRATION
Rizal Hall, Padre Faura Street, Manila, Philippines.

ECONOMIC DEVELOPMENT INSTITUTE
1818 H Street, N.W., Washington, D.C. 20433, U.S.A.

EUROPEAN FOUNDATION FOR MANAGEMENT DEVELOPMENT
51, rue de la Concorde, Bruxelles, Belgique.

EUROPEAN INSTITUTE FOR TRANSNATIONAL STUDIES IN GROUP AND ORGANIZATIONAL DEVELOPMENT
Viktorgasse 9, 1040 Vienna, Austria.

EUROPEAN INSTITUTE OF BUSINESS ADMINISTRATION
Boulevard de Constance, 77 Fontainebleau, France.

EUROPEAN RESEARCH GROUP ON MANAGEMENT
Predikherenberg 55, 3200 Kessel-Lo, Belgique.

INSTITUTE OF INTERNATIONAL LAW
82, avenue de Castel, 1200 Bruxelles, Belgique.

INTERNATIONAL AFRICAN INSTITUTE
210, High Holborn, London WCIV 7BW, United Kingdom.

INTERNATIONAL ASSOCIATION FOR METROPOLITAN RESEARCH AND DEVELOPMENT
Suite 1200, 130 Bloor Street West, Toronto 5, Canada.

INTERNATIONAL CENTRE OF RESEARCH AND INFORMATION ON PUBLIC AND CO-OPERATIVE ECONOMY
45, quai de Rome, Liege, Belgique.

INTERNATIONAL CO-OPERATION FOR SOCIO-ECONOMIC DEVELOPMENT
59-61, rue Adolphe-Lacombie, Bruxelles 4, Belgique.

INTERNATIONAL INSTITUTE FOR LABOUR STUDIES
154, rue de Lausanne, Case Postale 6, 1211 Geneve, Suisse.

INTERNATIONAL INSTITUTE FOR STRATEGIC STUDIES
18, Adam Street, London WC2N 6AL, United Kingdom.

INTERNATIONAL INSTITUTE OF ADMINISTRATIVE SCIENCES
25, rue de la Charite, Bruxelles 4, Belgique.

INTERNATIONAL MANAGEMENT DEVELOPMENT INSTITUTE
4, Chemin de Conches, 1200 Geneve, Suisse.

INTERNATIONAL SCIENCE FOUNDATION
2, rue de Furstenberg, 75006 Paris, France.

INTERNATIONAL SOCIAL SCIENCE COUNCIL
1, rue Miollis, 75015 Paris, France.

INTERNATIONAL STATISTICAL INSTITUTE
Prinses Beatrixlaan 428, Voorburg, Netherlands.

INTERNATIONAL TRAINING AND RESEARCH CENTER FOR DEVELOPMENT
47, rue de la Glaciere, 75013 Paris, France.

LATIN AMERICAN CENTRE FOR ECONOMIC AND SOCIAL DOCUMENTATION
Casilla 179-D, Santiago, Chile.

ORGANIZATION FOR ECONOMIC CO-OPERATION AND DEVELOPMENT
Chateau de la Muette, 2, rue Andre Pascal, 75775 Paris Cedex 16, France.

REGIONAL ECONOMIC RESEARCH AND DOCUMENTATION CENTER
B.P. 7138, Lome, Togo.

RESEARCH CENTRE ON SOCIAL AND ECONOMIC DEVELOPMENT IN ASIA--INSTITUTE OF ECONOMIC GROWTH
University Enclave, Delhi 7, India.

SOCIETY FOR INTERNATIONAL DEVELOPMENT
1346 Connecticut Avenue, N.W., Washington, D.C. 20036, USA.

SOUTHEAST ASIAN SOCIAL SCIENCE ASSOCIATION
Chulalongkorn University, c/o Faculty of Political Science, Bangkok, Thailand.

UNITED NATIONS INSTITUTE FOR TRAINING AND RESEARCH
801 United Nations Plaza, New York, NY, USA.

UNITED NATIONS RESEARCH INSTITUTE FOR SOCIAL DEVELOPMENT
Palais des Nations, 1211 Geneve, Suisse.

AUSTRALIA

AUSTRALIAN INSTITUTE OF INTERNATIONAL AFFAIRS
P.O. Box E181, Canberra, ACT 2600.

INSTITUTE OF ADVANCED STUDIES
The Australian National University, P.O. Box 4, Canberra ACT 2600.

STRATEGIC AND DEFENSE STUDIES CENTER
Research School of Pacific Studies, Australian National University, P.O. Box 4, Canberra, ACT 2600.

AUSTRIA

AUSTRIAN FOUNDATION FOR DEVELOPMENT RESEARCH (OFSE)
Turkenstrasse 3, 1090 Vienna, Austria.

VIENNA INSTITUTE FOR DEVELOPMENT
Karntner Strasse 25, 1010 Vienna, Austria.

BANGLADESH

BANGLADESH INSTITUTE OF DEVELOPMENT STUDIES
Adamjee Court, Motijheel Commercial Area, Dacca 2.

BELGIUM

CATHOLIC UNIVERSITY OF LOUVAIN
Center for Economic Studies, Van Evenstraat 2b, 3000 Louvain, Belgium.

FREE UNIVERSITY OF BRUSSELS
Department of Applied Economics, Avenue F-D Roosevelt 50, 1050 Brussels, Belgium.

UNIVERSITY OF ANTWERP
Centre for Development Studies, 13 Prinsstratt, 2000 Antwerp, Belgium.

BRAZIL

BRAZILIAN INSTITUTE OF ECONOMICS
Fundacao Getulio Vargas, Caixa Postal 4081-ZC-05, Rio de Janeiro, Brazil.

PROGRAMME OF JOINT STUDIES ON LATIN AMERICAN ECONOMIC INTEGRATION
Caixa Postal 740, Rio de Janeiro, Brazil.

BULGARIA

SCIENTIFIC RESEARCH CENTRE FOR AFRICA AND ASIA
Academy of Social Science, ul. Gagarin 2, Sofia 13, Bulgaria.

INSTITUTE FOR INTERNATIONAL RELATIONS AND SOCIALIST INTEGRATION
Bulgarian Academy of Sciences, Boul. Pencho Slaveicov, 15, Sofia, Bulgaria.

CANADA

CANADIAN ASSOCIATION OF AFRICAN STUDIES
Geography Department, Carleton University, Ottawa, K1S 5B6.

CANADIAN COUNCIL FOR INTERNATIONAL CO-OPERATION
75 Sparks Street, Ottawa 4, Ontario.

CANADIAN INSTITUTE OF INTERNATIONAL AFFAIRS
Edgar Tarr House, 31 Wellesley Street East, Toronto 284, Ontario.

CENTRE FOR DEVELOPING-ASIA STUDIES
McGill University, Montreal.

INSTITUTE OF INTERNATIONAL RELATIONS
University of British Columbia, Vancouver 8.

INTERNATIONAL DEVELOPMENT RESEARCH CENTRE
60 Queen Street, P.O. Box 8500, Ottawa K1G 3H9.

REGIONAL DEVELOPMENT RESEARCH CENTER
University of Ottawa, Ottawa 2, Ontario.

CHILE

CATHOLIC UNIVERSITY OF CHILE
Institute of Economics, Avda. Libertador Bernardo O'Higgins, No. 340, Santiago, Chile.

CATHOLIC UNIVERSITY OF CHILE
Centre for Planning Studies (CEPLAN), Avda. Libertador Bernardo O'Higgins, No. 340, Santiago, Chile.

UNIVERSITY OF CHILE
Planning Centre (CEPLA), Avda, Libertado Bernardo O'Higgins, No. 1058, Santiago, Chile.

COLOMBIA

UNIVERSITY OF ANTIOQUIA
Economic Research Centre, Apartado Aereo 1226, Medellin, Colombia.

CZECHOSLOVAKIA

INSTITUTE OF INTERNATIONAL RELATIONS
Praha 1 - Mala Strana, Nerudova 3, Czechoslovakia.

DENMARK

INSTITUTE FOR DEVELOPMENT RESEARCH
V. Volgade 104, DK-1552 Kobenhavn.

CENTRE FOR DEVELOPMENT RESEARCH
9, NY Kongensgade, 4K-1472 Copenhagen K, Denmark.

FRANCE

UNIVERSITY OF PARIS, INSTITUTE OF ECONOMIC AND SOCIAL DEVELOPMENT STUDIES
58 Boulevard Arago, 75013 Paris, France.

INSTITUTE FOR RESEARCH INTO THE ECONOMICS OF PRODUCTION
2 rue de Rouen, 92000 Nanterre, France.

INTERNATIONAL CENTRE OF ADVANCED MEDITERRANEAN AGRONOMIC STUDIES
Route de Mende, 34000 Montpellier, France.

INSTITUTE FOR ECONOMIC RESEARCH AND DEVELOPMENT PLANNING
B.P. 47, 38040 Grenoble Cedex, France.

GERMANY, FEDERAL REPUBLIC OF

INSTITUTE FOR DEVELOPMENT RESEARCH AND DEVELOPMENT POLICY
Ruhr-Universitat Bochum, 463 Bochum-Querenburg, Postifach 2148, Federal Republic of Germany.

INTERNATIONAL INSTITUTE OF MANAGEMENT
Wissenschaftszentrum Berlin, Criegstrasse 5-7, Berlin 33, D-1000.

GERMAN ASSOCIATION FOR EAST ASIAN STUDIES
Rothenbaumchaussee 32, 2 Hamburg 13.

GHANA

INSTITUTE OF AFRICAN STUDIES
University of Ghana, P.O. Box 73, Legon, Accra.

HUNGARY

INSTITUTE FOR WORLD ECONOMICS OF THE HUNGARIAN ACADEMY OF SCIENCES
P.O. Box 36, 1531 Budapest, Hungary.

INSTITUTE FOR ECONOMIC AND MARKET RESEARCH
P.O. Box 133, Budapest 62, Hungary.

INDIA

CENTRE FOR THE STUDY OF DEVELOPING SOCIETIES
29, Rajpur Road, Delhi 6, India.

INDIA INTERNATIONAL CENTRE
40 Lodi Estate, New Delhi 110003, India.

INDIAN COUNCIL FOR AFRICA
Nyaya Marg, Chankyapuri, New Delhi 21, India.

INDIAN COUNCIL OF WORLD AFFAIRS
Sapru House, Barakhamba Road, New Delhi 110001, India.

INDIAN INSTITUTE OF ASIAN STUDIES
23/354, Azad Nagar, Jaiprakash Road, Andheri, Bombay 38, India.

INDIAN SCHOOL OF INTERNATIONAL STUDIES
35, Ferozeshah Road, New Delhi 1, India.

INSTITUTE OF ECONOMIC GROWTH
University of Enclave, Delhi 7, India.

MADRAS INSTITUTE OF DEVELOPMENT STUDIES
74, Second Main Road, Gandhinagar Adyar, Madras 20, India.

INDONESIA

NATIONAL INSTITUTE OF ECONOMIC AND SOCIAL RESEARCH
Leknas, UC, P.O. Box 310, Djakarta, Indonesia.

ISRAEL

DAVID HOROWITZ INSTITUTE FOR THE RESEARCH OF DEVELOPING COUNTRIES
Tel-Aviv University, Ramat-Aviv, Tel-Aviv.

AFRO-ASIAN INSTITUTE FOR CO-OPERATIVE AND LABOUR STUDIES
P.O. Box 16201, Tel-Aviv.

ISRAELI INSTITUTE OF INTERNATIONAL AFFAIRS
P.O. Box 17027, Tel-Aviv 61170.

JAPAN

INSTITUTE OF DEVELOPING ECONOMIES
42 Ichigaya-Hommura-cho, Shinjuku-ku, Tokyo 162, Japan.

JAPAN CENTER FOR AREA DEVELOPMENT RESEARCH
Iino Building, 2-1-1 Uchisaiwai-cho, Chiyoda-ku, Tokyo, Japan.

KENYA

INSTITUTE FOR DEVELOPMENT STUDIES
University of Nairobi, P.O. Box 30197, Nairobi.

KOREA

INDUSTRIAL MANAGEMENT RESEARCH CENTRE
Yonsei University, Sodaemoon-ku, Seoul

INSTITUTE OF OVERSEAS AFFAIRS
Hankuk University of Foreign Studies, 270 Rimoon-dong, Seoul.

INSTITUTE OF THE MIDDLE EAST AND AFRICA
Rm. 52, Dong-A Building, No. 55, 2nd-ka, Sinmoonro, Chongro-ku, Seoul.

MEXICO

CENTRE FOR ECONOMIC RESEARCH AND TEACHING
Av. Country Club No. 208, Apdo. Postal 13628, Mexico 21, D.F.

NEPAL

CENTRE FOR ECONOMIC DEVELOPMENT AND ADMINISTRATION (CEDA)
Tribhuvan University, Kirtipur, P.O. Box 797, Kathmandu, Nepal.

NETHERLANDS

CENTRE FOR LATIN AMERICAN RESEARCH AND DOCUMENTATION
Nieuwe Doelenstraat 16, Amsterdam 1000, Netherlands.

INSTITUTE OF SOCIAL STUDIES
Badhuisweg 251, P.O. Box 90733, 2509 LS The Hague, Netherlands.

FREE UNIVERSITY, DEPARTMENT OF DEVELOPMENT ECONOMICS
De Boelelaan 1105, Amsterdam 1000, Netherlands.

CENTRE FOR DEVELOPMENT PLANNING
Erasmus University, Postbus 1738, Rotterdam, Netherlands.

DEVELOPMENT RESEARCH INSTITUTE
Hogeschoollaan 225, Tiburg 4400, Netherlands.

NEW ZEALAND

NEW ZEALAND INSTITUTE OF INTERNATIONAL AFFAIRS
P.O. Box 196, Wellington, New Zealand.

NEW ZEALAND INSTITUTE OF ECONOMIC RESEARCH
26, Kelburn Parade, P.O. Box 3749, Wellington, New Zealand.

NIGERIA

INSTITUTE OF AFRICAN STUDIES, UNIVERSITY OF NIGERIA
Univeristy of Nigeria, Nsukka, Nigeria.

NIGERIAN INSTITUTE OF INTERNATIONAL AFFAIRS
Kofo Abayomi Road, Victoria Island, G.P.O. Box 1727, Lagos, Nigeria.

NIGERIAN INSTITUTE OF SOCIAL AND ECONOMIC RESEARCH
Private Mail Bag No. 5, U.I. University of Ibadan, Ibadan, Nigeria.

NORWAY

INTERNATIONAL PEACE RESEARCH INSTITUTE
Radhusgt 4, Oslo 1, Norway.

NORWEGIAN AGENCY FOR INTERNATIONAL DEVELOPMENT (NORAD)
Planning Department, Boks 18142 Oslo Dep., Oslo 1, Norway.

THE CHR. MICHELSEN INSTITUTE (DERAP)
Fantoftvegen 38, 5036 Fantoft, Bergen, Norway.

PAKISTAN

DEPARTMENT OF INTERNATIONAL RELATIONS
University of Karachi, Karachi-32, Pakistan.

PHILIPPINES

ASIAN CENTER
University of the Philippines, Palma Hall, Diliman D-505, Quezon City, Philippines.

ASIAN INSTITUTE OF INTERNATIONAL STUDIES
Malcolm Hall, University of the Philippines, Diliman, Quezon City, Philippines.

INSTITUTE OF ECONOMIC DEVELOPMENT AND RESEARCH
School of Economics, University of the Philippines, Diliman, Quezon City, Philippines.

POLAND

RESEARCH INSTITUTE FOR DEVELOPING COUNTRIES
Rakowiecka 24, Warsaw, Poland.

CENTRE OF AFRICAN STUDIES
University of Warsaw, Al. Zwirki i Wigury 93, 02-089 Warsaw, Poland.

SINGAPORE

INSTITUTE OF ASIAN STUDIES
Nanyang University, Jurong Road, Singapore 22.

INSTITUTE OF SOUTH-EAST ASIAN STUDIES
Campus of University of Singapore, House No. 8, Cluny Road, Singapore 10.

SRI LANKA

MARGA INSTITUTE
P.O. Box 601, 61 Isipathana Mawatha, Colombo 5, Sri Lanka.

SUDAN

INSTITUTE OF AFRICAN AND ASIAN STUDIES
Faculty of Arts, University of Khartoum, P.O. Box 321, Khartoum, Sudan.

SWEDEN

INSTITUTE FOR INTERNATIONAL ECONOMIC STUDIES
Fack S-104 05, Stockholm 50, Sweden.

STOCKHOLM SCHOOL OF ECONOMICS, ECONOMIC RESEARCH INSTITUTE
Box 6501, 11383 Stockholm, Sweden.

UNITED KINGDOM

CENTRE FOR SOUTH-EAST ASIAN STUDIES
University of Hull, Hull HU6 7RX

CENTRE OF AFRICAN STUDIES
University of Edinburgh, Adam Ferguson Building, George Square, Edinburgh 8.

CENTRE OF LATIN AMERICAN STUDIES (CAMBRIDGE)
University of Cambridge, History Faculty Building, West Road, Cambridge CB3 9ES, England.

CENTRE OF LATIN AMERICAN STUDIES (OXFORD)
Oxford University, St. Antony's College, Oxford OX2 6JF, England.

CENTRE OF WEST AFRICAN STUDIES
University of Birmingham, P.O. Box 363, Birmingham B15 2TT.

INSTITUTE FOR THE STUDY OF INTERNATIONAL ORGANISATION
University of Sussex, Stanmer House, Stanmer Park, Brighton BN1 9QA, England.

INSTITUTE OF DEVELOPMENT STUDIES
University of Sussex, Falmer, Brighton BN1 9QN, England.

INSTITUTE OF LATIN AMERICAN STUDIES
University of London, 31 Tavistock Square, London WC1, England.

INSTITUTE OF LATIN AMERICAN STUDIES (GLASGOW)
University of Glasgow, Glasgow.

ROYAL INSTITUTE OF INTERNATIONAL AFFAIRS
Chatham House, St. James' Square, London SW1Y 4LE, England.

UNITED STATES

AFRICAN STUDIES CENTER (BOSTON)
Boston University, 10 Lenos Street, Brookline, MA 02146.

BROOKINGS INSTITUTION
1775 Massachusetts Avenue, N.W., Washington, D.C. 20036.

CENTER FOR ASIAN STUDIES
Arizona State University, Tempe, AZ 85281.

CENTER FOR COMPARATIVE STUDIES IN TECHNOLOGICAL DEVELOPMENT AND SOCIAL CHANGE
University of Minnesota, Minneapolis, Minnesota 55455.

CENTER FOR DEVELOPMENT ECONOMICS
Williams College, Williamston, MA 01267

CENTER FOR INTERNATIONAL AFFAIRS
Harvard University, 6 Divinity Avenue, Cambridge, MA 02138.

CENTER FOR INTERNATIONAL STUDIES
Massachusetts Institute of Technology, Cambridge, MA 02139.

CENTER FOR LATIN AMERICAN STUDIES, ARIZONA STATE UNIVERSITY
Arizona State University, Tempe, AZ 85281.

CENTER FOR LATIN AMERICAN STUDIES, UNIVERSITY OF FLORIDA
University of Florida, Room 319 LAGH, Gainesville, FL 39611.

CENTER FOR RESEARCH ON ECONOMIC DEVELOPMENT
506 East Liberty Street, Ann Arbor, MI 48108.

CENTER FOR STRATEGIC AND INTERNATIONAL STUDIES
Georgetown University, 1800 K Street, N.W., Washington, D.C. 20006.

CENTER OF INTERNATIONAL STUDIES, PRINCETON UNIVERSITY
Princeton University, 118 Corwin Hall, Princeton, NJ 08540.

HARVARD INSTITUTE FOR INTERNATIONAL DEVELOPMENT
Harvard University, 1737 Cambridge Street, Cambrdige, MA 02138.

INSTITUTE FOR WORLD ORDER
1140 Avenue of the Americas, New York, New York 10036.

INSTITUTE OF LATIN AMERICAN STUDIES
University of Texas at Austin, Sid. W. Richardson Hall, Austin, TX 78705.

STANFORD INTERNATIONAL DEVELOPMENT EDUCATION CENTER
P.O. Box 2329, Stanford, CA 94305.

UNIVERSITY CENTER FOR INTERNATIONAL STUDIES
University of Pittsburgh, Social Sciences Building, Pittsburgh, PA 15213.

WORLD FUTURE SOCIETY
4916 St. Elmo Avenue, Bethesda Branch, Washington, D.C. 20014.

UNIVERSITY OF HAWAII
Centre for Development Studies, Department of Economics, Porteus Hall, 2424 Maile Way, Honolulu, Hawaii 96822

URUGUAY

LATIN AMERICAN CENTRE FOR HUMAN ECONOMY
Cerrito 475, P.O. Box 998, Montevideo, Uruguay.

VENEZUELA

UNIVERSITY OF ZULIA
Department of Economic Research, Faculty of Economic and Social Sciences, Maracaibo, Venezuela.

YUGOSLAVIA

INSTITUTE FOR DEVELOPING COUNTRIES
41000 Zagreb, Ul. 8 Maja 82, Yugoslavia.

RESEARCH CENTRE FOR CO-OPERATION WITH DEVELOPING COUNTRIES
61 109 Ljubljana, Titova 104 P.O. Box 37, Yugoslavia.

INSTITUTE OF WORLD ECONOMICS AND INTERNATIONAL RELATIONS OF THE ACADEMY OF SCIENCES OF THE U.S.S.R.
Yaroslavskaya Ul. 13, Moskva I-243.

Appendix

COUNTRIES BY INCOME GROUP
(based on 1976 GNP per capita in 1976 US dollars)

INDUSTRIALIZED COUNTRIES

Australia
Austria
Belgium
Canada
Denmark
Finland
France
Germany, Fed. Rep. of
Iceland
Ireland
Italy
Japan
Luxembourg
Netherlands
New Zealand
Norway
South Africa
Sweden
Switzerland
United Kingdom
United States

DEVELOPING COUNTRIES BY INCOME GROUP
(Excluding Capital Surplus Oil Exporters)

High Income (over $2500)

American Samoa
Bahamas
Bermuda
Brunei
Canal Zone
Channel Islands
Faeroe Islands
French Polynesia
Gabon
Gibraltar
Greece
Greenland
Guam
Israel
Martinique
New Caledonia
Oman
Singapore
Spain
Venezuela
Virgin Islands (U.S.)

Upper Middle Income ($1136-2500)

Argentina
Bahrain
Barbados
Brazil
Cyprus
Djibouti
Fiji

French Guiana
Guadeloupe
Hong Kong
Iran
Iraq
Isle of Man
Lebanon
Malta
Netherlands Antilles
Panama
Portugal
Puerto Rico
Reunion
Romania
Surinam
Trinidad & Tobago
Uruguay
Yugoslavia

Intermediate Middle Income ($551-1135)

Algeria
Antigua
Belize
Chile
China, Rep. of
Colombia
Costa Rica
Dominica
Dominican Republic
Ecuador
Ghana
Gilbert Islands
Guatemala
Ivory Coast
Jamaica
Jordan
Korea, Rep. of
Macao
Malaysia
Mauritius
Mexico
Namibia
Nicaragua
Paraguay
Peru
Seychelles
St. Kitts-Nevis
St. Lucia
Syrian Arab Rep.
Trust Territory of the Pacific Islands
Tunisia
Turkey

Lower Middle Income ($281-550)

Angola
Bolivia
Botswana
Cameroon
Cape Verde
Congo, P.R.
El Salvador
Equatorial Guinea
Grenada
Guyana
Honduras
Liberia
Mauritania
Morocco
New Hebrides
Nigeria
Papua New Guinea
Philippines
Rhodesia
Sao Tome & Principe
Senegal
St. Vincent
Sudan
Swaziland
Thailand
Tonga
Western Samoa
Zambia

Low Income ($280 or less)

Afghanistan
Bangladesh
Benin
Bhutan
Burma
Burundi
Cambodia
Central African Empire
Chad
Comoros
Egypt
Ethiopia
Gambia, The
Guinea
Guinea-Bissau
Haiti
India
Indonesia
Kenya
Lesotho
Madagascar
Malawi

Maldives
Mali
Mozambique
Nepal
Niger
Pakistan
Rwanda
Sierra Leone
Solomon Islands
Somalia
Sri Lanka
Tanzania
Togo
Uganda
Upper Volta
Viet Nam
Yemen Arab Rep.
Yemen P.D.R.
Zaire

CAPITAL SURPLUS OIL EXPORTING DEVELOPING COUNTRIES

Kuwait
Libya
Qatar
Saudi Arabia
United Arab Emirates

CENTRALLY PLANNED COUNTRIES

Albania
Bulgaria
China, People's Rep. of
Cuba
Czechoslovakia
German Dem. Rep.
Hungary
Korea, Dem. Rep. of
Lao People's Dem. Rep.
Mongolia
Poland
U.S.S.R.

SOURCE : World Economic and Social Indicators
Document of the World Bank
Washington, D.C.

Index

About the Editor

Pradip K. Ghosh is Adjunct Associate Professor and Visiting Fellow at the Center for International Development at the University of Maryland, College Park. He is the author of *Thinking Sociology* and *Land Use Planning,* and editor of the International Development Resource Books series for Greenwood Press.